SCOTTISH LIFE AND SOCIETY

The Working Life of the Scots

Publications of the European Ethnological
Research Centre

Scottish Life and Society: A Compendium of Scottish Ethnology
(14 Volumes)

Already published:
Volume 3 *Scotland's Buildings*
Volume 4 *Boats, Fishing and the Sea*
Volume 5 *The Food of the Scots*
Volume 6 *Scotland's Domestic Life*
Volume 7 *The Working Life of the Scots*
Volume 9 *The Individual and Community Life*
Volume 10 *Oral Literature and Performance Culture*
Volume 11 *Institutions of Scotland: Education*
Volume 12 *Religion*
Volume 14 *Bibliography for Scottish Ethnology*

GENERAL EDITOR:
Alexander Fenton

Scottish Life *and* Society

A COMPENDIUM OF SCOTTISH ETHNOLOGY

THE WORKING LIFE OF THE SCOTS

Edited by
Mark A Mulhern
John Beech
Elaine Thompson

JOHN DONALD
in association with
THE EUROPEAN ETHNOLOGICAL RESEARCH CENTRE

First published in Great Britain in 2008 by

John Donald, an imprint of Birlinn Ltd
West Newington House
10 Newington Road
Edinburgh
EH9 1QS

www.birlinn.co.uk

ISBN 10: 0 904607 85 3
ISBN 13: 978 0 904607 85 4

British Library Cataloguing-in-Publication Data
A catalogue record for this book is available on request from the British Library

Typeset by Carnegie Publishing Ltd, Lancaster
Printed and bound in Britain by the Cromwell Press, Trowbridge, Wiltshire

Contents

List of Figures

List of Tables

List of Contributors

PROFESSOR TOM G BAUM
Professor of International Tourism and Hospitality Management, Strathclyde Business School, University of Strathclyde.

DR EWEN A CAMERON
Senior Lecturer, Scottish History subject area, University of Edinburgh.

DR E PATRICIA DENNISON
Director, Centre for Scottish Urban History, Scottish History subject area, University of Edinburgh.

DR RONA DOUGALL
Public Health Researcher, NHS Greater Glasgow and Clyde.

THOMAS J DOWDS
Independent scholar.

PROFESSOR JANET DRAPER
Professor of Education Studies, Hong Kong Baptist University.

PROFESSOR ALEXANDER FENTON
Founder, European Ethnological Research Centre.

ANNE M FINDLAY
Research Fellow, Centre for the Study of Retailing in Scotland, University of Stirling.

DR DAVID HAMILTON
Independent scholar and retired general practitioner.

DR RONNIE JOHNSTON
Reader, History Subject Area, School of Law and Social Sciences, Glasgow Caledonian University.

DR WILLIAM KENEFICK
Lecturer, History Department, University of Dundee.

DR WILLIAM W KNOX
Senior Lecturer, Department of Scottish History, University of St Andrews.

DR CATRIONA M M MacDONALD
Senior Lecturer, History Subject Area, School of Law and Social Sciences, Glasgow Caledonian University.

PROFESSOR ARTHUR McIVOR
Professor of Social History, Department of History, University of Strathclyde.

PROFESSOR CHARLES McKEAN
Professor of Scottish Architectural History, Department of History, University of Dundee.

PROFESSOR ALAN McKINLAY
Professor of Management, School of Management, University of St Andrews.

PROFESSOR RONALD McQUAID
Director, Employment Research Institute, Napier University.

DR ALASTAIR J MANN
Lecturer, Department of History, University of Stirling.

DR IRENE MAVER
Senior Lecturer and Head of Scottish Area, Department of History (Scottish), University of Glasgow.

JAMES MILLER
Independent scholar and author.

PROFESSOR CHARLES MUNN
Chief Executive of the Chartered Institute of Bankers in Scotland.

DR ANDREW NEWBY
Senior Lecturer, History, University of Aberdeen.

DR DENNIS P NICKSON
Reader, Department of Human Resource Management, Strathclyde Business School, University of Strathclyde.

DR CHRISTOPHER NOTTINGHAM
Reader in Contemporary History, Glasgow Caledonian University.

DR LINDSAY REID
Freelance medical historian and research midwife, Fife.

PROFESSOR RICHARD RODGER
Professor of Economic and Social History, University of Edinburgh.

TREVOR ROYLE
Historian, journalist and independent scholar.

PROFESSOR ADRIENNE SCULLION
Professor of Drama and Head of Theatre Studies, Film and Television Studies at the University of Glasgow.

PROFESSOR LEIGH SPARKS
Professor of Retail Studies, Institute for Retail Studies, University of Stirling.

PROFESSOR STEPHEN P WALKER
Professor in Accounting, Cardiff Business School, Cardiff University.

DR ALAN B WATSON
Senior Lecturer, Caledonian Business School, Glasgow Caledonian University.

WILLIAM G D WATSON
Retired lecturer.

PROFESSOR CHRISTOPHER A WHATLEY
Vice Principal and Professor of Scottish History, University of Dundee.

Foreword

In his Introduction to this volume, editor Mark Mulhern rightly stresses the importance of work in shaping identity, whether at personal, local, regional or national level. Understanding individual experience is at the heart of the ethnological approach and *The Working Life of the Scots* offers insights on the contexts in which the work of men, women and children has been carried out through the centuries. It may be as a family tradition, or a learned craft, trade or profession, or an avocation. At all times the emphasis is on the individual within the work group or the work place, and lived experience is highlighted. This volume helps answer the question, 'What was it like to … ?'.

The nature of Scotland, its topography and geology, its regional variety, its resources both material and human, its economic, political and social structures and institutions, has ensured a formidable diversity in the working lives of its people for a nation of its size. Attaining the basic necessities of life – sustenance, shelter and protection, and social contact – calls for food production, buying and selling, buildings and textiles, individual and communal effort. Travel and social contacts elsewhere influenced agricultural improvement, while the impact of scientific and economic thinkers as well as our philosophers and historians made Scotland a laboratory for the practical application of Enlightenment ideas, with implications for education, personal health and the well-being of society. The fight for improved working conditions and combat in the military sense are part of the story too. Concentrations of coal, iron and water, oil and gas, have brought Scotland to the fore in industrial and engineering innovation over centuries and into the present.

The *Compendium* provides a resource for Scotland which is engaging scholars from many specialisms and, while this volume has a special concentration, the working lives of the Scots are in fact reflected in all its volumes, those which have already appeared on education, religion, food, buildings, domestic life, the individual and community life, oral literature and performance culture, boats, fishing and the sea, not forgetting the invaluable *Bibliography*, and those to be published shortly on farming and rural life, the law, transport and communication, and the final volume with its overview.

Identity is one of the central issues in our world today, just as energy and communications are major concerns, and this volume contributes Scottish experience to a wider picture. It is in itself a tribute to the work of the Scots in different periods, places and social contexts, to the lives shaped by work as explored here, and the work of the contributors and editors. As Ecclesiastes 5:12 has it, 'Sweet is the sleep of the labourer'. Well done, all!

Margaret A Mackay

Acknowledgements

The editors wish to thank the many people involved in bringing this volume to fruition. This has been a truly collaborative exercise dependent upon the best of scholastic endeavour on the part of those who have many competing demands placed upon their time. In addition to those who contributed the chapters that follow, there were many others who contributed to the volume by acting as readers, and to them the editors would like to record their thanks.

We are indebted to the Scottish Life Archive of National Museums Scotland and its staff, especially Dorothy Kidd and Kate MacKay, and to Ian Mackenzie of the School of Scottish Studies Archive, for many of our photographs. The staff of Perth Museum and Art Gallery, the National Library of Scotland and many other repositories were helpful in identifying and supplying many of the images used here. All images from the Scottish Life Archive are reproduced by kind permission of the Trustees of National Museums Scotland.

In addition, grateful thanks are due to the following bodies:

The Scotland Inheritance Fund, which provides substantial support for the preparation and production costs of the *Compendium of Scottish Ethnology*.

The Trustees of National Museums Scotland, who provided support for the European Ethnological Research Centre during the early production of this volume.

Abbreviations

ACCA	Association of Chartered Certified Accountants
AIS	Architectural Institute of Scotland
BMA	British Medical Association
BTEC	Business and Technology Education Council
CA	Chartered Accountant
CABIN	Campaign against Nationalisation of the Building Industry
CBI	Confederation of British Industry
CIMA	Chartered Institute of Management Accountants
CIPFA	Chartered Institute of Public Finance and Accountancy
CMB	Central Midwives Board
CPD	Continuing Professional Development
CSA	Clyde Shipbuilders' Association
CWS	Co-operative Wholesale Society
DOMINO	Domiciliary In and Out
DP	Displaced Person
DTP	Desktop Publishing
DWRGWU	Dock, Wharf, Riverside General Workers' Union
EGAMS	Expert Group on Acute Maternity Services
EIS	Educational Institute for Scotland
EOS	Edinburgh Obstetrical Society
FBI	Federation of British Industry
FIRE	Finance, Insurance and Real Estate
GDP	Gross Domestic Product
GHLU	Glasgow Harbour Labourers' Union
GHMWU	Glasgow Harbour Mineral Workers' Union
GPMU	Graphical, Paper and Media Union
GTC	General Teaching Council
HGV	Heavy Goods Vehicle
IAAG	Institute of Accountants and Actuaries in Glasgow
ICAEW	Institute of Chartered Accountants in England and Wales
ICAI	Institute of Chartered Accountants in Ireland
ICAS	Institute of Chartered Accountants in Scotland
IMF	International Monetary Fund
LSA	Local Supervising Authority
MMR	Measles, Mumps and Rubella vaccine
MRSA	Methicillin-resistant *Staphylococcus aureus*
NASD	National Amalgamated Stevedore and Dockers

NBS	National Board for Scotland for Nurses, Midwives and Health Visitors
NCB	National Coal Board
NCEO	National Conference of Employers' Organisations
NGA	National Graphical Association
NHS	National Health Service
NJC	National Joint Council for the Port Transport Industry
NMW	National Minimum Wage
NTWF	National Transport Workers' Federation
NUDL	National Union of Dock Labourers
NUM	National Union of Mineworkers
NVQ	National Vocational Qualifications
NWETEA	North Western Engineering Trades Employers' Association
OECD	Organisation for Economic Co-operation and Development
PBR	Premium Bonus Rate
PC	Personal Computer
PoW	Prisoner of War
PPITB	Printing and Publishing Industry Training Board
PTSD	Post-Traumatic Stress Disorder
QNI	Queen's Nursing Institute
QNIS	Queen's Nursing Institute of Scotland
RCAHMS	Royal Commission on the Ancient and Historical Monuments of Scotland
RIAS	Royal Incorporation of Architects in Scotland
RIBA	Royal Institute of British Architects
RSI	Repetitive Strain Injury
SAA	Society of Accountants in Aberdeen
SAE	Society of Accountants in Edinburgh
SBEF	Scottish Building Employers' Federation
SCOTMID	Scottish Midland Co-operative Society
SCOTVEC	Scottish Vocational Education Council
SCWG	Scottish Co-operative Women's Guild
SCWS	Scottish Co-operative Wholesale Society
SGF	Scottish Grocers' Federation
SHAW	Stress and Health at Work Study
SLA	Scottish Life Archive
SME	Small and Medium-sized Enterprises
SMMDG	Scottish Multiprofessional Maternity Development Group
SMMDP	Scottish Multiprofessional Maternity Development Programme
SNP	Scottish National Party
SOGAT	Society of Graphical and Allied Trades
SPEF	Scottish Print Employers Federation
SSLPL	South Side Labour Protection League
STA	Scottish Typographical Association
STGWU	Scottish Transport and General Workers' Union
STUC	Scottish Trades Union Congress

SUDL	Scottish Union of Dock Labourers
TGWU	Transport and General Worker's Union
TQM	Total Quality Management
TUC	Trades Union Congress
UCS	Upper Clyde Shipbuilders
UKCC	United Kingdom Central Council for Nurses, Midwives and Health Visitors

Glossary

This selective list contains only those terms found in the text that may need explanation.

asbestosis, a disease of the lungs caused by inhaling particles of asbestos.

custumar, an official having charge of the customs; a customs officer.
cut, a passage or channel, an artificial watercourse cut or dug out; a channel, canal, cutting.

forme, in printing, the item that imparts the text/image on to the printed substrate.

howdie, a midwife, untrained nurse; a woman who laid out the dead.

liner, an official whose duty was the tracing of the boundaries of properties in burghs.
longwall, a method of coal extraction in which the whole seam of coal is removed in one working.

Minnitt apparatus, a device which delivered a mixture of air and the anaesthetic nitrous oxide to safely provide pain relief, especially during labour.

pillar and stoop (also known as post and stall, board and pillar, stoop and room), a method of coal extraction which left a substantial pillar of coal in place to support the roof of the working.

navigation, canal, an artificial waterway; a river or channel that has been dredged, straightened, etc., to make it navigable.
navvy, construction worker; especially a labourer employed in the construction of (originally) a canal, (now frequently) a road, railway, etc.
Nissen hut, a prefabricated shelter consisting of a sheet of corrugated steel bent into half a cylinder and planted in the ground with its axis horizontal. The semicircular ends are closed with masonry walls.

soutar, a maker or mender of shoes; a shoemaker or cobbler.

toft, originally, a homestead, the site of a house and its out-buildings, including an attached piece of arable land.

tron, a weighing machine; a pair of scales or other machine for weighing merchandise; a public weighing apparatus in a city or burgh.

Introduction

Introduction

MARK A MULHERN

The title of this volume is concise, and at the same time ambitious. The meaning of 'work' is apparently understood by all. However, 'work' carries many different meanings. Work is a measure of the expenditure of energy and the effects of that expenditure. There is some value in thinking of work as a description of energy expended. Such a description allows us to understand that the preparation of a meal in the home, the repair of a shoe, the tending of a garden are all work.

This description denotes effort, but it fails to define the 'world of work'. Work is also that which is undertaken by human beings to secure for themselves and their families a proportion of available resources – that is, employment and the workplace. The workplace creates distinct environments. These environments lead to the creation of workplace associations and behaviour patterns that result in the development of distinct workplace cultures. These cultures then permeate the wider regional, national and supra-national culture. In other words, work and the workplace play a key role in the formation of identities at all levels of society.

Whilst very much a group activity, work, as in all human endeavour, is composed of individual effort. The role of the individual in society is central to the ethnological method and has been key in forming the structure and content of this volume.

The ubiquity and importance of work in human relations is such that it has come within the ambit of investigation of many scholars over the years. Some, investigating the imbalance of power relations in society, look to the world of work as a means of providing explanation. Others are interested in the history of forms of economic endeavour and therefore analyse the history of particular groups of workers. Yet others seek to explain wider historical trends by studying whole cadres of work such as 'crafts', 'trades' or 'professions'. This last approach is the one that most closely approximates to the method adopted in structuring this volume.

In Scotland, if the preponderance of work is considered, two organising themes suggest themselves – 'craft' and 'service'. These themes are capable of including work involved in 'trade'. As to 'professions', the role of the professions is notable in Scotland as elsewhere, and considerations of professions in the working life of the Scots are included here. With the development of economic activity and other human endeavour, certain work roles have been created, others have been elaborated and others have disappeared.

As will be seen, 'service' is descriptive of both a notable sector of employment in Scotland and of an aspect of the working life of the Scots across a range of occupations. Likewise 'craft' forms a part of work across a number of sectors of employment. For example, the building of ships or manufacture of semi-conductor components utilise many craft skills. Craft is not the reserve of the self-employed artisan making bespoke goods. Craft is that which is borne of skill developed over a period of time, in which the worker takes pride.

This project is ambitious in seeking to be as comprehensive as possible, but it is also highly focused in its approach. The editors have engaged those with detailed knowledge of the particular topic they have been asked to address. The editors have not necessarily sought only those following an ethnological approach to contribute to this volume. This is a reflection of the extent to which ethnological study in Scotland and elsewhere has tended to focus on aspects of rural life and forms of 'traditional' cultural milieu embodied in folklore and folklife. Volumes such as this seek to encourage the participation of social scientists from various fields in an ethnological pursuit, and it is to their great credit that so many extremely busy and able scholars have so willingly involved themselves here.

This volume forms part of a fourteen-volume series which takes as its subject the lives of those who live and have lived in Scotland. The series approaches this study by examining discrete elements of those lives and of the institutions within Scotland that determine the ways in which those lives are lived, thereby contributing to a sense of identity – personal, local, regional and national. The place of work in this matrix is of clear importance and is the reason why this subject is examined in a separate volume. In addition, studies of work tend to concentrate on structural or economic aspects of working life. There is therefore space for an account to be given of working life in Scotland that focuses on the range of individual experience. Such an approach has been undertaken in the past, most notably by the Scottish Working People's History Trust, and this volume makes a further contribution to this effort.[1] The distinction in the approach taken here is that of the ethnological approach in general. Where economic and, to a lesser extent, historical analyses draw upon information on large groups of people to give an account of aggregate experience, ethnology looks to individual experience and makes comparisons across time, place and milieu in pursuit of an account of aggregate experience.

Ethnology is a discipline that seeks to find the truth of the lived experience through the object of the individual. By recording what people do and the ways in which they do what they do, ethnologists seek to establish a view of what society is with reference to all those who compose that society – not just the privileged or the vocal. Whilst simple in conception, the methodology of ethnological enquiry is difficult in practise and the analysis of findings is as fraught with the problems of bias and teleology as that of any related discipline. The key analytical tool employed by the ethnologist (and by all social scientists) is that of comparison. In this volume some authors make greater use of this tool than others, but it is left open to the reader to consider each of these contributions and compare what one author is revealing about an aspect of one type of work with that of another author about another type of work.

This approach has informed the structure and content of this volume. Within the broad demarcation of 'craft' and 'service', refinement was introduced by taking broad themes that gave an account of what was and is the working life of the Scots. This account has been given by considering the historical context of the workplace and by examining the historical pattern of work in Scotland from the twelfth century onwards. Following this, work undertaken to meet the necessities of life was considered in order to assess both what forms of work are undertaken in fulfilment of the necessities of life and what those types of work tell us about the people of Scotland. However, life is composed of more than the necessities of an individual's own life or that of their family. At times people come together to strive for the communal good, either of their immediate community or of the wider community, a motivation encapsulated in Scotland in the long-standing concept of 'the Commonweal'.

Such co-operative human activity is evident in many different forms, and this volume seeks to examine areas whereby such activity has involved the greatest number of individuals working for the collective good. Whether in trade unionism or cooperative retailing, people have striven in the workplace to secure for others – and themselves – a more beneficial bargain than might otherwise have pertained. Less prosaically, this type of workplace endeavour demonstrates the ability of individuals to identify a commonality of interest and an aspiration to do what is right for its own sake. This aspect of work is one in which individuals come together to form community and, to a certain degree, act according to a morally framed world view rather than according to economic self-interest.

An area of work where there exists a self-conscious awareness of regional and national identities is that of words and imagination. Those whose work is, in part, the communication of identities to others, both in Scotland and beyond, are perhaps most alive to the impact of an individual's work upon their sense of identity. However, this should not obscure the fact that people who work in these areas are every bit as exposed to the vicissitudes of economic necessity as those in any other type of employment. Indeed many in these fields are more exposed to the harsh realities of making a living though insecure and seasonal work than others in more 'necessary' types of work.

These more 'necessary' types of work include the public services and industry. To many, these types of work epitomise the Scottish economy and therefore the working experience of the Scots. Whilst these types of work are most definitely important as part of the story of the working life of the Scots, they in no way tell of the entirety of that experience. Indeed, such is the variety of working life that the thirty-one chapters in this book do not cover the entirety of that experience – no single volume could hope to.

The difficulty in researching the working life of the Scots is not really one of availability of source material. Even where such material does exist, it is preserved in inchoate form and is scattered across a range of repositories, both formal and personal. In seeking the experience of the individual in pursuit of a broader analysis of a theme, one is hampered by the fact that such source material, where it exists, is atomised and difficult to collect.

This volume seeks to bridge the approaches of differing disciplines by incorporating differing analyses, such as historical or sociological, within a work conceived to provide an account of the lived experience across Scotland. Of course, neither historical nor sociological approaches are antagonistic to the ethnological approach, but these disciplines differ markedly from ethnology in that they, broadly speaking, take as their object of enquiry 'groups'.

Another function of a work of this type is to highlight areas worthy of further study. From this volume it is clear that work in the insurance industry in Scotland is at the earliest stages of investigation. Likewise, a comprehensive study of the construction of the schemes of the North of Scotland Hydro-Electric Board is long overdue. This work has begun with, amongst others, the works of Miller and Payne, but much remains to be done.[2] These schemes represent the most important civil engineering project undertaken in Scotland in the twentieth century, and a detailed study of their construction would reveal an interesting and important aspect of Scottish life and society. Such a study may also be helpful in the work of policy formation at a time when renewable energy production and supply are the focus of so much attention.

Another area worthy of more detailed study is that of worker remuneration. We know surprisingly little about the ways in which workers were compensated for their labours. True, we know about wage levels for certain types of work during certain periods. We also know that for many, payment was often made, in part or wholly, in kind as well as in cash and that the feudal basis of Scottish land ownership played a role in preserving such systems of remuneration. What is missing from these aspects of understanding, however, is any sense of comprehensiveness and locality. To be able to make valid comparisons we need to be assured that we are comparing like with like. In the field of remuneration, the work has yet to be undertaken to allow such comparisons.

This is partly a reflection of the survival of relevant source material and partly a reflection of the difficulty of interpreting the source material that does exist. The simpler material with which to make comparisons is that which tells of cash payments. However, there is a limited range of this type of information available. One of the most useful sources, the statistical accounts, are not without their limitations (limited data, inconsistent coverage, etc.). More problematically, we do not currently have a sufficient range of material available to allow for the compilation of comprehensive tables of data on wage levels. We are therefore left in the position whereby it becomes necessary to interpolate from those sources that we do have.

Some examples of this approach do exist, but this approach needs to be adopted more fully.[3] Notwithstanding the problems of researching cash wages, research into payment in kind proves all the more elusive. Records of such payments were not kept in any comprehensive manner. Such payments could form the part of a contract of employment, but by no means all who were employed had such contracts or offers of employment.

As Fenton has shown, such interpolation is possible by making judicious use of evidence drawn from as wide a base as possible (see chapter 5). For example, advertisements in newspapers can indicate the jobs on offer, the remuneration offered and the status of the job being advertised. Making use of material such as this can allow the many gaps that exist in this database to be spanned. With proper caution it becomes possible to draw closer to a better understanding of the general pattern of remuneration in Scotland over time. At the same time, these analyses allow us to identify the gaps in the database and can therefore help to refine the approaches of future researchers.

It is hoped that this volume will provide the reader with a rounded understanding of the working life of the Scots. There is much in what follows that will be true of the working life of those outwith Scotland – aspects of being that are common to the experience of work. There are, however, aspects of the working life of the Scots that arise because of the fact that the life is or has been lived in Scotland. For example, the scale of work in the 'service' sector has been an increasingly notable feature of working life in Scotland. Similarly, the shift away from work in agriculture is often characterised as an acute shift which rendered agriculture a marginal aspect of the working life of the Scots. In relative terms, such a conclusion may be true. However, this is, to an extent, a reflection of a marked increase in the total size of the working population. In absolute terms, the number of people working in agriculture in 1841 was approximately 230,000; by 1931, there remained just under 200,000 people employed in agriculture. That is, agriculture continued to be the basis of family and communal life for a similar number of people for a period of ninety years – a period when there was marked change in employment patterns and wider society.

In drawing out these points, this volume acts to refine our understanding of not only working life in Scotland but also the pattern of work. The place of work in the formation of identity of the individual, a locality and communities is well attested. Given this, clearer understanding of what that working life is and has been enables a better understanding of what these identities were and are. In so doing, this volume makes a further contribution to the series as a whole and to more general study of Scotland and the Scots.

NOTES

1. For example, MacDougall, I. *'Hard work, ye ken': Midlothian Women Farmworkers*, Edinburgh, 1993; MacDougall, I. *Hoggie's Angels – Tattie Howkers Remember*,

East Linton, 1995; MacDougall, I. *Onion Johnnies – Personal Recollections by Nine French Onion Johnnies of Their Working Lives in Scotland*, East Linton, 2002.

2. Miller, J. *The Dam Builders: Power from the Glens*, Edinburgh, 2002; Payne, P. *The Hydro: A Study of the Development of the Major Hydro-Electric Schemes Undertaken by the North of Scotland Hydro-Electric Board*, Aberdeen, 1988.
3. See Gibson, A J S and Smout, T C. *Prices, Food and Wages in Scotland, 1550–1780*, Cambridge, 1995.

PART ONE

•

Timeline and Employment

1 The Occupational Structure of Towns, *c*1100–1700

E PATRICIA DENNISON

EARLY TOWNSPEOPLE AS TRADERS

Written sources are elusively vague on the occupational structure of towns in the twelfth century – even though towns are first documented at this time. They were probably in existence before this, however.[1] The early sources available are essentially formal, legal rulings: the *Leges Burgorum*, reputedly of the reign of David I (1124–53), drawn up perhaps for Berwick, which were to become the basic foundation for specific Scottish burghs; the *Assise Regis Willelmi* (laws of William the Lion, 1165–1214); the *Statuta Gildae*, the statutes of the guild of Berwick, the earlier part of which is attributable to 1249 and the later specifically dated 1281 and 1294; along with charters and ecclesiastical records in the form of cartularies of religious houses. The more intimate records of burgh and guild courts do not survive until as late as the fifteenth century, apart from those of Aberdeen, which start at the end of the fourteenth century, and even this record has a loss of registers from 1413 to 1433. But, in spite of such paucity, these early sources do give a certain insight into the workings of Scotland's first burghs.

Burghs were settlements deliberately created by the crown, or with royal permission by a notable ecclesiastic or aristocrat, with specific legal rights. Other settlements that displayed urban characteristics may have existed but have escaped recognition simply because they were never raised to burgh status. For the purposes of this essay, therefore, 'burgh' and 'town' are treated as synonymous.

What emerges with great clarity is that Scotland's early burghs were primarily markets, founded for trading purposes. The burgh was, in effect, a community organised for trade.[2] Such a recognition was made as early as the twelfth century, judging from the percentage of clauses dealing with mercantile matters in early burghal legislation.[3] Of great significance to the burgesses, the freemen of the burgh, was the right this gave of non-payment of toll to the owner of a market, so enabling the burgess to travel at will around the country buying and selling. Equally importantly, the burgh community gained the right to hold its own market and exact tolls on those attending it.

In some burghs, guilds merchant were founded, enabling the bestowed burgh to gain maximum advantage from its newly gained economic privileges. Such a right was noted in clause xciv of *Leges Burgorum* and again in clause xxxix of *Assise Regis Willelmi*, when guilds merchant received official sanction; it was decreed that merchants of the realm were to have their guild with the liberties to buy and sell in all places within the bounds of liberties of burghs, to the exclusion of all others. This would have a profound effect on the occupational structure of towns throughout the Middle Ages.

It cannot be concluded with certainty that all early burghs had a guild merchant, but where it is known that they existed the guild is seen to have had a dominant role in both the organisation and the occupational hierarchy of the burgh. Both Perth and Roxburgh, for example, had guilds at least by pre-1189x1202.[4] St Andrews' guild merchant was granted in 1189x1202, whereas Edinburgh's possibly dates to 1209 and those of Inverness and Inverkeithing may be dated to 1165?x1214. Aberdeen's and Stirling's guilds merchant were granted in 1222, as was possibly also that of Ayr, that of Stirling in 1226 and that of Elgin in 1234. Dundee's guild merchant existed at least by 1249x1286 and possibly as early as 1165x1214. Berwick had a guild by 1249 at the latest, if not considerably earlier.[5] Although there was

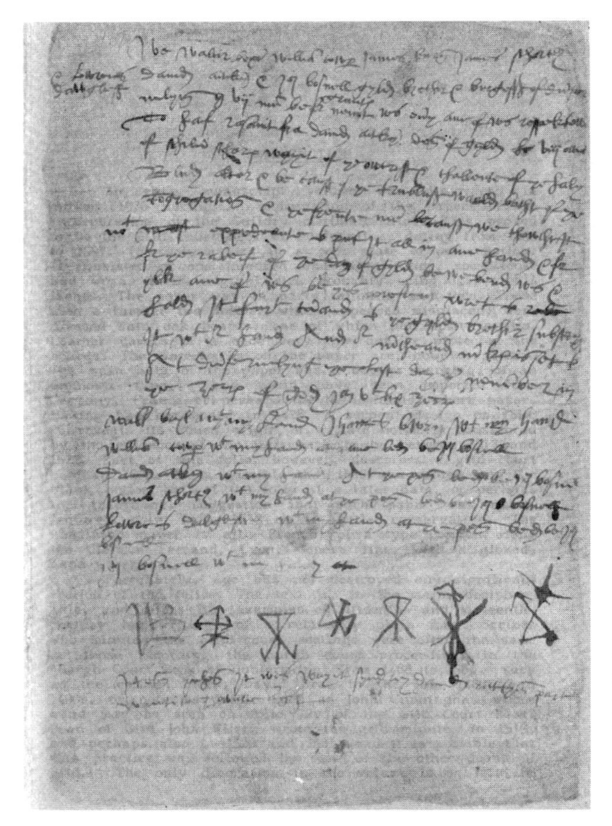

Figure 1.1 Merchant marks in Dunfermline Gild Court Book. Source: Reproduced by permission of The Incorporation of the Guildry of Dunfermline.

an intermingling of the functions of local government between the guild merchant and the burgh assize or burgh council, the guild became particularly concerned with trading issues. It was the resultant control of the market and related issues that gave the guild economic and, consequently, political and occupational dominance in the urban hierarchy.

But who formed this élite of the guild merchant? The *Leges Burgorum* specifically excluded from entry dyers, fleshers and soutars (cobblers), who practised their crafts with their hands.[6] The burgh charters of Perth, Stirling and Aberdeen likewise barred weavers from entering guild merchant membership. The members of the guild merchant of Dunfermline, the burgh with the earliest extant guild book,[7] however, reveals that their fraternity included men of very different social standing – gentry, churchmen, graduates, merchants, town clerks, tavern keepers, cordiners (shoemakers), plumbers, weavers, masons and skinners.

The Perth 'Guildrie Book'[8] indicates, in spite of the ruling in its foundation charter, that not only weavers but also a high proportion of other craftsmen – such as baxters (bakers), dyers, furriers and skinners – were accepted into the guild, where the various branches of the leather trade underpinned much of the local economy. In practice, then, craftsmen were accepted into the early guild merchant, no doubt partly depending on the specialisations of the town. It was not the sole preserve of the wealthy élite merchant.

A primary function of the guild merchant was to run the town market efficiently. To ensure that all transactions were open and fair, selling was to take place at the market cross or at official booths at the market place, and not at the quayside, where shady deals might take place surreptitiously, or in 'myrk howsis and quiet loftis'.[9] All weighing had to be overt, at the town's official weighing machine, or tron, and the town's official weights were the only ones approved, to ensure a common standard and discourage cheating.

Punishment for the use of false measures was severe: often banishment from the town.[10] Two of the most serious forms of malpractice at the market were forestalling and regrating. The former – the purchase of goods

Figure 1.2 A tron or weighing beam as depicted on the John Geddy map of St Andrews, *c*1580. Source: Reproduced by permission of the trustees of the National Library of Scotland.

before they reached the open market, so avoiding the payment of toll for use of the market – was heavily punished, as was regrating – the buying in bulk with a view to hoarding and selling at an advantageous time when prices were high. Restrictions were being placed on wholesalers.

The authorities also imposed a firm policy of quality and quantity control. Burgh officers, the *appressiatores carnium*, were appointed to appraise the standard of meat; ale tasters, *appressiatores cervicie* or 'cunnaris', controlled both the quality and price of ale sold at the market; and the price, weight and quality of bread was likewise monitored by the burgh officials. Again, contravention of any of these regulations brought punishment: in 1499, in Dunfermline, anyone found selling ale for more than 8d. (approximately 3 new pence) per gallon would sustain the removal of their 'caldronis and veschallis and dinging out of the bodumis at the Mercat Cros'.[11]

TOWNSMEN AS PRODUCERS – AND SELF-SUFFICIENT?

The guild merchant might be seen as the highest rung of the occupational ladder. But, no matter how efficiently an élite ran the burgh market, its functioning was impossible without one essential factor: people to produce and sell their goods. This sector was the very basis of the occupational structure of the medieval town. But, here again, the guild held certain monopolies. The traffic in staple goods – hides, furs, skins, wool and woolfells – was officially carried out solely by guild members. Some items were also imported from abroad, and merchants travelled regularly overseas. This was not the sole prerogative of the merchants of royal burghs: the merchants of Dunfermline, an abbatial burgh, for example, traded as freely into continental Europe. Furs, wines, spices, exotic fruit, dyes and luxury manufactured goods might reach the weekly town markets and certainly would feature in annual fairs. But these commodities were not the produce of the average townsman. The majority of goods on sale at the weekly markets were produced locally.

It is quite possible that in the early years of burgh life there was little specialisation. Many families would have been almost self-sufficient, selling only any surplus at the town market. Burgh life was initially very much geared to self-sufficiency. With the family home built usually on the front of the burgage plot, or toft, there was an extent of land at the rear – the backlands – where animals might be kept for slaughtering or milking, fowl for eggs or food, and crops grown. This sometimes relatively extensive agricultural space was supplemented by a share of grazing and peat-digging for fuel, and heather and turf extraction for roofing, in the common lands outside the burgh's immediate precinct. Medieval Scottish towns were essentially rural, and town dwellers were often closely involved in agricultural pursuits. Many, indeed, probably had more than one job, taking seasonal agricultural employment. The population of Perth is known to have doubled in the time of harvest in 1584, thus adding to the problems of urban poverty and order.[12]

For how long this self-sufficiency remained the norm for the ordinary people of the town is unclear. There was no reason why early town dwellers might not grow, spin, weave and tailor their own clothes, dress their own leather, supply their own food and drink, and even build their own houses. Froissart chronicles (doubtless with an element of exaggeration) that the Scots found no great problem in rebuilding their homes, since 'with four or five stakes and plenty green boughs to cover them' they were remade almost as soon as they were destroyed or burnt.[13]

Rural pursuits were commonplace in Scottish towns. The elaborate corn exchanges built in the nineteenth century in locations such as Dunbar and Dalkeith are testimony not only to the town's function as a market, but also to the town as a producer of foodstuffs well into modern times.

SPECIALISATION OF CRAFTS

Written sources reveal that gradually specialisation became the norm for many in urban life. Burghs were not merely formally created and artificial, as opposed to a naturally evolving organism; they were also formally laid out. This required, from the outset of burgh life, a specialised skill: that of the town planner. In many burghs there was a quite deliberate planning of streets and burgage plots, often respecting natural features such as rivers, marshes and hills. This, in itself, was little different from the countryside where agricultural activity was determined by geography, geology and climate. In burghs, however, there was established a formal system of territorial planning, which was reinforced and maintained by legal authority. St Andrews was laid out by Mainard the Fleming, who had in all probability already planned Berwick; Ranulf of Haddington left that town to lay out the ecclesiastical burgh of Glasgow.[14] Once formally planned and laid out, the maintenance and subsequent later delineations of the town were supervised by town officers, skilled in measuring, who were specifically appointed to this task: the liners.

From the first days of burghs also, masons, probably peripatetic, required very precise skills to assist with the construction of prestigious buildings such as cathedrals, friaries and castles.[15] Housebuilding became more elaborate in the later Middle Ages and required specific abilities of craftsmen. Slaters became more common as roofing gradually evolved from thatching with heather, turf or water-resistant plants. Documentary evidence suggests that such specialisation was not unique. Family names such as Baxter, Litster, Wright, Walcar and Barker began to appear in the written record. Street names also indicate a congregation of people practising a single craft, whether out of common interest or for reasons of safety – when, for example, craftsmen such as candle makers were encouraged to settle on the outskirts of towns to lessen the risk of fire; or when noxious crafts, such as tanning, were banned to the edges of settlement. Candlemaker Row in Edinburgh, Walkergate in Glasgow and Fisherrow in Musselburgh all illustrate the principal local occupations. Archaeological evidence suggests,

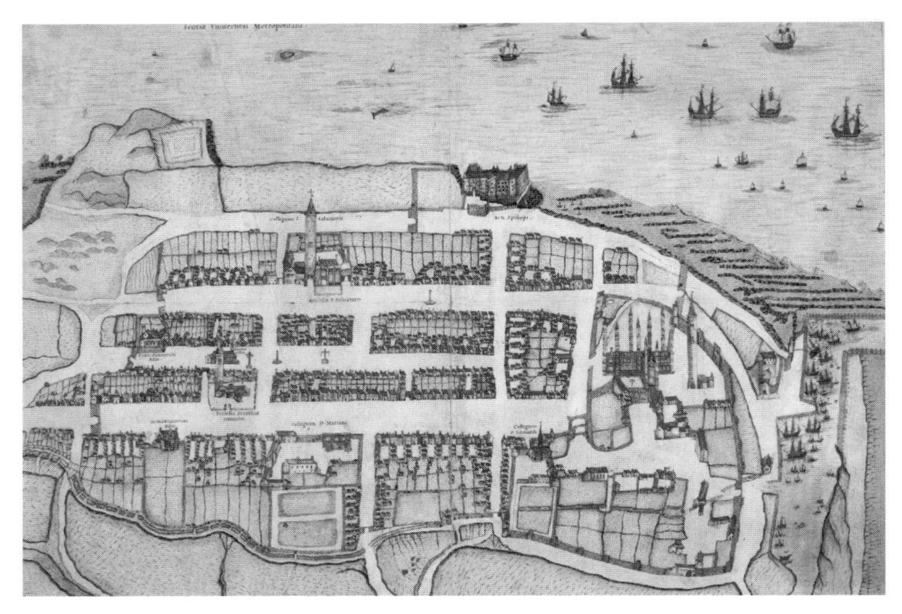

Figure 1.3 St Andrews, *c*1580, showing the east end as laid out by Mainard the Fleming. Source: Reproduced by permission of the trustees of the National Library of Scotland.

also, that an element of specialisation began to creep into towns. Not only was this more cost-efficient, but town life, with people of varied skills living in close proximity, opened the door to the growth of specialised trades or crafts. Initially, much of this manufacturing would have taken place in the backlands of the tofts, and was essentially a home-based activity.

Once specialisation was the norm, supervision of a trade or craft was necessary. Precisely how this was effected depended on the town. In Dunfermline, the responsibility for overseeing the lucrative cloth and leather industries lay with the guild merchant until the early sixteenth century. This virtual economic dominance of a town by one group may have been resented, but the solution was found not by setting up rival craft organisations, but rather by joining the guild.[16] This may, however, have been the reaction of an essentially small town where each would know his neighbours well.

In larger towns, the three-quarters of a century after 1450 were to see radical changes in the urban economic structure. Various crafts were gradually granted a formal status, independent of the guild merchant, as incorporated craft guilds. In Edinburgh, the earliest craft incorporation was that of the skinners, founded in 1474; although, significantly, they had maintained their own altar in the burgh church since 1451, a clear indication of an already existing sense of separateness and specialisation. By 1523, Edinburgh had fourteen craft guilds. These were, however, conglomerate organisations: the hammermen craft, for example, was not merely a guild of smiths, but included under its umbrella seven differing trades, including workers in leather, pewter, gold and tin.[17] Dundee and Perth had nine incorporated crafts each, and Aberdeen and Stirling seven each by 1540.[18] This was a pattern followed by smaller burghs at a more modest rate.

Craft guilds enabled the craft masters to take closer control not only over the quality of goods produced but also over their colleagues, their working practices, and their journeymen and apprentices. The system also guarded against encroachment into the craft by those unskilled or not a member of the craft guild. Through collective interest, mutual aid and fraternity the craft guilds polarised into exclusive sects of tradesmen within the wider burgh community. Production, wholesaling and retailing were, in this medieval system, not specialised operations; there was little distinction between them. What did emerge in the guild system was a close awareness of the relative importance of specific craft guilds. Large crafts, such as the hammermen and the baxters, in particular, had a strong sense of their own importance, displayed clearly in their seat placement at the parish church of St Giles in Edinburgh and openly on saints' days and at the celebration of major feast days such as Corpus Christi, when the crafts paraded through the streets of the town in order of importance.[19]

The production of goods could not, however, be an exclusive, introspective activity. Although the burgh might increasingly be putting down rules to monitor the production and quality of craft goods and to monopolise that activity, it could not exist in a vacuum. Craftsmen were often dependent for a supply of their raw materials from outsiders, be they local or from overseas. This in its turn brought the need for carters and carriers. In Aberdeen, porters – the forerunners of the shore porters – not only employed women as well as men but even paid them the same wage: a sure comment on the women's strength and stamina! The burgh's rural hinterland, in particular, played a pivotal role. Indeed, it might even be argued that without the hinterland the town was incapable of functioning economically. Not only were the occupations of the town dependent on supplies of raw material, the surrounding areas also provided a workforce. Piece-work became more prevalent, a process that was intensified as the Scottish economy moved from a concentration on wool to that of linen. Many craft produces, in reality, ended up as town goods that had been part-made in the countryside.

OTHER OCCUPATIONS

Not all townspeople, however, were members of a craft guild. In ports, various specialised occupations associated with the sea were more common than inland. Ropemakers, net menders, skippers, boatmen and fishermen plied their trade and were to be found in all coastal towns. Kirkcaldy and Dumbarton, along with other seafaring burghs, were the homes of shipbuilders.[20] The admittance of Irishmen as navigators to the status of burgesses in Stranraer in 1689 is a fine comment on that town's role within the Irish Sea context.[21]

There were many, also, found both inland and in coastal towns, with lowlier occupations, such as labourers, carters, servants, cleaners and beggars. A small number functioned as carriers of information and goods in overland communications. Large towns also served as a market for service skills, farming out the more menial tasks to an underclass, many of whom migrated from smaller towns or rural areas to the larger regional centres. Whatever the size of town, supplies of food and drink were essential. Although many of those working in this sector were members of the guild system, there was a vast sub-culture of more menial workers employed both in this field and in agriculture. As mentioned earlier, the population of Perth is known to have doubled in the time of harvest in 1584.[22]

At times of plague or other infectious diseases, specialised services were required. In Dumbarton, for example, in 1604 when plague threatened, watchmen were placed on the town's bridge at the Knowle Burn and at the college church to prevent strangers bringing disease into the town. Forty years later, when plague or possibly typhus did hit, 'cleingers' or cleaners were hired from Paisley to fumigate the house of a deceased woman, the body being handled as little as possible as it was dragged by ropes to the grave.[23] In Aberdeen in 1514, goods suspected of being infected with plague were burned, and cleansers were appointed to clean the houses and gear of victims. When plague struck the town between May and December of 1647, 1,400 died – possibly a fifth of the population. Aberdeen town council estimated that this outbreak cost £30,000 Scots through loss of trade, hiring cleansers, burying the dead, building huts and a gibbet to house and control the infected, and purchasing food for the poor.[24]

Ecclesiastics were an intrinsic part of life in medieval towns. The parish church was a focal point of burgh life, all towns being single-parish in the Middle Ages. From birth and baptism until death and the last rites, church life and its practitioners pervaded urban living. The church bell called the sleeping to work and tolled the curfew at night, and the church cycle determined the weekly and the seasonal routine through holy days and religious processions. Clerics were the teachers, the source of much medical knowledge, and the notaries public, assisting not only with the writing of letters but also the signing of names, with a hand held by the notary, and verifying merchant marks. Towns also attracted friaries to their outskirts. The friars' teaching and preaching brought knowledge and entertainment to townspeople. Although a minority, churchmen were a vital part of urban life. And sometimes, as in Old Aberdeen, ecclesiastics were a sizable minority. Here, almost one adult in four was a cleric. In nearby New Aberdeen in 1500, with a population of about 4,500, one adult male in twenty would have been a priest, chaplain or friar.[25]

A large section of urban society appears scantily in the records but had a crucial role to play in the occupational and economic structure of the town – women. Women rarely feature in guilds merchant or craft guilds. Indeed, they had little or no official role within the burgh hierarchy. Their presence, however, was one that was keenly felt within burgh society. The records suggest that much of the brewing within towns was the remit of women.[26] In Aberdeen, women are noted as brewers, bakers and shopkeepers, one woman even being described as a merchant in the 1690–1 customs book.[27] Simpler crafts, such as baking, brewing and candle making, might be undertaken by women at home with the intention of supplying the family. A surplus could then be sold at the market.[28] Women also, on occasion, feature as ship owners, custumars (customs officers) and money lenders.[29] One occupation that fell to women was that of prostitution.[30] Before 1750 it was the custom throughout Europe for well-to-do women not to breastfeed their own children. Only the very wealthy could afford live-in wet nurses; the majority would hand over their children to a married woman in a rural village who took urban youngsters into her home. Those who did succeed in obtaining employment as a wet nurse in the town might receive £20 cash per year, double what a woman might earn as a domestic servant.[31] Women, particularly widows, played an important role in the economy. This somewhat gives the lie to the theory that women were an underclass in society. Certainly, men dominated the more prestigious occupations, but many women were capable of running the family business during the absence of a husband overseas.

The changing role of women is witnessed in the rise of schools for girls, particularly in the 1600s. Girls were often taught by women teachers. Dumbarton, Dundee, Haddington, Kirkcaldy, Leith South and North, Montrose and Paisley all have recorded dame schools. The first official, regulated school for girls in Aberdeen came in 1642, with the mortification of £1,000 Scots by Catherine Forbes, Lady Rothiemay. The rhyme in the commonplace book of the master of the music school in Aberdeen would suggest that some at least in that town felt that women were as able as men to educate:

> The weaknes of a womanis witt
> Is not to natures fault
> Bot laike of educationne fitt
> Makes nature quhylls to halt.[32]

CHANGING PATTERNS OF OCCUPATIONS?

Medieval towns may, then, be seen as market centres which provided both goods and services. But these goods and services were worthless without purchasers. If the town was to fulfil its primary occupation – that of trader – it had to attract buyers. The rural hinterland had, yet again, a prominent role in the survival of the town. Some burgh charters contained a radical and far-reaching privilege for the endowed town. Many were granted the sole right to trade over an extended rural hinterland; and, importantly, it required all the inhabitants within this specified landward area to market their goods in the burgh of their locality. All who attended the market were obliged to pay tolls for the privilege, unless a free burgess. These dues might be collected at the town gates, or ports, or at the tolbooth. It was here that the town weights were normally housed; and the tolbooth, with the market cross, became the symbol of the burgh's secular status as a trading centre. Transport into the town market was regulated also by the burgh officers, particularly in the case of meat, which had to be visibly fresh. The consequent relationship of a town with its surrounding countryside and the extent and prosperity of this rural neighbourhood were to have a profound influence on the economic success and the occupational structure of the town.

However, the occupational makeup of individual towns varied from town to town, and from time to time. But it is fair to say that increasing size and material wealth brought about both economic diversification and stratification within urban occupations. As bodies formalised, internal dissensions might appear, and in the late fifteenth and early sixteenth centuries there was in larger towns a hardening of roles and the emergence of a certain élitism in both craft and merchant organisations. Moreover, within Scotland itself, the development of a pyramid, with a shifting foundation, a jockeying for middling status, and an increasingly constricting stranglehold by Edinburgh at the summit would play havoc with the occupational structures of many lesser towns. Allied to this, towards the end of the Middle Ages and into the seventeenth century, a proliferation of new burghs was created. The traditional occupational structure of the medieval burgh was, yet again, challenged.

War also affected life in towns. Although many burghs in the Middle Ages, particularly in the Borders region, had suffered from the attack of unfriendly neighbours, the seventeenth century was to witness an increasing phenomenon, which was to reach its zenith in the next century: that is, occupation by soldiers, whether friend or foe. Linlithgow, Dalkeith, Dumbarton and Ayr all played host to Cromwellian troops, who were billeted on them for quite extensive periods of time. The impact on the social and occupational balance of the town could be profound, with the demands for food, candles, blankets and bandages not only creating a new market, but also a dearth for the local people. In some cases, the burgh became a garrison town. Stornoway virtually disappeared as a town, becoming, in effect, one large military camp.[33]

When an analysis is made of the occupational structures of different towns at the end of the period under discussion, it is not surprising to find that Edinburgh distinguished itself clearly from other burghs, containing a prominence of professions. Practitioners of law accounted for some 62 per cent of the professions, and medicine a further 22 per cent.[34] Surgeons, physicians and apothecaries all played an important role in Edinburgh medicine.[35] The church, education and the army constituted a significant proportion of the professions, and landowners, with their own urban households, were an important minority of the population.[36] In the capital of Scotland this is not surprising, nor is the vast array of back-up services that the capital and its professions required. From domestic service, both male and female, to the production of luxury items, to prostitution, Edinburgh was a draw for skilled and unskilled workers. The existence of goldsmiths, silversmiths, wig makers, coach builders, printers, booksellers, pistol manufacturers and clock makers is a reflection of the demands of a capital city.

The professions featured prominently also in larger burghs that were heads of sheriffdoms, such as Aberdeen and Perth. Certain other relatively large burghs, however, such as Dalkeith, Musselburgh and Leith, had few professional occupations, their proximity to Edinburgh negating the need for such facilities. Similarly, such burghs had fewer merchants proportionately than other burghs. Again, this resulted from proximity to Edinburgh, whose merchants dominated the trade of nearby towns. The impact of larger burghs on their smaller neighbours and the networking of satellite towns is also evident by the end of the seventeenth century. Musselburgh, a satellite town of Edinburgh, took advantage of the market that the capital offered. Vegetables were grown and salt panned, and both were then carried in creels on the backs of women to the city. On their return, the women brought back washing, which in its turn boosted the sales of the several soap boilers and starch makers in the Musselburgh district.[37] Water carriers also found a market from these occupations. Dalkeith had a high proportion of fleshers, which may well have come about because the slaughtering of animals in Edinburgh had been banned in 1695,[38] and Dalkeith was clearly supplying meat to the capital. Interestingly, the near neighbour of Dalkeith, Musselburgh, had a high degree of concentration of leather manufacturers:[39] Dalkeith's hides were in all probability being processed in Musselburgh.

Figure 1.4 The Edinburgh Trades in front of Holyrood Palace, *c*1685–1744, by Roderick Chalmers. Source: Reproduced with permission of the trustees of the National Galleries of Scotland.

Figure 1.5 Musselburgh High Street in the late nineteenth century, with sixteenth-century tolbooth. The town is still essentially rural – with the town herd leading sheep along the High Street to the Common Green. Source: Reproduced by permission of RCAHMS, Crown Copyright.

The use of the general term 'merchant' until this time suggests that, as in medieval times, trading was still unspecialised. Interestingly, however, certain sources indicate that some craftsmen had their own shops and dealt, at least in part, with the distribution of their own manufactures. Production, wholesaling and retailing were, as in medieval times, clearly still not distinctive occupations.[40]

The marked difference between the occupations of coastal and inland towns remained constant, as did the paucity of people involved in overland communications. Interestingly, Bo'ness, Grangepans, Newark and Greenock had large proportions of the pollable population involved in shipping and transport: 18.8 per cent, 23.8 per cent, 26.8 per cent and a staggering 38 per cent respectively.[41]

A NEW ECONOMY AND AN OLD?

The seventeenth century was to see the rise of new manufactories in larger towns, with encouragement from parliament to establish such specialised industries, as well as the fostering of overseas trade. Colonies were established in New Jersey and South Carolina by the 1680s, at which time the West Indies were also a vital trading partner. Such commercial contacts resulted in soap works and sugar refining, in Glasgow in particular. Here, in 1688, a 'soaperie' was established, with permission for another in the

Candleriggs five years later.[42] As a direct result of trade with the West Indies, sugar houses were set up; the first, the Wester Sugar House, being established in 1667 and extended in 1675. Two years later it was again enlarged, and a further process was added to sugar refining, rum being distilled from the waste molasses. In 1669 the Easter Sugar House was set up, and by the turn of the century a further two houses had been established to boil sugar, make candy and distil spirits.[43]

Woollen manufactories and glassworks were also established, alongside traditional crafts such as tanning. The Incorporation of Weavers in Glasgow had by 1700 three joint-stock companies for weaving both woollen and linen cloth, one employing as many as 1,400 people.[44] There is evidence also of a new commodity appearing on the market: tobacco. This would also affect the employment structure, as by 1668 a pipemaker was established in Glasgow, and two years later the magistrates issued licences to dig for pipe clay.[45]

At the end of this period, throughout all towns manufacturing was still the mainstay of burgh life, employing half of the recorded male workforce in many cases. Food and drink production were still essential urban occupations, usually accounting for somewhere between 15 and 20 per cent of the workforce.[46] And, as ever, the staples of urban life – prostitutes and beggars – gravitated to towns. Interestingly, the medieval rural characteristics of town life survived strongly in smaller burghs. Where the backlands of tofts had not been developed with outbuildings or dwellings – burgage repletion – the traditional use of the land for growing food and housing animals, continued. In smaller towns, the medieval culture of self-sufficiency had not been totally eradicated, and their occupational structure in 1700 remained broadly similar to that of 1100.

NOTES

1. Dennison and Simpson, 2000, 719–20.
2. Dickinson, 1945–6, 224; Maitland, 1897, 193.
3. Ballard, 1916, 16.
4. When 'x' is used in a date range it indicates that the topic under discussion occurred sometime between the range given.
5. Dennison, 1996, 215; Torrie, 1988, 245.
6. *Leges Burgorum*, cl. xciv.
7. Torrie, 1988, passim.
8. Stavert, 1993, passim.
9. MS Dundee Burgh and Head Court Book, 4 October 1521.
10. MS Dundee Burgh and Head Court Book, 9 October 1523.
11. Beveridge, 104.
12. Lynch, 1988, 263 and 282.
13. Froissart, 1812, ii, 7.
14. Lawrie, 1905, no. 169; Marwick, 1894–97, I, pt ii, 5.
15. Dennison and Coleman, 1998, 97.
16. Torrie, 1988, 252–3.
17. Lynch, 1988, 262.

18. Lynch, 1988, 261; Marwick, 1909, 47–73; Bain, 1887, 308; Murray, 1924, I, 359; Renwick, 1887–9, 274–7.
19. Bain, 1887, 56.
20. Torrie and Coleman, 1995(a); TA, ii, 278, 279; *TA*, iv, 97, 458; Dennison and Coleman, 1999, *TA*, iii, 368.
21. Torrie and Coleman, 1995(b), 18; MS Stranraer Burgh Court Book, 1596–1672, 9 May 1689.
22. Lynch, 1988, 263.
23. Dennison and Coleman, 1999, 31; Dumbarton Burgh Archives, MS1/1/0, 49; Roberts and MacPhail, 1972, 158.
24. Tyson, 2002, 114.
25. Lynch et al., 2002, 289.
26. Lynch, 1988, 277.
27. Tyson, 2002, 122.
28. Blanchard et al., 2002, 139.
29. Dundee Burgh and Head Court Book, 6 October? (insert in 1523 records); ER, i, pp.xcv, 76, 95, 170–2, 275, 316–17, 36; Sanderson, 1987, 91–102.
30. Dennison et al., 2002, 75–6.
31. DesBrisay, Ewan and Diack, 2002, 57.
32. Vance, 2002, 319.
33. Dennison and Coleman, 1997, 24–5.
34. Whyte, 1987, 222.
35. Dingwall, 1995, 3 and passim.
36. Whyte, 1987, 222.
37. Whatley, 1990, 57; Paterson, 1861, 61.
38. *APS*, ix, 361.
39. Whyte, 1987, 233; Dennison and Coleman, 1996, 36.
40. Whyte, 1987, 226.
41. Whyte, 1987, 224.
42. Marwick, 1876–1916, iii, 112 and 173.
43. Smout, 1961–2, 241–3.
44. Gibb, 1983, 45.
45. Marwick, 1876–1916, iii, 112 and 147.
46. Whyte, 1987, 227–8.

BIBLIOGRAPHY

Primary Sources

Thomson,T and Innes, C. *Acts of the Parliaments of Scotland*, Edinburgh, 1814–75.

Assise Regis Willelmi. In Innes, C, ed, *Ancient Burgh Laws*, Edinburgh, 1868, 60–62.

Beveridge, E, ed. *Burgh Records of Dunfermline*, Edinburgh, 1917.

Dickson, T, Paul, J B and McInnes, C T, eds. [*TA*] *Accounts of the (Lord High) Treasurer of Scotland*, Edinburgh, 1877–1913.

Dundee Burgh and Head Court Book, Dundee and District Archive Centre.

Dumbarton Burgh Archives, MS1/1/0, Dumbarton Public Library MS collections.

Stuart, J, Burnett, G, Mackay, Ae. J G and McNeill, G P, eds. [*ER*] *Exchequer Rolls of Scotland*, Edinburgh, 1878–1908.

Froissart, J. *Chronicles of England, France, Spain, Portugal, Scotland, Brittany, Flanders and the Adjoining Countries*, from the French by J. Bourchier. Reprinted from Pynson's edition of 1523 and 1525, London, 1812.

Lawrie, A C, ed. *Early Scottish Charters prior to 1153*, Glasgow, 1905.
Leges Burgorum. In Innes, C, ed, *Ancient Burgh Laws*, Edinburgh, 1868, 4–58.
Marwick, J D and Renwick, R, eds. *Charters and Other Documents Relating to the City of Glasgow*, Glasgow, 1894–97.
Marwick, J D, et al., eds. Extracts *from the Records of the Burgh of Glasgow*, 11 vols, Edinburgh, 1876–1916.
Renwick, R, ed. *Extracts from the Records of the Royal Burgh of Stirling*, Glasgow, 1887–9.
Roberts, S and MacPhail, I M M, eds. *Dumbarton Common Good Account*, Dumbarton, 1972.
Statuta Gildae. In Innes, C, ed, *Ancient Burgh Laws*, Edinburgh, 1868, 64–88.
Stavert, M L, ed. *The Perth Guildry Book, 1452–1601*, Edinburgh, 1993.
Stranraer Burgh Court Book, 1596–1672 (MS ST1/1/0).
Torrie, E P D, ed. *The Gild Court Book of Dunfermline, 1433–1597*, Edinburgh, 1986.

Secondary Sources

Bain, E. *Merchant and Craft Guilds: A History of the Aberdeen Incorporated Trades*, Aberdeen, 1887.
Ballard, A. The theory of the Scottish burgh, *Scottish Historical Review*, xiii (1916), 16–29.
Blanchard, I, Gemmill, E, Mayhew, N and Whyte, I D. The economy: Town and country. In Dennison E P, Ditchburn, D and Lynch, M, eds, *Aberdeen before 1800: A New History*, East Linton, 2002, 129–58.
Dennison, E P. Burghs with gilds merchant by 1550. In McNeill, P G B and MacQueen, H L, eds, *Atlas of Scottish History*, Edinburgh, 1996, 215.
Dennison, E P and Coleman, R. *Historic Melrose: The Archaeological Implications of Development*, Edinburgh, 1998.
Dennison, E P and Coleman, R. *Historic Stornoway: The Archaeological Implications of Development*, Edinburgh, 1997.
Dennison, E P and Coleman, R. *Historic Musselburgh: The Archaeological Implications of Development*, Edinburgh, 1996.
Dennison, E P and Coleman, R. *Historic Dumbarton: The Archaeological Implications of Development*, Edinburgh, 1999.
Dennison, E P and Simpson, G G. Scotland. In Palliser, D, ed, *The Cambridge Urban History of Britain* (vol i, 600–1540), Cambridge, 2000, 715–37.
Dennison, E P, Ditchburn D and Lynch, M, eds. *Aberdeen before 1800: A New History*, East Linton, 2002.
Dennison, E P, DesBrisay, G and Diack, L. Health in the two towns. In Dennison, Ditchburn and Lynch, 2002, 70–96.
DesBrisay, G, Ewan, E and Diack, L. Life in the two towns. In Dennison, Ditchburn and Lynch, 2002, 44–69.
Dickinson, W C. Burgh life from burgh records, *Aberdeen University Review*, xxi (1945–6), 214–26.
Dingwall, H. *Physicians, Surgeons and Apothecaries: Medical Practice in Seventeenth-century Edinburgh*, East Linton, 1995.
Gibb, A. *Glasgow: The Making of a City*, Glasgow, 1983.
Lynch, M, Spearman, M and Stell, G, eds. *The Scottish Medieval Town*, Edinburgh, 1988.
Lynch, M. The social and economic structure of the larger towns, 1450–1600. Lynch, M, Spearman, M and Stell, G, eds. *The Scottish Medieval Town*, Edinburgh, 1988, 261–86.

Lynch, M, Desbrisay, G and Pittock, M G H. The faith of the people. In Dennison, Ditchburn and Lynch, 2002, 289–308.
Maitland, F W. *Domesday Book and Beyond*, Cambridge, 1897.
Marwick, J D. *Edinburgh Crafts and Guilds*, Edinburgh, 1909.
Mill, A J. *Mediaeval Plays in Scotland*, Edinburgh, 1927.
Murray, D. *Early Burgh Organisation in Scotland*, Glasgow, 1924.
Paterson, J. *The History of the Regality of Musselburgh*, Musselburgh, 1861.
Sanderson, M H B. Janet Fockart: Merchant and money-lender. In Sanderson, M H B, *Mary Stewart's People: Life in Mary Stewart's Scotland*, Edinburgh, 1987, 91–102.
Smout, T C. The early Scottish sugar houses, 1660–1720, *Economic History Review*, xiv, 1961–62, 240–53.
Thomson, D. *The Dunfermline Hammermen: A History of the Incorporation of Hammermen in Dunfermline*, Paisley, 1909.
Thomson, D. *The Weavers Craft, Being a History of the Weavers Incorporation of Dunfermline*, Paisley, 1903.
Torrie, E P D. The guild in fifteenth-century Dunfermline. In Lynch, Spearman and Stell, 1988, 245–60.
Torrie, E P D and Coleman, R. *Historic Kirkcaldy: The Archaeological Implications of Development*, Edinburgh, 1995(a).
Torrie, E P D and Coleman, R. *Historic Stranraer: The Archaeological Implications of Development*, Edinburgh, 1995(b).
Tyson, R E. People in the two towns. In Dennison, Ditchburn and Lynch, 2002, 111–28.
Vance, C. Schooling the People. In Dennison, Ditchburn and Lynch, 2002, 309–26.
Whatley, C. *The Scottish Salt Industry, 1570–1850*, Aberdeen, 1990.
Whyte, I D. The occupational structure of Scottish burghs in the late seventeenth century. In Lynch, M, ed. *The Early Modern Town in Scotland*, London, 1987, 219–44.

2 The Making of Industrial Scotland, 1700–1900: Transformation, Change and Continuity

CHRISTOPHER A WHATLEY

The 200 years from around 1700 to the start of the twentieth century witnessed a spectacular transformation in Scotland's economy. Accompanying this were profound changes in Scottish society. Amongst these were radical developments in the nature of work. At the start of the period, around eight out of ten Scots lived and worked in the countryside, which was characterised by scattered farmsteads, cottages and hamlets, and a few nucleated villages. Although rural society was becoming more commercialised, most rural dwellers spent at least some part of their time producing food, fuel, tools and furnishings for their own use. Nevertheless, as shown in the previous chapter, there was increasing demand for the services of skilled craftsmen such as masons, wrights and thatchers, as well as for the labour of the semi- and unskilled – for ditching and harvesting, for instance – normally on a daily basis. Longer contracts were used to bind the thousands of young females who worked on a year-long basis in farm service in Scotland prior to marriage.

Reference to workers by a single occupation is misleading however: few people made a living by sticking to a single trade; adaptability was crucial, and remained so well into the nineteenth century.[1] Remuneration would usually include a sum of money but also payment in kind, for instance in the form of shelter, food and drink, although there were enormous variations in what was a highly unstandardised system, despite wage rates being set by the justices of the peace in several counties. The pattern of work was largely determined by the seasons, the needs of an employer for labour to perform a specific task, and the weather. Wages were usually lower in the winter, when there were fewer daylight hours. Estimates vary, as did experience, but it is highly unlikely that many day labourers found work on even two-thirds of the three hundred and sixty-five days of the year; in some cases far fewer.[2] For most households subsistence living was the norm, with periodic windfalls bringing about brief periods of respite. It was easier and more common to slip into poverty, which was endemic.

It was in the towns that most full-time work was carried on, with regulated hours heralded at the commencement of the day at 5 a.m. or

6 a.m. by the burgh's piper or drummer. Colliers were expected to start at 4 a.m. or even earlier. Between twelve and fourteen hours' work was the norm. Employment as a journeyman in building, tailoring, shoemaking, candle making or the myriad of urban crafts that were found in most burghs required a lengthy apprenticeship, although, as in the countryside, there was a reserve of unskilled men and women – perhaps a third of the urban population – who eked out a living through a combination of casual work, begging, occasional recourse to poor relief and, periodically, petty crime. Domestic service was a source of regular employment for females, many of whom travelled long distances to secure work in the bigger towns of Edinburgh, Glasgow, Dundee and Perth, although they were particularly numerous in St Andrews as well.[3] Edinburgh had its shopkeepers and retailers, women as well as men, along with sick-nurses, wet nurses and washerwomen, although it is unlikely that these occupations were confined to the capital.[4]

A degree of regional and local specialisation had occurred by the end of the seventeenth century. Edinburgh and regional centres such as Aberdeen housed higher proportions of professionals such as lawyers. Towns in which manufacturing employed as many as half of the resident adult males were also to be found, as in Perth (glove making), for example, and Paisley, which by the 1690s had already established itself in textiles.[5] In and around some

Figure 2.1 Housemaid brushing the stairs, 19 Corrennie Gardens, Edinburgh, 1903. Source: SLA W564418.

of the older burghs, but also the newer burghs of barony, clusters of coal workers were to be found, ranging from fewer than a handful at the smallest pits to bigger numbers plus their female bearers at the larger collieries like Wemyss in Fife or Woolmet in Midlothian. Along the shores of the firth of Forth, and in Ayrshire, salt manufacturing had become sufficiently prominent to give some places their names, as in Prestonpans, Grangepans and Saltcoats.[6]

Whilst a range of goods was processed or manufactured in Scotland, the scale of operations was small and quality was generally poor, apart perhaps from the work of some silversmiths and gun makers. Architects like William Bruce and James Smith heralded a new age in Scottish building, aided by master masons from the Mylne family, although in keeping with a society that maintained strong ties with Europe, French and Dutch influences were plainly evident.[7] Yet it is striking that even though the country's most important manufactured commodity was linen cloth, accounting for some 22 per cent of total exports in 1704, the value of imported linens, muslins and cotton was even greater.[8] The Scottish parliament and privy council was in the practice of granting special privileges to new enterprises in order to relieve the strain on the country's balance of payments, but few survived for long. Of forty-seven joint-stock companies formed between 1690 and 1695, only twelve appear to have been in business in 1700. One state-supported industry was glass making, but in 1695 there was only a single glasshouse in Scotland, compared to almost ninety in England.[9] Scotland's was an underdeveloped economy, dependent for roughly 60 per cent of its export income from raw materials, minerals, foodstuffs, semi-manufactures and low-grade manufactured commodities like marine salt, made by boiling sea water in iron (imported from Sweden) pans.[10] Discerning, fashion-conscious consumers, the proprietors of new or refashioned houses, imported their furniture, mirrors, clocks and crockery, much of it from China or Japan, although the products and culture of the Low Countries were deeply admired too.[11]

By 1900 Scotland had become one of Europe's most densely urbanised regions, its economy dominated by heavy industries such as shipbuilding and engineering, mining, iron and steel making and the production of textiles. The transformation, which no one had predicted for an ailing economy which in 1700 lagged behind that of Ireland,[12] had been remarkably rapid. Indeed as early as 1850 Scotland had become more industrialised than the rest of Britain.[13] The pace had been set by cotton, production of which soared between 1780 and 1830. Around 1826, nine out of ten workers in manufacturing in Scotland had been employed in textiles, and 60 per cent of these were in cotton.[14] Although cotton lost its place in the vanguard of Scottish industrialisation, it continued to enjoy long periods of success, intermingled with shorter bouts of depression; long-term decline seems to have set in only at the end of the 1860s.[15] Linen, too, continued to grow, although more slowly than it had in the seventeenth and eighteenth centuries, when it had led manufacturing in Scotland, and in terms of entrepreneurial stock,

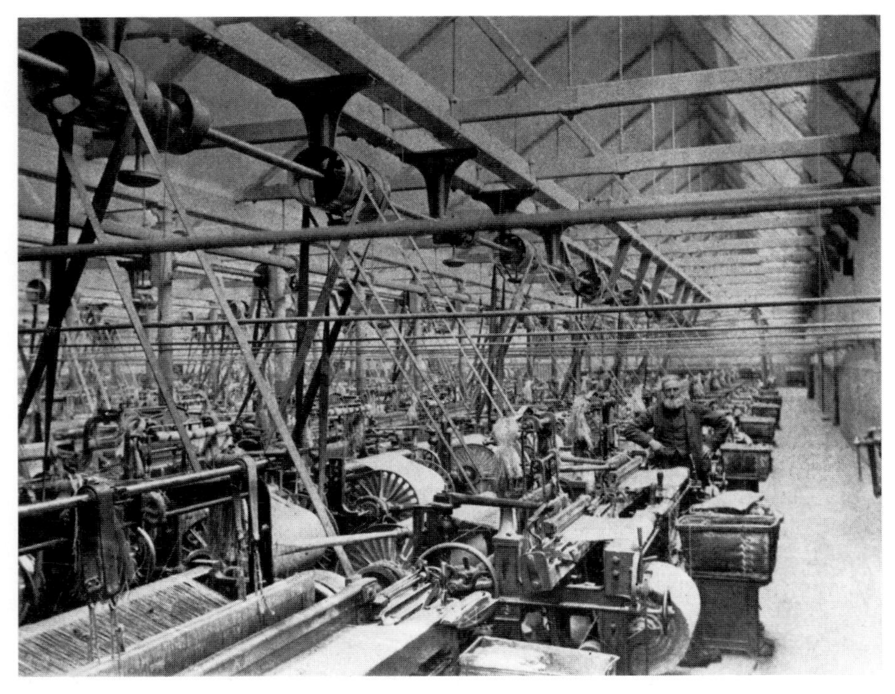

Figure 2.2 Weaving department at Caldrum Works, Dundee, 1875. Reproduced by permission of Dundee University Archives.

capital and skills laid the foundations of the cotton industry.[16] In Dundee it was partly superseded by jute, in which commodity the city led world production for a time in the 1850s, '60s and '70s. Although in output terms the jute industry peaked in the early 1900s, international competition had become intense in the preceding two decades.[17]

Woollens, another Scottish manufacture with roots stretching back to medieval times but which was struggling to survive around the time of the Union of 1707 (after which it suffered further, due to English competition), expanded substantially from the end of the eighteenth century, although not without a sharp check at the end of the 1820s. The Borders region, as well as the southern reaches of the Ochil hills, provided ample running water as well as supplies of wool from Cheviot sheep, reared increasingly enthusiastically in the Southern Uplands from the 1790s – although they had been introduced earlier. With a shift in favour of tweed cloths, checks and tartans partly inspired by Sir Walter Scott's writings, the number of mills doubled between 1835 and 1850. By 1871 there were some 218 woollen mills in Scotland, mainly in the south, but including the sizable works of the makers of high quality overcoatings, J & J Crombie of Abrdeen.[18] Although less dependent upon exports than the other Scottish textiles, that half of the woollen cloth made in Scotland was sold overseas in the nineteenth century is an outstanding achievement, given that virtually none could be sold abroad in the early 1700s.

Metals, primarily iron, had been made in Scotland from the sixteenth century, perhaps earlier, but on a very small scale. Lead mining and smelting date back to the thirteenth century, but even when lead production was at its height, in the early 1800s, no more than five or six thousand tons was extracted in a year, and probably no more than a thousand or so miners and process workers were engaged.[19] Iron production had risen in the eighteenth century, most notably with the foundation of what would become the world-renowned ironworks at Carron, in 1759, although other works were opened in Lanarkshire and Ayrshire prior to the end of the century.[20] From around 1830, following the introduction of Neilson's hot-blast furnace, which had a dramatic effect on production costs, pig iron production rose rapidly to alter the character of the Scottish industrial economy. From 5 per cent of UK iron output in 1825, Scotland's share rose to 25 per cent in 1840. Production peaked in 1869–70 at a spectacular 1.2 million tons.

Demand for coal (rising anyway as the number of domestic hearths and stoves grew) for conversion into gas for heat and lighting, for numerous industrial uses at home, and for export, surged. Output, which was around a quarter of a million tons in 1700, rose to a staggering 35.4 million at the turn of the twentieth century, close to its all-time peak.[21] Over 100,000 miners were employed, a number that would rise still further before long-term decline set in after 1925. Coal and iron provided the raw materials for the shipbuilding and engineering industries, although craft skills and the labour of the unskilled had to be bought at lower costs to achieve Scottish ascendancy in these sectors.[22] Steel making – for ships' plates – took off later, using imported ores, but by 1900 the Scottish steel industry was turning out over 960,000 tons each year.[23] From a country with few manufactured exports, Scotland became one that was arguably over-dependent upon a limited range of interconnected industries. Scotland dominated the world's supply of locomotives – most of which were made at Springburn, Glasgow – and was the source of thousands of steam engines and boilers as well as iron bridges,

Figure 2.3 Airdrie-built steam locomotive manufactured by the Airdrie Iron Company, 1882. Reproduced by permission of North Lanarkshire Archives.

gas-work machinery and machine tools that were exported throughout the British empire. The world's cheapest pig iron was also made in Scotland.[24] In shipbuilding, which became the unsteady barometer of the economic health of the West of Scotland, Clydeside replaced London as the centre of British ship construction and accounted for 18 per cent of the world's output; three firms with Scottish connections, using mainly Scottish-built ships, secured large swathes of the world's shipping lanes: P & O, Cunard and William Mackinnon's British India Steam Navigation Company. Several smaller shipping companies also became household names.[25]

Economic change in Scotland in the eighteenth and nineteenth centuries was not simply a matter of expanding markets and output. Regional specialisation was part and parcel of the process. Thus, while in the middle of the eighteenth century flax was spun virtually everywhere in Scotland, a century later large-scale linen production was confined to the counties of Fife and Angus, which included Dundee. The former gained a reputation for the manufacture of fine linen cloth – damasks – in Dunfermline (but, by contrast, linoleum was made at Kirkcaldy), while Dundee and surrounding towns such as Forfar reigned supreme in the production of coarse linen and jute.

Perthshire, which had probably been the main Scottish centre for linen yarn and cloth in 1700, with major markets for both in Perth and Coupar Angus, had few links with the industry by 1900, although very early in the eighteenth century there had been indications that Dundee had begun to establish itself as a rival centre, partly by reducing its customs duties. As has been seen already, Paisley had also confirmed its place in the West of Scotland

Figure 2.4 Canmore Linen Works, Forfar, Angus, built 1867. Reproduced by permission of RCAHMS, crown copyright.

linen trade at an early date, later strengthening its grip in fine textiles through silk weaving, cotton spinning and thread making. Yet the concentration of industry in particular regions or districts was not a straightforward process. Cotton spinning, for example, which by 1840 was virtually restricted to Glasgow and Paisley (a handful of important water-powered mills such as Stanley in Perthshire excepted), had in the 1770s and 1780s spread widely, with the early entrepreneurs setting up mills on streams and rivers as far apart as Gatehouse-of-Fleet in Galloway, Spinningdale in Sutherland in the north, west as far as Rothesay on the island of Bute, and in the east at Aberdeen and Penicuik, the first in Scotland, which commenced in 1776.[26] By the middle of the nineteenth century, however, most of these locations had seen the last of cotton; Aberdeen, where textile manufacturing had been the main industry in 1800, was suffering from prolonged depression as the town struggled to compete with better-placed rivals to the south and west.[27]

In the extractive industries, shipbuilding and engineering, similar patterns occurred, although by no means identically. Thus in shipbuilding, with the increased use of iron and steam power, and rising capital costs, the main yards became concentrated on the Clyde, where supplies of key components were more easily obtained. Ports such as Aberdeen (perhaps the main employer of shipyard labour in 1831), Dundee and Leith were restricted to the construction of smaller iron-hulled boats, or wooden vessels, for specialist uses like fishing (Aberdeen) and whaling (Dundee), although in some cases these ventures were highly successful.[28] Indicative of the general trend was the firm of Alexander Stephen, which had begun business in Aberdeen but left in 1830 for Dundee and made wooden vessels, opened an iron shipyard at Glasgow in 1851 and later ended up further down the Clyde as important builders of ships' engines.[29]

Coal mining, of course, was restricted to those districts where workable coal lay underground. By the early 1870s, however, around 80 per cent of Scottish coal mined was coming from west central Scotland. But as former markets declined and new opportunities arose, and the coal and iron companies moved in search of new and thicker seams – following the railway south into Ayrshire during the 1840s, for example – the locus of operations shifted. The Fife coalfield grew in importance from the 1880s, with markets being tapped in the Baltic, northern Europe and Russia. Ayrshire's importance, on the other hand, declined as the most productive deposits were worked out.[30] Other parts of the country with particular natural advantages saw them exploited, too, as in the case of Aberdeen granite, which had been used for London pavements from the 1760s.[31] Activity of this sort was not confined to the Scottish mainland. As far north as Shetland, copper and iron ore were mined in the 1870s, while in the mid-1880s unusually heavy demand for granite inspired an Aberdeen-based company to open a quarry at Hildasay.[32] The tentacles of the Industrial Revolution reached the Western Isles and Orkney, too, with demand for kelp drawn from the coastal shores and sea lochs intensifying towards the end of the eighteenth century. It was used by glass manufacturers and soap makers in the south, as well as for dyes, and although demand collapsed in the 1820s, with drastic consequences for the thousands of crofter–kelpers who depended on the cash income provided by kelp gathering and burning, its use as a source of iodine led to a partial revival in the second half of the nineteenth century.[33] With woodland unusually scarce in Scotland (in comparison to the rest of Europe), proprietors of wooded estates improved their management from around the mid-1700s and seized the available opportunities to exploit the market for timber, thereby generating localised sawmill and related work, although the pace of cutting slowed with a fall in timber prices some time after 1866.[34]

As market opportunities expanded, and in order to better exploit these, at the level of the individual entrepreneur or firm ways were sought to increase output and, where possible, improve quality. There were acute skills shortages in Scotland, and in many sectors including mining, iron making, glass working, pottery and textiles, workers had to be brought north from England or further afield to impart their knowledge.[35] Profit levels and a company's ability to compete were affected by production costs, and employers looked for ways to reduce these.

It had been rare for more than a handful of workers to be employed under a single roof at the turn of the eighteenth century, although some large workshops or proto-factories had been erected.[36] One of the most notable was the woollen manufacture at Newmills, founded in 1683, with another being established in Glasgow in 1699, which employed up to 1,400 people.[37] Expansion of output, however, did not necessarily mean concentration of production. The costs involved in the construction of large premises could be high. During the first half of the eighteenth century employers could first of all utilise what access they had to credit rather than resorting to fixed capital, and tap pools of underemployed labour in Scotland, offering paid work. Women in particular were an underused resource. There were reports that females in some rural districts had little to do for half of the year; at the lead-mining village of Leadhills in 1747, it was the existence of 400 families whose womenfolk were 'quite idle' that persuaded the British Linen Company to put out flax to be spun there, in return for which it was estimated that those so employed might earn between 1s. 4d. (7 pence) and 2s. 4d. (12 pence) a week.[38] While the main user of domestic labour was the textile industry, work was given out in countless other trades, too, including the partners of Carron Company, who at the time they were consolidating the works near Falkirk, distributed rod iron to men and boys in parts of Stirlingshire and Fife to be made into nails.

This system – putting out – had obvious advantages but also drawbacks, including lack of quality control, theft by workers of raw materials, and irregular working. These persuaded some masters to gather their employees together in to a place where they could be better supervised. It was largely in order to create an environment in which they could more easily monitor their workers' timekeeping and workplace practices that the proprietors of the Carron ironworks built a wall round the works and appointed porters to man the gates.[39]

Figure 2.5 Falkirk, Stirlingshire, showing the Carron ironworks in the distance, *c*1790. Source: SLA S06026.

The appearance in Scotland in the 1760s and 1770s of mechanical spinning devices from England – Hargreaves' spinning jenny, Crompton's mule and Richard Arkwright's water frame; the last two requiring water or animal power to drive them – provided another incentive to begin to invest in the prime symbols of the Industrial Revolution, the cotton mills. All three machines were adopted, with jennies and mules being powered by hand, although horse gins, too, were used to drive mules. By the 1790s, carding and spinning machinery in cotton, linen and wool could be driven by steam. Scotland was well-endowed with running water in suitable locations such as the Cart basin in Renfrewshire, which was endowed with superb sites for mills, into which raw cotton could readily be transported and from where manufactured goods were easily exported. Considerable use was made of water power by Scottish industrialists long into the nineteenth century.[40] However, the country's rich geological resources freed Scottish producers from the constraints imposed by an advanced organic energy regime based on wind, water and muscle power.[41] Steam revolutionised Scottish industry but made its mark in agriculture, too: with steam threshing mills, for example, although wind and water could also be and were used to power threshing machinery up to the end of the nineteenth century.[42] Its effects were often profound and unpredictable, leading for example to the decline in Orkney of long-line fishing for cod off the Faroes after the 1880s, when steam trawlers began to sweep the banks clean.[43] Yet while steam trawling would triumph after 1900, it is notable that the two great booms in herring fishing in the north of Scotland (1880–6 and 1895–1914) were in part based on investment in large numbers of sailboats, although steam capstans were used to haul the nets.[44]

Cotton mills were not the first segment of the textile industry in which large sums of fixed capital were invested. Relatively large sums had been required to lay out bleachfields from 1727, when determined efforts began to be made to improve the notoriously poor finish of linen cloth exported from Scotland.[45] In the region of 250 fields were established between 1728 and 1830, mainly in east-central Scotland but also to the west and south, as well as in Inverness-shire and even Shetland. Costs could range from a few hundred pounds to several thousands. The linen industry, too, saw the erection of innumerable lint mills for the mechanical preparation of flax, prior to spinning. Even though lint mills and bleachfields brought together numbers of workers in a single location – as had, for example, wooden shipbuilding yards, glassworks and collieries – it was the cotton mills, followed by other large, usually multi-storeyed, buildings for spinning flax and wool, that heralded the revolution in labour management and production.

Yet the march to industrialisation was not one that followed a single route. The mechanisation of spinning was accompanied by a massive extension in the numbers of outworkers, specifically handloom weavers, whose numbers rose from around 25,000 in 1780 to a peak of over 84,000 in 1840, after which came inexorable decline.[46] Muslin sewing – embroidery and plain stitching – employed many thousands of females in their own homes

Figure 2.6 Winders at Victoria Spinning Company Ltd, Queen Victoria Works, Dundee, *c*1902. Reproduced by permission of University of Dundee Archives.

after around 1780, in the early decades at higher wages than were available in the mills or bleachfields.[47]

Attempts to exert tighter control over workers in Scotland were not new. The proprietors of the Newmills woollen mills experienced labour difficulties throughout the works' existence; the Clerks of Penicuik attempted for the best part of the century after 1650 to encourage their colliers to work regularly and embrace Christian values. In many trades, customs such as heavy drinking or taking a holiday on Mondays were strictly adhered to. On the land and in trade and industry, employers recognised that one of the keys to success was a disciplined labour force. One of the principal challenges the partners in the Carron works faced in their early years was that they had set up business in a 'Country of Idleness'.[48] Results were mixed: there are indications that the numbers of days worked each year by agricultural labourers rose from around the middle of the eighteenth century.[49] Skilled workers, on the other hand, or those who, collectively, were able to play the market system, restricted their hours of work, and in the case of coal miners in Ayrshire at the end of the eighteenth century, limited the number of days they worked each week to as few as four.[50] It was the desire on the part of linen manufacturers to make it easier to recruit men (and later women) who had not served the lengthy apprenticeships as handloom weavers required by the urban trade incorporations that caused them to seek legislation outlawing this perceived impediment.

Struggles between masters and workers over entry requirements, new technology and the pace of work were particularly intense during the first decades of the economic revolution and resulted in fierce and widespread industrial conflicts during the 1820s and 1830s.[51] Disputes continued, however, through the rest of the century, as employers strove under conditions of growing competition to reduce labour costs.[52] Their position was strengthened by legal judgements, favourable changes in employment law and, above all, by the demographic variable turning in their favour. Rapid population increase after 1800 and increased rural–urban migration meant that the labour supply situation in Scotland, which had been relatively tight towards the end of the eighteenth century, loosened substantially.[53] Where methods used to regulate the behaviour of working people within the workplace failed, some employers adopted paternalist strategies, providing housing, schools and other welfare benefits as a means of exercising social control.[54] In such ways the hope was that affection for an employer and perhaps even industrial society could be inculcated, although paternalism had its limits and a downturn in the trade cycle and wage reductions appear to have had the effect of cooling working-class ardour.[55]

A crucial factor in easing the transition to mill and factory work and the world of disciplined labour was the employment on a regular basis of women and children. Women and children formed the 'shock troops' of modern industry in Scotland.[56] Although they were by no means docile, as the owners and managers of mills and factories in towns like Dundee, Glasgow and Paisley were to discover,[57] women were perceived to be more easily managed, cheaper to employ and better suited for tasks that required manual dexterity. Children were favoured, not only for their small size and because they were 'tractable' (the more so the better), but because child paupers and those children whose parents were anxious either to reduce household costs or to maximise the earning potential of the family, could do little to resist mill or factory employment.[58] Accordingly, women and children predominated in the early mills, much more so than in England. In cotton spinning in 1834 the male–female ratio in Lancashire was 100:102, in Glasgow it was 100:160 and for the Scottish cotton industry as a whole perhaps even 100:209.[59] Young women were not only operating the new spinning machinery, but as time went by – beginning in the eighteenth century – female labour was used to replace or supplement that of males in dressmaking and bleaching, for example.[60] The rise of heavy industry after *c*1830, however, created much more employment for men, and as a proportion of the total labour force, women's share fell, from just over 36 per cent in 1851 to 30 per cent in 1901. These figures record only 'regular' paid work, however, and exclude much of the casual, temporary and seasonal paid employment necessarily taken on by females, whether they were spinsters, married or widowed.[61]

There is no doubt that mechanisation, the centralisation of work and greater workplace discipline combined to alter radically the nature of work. Initial problems with new machinery and uncertain water supplies were overcome, and by the early 1830s most Scottish mills were operating for at least 12 hours a day, 6 days a week, around 306 days a year.[62] Long hours of hard work carried out under the close supervision of a master, manager or foreman were the norm in most occupations, however, whether in a mill, factory, workshop or in the open air. The adoption of piece-work methods also helped keep workers' noses to the grindstone. In unsupervised trades like handloom weaving, often carried on in loom sheds attached to or nearby the weaver's place of residence, the whip hand was held by necessity, with earnings falling to desperately low levels other than in the fine trades and woollens.[63] In some types of weaving, women had been employed, swelling the pool of labour and further reducing wages. Even in handloom weaving, though, which in some parts of the country was still treated as a part-time calling (or had become so, allowing for, or necessitating, breaks to be taken during spring and harvest time), some employers saw advantage in gathering their workers under one roof. When weaving prices were low, some simply refused to work at the loom.[64] At the same time there were more independent weavers, whose circumstances became increasingly desperate in the 1830s and '40s as webs distributed by agents in cut-throat competition with each other became more and more scarce, and their ability to earn a decent wage a distant memory.[65]

But while the large, centralised workplaces – notably in textiles – rightly catch the historical imagination, as late as 1851 half the firms in the main urban centres employed fewer than five workers.[66] In the countryside too, in the expanded and planned villages that had grown throughout

Lowland Scotland as cottars and sub-tenants were removed from the larger consolidated farms beloved of improvers, there were substantial numbers of craftsmen working on their own. Small-scale craft producers, blacksmiths and farriers, wrights (carpenters and joiners) and millwrights, sawyers and the like were essential in order to sustain not only the rural villages but also to serve the needs of agriculture, which had become more specialised. The 'jack of all trades' was becoming a thing of the past. Carts had to be built and repaired, as did threshing mills, ploughs and other farming machinery, while increased numbers of horses had to be shod. As self-provisioning amongst agricultural employees became rarer, even butchers were needed in the rural villages.[67]

In the middle of the nineteenth century, almost one in three workers in Scotland remained in agriculture; the figure was slightly less than one in five by the end of Queen Victoria's reign. There was much greater regional variation than there had been in the past, with only a small proportion of the working population in west-central Scotland registered as agricultural workers in 1901. Rural employment continued to be much more important in the Borders, Aberdeenshire and the Highlands and Islands. Many of these workers – at least 26 per cent in 1871 – were women.[68] Unlike England, Scotland had retained a relatively heavy dependence on female farming labour, with women carrying on the same kinds of tasks they had been familiar with prior to the revolution in agriculture that had commenced north of the border around 1760, with widespread enclosure and the use of new farming techniques. The family labour unit survived, as did the concept of the farm servant (as opposed to the farm labourer, more common south of the border, but also in parts of Scotland such as the Lothians),[69] with labour contracts lasting for six months or a year, depending upon marital status; payment in kind, at least for part of the wage, also remained.[70] But the type of farming had altered: largely pastoral across the board in the early eighteenth century, with a heavy emphasis on black cattle and sheep, and oats, over the course of the following two centuries a process of regional specialisation took place, with heavy, labour-intensive cropping dominating in the arable south-east, where the largest farms were located.[71] In sharp contrast, pastoral farming, and above all dairying, characterised the industry in Ayrshire and the south-west, as it had from the second half of the eighteenth century.[72] In the Borders and Highlands, landlords cleared vast acreages of their estates for sheep, as wool and meat prices soared prior to 1815 and held up better thereafter than for black cattle and kelp.[73]

Economic revolution brought with it a massive shift in the distribution of the Scottish population. In the mid-eighteenth century just over half of the population of Scotland (about 1.26 million) lived north of the River Tay. Perthshire was the most populous county, with just over 120,000 inhabitants. Reflecting the economic changes outlined above, by 1801 Lanarkshire, now with a population of 147,692, had seized Perthshire's mantle.[74] It was the population of Renfrewshire, however, the heartland of the burgeoning cotton industry, that grew fastest, at 195 per cent between 1755 and 1801.[75]

Figure 2.7 Man burning seaweed at Holm Island, Orkney Isles, early twentieth century. The calcinated ashes of this seaweed were sold as a raw ingredient to a range of manufacturers. Source: SLA C7728–36.

The central belt drew people from both the Highlands and the Borders, so that by the final decades of the nineteenth century two-thirds of Scots were to be found crammed within the counties that lay along the Forth–Clyde axis. In 1911, three counties – Renfrew, Lanark and Ayr – accommodated 46 per cent of the Scottish people.[76] Depopulation occurred in what now were the fringe areas of the Borders and the north, although this was not immediate. Population had continued to grow throughout Scotland well into the nineteenth century. Decline was slow and uneven, beginning first in Argyll and Perthshire after 1831.

What, historically, had been a mobile population became an overwhelmingly urban one. From one of the least urbanised societies in Europe around 1700, by 1900 Scotland lay behind only England and Wales in the European urban league, with almost 50 per cent of the country's inhabitants residing in towns of more than 20,000 people. By 1911, one in three Scots lived in one of the four big cities, led by Glasgow and Edinburgh, with 1.4 million inhabitants between them, and Dundee (165,000) and Aberdeen (164,000) a long way behind.[77] Important, too, were the many smaller towns: regional centres like Haddington and Dumfries, mining and industrial towns such as Coatbridge or Clydebank, and those where the leading activity was fishing or the provision of spa facilities. Visitors in search of cures or alleviation of their ills provided a growing market from the eighteenth century. By the later nineteenth century, with omnibus and rail links established, Moffat was drawing in over 2,000 temporary inhabitants each year, mainly from central Scotland. Bridge of Allan was even more successful, with many new jobs being created in the hotels and lodging houses required to accommodate a fairly steady stream of guests.[78]

As this suggests, while Scottish towns had much in common, each was distinctive in ways that are too numerous to deal with here.[79] Glasgow and Dundee were both major manufacturing cities, but the former had a better balance and was considerably stronger in engineering, chemicals, printing and publishing than Dundee, which was dangerously dependent upon coarse textiles. Aberdeen had thrown off its dependence upon cotton, linen and wool, and had developed a more rounded portfolio of interests. As the jute and linen trades employed many more women than men, Dundee's social structure – and character – also differed from that of Glasgow, where there were 95 males to 100 females (the ratio in Dundee was 74:100). At Clydebank, a metals and engineering town where male labour predominated, there was little female employment and the sex ratio was 123:100. Yet the labelling of towns by their principal economic characteristic can mislead: like Dundee, Galashiels was a textile town, but there were more males than females employed in the town's main industry, and a higher proportion of women were without occupations.

What should not be overlooked, however, is that most towns were service providers, too, and had in some cases in the eighteenth century made deliberate efforts to draw people to them, by enhancing the appearance of civic buildings and through the provision of, for example, municipal race meetings. Selkirk attempted this in the early 1700s. Edinburgh is the prime example, although it was also an important manufacturing centre, with brewers such as Youngers and McEwan's the most prominent in an industry that was employing over 1,000 of the city's inhabitants by 1900.[80] Edinburgh, however, with its legal and civil courts, was the main provider in Scotland of employment for lawyers as well as the professional salaried classes generally, with higher proportions of men in medicine, the arts, the civil service, the church and teaching than in either Glasgow, Aberdeen or Dundee.[81] Professional employment had risen to around 11 per cent of the total in the four main cities by the early twentieth century. Not surprisingly, Edinburgh also boasted the biggest number of domestic servants, accounting for 70 per cent of the city's employed females in 1841. In Glasgow and Dundee, the figures – even later in the century – were around 20 and 10 per cent respectively. The stable demand that salaried and professional employment brought also had favourable effects on the food and drink trades as well as artisanal producers of high-quality goods, like jewellery.

The emergence of salaried employment in Scotland reflected the growing maturity of the Scottish economy at the end of the nineteenth century. Living standards were considerably higher for most Scots, who basked in the warm glow of satisfaction that success in international markets had brought. Skilled male workers, often staunchly Protestant and liberal in their politics, were in their element, enjoying to the full the benefits of Victorian society without conceding their right to dissent, or in the workplace to form trade unions and withdraw their labour.[82] Women were capable of acting collectively as well, although without the cultural or institutional support open to men, fewer did. The descendants of Catholic Irish immigrants who had flocked to mining towns like Blantyre in the 1830s and '40s adapted to life in Scotland, bolstered by an increasingly active and confident Roman Catholic church.[83] Highlanders too, found their feet in urban society. Yet economic change had not eliminated uncertainty or the prospect of rapid descent into poverty for all. Indeed even for those in employment, real wages in Scotland generally were still lower than in England and Wales, although the differential was less than it had been in 1700.[84] In part, Scottish economic success was dependent upon skill and technical expertise. Equally significant, however, was the fact that Scotland was a low-wage economy. This was one of a number of factors that caused so many Scots to emigrate in search of better opportunities overseas: between 1830 and 1914 some 2 million people departed from Scotland's shores.[85]

For the majority who stayed, the nature of work was different, but as has already been noted in relation to agriculture, the transformation of Scotland into an industrial, urbanised society had not severed all links with the past. The process of making salt, for example, had altered little since the sixteenth century, imperceptibly at Saltcoats where only seawater was used (elsewhere, rock salt from Cheshire was imported and poured into the pans, to strengthen the brine).[86]

Figure 2.8 Saltworks, High Street, Prestonpans, East Lothian. This was the last salt works in Scotland. Reproduced by permission of RCAHMS, crown copyright.

From the mid-eighteenth century, coal masters had tried to subordinate their hewers, through the introduction of longwall working, Irish strike breakers, fines and day contracts. The unique nature of the work, however, knowledge and understanding of which was passed from father to son, meant that coal workers in the early twentieth century maintained a degree of independence allied to habits of work that had been in evidence a century and more earlier. Although steam power had made it possible to pump water from deep mines that would have been inaccessible in the early 1700s, 98 per cent of the coal mined in Scotland in 1900 was cut by hand. The rapid advance of mechanisation lay some way in the future.[87]

Most Scots in employment had become more or less full-time waged workers, even those who continued to live and gain employment in the countryside. On a day-to-day basis, the clock and the dictates of employers determined patterns of work in and around the towns, and on the factory farms of the Lothians. The nature of the technology in use had a major influence too, while the availability of employment was dependent on the trade cycle and shifts in the demand for products. Working hours were almost certainly longer – or at least the working day, week or year was less likely to be punctuated by stoppages. Early starts were still commonplace. The weather and seasons could still exert their influence in some occupations – examples are building work and dock labour, even in the last case with the appearance of the steamship. Moffat's hospitality workers had little to do before May or after September, in common with most the rest of the country that had begun to develop the tourist trade, as visitors tended to come in the summer. There was also a marked seasonal element to the clothing trades, with casual workers in particular being prone to sharp shifts in demand for their services. This most heavily exploited but sizable segment of the labour market comprised more females than males, but along with juveniles of both sexes, and the old, they continued, as they had two centuries beforehand, to depend on what work they could get, in return for scanty wages.[88]

NOTES

1. Carswell, 1951, 52–4; Young, 1996, 32.
2. Gibson and Smout, 1995, 280–5; Whatley, 1987, 15–17.
3. Whyte and Whyte, 1988, 92–8.
4. Sanderson, 1996, 1–4.
5. Whyte, 1987, 222–33
6. Whatley, 1987, 3.
7. Glendinning, MacInnes and MacKechnie, 1996, 71–146.
8. Saville, 1996, 60.
9. Turnbull, 2001, 283.
10. Whatley, 1987, 9–32.
11. Whatley, 2000, 33.
12. Mitchison, 1983, 2–11.
13. Whatley, 1997, 35.
14. Knox, 1999, 34.

15. Knox, 1995, 17–38.
16. Butt, 1977, 116–28.
17. Lenman, Lythe and Gauldie, 1969, 23–42.
18. Gulvin, 1973, 70–132.
19. Smout, 1967, 103–7.
20. Butt, 1976, 68.
21. Campbell, 2000, 22.
22. Moss and Hume, 1977, 91.
23. Lythe and Butt, 1975, 215; Slaven, 1975, 10–11.
24. Knox, 1999, 85.
25. Fry, 2001, 260–2.
26. Donnachie, 1987, 19–36.
27. Perren, 2000, 80–2.
28. Pollard, 1979, 56; Perren, 2000, 91–2.
29. Perren, 2000, 76; Moss and Hume, 1977, 40, 97.
30. Campbell, 2000, 20–5.
31. Perren, 2000, 77.
32. Smith, 1984, 189.
33. Thomson, 1983, 45–53; Smith, 1984, 189–90.
34. Smout and Watson, 1997, 92–100; Dunlop, 1997, 176–9.
35. Whatley, 1995, 364; Whatley, 1997, 47–8; Duckham, 1976, 28–45.
36. Lenman, 1977, 42.
37. Whatley, 2000, 34.
38. Whatley, 2000, 129.
39. Whatley, 1988, 235.
40. Shaw, 1984, 494–537.
41. Wrigley, 1988, 34–57.
42. Fenton, 1997, 366–9.
43. Fenton, 1997, 602.
44. Smith, 1984, 171, 177; Fenton, 1997, 613.
45. Durie, 1987, 2.
46. Murray, 1978, 23.
47. Collins, 1988, 242–4.
48. Whatley, 1988, 228–31.
49. Gibson and Smout, 1995, 283.
50. Whatley, 1988, 239.
51. Whatley, 1995, 363, 377; Knox, 1999, 47–51.
52. Knox, 1995, 164–72.
53. Whatley, 1994, 22.
54. Knox, 1990, 143–5.
55. Whatley, 2000, 85–6; Campbell, 2000, 256–76.
56. Whatley, 1988, 244; Butt, 1987, 140–2.
57. Gordon, 1991, 102–36.
58. Whatley, 1988, 244–5.
59. Whatley, 1994, 29–30.
60. Knox, 1995, 19–20; Whatley, 1994, 23–9.
61. Gordon, 1991, 20.
62. Whatley, 1988, 236.
63. Murray, 1978, 114–6.
64. Whatley, 1992, 86–92.

65. Murray, 1994, 226–9.
66. Knox, 1999, 87.
67. Young, 1996, 23–33.
68. Devine, 1984, 98.
69. Orr, 1984, 32.
70. Campbell and Devine, 1990, 56.
71. Anthony, 1997, 18.
72. Campbell, 1984, 59.
73. Richards, 1982, 192; Devine,1994, 57.
74. Lythe and Butt, 1975, 88, 94.
75. Whatley, 2000, 222.
76. Anderson and Morse, 1990, 8.
77. Morris, 1990, 73.
78. Durie, 2003, 89–100.
79. Morris, 1990, 73–102.
80. Donnachie, 1979, 166–7, 199–200.
81. Rodger, 1993, 83.
82. Whatley, 1996, 1–26; Knox, 1999, 122–36.
83. Campbell, 2000, 323–6.
84. Campbell, 1990, 80.
85. Anderson and Lee, 1990, 15.
86. Whatley, 1987, 56.
87. Campbell, 2000, 73–107, 110,
88. Treble, 1979, 51–90.

BIBLIOGRAPHY

Anthony, R. *Herds and Hinds: Farm Labour in Lowland Scotland, 1900–1939*, East Linton, 1997.

Anderson, M and Morse, D J. The people. In Fraser and Morris, 1990, 8–45.

Butt, J. Capital and enterprise in the Scottish iron industry, 1780–1840. In Butt, J and Ward, J T, eds, *Scottish Themes*, Edinburgh, 1976.

Butt, J. The Scottish cotton industry during the industrial revolution, 1780–1840. In Cullen, L M and Smout, T C, eds, *Comparative Aspects of Scottish and Irish Economic and Social History, 1600–1900*, Edinburgh, 1977, 116–28.

Butt, J and Ponting, K, eds. *Scottish Textile History*, Aberdeen, 1987.

Butt, J. Labour and industrial relations in the Scottish cotton industry during the industrial revolution. In Butt and Ponting, 1987, 139–60.

Campbell, A. *The Scottish Miners, 1874–1939, Volume One: Industry, Work and Community*, Aldershot, 2000.

Campbell, R H. Agricultural labour in the south-east. In Devine, T M, ed, *Farm Servants and Labour in Lowland Scotland, 1770–1914*, Edinburgh, 1984, 55–70.

Campbell, R H and Devine, T M. The rural experience. In Fraser and Morris, 1990, 46–72.

Carswell, J. *The Autobiography of a Working Man*, London, 1951.

Collins, B. Sewing and social structure: The flowerers of Scotland and Ireland. In Mitchison, R and Roebuck, P, eds, *Economy and Society in Scotland and Ireland, 1500–1939*, Edinburgh, 1988, 242–54.

Devine, T M, ed. *Farm Servants and Labour in Lowland Scotland, 1770–1914*, Edinburgh, 1984.

Devine, T M. Women workers, 1850–1914. In Devine, 1984, 98–123.
Devine, T M. *Clanship to Crofters War: The Social Transformation of the Scottish Highlands*, Manchester, 1994.
Donnachie, I. *A History of the Brewing Industry in Scotland*, Edinburgh, 1979.
Donnachie, I. The textile industry in south west Scotland, 1750–1914. In Butt and Ponting, 1987, 19–36.
Duckham, B. English influences on the Scottish coal industry, 1700–1815. In Butt and Ward, 1976, 28–46.
Dunlop, B M S. The woods of Strathspey in the nineteenth and twentieth centuries. In Smout, T C, ed, *Scottish Woodland History*, Edinburgh, 1997, 176–89.
Durie, A J. Textile finishing in the north east of Scotland, 1727–1860. In Butt and Ponting, 1987, 1–18.
Durie, A J. *Scotland for the Holidays: Tourism in Scotland, c1780–1939*, East Linton, 2003.
Fenton, A. *The Northern Isles: Orkney and Shetland*, East Linton, 1997.
Fraser, W H and Morris, R J, eds. *People and Society in Scotland, Volume II: 1830–1914*, Edinburgh, 1990.
Fry, M. *The Scottish Empire*, East Linton, 2001.
Gibson, A J S and Smout, T C. *Prices, Food and Wages in Scotland, 1550–1780*, Cambridge, 1995.
Glendinning, M, MacInnes, R and MacKechnie, A. *A History of Scottish Architecture From the Renaissance to the Present Day*, Edinburgh, 1996.
Gordon, E. *Women and the Labour Movement in Scotland, 1850–1914*, Oxford, 1991.
Gulvin, C. *The Tweedmakers: A History of the Scottish Fancy Woollen Industry, 1600–1914*, Newton Abbot, 1973.
Knox, W W. The political and workplace culture of the Scottish working class, 1832–1914. In Fraser and Morris, 1990, 138–66.
Knox, W W. *Hanging By A Thread: The Scottish Cotton Industry c1850–1914*, Preston, 1995.
Knox, W W. *Industrial Nation: Work, Culture and Society in Scotland, 1800–Present*, Edinburgh, 1999.
Lenman, B, Lythe, C and Gauldie, E. *Dundee and its Textile Industry, 1850–1914*, Dundee, 1969.
Lenman, B. *An Economic History of Modern Scotland*, London, 1977.
Lythe, S G E and Butt, J. *An Economic History of Scotland, 1100–1939*, London, 1975.
Mitchison, R. Ireland and Scotland: The seventeenth-century legacies compared. In Devine, T M and Dickson, D, eds, *Ireland and Scotland, 1600–1850: Parallels and Contrasts in Economic and Social Development*, Edinburgh, 1983, 2–11.
Morris, R J. Urbanisation and Scotland. In Fraser and Morris, 1990, 73–102.
Rodger, R. Employment, wages and poverty in the Scottish cities, 1840–1914. In Morris, R J and Rodger, R, eds, *The Victorian City: A Reader in British Urban History, 1820–1914*, Harlow, 1993, 73–113.
Moss, M and Hume, J R. *Workshop of the British Empire: Engineering and Shipbuilding in the West of Scotland*, London, 1977.
Murray, N. *The Scottish Hand Loom Weavers, 1790–1850*, Edinburgh, 1978.
Murray, N. The regional structure of textile employment in Scotland in the nineteenth century: East of Scotland hand loom weavers in the 1830s. In Cummings, A G J and Devine, T M, eds, *Industry, Business and Society in Scotland Since 1700*, Edinburgh, 1994, 218–33.

Orr, A. Farm servants and farm labour in the Forth valley and south-east Lowlands. In Devine, 1984, 29–54.
Perren, R. The nineteenth-century economy. In Fraser, W H and Lee, C H, eds, *Aberdeen, 1800–2000: A New History*, East Linton, 2000, 75–98.
Pollard, S and Robertson, P. *The British Shipbuilding Industry, 1870–1914*, Harvard, 1979.
Richards, E. *A History of the Highland Clearances: Agrarian Transformation and the Evictions, 1746–1886*, London, 1982.
Sanderson, E. *Women and Work in Eighteenth-Century Edinburgh*, London, 1996.
Saville, R. *Bank of Scotland: A History, 1695–1995*, Edinburgh, 1996.
Shaw, J. *Water Power in Scotland, 1550–1870*, Edinburgh, 1984.
Slaven, A. *The Development of the West of Scotland, 1750–1960*, London, 1975.
Smith, H D. *Shetland Life and Trade, 1550–1914*, Edinburgh, 1984.
Smout, T C. Lead mining in Scotland, 1650–1850. In Payne, P L, ed, *Studies in Scottish Business History*, London, 1967, 103–35.
Smout, T C and Watson, F. Exploiting semi-natural woods, 1600–1800. In Smout, T C, ed, *Scottish Woodland History*, Edinburgh, 1997, 86–100.
Thomson, W P L. *Kelp Making in Orkney*, Stromness, 1983.
Treble, J. *Urban Poverty in Britain*, London, 1979.
Turnbull, J. *The Scottish Glass Industry, 1610–1750*, Edinburgh, 2001.
Whatley, C A. *The Scottish Salt Industry, 1570–1850*, Aberdeen, 1987.
Whatley, C A. The experience of work. In Devine, T M and Mitchison, R, eds, *People and Society in Scotland, Volume 1: 1760–1830*, Edinburgh, 1988, 227–51.
Whatley, C A. *Onwards from Osnaburgs: The Rise and Progress of a Scottish Textile Company, Don & Low of Forfar, 1792–1992*, Edinburgh, 1992.
Whatley, C A. Women and the economic transformation of Scotland, *c*1740–1830, *Scottish Economic & Social History*, 14 (1994), 19–40.
Whatley, C A. Labour in the industrialising city, *c*1660–1830. In Devine, T M and Jackson, G, eds, *Glasgow, Volume 1: Beginnings to 1830*, Manchester, 1995, 360–401.
Whatley, C A. *The Diary of John Sturrock, Millwright, Dundee, 1864–65*, East Linton, 1996.
Whatley, C A. *The Industrial Revolution in Scotland*, Cambridge, 1997.
Whatley, C A. *Scottish Society, 1707–1830: Beyond Jacobitism, towards Industrialisation*, Manchester, 2000.
Whatley, C A. Altering images of the industrial city: the case of James Myles, the 'Factory Boy', and mid-Victorian Dundee. In Miskell, L, Whatley, C A and Harris, B, eds, *Victorian Dundee: Image and Realities*, East Linton, 2000, 70–95.
Whyte, I D. The occupational structure of the Scottish burghs in the late seventeenth century. In, Lynch, M, ed, *The Early Modern Town in Scotland*, London, 1987, 219–44.
Whyte, I D and Whyte, K A. The geographical mobility of women in early modern Scotland. In Leneman, L, ed, *Perspectives in Scottish Social History*, Edinburgh, 1988, 83–106.
Wrigley, E A. *Continuity, Chance and Change: The Character of the Industrial Revolution in England*, Cambridge, 1988.
Young, C. Rural independent artisan production in the east-central Lowlands of Scotland, *c*1600–1850, *Scottish Economic & Social History*, 16 (1996), 17–37.

3 Work in Twentieth-Century Scotland

WILLIAM W KNOX and ALAN MCKINLAY

From the Industrial Revolution onwards work has dominated our lives, monopolising our creativity and time, and, through the wage bargain, creating a dependency relationship between worker and employer hitherto unknown in history. A skilled worker in 1900 had a basic working week of fifty-four hours, railway engine drivers and guards about seventy. Even today British workers spend more time in the workplace than workers in comparable European Union member states. However, in spite of its centrality to life, the development of work and working practices and the impact this has had on skill is something that has until the last two or three decades largely been ignored by historians and social scientists alike. It was the student and worker unrest in France and Italy in the late 1960s and early 1970s which stimulated academics to reinvestigate the relationship of the workplace to politics.[1] As so often in these decades, Marx provided the starting point for analysis with his path-breaking study of the capitalist labour process. Marx argued that to ensure the continuous flow of profit the capitalist must not only provide the materials for labour to work on, but he or she must also acquire control of the conditions under which the speed, skill and dexterity of the worker operates. Once that is realised, the goal of capital becomes the subordination of labour. In the process of subordination, science and technology are used to break down complex skills into routinely performed operations by unskilled labour. In this way capital and its agents in the system of production gain control over the labour process and effectively destroy any resistance coalesced around the maintenance of skill. Deskilled labour becomes homogenised and easily exploited and manipulated as recalcitrant workers are dismissed and replaced by more passive ones without significantly disrupting production.[2] Marx's work on the labour process was adopted and enriched by Harry Braverman, who provided an extended critique of scientific management, particularly as propounded by Frederick W Taylor in *The Principles of Scientific Management* (1914).[3] Braverman arrived at two main conclusions not too dissimilar from those of Marx. First, crucial to understanding the development of the labour process in a capitalist society is the desire to cheapen the cost of production by substituting unskilled for skilled labour; second, and more important, is the desire to guarantee effective employer control over the labour process by, as Tony Elger puts it, 'dissolving those esoteric skills which underpinned effective craft control and reorganising production in the hands of capital'.[4]

Thus, a clear lineage of deskilling was established from the breakdown of traditional crafts during the period of industrialisation through to the establishment of the assembly line and homogenisation of labour in the twentieth century.

This view was, however, not without its critics, and a powerful revisionist critique of Braverman, from both the right and the left of the political spectrum, was quickly mounted. Drawing on an important article by Raphael Samuel on the uneven development of nineteenth-century industrial capitalism,[5] the revisionists argued that there was no linear trend towards deskilling, as the nature of work patterns remained disjointed and haphazard. Far from being deskilled, many workers were able to retain a large measure of control over the work process. The labour-intensive methods favoured by employers, and the highly differentiated product markets in which they operated, ensured that skill remained at a premium throughout the nineteenth century. Consequently, the scope that employers had for deskilling was limited, and because of this, instead of acting to destroy skills employers had a vested interest in nurturing them.[6] Underscoring this view, Patrick Joyce argued that the Victorian workplace was a terrain of compromise rather than one of conflict, with 'capital often ced[ing] to labour control [over the labour process]'.[7] Although the twentieth century has seen major changes in the nature of technology and the organisation of the workplace, revisionists have maintained that the tendency has not been towards the destruction of skill but towards its recomposition. Some skills have disappeared, increasingly so with the onset of computerised techniques in industry: riveting in the shipbuilding industry has given way to welding; muscle power in dockwork gave way to automatic methods of cargo handling and containerisation; and so on. Revisionists would argue that the disappearance of older skills and competencies has made way for the emergence of new skills. New technologies, it has been argued, have not degraded labour; rather they have increased the technical expertise of workers, encouraged greater flexibility, increased the sense of relative autonomy in making decisions which reduced the level of alienation, and allowed workers more freedom within a production regime that was challenging and interesting rather than merely routine.[8]

How far does the experience of Scottish workers correspond to these diametrically opposing views? Before examining these debates in detail, it must be remembered that Scotland's drive to industrialisation developed over a much longer period than that of England and that the period of deindustrialisation has also been a far more protracted process. Moreover, England possessed a much more diversified economic base than Scotland, where a narrowly integrated industrial structure based on shipbuilding, engineering, coal, iron and steel inhibited the emergence of new industries for most of the twentieth century. The economy, because of the nature of the product – ships, railway engines, various kinds of machinery – was also much more export-orientated and thus more limited in its capacity to develop mass, standardised production. This peculiar economic structure led to the creation of a formidable sectarian masculine culture, which, due to its equally powerful political connections in the Scottish labour movement, lingered longer than the economic realities allowed for. The relationship between economic and technological change and the nature and character of work in Scotland was more complex than elsewhere, and the scope for change was more restricted until the last few decades of the twentieth century.

Part of the reason for this was timing. The opening decade of the twentieth century witnessed far-reaching technological change, as were the closing decades. A series of changes beginning around 1880 led Michael Mann and others to talk of a 'second industrial revolution'.[9] The chief characteristics of the 'second industrial revolution' were first, the introduction of new semi-automatic machinery in coal mining, shipbuilding and engineering; second, the increasing use of unskilled and semi-skilled labour in tasks hitherto the preserve of time-served men; third, the adoption of a rudimentary system of standardised and interchangeable parts; fourth, the predominance of the factory over the workshop as the primary unit of production; and, last, the introduction of aspects of scientific management, particularly payment by results, and new specialist categories of management concerned with the design, planning and supervision of production. It is important to note that it was not just heavy industry that was affected by changes in this period: smaller-scale trades such as granite polishing[10] and trawling were also affected, although not as profoundly as, for example, coal mining.

In engineering, the introduction of the capstan or turret lathe at the end of the nineteenth century, as well as specialised boring and grinding machines, reduced much of the work of skilled workers to that of preparation. The Amalgamated Society of Engineers claimed that as a result of these new machine tools there were only 'seven out of 46 federated districts' employing the turret lathe which were not 'manned by handymen'.[11] One contemporary felt that the changes in engineering had specialised the workman to such an extent that 'every morning each man knew the job he was going to do during the day. The jobs were so ridiculously simple that anyone could do them.'[12]

While the restructuring of engineering allowed employers the luxury of dreaming of the transition from workshop to assembly-line production, shipbuilding offered much less scope for innovative methods of rationalisation. Trade fluctuations made employers reluctant to invest in new and expensive machinery, and this led to a continued emphasis on labour-intensive production methods. However, in the larger Clyde yards, the increasing size of ocean-going liners made the construction of the hull by handwork extremely difficult and expensive. These problems were in large measure overcome by the introduction of pneumatic rivet machines and electrically powered drills. The impact of these new machines was most keenly felt by the least skilled handworkers – riveters and caulkers – and this allowed employers to make greater use of apprentice labour in these

Figure 3.1 Electric coalcutter at the coal face with operator, Lanarkshire, *c*1960. Reproduced by permission of North Lanarkshire Archives.

trades. Noting this change, a Glasgow factory inspector remarked: 'Jobs formerly done by journeymen can now with [pneumatic] tools be undertaken by apprentices.'[13] The change from iron to steel shipbuilding also led to a significant reduction in the levels of skill required by both platers and their helpers, as manipulation of cold steel plates proved easier than dealing with heated iron plates.[14]

Similar encroachments into skilled work were experienced in other trades. Coal mining was increasingly subject to mechanisation, with 22 per cent of coal cut by machine in Scotland by 1913, a figure higher than that of any other mining district in Britain.[15] In addition, there was the installation of conveying machinery and abandonment in many areas of the 'pillar and stoop' method of coal-getting in favour of the longwall method, which allowed for the more intensive supervision of the collier and other underground workers.

The lesser populated trades were not excluded from the trend towards deskilling. Woodworking trades in the construction industry saw greater use being made of ferro-concrete in the building of floors and beams, developments that threatened to abolish the rougher carcass work of the carpenters. In stonework, the practice of dressing stone at the quarry and the arrival of the pneumatic chisel and other cutting devices undermined the work of the mason by no small degree.[16] Coopers also found themselves 'not so skilled' as the 'Division of Labour has come in'.[17] Similarly, the invention of the linotype composing machine in the newspaper trade removed from the compositor the skill of producing justified lines of type and was replaced with the less demanding skill of keyboard operation. Monotype had the same impact in the book trade.[18]

Outside the centres of industry and population, the cold draught of technological change was also being felt. The white fishing industry of Aberdeen underwent a transition in social relations with the introduction of the steam trawler. Prior to this the boats and tackle belonged to the fishermen, but steam technology put the cost of ownership beyond the reach of the small enterprise. The larger trawlers needed to encroach into the rich fishing grounds of the northerly waters, which added to the time at sea and the cost of fuel, further reinforced the trend towards capitalist ownership of the industry. Fishing, therefore, became a huge commercial operation requiring a high level of finance. This led to the proletarianisation of large numbers of previously independent fishermen.[19]

The reorganisation of production and the encroachments by machinery into the realm of skilled handwork can also be seen as part of a general process in the early twentieth century of intensifying the exploitation of labour by capital. This was facilitated by the introduction of electric

Figure 3.2 Man operating the keyboard of a Linotype composing machine. Reproduced by permission of Scottish Centre for the Book, Napier University, Edinburgh.

lighting, which made shiftworking more common and led to excessive overtime working in many trades. Tighter work discipline also followed. In engineering, the failure of the 1897 strike over the introduction of new technology, which threatened to displace skilled workers, won for management the right not only to control the labour process but also to introduce 'new systems of supervision'.[20] On the railways in 1900, a new system of control was introduced which related the movement of rolling stock to the availability and need for labour. To operate the new system, a 'strictly enforced hierarchy of obedience and accountability' on the part of the workforce was necessary.[21] However, it was the extension of the incentive payments systems that was most effective in raising the work effort and intensifying industrial discipline. By 1914, 46 per cent of fitters and 37 per cent of turners in the British engineering industry were on piece-rate payment, compared to only 5 per cent of all engineers and boilermakers in 1886.[22]

The new working regimes, combined with the greater specialisation of skill, led to changes in the customs and habits of the workplace. The ceremonials and rituals which were an integral part of workplace culture, and the social supports of artisan solidarity, disappeared in the larger works towards the end of the nineteenth century. The initiation ceremony for apprentices, which was ritualised and involved copious amounts of alcohol, was one casualty of the tighter work disciplines. Work had become too regimented and serious for such horseplay. Even in the smaller workshops the old forms of apprentice socialisation were being stripped of their intrinsic value. The washing of the stonemason's apron, an important event in the rites of passage of the apprentice, was abandoned, and the 'more modern masons were satisfied by taking the youth to a public house and making him drink a pint of beer'.[23] True, the ritual smearing of the genitals with grease, oil or ink, depending on the trade, continued, but these actions were more the product of a closed male environment than the surviving remnants of a once powerful artisan culture of rituals, signs and ceremonials.

The changing state of social relations in the workplace also affected managerial agencies for controlling labour as managers tried to assume total control of the labour process. As a result, gangs of subcontractors began to disappear in shipbuilding and other trades,[24] a development that simplified wage disputes into general struggles between workers and employers, rather than personal ones between men and contractors. The direct control over workers by foremen was also being eroded thanks to the introduction of complicated new payment systems and the time-measurement of jobs as well as, in the larger works, the establishment of the personnel department, which dealt with the hiring and firing of labour.[25] This harder edge to workplace relations also manifested itself in the decline of employer paternalism. In the coal industry, the decline of aristocratic ownership and the increasing role of more entrepreneurially minded employers brought an end to paternalism in the Midlothian coalfields.[26] On the railways, by 1912 the 'system of [paternalism] had been gutted of any meaningful reciprocity'.[27] Even in the smaller towns, such as Paisley, dominated by a single industry, paternal relations were in sharp decline. The expansion of the thread trade, thanks to the invention of the sewing machine, saw personal ties between workers and employers loosen, as evidenced in the growing number of industrial disputes.[28]

The drive towards deskilling and specialisation in the decades running up to World War I is inescapable. However, there were limits to this trend, determined by the nature of the product and the market. The differing technical demands of international markets meant that it made little commercial sense to implement techniques of mass production: ships, machines, railway engines, boilers and so on had to be made according to the needs of the purchaser. In the consumer-orientated trades, such as construction and jobbing printing, the one-off nature of much of the work also imposed limits on the use of labour-saving technology and, as a result, workers retained a measure of craft control. The subordination of labour to capital was, thus, never quite complete during this rapid phase of technological and organisational change in industry, since skill remained at a premium in many trades.

The industrial situation in the post-war years was very different to the pre-1914 era as mass unemployment led to a huge increase in employer power and a significant weakening of workers' ability to resist the imperatives of capital and the impact of technological change on the distribution of skills in the workplace. Employers took advantage of favourable conditions to reorganise working methods and introduce scientific management systems that threatened to kill off any remaining elements of craft control over production and output. All trades were subject to an intensification of industrial discipline and to general speed-ups in production. However, like other periods of change in the labour process, the impact of those changes was uneven.

Among the larger employers of labour, coal mining and engineering probably experienced the greatest number of changes, although jute also underwent a profound reorganisation of production processes in the interwar period. However, because of the high levels of industrial conflict which characterised workplace relations in coal mining and engineering in the 1920s, historians have been drawn to these industries. Coal mining has been the subject of numerous studies into the changing nature of skill and work practices. All of them point to the increased mechanisation of coal-getting and the impact this had on the traditional skills of the hewers. In response to increased competition and falling profits, Scottish coal owners sought to lower labour costs by increasingly making use of mechanisation. By 1935, 71.6 per cent of coal output in Scotland was cut mechanically, while the figure for Britain as a whole was 59 per cent.[29] The gains in productivity did little to increase the profits of the Scottish coal companies. However, the way they were engineered had a devastating impact on the health and skill of underground workers. The elimination by machines of the old hewing skills of holing and undercutting coal, and the widespread introduction of a twenty-four hour production regime, as Stuart MacIntyre has argued, allowed the coal companies to 'replace older miners with unskilled, but more vigorous men, and lent itself to speed ups and general intensification of the work process'.[30] Such pessimism has been challenged by Barry Supple, who argues in his official history of the coal industry that mechanisation did not so much reduce mining skills as it led to their recomposition,[31] as the operation of expensive new machines, such as coal cutters, involved a fair amount of training. Moreover, mechanisation was only really feasible in a large pit. The small size of many pits in the West of Scotland made it uneconomical to adopt the new technology of coal-getting, thus traditional pick-and-shovel methods still found favour with employers.

In spite of the incomplete and incremental nature of technological change, Supple goes too far in playing down the degeneration of skill and health encouraged by mechanisation. Even he admits that mechanisation altered the character of pit work, saying that 'strength rather than dexterity or experience was becoming paramount'.[32] It also created a new division of labour. Under traditional methods of coal-getting, small teams of workers were organised under the leadership of the independent collier and were responsible for the whole operation. Mechanisation called for a multiple-shift system employing larger teams of workers, each performing

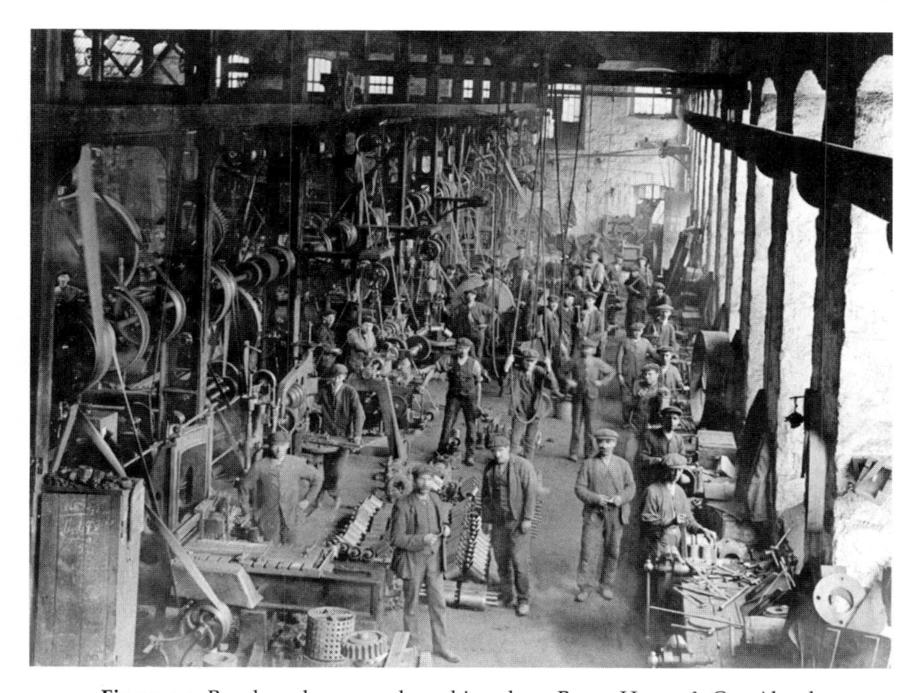

Figure 3.3 Benchwork area and machine shop, Barry, Henry & Co., Aberdeen, 1911. This picture shows that the engineers were using modern tools such as lathes, shapers and boring machines. Reproduced by permission of Aberdeen City Council.

a specialised part of the mining process. The destruction of all-round skills was accompanied by deteriorating conditions underground as the new machines increased the levels of coal dust that miners were exposed to and the new working regimes had an adverse effect on their psychological well being. Mechanisation turned miners and their families from being among the healthiest sections of the working class to one of the least. Health reports for the period 1932–9 revealed abnormally high incidences of acute and chronic sickness among the workforce. Doctors also reported unusually high levels of psychosomatic and psychoneurotic illness among the miners.[33]

Engineering also experienced a redistribution of skills, a process that had begun during World War I. Employers realised that, in the postwar era, engineering would require a new division of labour, which involved 'highly specialised machinery and a staff of skilled supervisors and tool makers ... total operations are [to be] reduced to so simple a character that they can be performed by unskilled labour'.[34] This production strategy was put into operation during the 1920s and '30s by the larger firms and significantly altered the balance between skilled and semi-skilled workers, and also led to a new division of labour among the former. In 1914 skilled workers accounted for 60 per cent of the total engineering workforce, while the semi-skilled operatives made up 20 per cent and the unskilled the rest. By 1931 the respective figures were 32 per cent skilled, 57 per cent semi-skilled and 15 per cent unskilled. Additionally, women workers were more prominent, rising from 3 per cent of the total engineering workforce in 1907 to 13.2 per cent in 1935.[35] Complementing the increase in less skilled workers was the greater degree of specialisation that time-served men were subjected to. Long-run batch work encouraged greater specialisation among fitters and limited the need for the all-round skills of fitting. In turning, the introduction of better machine tools after 1918 narrowed the range of skills and severely reduced the number of lathe operations performed by the turner. These changes had two important knock-on effects: first on apprenticeship, and then on the hierarchy of the workplace.

The movement towards simplification and subdivision of labour discouraged employers from engaging apprentices. An inquiry in the 1930s by the Amalgamated Engineering Union found that only 16 per cent of 1,332 'fair sized' firms engaged apprentices.[36] Moreover, the increase in the numbers 'picking up' the trade accompanied the creation of a new hierarchy. In the fitting department, an élite cadre of workmen, comprising around a third of fitters, was evolving, centred on the tool room. Below this élite group, fitting became a trade divided into 'a multitude of tasks none of which utilised the full range of fitting skills and techniques'.[37] In the turning department a distinction was made between the 70 per cent of 'rough' turners and the 30 per cent of 'finishers'.[38] The progressive deskilling of the engineering workers was underscored by a more intensive regime of industrial discipline. After the 1922 lock-out over the prerogative of management to allocate overtime working, the victorious employers abolished the tradition of 'minutes of grace' at starting times and meal breaks, and in Edinburgh 'speeding up and time checking even for the lavatory has been brought to a fine art'.[39] Buttressing the discipline was the increasing use of payments' systems, such as Premium Bonus Rate (PBR), to break collective controls by promoting individual incentives. By 1927 just over 63 per cent of turners and 51 per cent of fitters were on piece rates; by 1941 this had increased to nearly 80 per cent for turners and 70 per cent for fitters.[40]

Engineering workers experienced a greater levelling of skill and tighter industrial discipline at a time when they were incapable of registering any meaningful protest against the changes in the labour process. Much the same could be said of smaller trades. The collapse of the Dundee jute strike in 1923 provided the employers with a free hand to rationalise and modernise the production processes. Spinners were forced to work two frames rather than one, and high-speed spinning was further advanced in the 1930s. In the same decade, weavers were subject to much higher levels of exploitation. In 1931 they were operating two eighty-inch looms; two years later they were operating two three-yard looms. As the new machinery had to be run continuously to obtain maximum returns on the investment, a three-shift system was introduced. Legislation prevented the use of women and juveniles on night shift, which meant that there was an increase in the number of male operatives and a corresponding reduction in the amount of juvenile labour employed as piecers and shifters. By these methods a restructuring of the labour force was achieved and a 10 per cent increase in output obtained from a much smaller workforce.[41]

Of the major industries to be crippled by the collapse of world markets in the interwar period, shipbuilding was the least able to reorganise production on the basis of new technology and/or scientific management techniques. Shipbuilding employers were consistently opposed to any significant capital investment for fear of being burdened by overheads during one of the industry's periodic downturns. However, this by no means meant technological paralysis. An assault was made on the skills of the platers, and there were attempts to introduce welding on a more extensive scale. The piano punch held out the possibility of increasing the level of prefabrication, since the plates could be assembled under cover, a development that would have almost certainly raised the level of direct supervision and threatened the platers with semi-skilled status. In an atmosphere of intense international competition, few members of the Clyde Shipbuilders' Association were prepared to challenge the Boilermakers' Union over the right to determine who worked the piano punch, and the men were, as they had been in the past with other technologies, allowed to appropriate the new machine. The only men to suffer were the platers' helpers, who saw their numbers decline and their wages being reduced.[42] The 1930s witnessed a more important challenge to union control of the production process in shipbuilding. Simpler, more standardised ships began to dominate world production. Prefabrication called forth the skills of the welder, a prospect that threatened to undermine the positions of platers and riveters alike. A direct indication of this was given in the early 1930s, when the Shipbuilding Employers'

Federation reclassified the welding of plates to the ship frame as semi-skilled work. However, the stranglehold that the Boilermakers' Union had imposed on the labour process meant that the introduction of welding was significantly delayed in the Clyde yards. By 1939 welders only accounted for 2.8 per cent of the total workforce on the Clyde.[43] Dockwork was also seriously affected by technological change in shipbuilding, principally the shift from sail to steam. Steam ships exacerbated casual employment on the waterfront as more cargo was handled in shorter periods. Manual strength replaced skill as the key requirement for quayside labour. Dock labour was unable to halt or significantly negotiate mechanisation and work organisation.[44]

All this points to the fact that mass unemployment and the general weakness of trade unions in the interwar depression did not result in the universal domination of the labour process by employers, and neither did it lead to sweeping technological change. Worker resistance continued to impede change in the workplace, but employers were arguably more restricted in modernising plant and machinery because of their reliance on labour-intensive production methods. It was only in those trades amenable to technological change and increased supervision that employers were able to assume the unrestricted right to manage their enterprises as they saw fit. However, even here, as engineering shows, the sheer variety of the product market impeded the march towards Taylorist goals of rationalisation and standardised production regimes. Deskilling and managerial control of the labour process remained distant but as yet unrealised ambitions. However, almost all workers experienced some loss of control over their labour and were subject to a more intensive form of discipline and working rhythm.

The depressed economic conditions of the interwar decades placed a brake on attempts by employers to improve the methods and machinery of production, although trends were observable. The post-World War II boom, however, created a more optimistic climate among business leaders, and investment in new technology grew rapidly in all sectors of manufacturing industry, affecting both small and large enterprises, and it also encroached into the realm of service industries.

In the construction industry, the process of prefabrication had developed so extensively that as early as 1947 the president of the Amalgamated Woodworkers' Society considered that many jobs could be done by 'mere process workers'.[45] As oil tanker building began to dominate the output of the Clyde yards in the 1950s, flowline methods of construction using the maximum amount of prefabrication were called for. This increased the demand for welding skills, a development that led to the rapid decline of riveting and plating. The simplicity of tanker construction meant that there were few opportunities to exercise traditional skills.[46] Faced with tight margins and weak unions, shipyard managers relentlessly reordered the labour process and reduced stable employment to a small core of essential skilled workers complemented by contract labour.[47] The allied steel industry saw key processes fully automated in the 1960s; this effectively undermined the personal system of steel making, which in the past had depended on 'the closely guarded skill and judgement of the men who ran the furnances'.[48] Similarly, in the other great jewel in the industrial crown – coal – there was a profound reorganisation of production methods with the introduction in the 1950s of power-loading machinery. The new machines stripped the coal off the face layer by layer and simultaneously loaded it on to an armoured conveyor, which followed along the coal face. Coal could now be cut on all shifts, whereas in the interwar period only one shift cut and others prepared and cleared. By the 1960s the armoured flexible cutter had been installed in all longwall faces in Britain.[49] Even industries such as whisky experienced the widespread introduction of process technologies in areas such as distilling and bottling, often driven by multinational management teams.[50]

These innovations did nothing to halt the decline of heavy industry in Scotland. Equally, the shift from heavy to light manufacturing also did nothing to enhance the skill base of Scottish workers. The branch plants established in 'Silicon Glen' by incoming overseas firms, mainly from the USA, placed emphasis on semi-skilled assembly work with little need for technical and skilled workers, as much of the work involved the hand-preparation of printed circuit boards, a task seen as eminently suitable for women.

These firms accelerated the deskilling process already at work in the economy, and their impact was felt even among those technical workers, such as draughtsmen and designers, who were thought to be outside the industrial proletariat. In 1963 the Technical and Salaried Staff Association listed seven broad categories of technical worker. Ten years later this had grown to 486 different categories, which is evidence of the degree to which the division of labour had affected this once privileged group of 'office' workers.[51]

The service sector, too, experienced change through the implementation of new technological regimes that revolutionised working methods. The growth of large retail outlets reduced much of shop work to filling shelves and running tills, and the increased use of prepackaged goods led to the abandonment of the six-year grocery apprenticeship. As point-of-sale equipment became more sophisticated, even simple addition skills became unnecessary. As shop workers were progressively deskilled, the administration of their activities, and those of other workers, became subject to greater specialisation and subdivision of labour. Clerical work has proved to be more easily deskilled than manual, as it is conducted through the medium of paper. This makes it easier to arrange and rearrange the office on the principles of scientific management. The advent of wordprocessors and sophisticated computer software has reduced the skill content of the tasks performed by the majority of clerical workers, and this has done much to encourage functional specialisation. The effect has been to narrow both the range and scope of clerical work and, at the same time, enhance the power that the organisation has over the individual.[52] An example of this was provided by a female watch assembler in the Timex factory in Dundee when

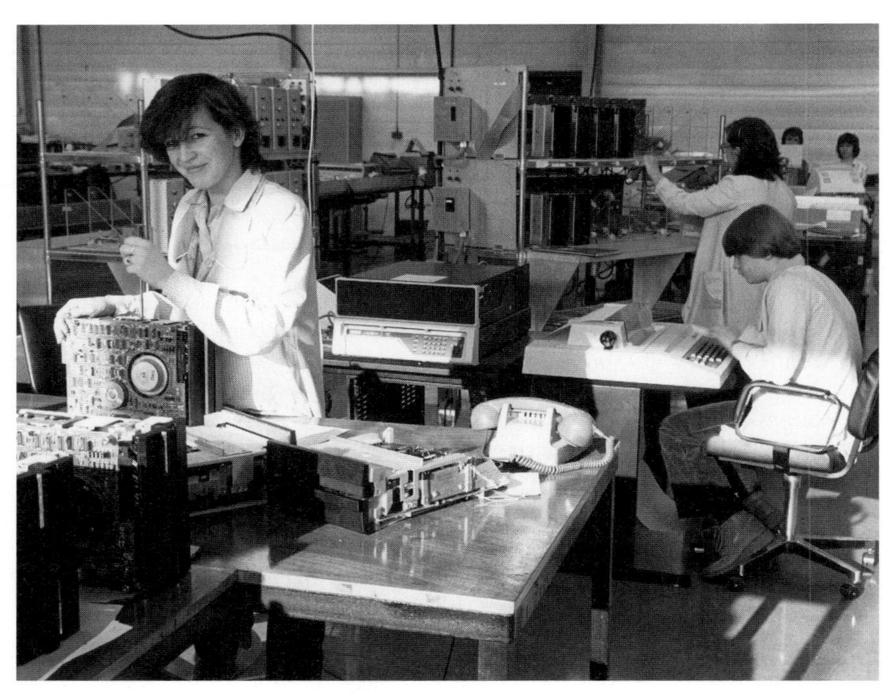

Figure 3.4 Employees at MFE Ltd, manufacturer of floppy disk drives, Brucefield Industrial Estate, Livingston. Source: SCRAN.

she remarked that: 'You were like a battery hen ... the watches were coming down in containers. It was never ending. The atmosphere was ... very strict – you daren't look up or get talking.'[53]

The motives behind the introduction of new technologies were primarily economic, as employers hoped that it would lower production costs and/or increase output. However, the introduction of new technology also created opportunities to challenge labour's independence in the workplace and open skilled work to intrusion from semi- and unskilled workers. New systems of control were introduced as older forms proved incompatible with greater bureaucracy and increased workers' power. Some employers, such as IBM,[54] relied on sophisticated variations of paternalist methods in manufacturing consent from the workers, but since 1945 the majority of firms have used the wage packet as the best way of securing power over the labour force. In the 1940s and '50s piece-rate payment was widespread in industry in Scotland. Half the workforce in coal mining, including all face workers, was on piece-rate, and in engineering the system was almost universal. However, the emergence of full employment and a stronger trade union movement saw employers move away from this system to measured day work (based on hourly rates and performance standards established by work-measurement techniques), first in coal mining in the 1960s, then in engineering, shipbuilding, dockwork and the motor car industry. The new system produced two contradictory effects: first, it encouraged the growth of collective bargaining by establishing uniform national rates for all workers in an industry; and second, it increased workplace negotiations and the degree of supervision to which workers were subjected.[55] The motor car industry saw the number of supervisors increase from three to run the plant along with a manager, superintendent and foreman, to a supervisor for every twenty-five men. This was also the experience of the coal industry.[56]

There were, however, limits to the progressive deskilling of the Scottish labour force. The nature of the product, the structure of the market and the degree of worker resistance all influenced the extent and speed of the process. In the shipbuilding industry, the historic reluctance of employers to add substantially to fixed overheads because of the vulnerability of the trade to shifts in world demand for ships meant that they continued to rely on skill-intensive methods of production. As the Shipbuilding Employers Federation noted in the 1970s: 'The present organisation of shipyard labour so far as skilled trades are concerned … is largely based on the pattern set when iron and, later, steel ships were first built in this country.'[57] Where management was successful in introducing new machinery and working methods, they were largely appropriated by the Boilermakers' Union. Similarly, the unstandardised nature of much engineering work meant that new techniques were inoperable in many establishments. Moreover, in the modern and increasingly computerised sector of the economy, research has shown that workers have greater freedom to organise their work and the speed at which it is performed. Computerisation has increased the demand for high-level skills in the area of programming and systems analysis, as well as in management information systems. Another consideration is that while technological change can result in the destruction of traditional skills, at the same time it can change the nature of existing skills or create new ones. In coal mining, progressive technical change made the old collier's skills redundant, but it created a greater demand for skilled maintenance staff and electricians. In 1957 maintenance craftsmen accounted for only 6 per cent of the total mining workforce in Britain; by 1981 this had increased to 20 per cent.[58] Even in the highly automated electronics industry there was an appreciable, if short-lived, rise in the level of skill in the mid-1970s, a phenomenon that posed a major threat to the jobs of semi-skilled workers as the hand-preparation of printed circuit boards declined sharply.[59] Thus the impact of technological change is contradictory and does not easily conform to Braverman's model of an irresistible shift towards an increasingly deskilled and defenceless proletariat.

However, what is clear from the foregoing historical sketch is that traditional craft skills have disappeared over the course of the twentieth century, and with them the existence of a culture and a system of values to which they gave rise. The independent craftsman, symbolised in the ownership of tools and the extensive system of workplace rituals and ceremonies, which served to emphasise his status in the workplace and underpin the values of craft pride and solidarity, disappeared with the arrival of the stopwatch, quality control, planning offices and modern technology. In his place emerged the semi-skilled assembly line worker and the technician, more specialised and subject to greater managerial control. Workers outside the realm of skilled employment were not unaffected by the profound restructuring of skill, which has been discussed above. Indeed, such were the reverberations of this alteration of skill in the workplace that all were affected by it; indeed, some had their whole way of life changed. The National Dock Labour Scheme and the decline in port transport destroyed the culture and community of dockworkers as family networks dispersed. The collapse of employment in the decade before the 1989 dock strike was paralleled by the rapid erosion of union bargaining powers. Wholesale changes in work organisation and employment practices, with little meaningful negotiation, have returned the docker to the status of a casual labourer similar to that of his Victorian forebear.[60] Similarly, offshore oil labour has failed to establish a durable and robust presence, even in areas such as health and safety.[61]

The same point could be made of coal mining communities in Scotland in the 1980s. The disappearance of traditional working-class communities, however, should not be seen as providing justification for nostalgic longings for the 'old days'. The ownership of skill was always exclusive and sectional. Outsiders, such as women and Catholic Irish, were always denied access to apprenticeships through some form or other of trade practice. Deskilling in many ways increased opportunities for upward mobility, if measured in terms of the wage bargain, for these previously marginalised sections of the Scottish labour force. Although specific figures do not exist for Scotland to any great extent, within a British context the twentieth century witnessed a narrowing of differentials between skilled and unskilled workers. If we take engineering, time-working fitters and turners in 1906 were earning 58 per cent more than time-working engineering labourers; in 1970 the differential decreased to 39 per cent; and between June 1970 and June 1978 there was a further reduction to 28 per cent. In the building trade the change was even more pronounced. The differential between bricklayers and their labourers fell from 50 per cent in 1906 to 13 per cent in 1978.[62] The main wage divide nowadays is between men and women rather than between skilled and unskilled. In 1981, the average earnings of a working woman in Scotland were 60 to 62 per cent of the male average; in 1991 the gap had narrowed to 68.1 per cent,[63] where it roughly remains at the time of publication.

The destruction of the once all-powerful sectarian masculine culture of the skilled worker in Scotland has allowed a new workplace culture to emerge that on one level is more democratic, less misogynist and less anti-Catholic, but on another less cohesive and disciplined. In spite of its weaknesses and flaws, the older workplace culture was capable of mobilising workers around commonly agreed interests and values, which had they been absent may have brutalised workers and their families to a greater extent than that to which their suffering and hardship over the course of the last century bears eloquent testimony. However, as the world of work was afforded primacy of concern in this male Protestant culture, other levels of experience, including gender and sexuality, were frozen out. The changes in the labour process, deindustrialisation and the growth of the branch-plant economy in general have altered these perceptions, but at a price. The trade union movement has declined rapidly, although this is as much due to government legislation as technological change, and the mentality of collectivism has been almost eradicated. Braverman saw deskilling as an historical process that would, over time, result in a homogeneous and revolutionary working class no longer distracted by issues of skill and sectionalism. The outcome has been rather the opposite; far from the social structure becoming simplified, it has become more complex.

NOTES

1. See Gorz, 1976.
2. Marx, 1969.
3. Braverman, 1974.
4. Elger, 1982, 25–53.
5. Samuel, 1977, 6–72.
6. See for instance More, 1980; Joyce, 1984, 67–76; Zeitlin, 1979, 263–74.
7. Joyce, 1984, 69.
8. Marshall, 1987, 37–40.
9. Mann, 1993, 597–627.
10. Donnelly, 1994.
11. *ASE Monthly Journal*, September 1906, 30.
12. McShane and Smith, 1978, 59–60.
13. Quoted in Levine, 1954, 431.
14. McClelland and Reid, 1985, 173–4.
15. Slaven, 1975, 168.
16. Working Man, 1908, 255.
17. The Webb Collection on Trade Unions, British Library of Political and Economic Science, London, XVII, f.115.
18. Zeitlin, 1979, 267–8.
19. Thompson, 1983, 116–8.
20. Zeitlin, 1979, 38.
21. Price, 1984, 134–5.
22. Hobsbawm, 1964, 320.
23. Gilchrist, 1940, 21.
24. Reid, 1980, 73–6.
25. Littler, 1982, 87–8.
26. Hassan, 1980, 90.
27. Price, 1986, 136.
28. Knox, 1995, 159–61.
29. Long, 1978, 124.
30. MacIntyre, 1980, 63.
31. Supple, 1987, 437–8.
32. Supple, 1987, 437–8.
33. Morris, 1974, 138.

34. *Mavor and Coulson Magazine*, December 1918, 160–1.
35. Zeitlin, 1979, 270.
36. Penn, 1983, 50.
37. McKinlay, 1986, 131.
38. McKinlay, 1986, 137.
39. *Labour Standard*, 9 May 1925.
40. Kibblewhite, 1979, 269.
41. Kiblewhite, 1979, 252–4.
42. McKinlay, 1986, 279.
43. McGoldrick, 1980, 200.
44. Kenefick, 2000, 128–37.
45. Cockburn, 1983, 109–10.
46. Robertson, 1954, 13.
47. McKinlay and Taylor, 1994, 293–304.
48. Pagnamenta and Overy, 1984, 76–101.
49. Ashworth, 1986, 74–8.
50. Findlay et al., 1998, 209–26.
51. Smith, 1987, 91.
52. Crompton and Jones, 1984, 47.
53. Timex assembler, October 1994, T/94/DM.OR.47, Dundee City Council, Museum Collections.
54. Cressey et al., 1985, 91; Dickson et al., 1982, 506–20.
55. Findlay and McKinlay, 2004, 2–28.
56. Price, 1986, 235.
57. McGoldrick, 1983, 200.
58. Penn and Simpson, 1986, 339–40.
59. Walker, 1987, 66; Goldstein, 1992, 269–84.
60. Turnbull et al., 1992, 30–1.
61. Woolfson et al., 1996, 348–54, 405.
62. Routh, 1980, 127–8.
63. Engender, 1994, 19.

BIBLIOGRAPHY

Ashworth,W. *The History of the British Coal Industry, Vol. 5, 1946–1982: The Nationalised Industry*, Oxford, 1986.

ASE Monthly Journal, September, 1906.

Braverman, H. *Labor and Monopoly Capital: The Degradation of Work in the Twentieth Century*, New York and London, 1974.

Cockburn, C. *Brothers: Male Dominance and Technological Change*, London, 1983.

Cressey, P, Eldridge, J and MacInnes, J, eds. *Just Managing: Authority and Democracy in Industry*, Milton Keynes, 1985.

Crompton, R, Jones, G. *White-Collar Proletariat: Deskilling and Gender in Clerical Work*, London, 1984.

Dickson, T, McLachlan, M V, Prior, P and Swales, K. Big Blue and the Unions: IBM, individualism and trade union strategy, *Work, Employment and Society*, 2, 1980, 506–20.

Donnelly, T. *The Aberdeen Granite Industry*, Aberdeen, 1994.

Elger, T. Braverman, Capital accumulation and deskilling. In Wood, 1982, 25–53

Engender, *Gender Audit 1993*, Edinburgh, 1994.

Findlay, P, Marks, A, Hine, J A, McKinlay, A and Thompson, P. The politics of partnership? Innovation in employment relations in the Scottish spirits industry, *British Journal of Industrial Relations*, 36/2, 1998, 209–26.
Findlay, P and McKinlay, A. Restless factories: Shop steward organisation on Clydeside, c1945–70, *Scottish Labour History*, 39, 2004, 2–28.
Gilchrist, A. *Naethin' at a': Stories and Reminiscences, etc.*, Glasgow, 1940.
Goldstein, N. Gender and the restructuring of the high-tech multinational corporations: New twists to an old story, *Cambridge Journal of Economics*, 3, 1992, 269–84.
Gorz, A, ed. *The Division of Labour: The Labour Process and Class-Struggle in Modern Capitalism*, Hassocks, 1976.
Hassan, J A. The landed estate, paternalism and the coal industry in Midlothian, 1800–1880, *Scottish Historical Review*, LIX, 1980, 73–91.
Hobsbawm, E J. *Labouring Men*, London, 1964.
Joyce, P. Labour, capital and compromise: A reply to Richard Price, *Social History*, 9, 1984, 67–76.
Kenefick, W. *'Rebellious and Contrary': The Glasgow Dockers, 1857–1932*, East Linton, 2000, 128–37.
Kibblewhite, E. The impact of unemployment on the development of trade unions in Scotland, 1918–1939. Unpublished PhD thesis, University of Aberdeen, 1979.
Knox, W W. *Hanging by a Thread: The Scottish Cotton Industry, c1850–1914*, Preston, 1995.
Labour Standard, 9 May 1925.
Long, P. The economic and social history of the Scottish coal industry, 1925–1939: With particular reference to industrial relations. Unpublished PhD thesis, University of Strathclyde, 1978.
Levine, A L. Industrial change and its affects upon labour, 1900–1914. Unpublished PhD thesis, University of London, 1954.
Littler, C. *The Development of the Labour Process in Capitalist Societies: A Comparative Study of Work Organisation in Britain, Japan and the USA*, London, 1982.
MacIntyre, S. *Little Moscows: Communism and Working Class Militancy in Inter-War Britain*, London, 1980.
McClelland, K and Reid, A. Wood, iron and steel: Technology, labour and trade union organisation in the shipbuilding industry. In Harrison, R and Zeitlin, J, eds, *Divisions of Labour: Skilled Workers and Technological Change in Nineteenth Century England*, Brighton, 1985, 151–84.
McGoldrick, J. A profile of the Boilermakers' Union. In Kruse, J and Slaven, A, eds, *Scottish and Scandinavian Shipbuilding Seminar: Development Problems in Historical Perspective*, n.p., 1980, 197–219.
McGoldrick, J. Industrial relations and the division of labour in the shipbuilding industry since the war, *British Journal of Industrial Relations*, 31(2), 1983, 197–220.
McKinlay, A. Employers and skilled workers in the inter-war depression: Engineering and shipbuilding on Clydeside, 1919–1939. Unpublished DPhil thesis, University of Oxford, 1986.
McKinlay, A and Taylor, P. Privatisation and industrial Relations: British shipbuilding, 1970–1992, *Industrial Relations Journal*, 25/4, 1994, 293–304.
McShane, H and Smith, J. *No Mean Fighter*, London, 1978.
Mann, M. *The Sources of Social Power, Vol. II: The Rise of Classes and Nation-States, 1760–1914*, Cambridge, 1993.
Marshall, G. What is happening to the working class?, *Social Studies Review*, 2, 1987, 37–40.

Marx, K. *Theories of Surplus Value: 2 Vols*, London, 1969.

Mavor and Coulson Magazine, 1918.

More, C. *Skill and the English Working Class, 1870–1914*, London, 1980.

Morris, J N. Coalminers, *Lancet*, CCLII, 1974, 138.

Pagnamenta, P and Overy, R. *All Our Working Lives*, London, 1984, 76–101.

Penn, R. Trade union organisation and skill in the British cotton and engineering industries, 1850–1960, *Social History*, 8, 1983, 37–55.

Penn, R and Simpson, R. The development of skilled work in the British coalmining industry, 1870–1985, *Industrial Relations Journal*, 17, 1986, 339–49.

Price, R. Structures of Subordination in Nineteenth Century British Industry. In Thane, P, ed, *The Power of the Past: Essays for Eric Hobsbawm*, Cambridge, 1984, 134–5.

Price, R. *Labour in British Society*, London, 1986.

Reid, A. The division of labour in the British shipbuilding industry, 1880–1920: With special reference to Clydeside. Unpublished PhD thesis, University of Cambridge, 1980.

Robertson, D J. Labour turnover in shipbuilding, *Scottish Journal of Political Economy*, 1, 1954, 9–32.

Routh, G. *Occupation and Pay in Great Britain, 1906–1979*, London, 1980.

Samuel, R. Workshop of the world: Steam power and hand technology in Victorian Britain, *History Workshop Journal*, 3, 1977, 6–72.

Slaven, A. *The Development of the West of Scotland, 1750–1960*, London, 1975.

Smith, C. *Technical Workers: Class, Labour and Trade Unionism*, Basingstoke, 1987.

Supple, B. *The History of the British Coal Industry, Vol. 4, 1913–1946: The Political Economy of Decline*, Oxford, 1987.

Thompson, P. *Living the Fishing*, London, 1983.

Timex Assembler, October 1994, T/94/DM.OR.47, University of Dundee, Museum Collections.

Turnbull, P, Woolfson, C and Kelly, J. *Dock Strike: Conflict and Restructuring in Britain's Ports*, Aldershot, 1992.

Walker, J. The Scottish 'electronics' industry, *Scottish Government Year Book 1987*, 1987, 57–80.

Webb Collection on Trade Unions, British Library of Political and Economic Science, London, XVII, f.115.

Wood, S, ed. *The Degradation of Work? Skill, Deskilling and the Labour Process*, London, 1982.

Woolfson, C, Foster, J and Beck, M. *Paying for the Piper: Capital and Labour in Britain's Offshore Oil Industry*, London, 1996.

Working Man, *Reminiscences of a Stonemason*, London, 1908.

Zeitlin, J. Craft control and the division of labour: Engineers and compositors in Britain, 1890–1930, *Cambridge Journal of Economics*, 111, 1979, 263–74.

4 Employment and Employability

RONALD W MCQUAID

INTRODUCTION

The changing nature of employment has profoundly influenced the social, economic and political development of Scotland. In particular, historically there have been severe contractions first in agricultural and then in manufacturing employment, with large rises in various parts of the service sector. These, together with the changing nature of jobs within each of the sectors, have resulted in the transformation of important social, economic and cultural characteristics of work and of the communities in which people live and work.

Many of the general improvements in employment have made Scotland a more attractive place to live in, or move to, with higher living standards. This chapter considers aspects of these societal and economic changes – particularly looking at some of the issues surrounding employment and unemployment. It starts by considering the broad trends of employment over the last century and a half in Scotland and elsewhere, especially the rise of the service sector and some of the reasons for this. It then considers the ebb and flow of employment change as different technologies, consumption patterns and industries appear to dominate for periods, with different effects on employment and society. Finally, the chapter discusses the employability of people, specifically of those out of work, and the impact of this on an individual's sense of belonging to the world of work and, by extension, wider society.

EMPLOYMENT CHANGE

The 1841 census in Scotland shows agriculture employing 23 per cent of the working population (that was the occupied population over ten years old).[1] Agriculture comprised 231,629 workers out of a total of 985,907 workers (Figure 4.1).[2] At this time 36 per cent of jobs (352,016) in Scotland were in manufacturing, 3 per cent in mining, 7 per cent in utilities (including transport and communications, such as posts and telegraphs and water) and 5 per cent in construction. Only a quarter (26 per cent), or 252,897 workers, were in services (including shops, education, health, domestic servants, etc). In Britain as a whole the figures were similar, with 22 per cent in agriculture (covering 27 per cent of working males), relatively fewer in manufacturing at

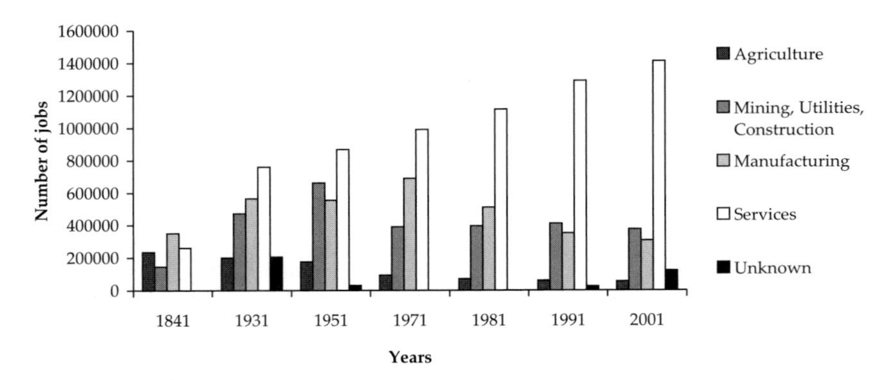

Figure 4.1 Total employment by sector in Scotland, 1841–2001. Source: Vision of Britain (2007) and 2001 census of employment, available online at: http://vision.edina.ac.uk/theme_sources.jsp?u_id=10039757&c_id=10090283&data_theme=T_IND, accessed January 2008.

31 per cent, but relatively more (30 per cent) in services. There were similar shares in mining (3 per cent), utilities (8 per cent)[3] and construction (6 per cent).

Moving on in time, one of the most notable features of employment change was the absolute and relative decline in agriculture to 9 per cent of employment in 1931, 4 per cent in 1971, 3 per cent in 1981 and 2 per cent in 2001. The long-term decline of agriculture and industrial change is poignantly described by Lewis Grassic Gibbon in his series of novels gathered together as *A Scots Quair*, first published in the 1930s:

> These were the last of the peasants, the last of the Old Scots folk. A new generation comes up that will know them not, except as a memory in a song ... And the land changes, their parks and their steadings are a desolation where the sheep are pastured, and we are told that great machines come soon to till the land, and the great herds come to feed on it, the crofter is gone.[4]

These manual agricultural jobs were soon to disappear, but as we have seen the decline was by then over a century old. In the last quarter of the twentieth century some of Scotland's manufacturing centres may arguably have experienced a similar reluctance to accept that 'blue-collar' physical manufacturing work was largely disappearing, as direct manufacturing jobs declined and those jobs that remained were becoming less physical.

In the first two-thirds of the twentieth century, manufacturing showed little significant change, accounting for 26 per cent of all jobs in 1931, 25 per cent in 1951 and 32 per cent in 1971. However, since the 1970s manufacturing has showed a dramatic decline, in absolute and relative terms, to 24 per cent in 1981 and 13 per cent in 2001 (representing a fall in absolute employment from 696,850 in 1971 to 299,205 in 2001). Hence many of the descriptions and analysis of changing 'blue-collar' and other employment over the last third of a century should be seen in the context of rapidly declining employment opportunities, some of the causes of which are considered below.

Meanwhile, employment in service industries (excluding transport) rose to over a third (34 per cent) of employment in 1931, 38 per cent in 1951 before rising inexorably to around 46 per cent in 1971, 54 per cent in 1981 and 63 per cent in 2001.[5] This represented a rise in absolute employment from 984,510 in 1971 to 1,415,389 in 2001.

Compared to Britain as a whole, the pattern in Scotland is similar. The main, relatively minor, difference was that employment in manufacturing, although 1 per cent below the Scottish figure in 1931, had 'only' fallen to 13 per cent in 2001 – indicating a slightly greater initial relative industrialisation, in terms of employment, in Scotland in the nineteenth century, but a more rapid deindustrialisation over the last 70 years. Services, and the other sectors, represented the same, or very similar, shares of employment in 2001, other than mining, where Scotland had around 1 per cent, with Britain having a negligible share of employment. Given this, the scale of the decline of manufacturing – and previously agriculture – and rise of services during the last third or quarter of the twentieth century can be seen as parts of much longer-term processes.

INDUSTRIAL CHANGE FROM 1985 TO 2005

The growth of services has been far from consistent, with some declining and others growing in different periods. It is worth noting that the term 'service sector' is really a bit of a misnomer. It actually incorporates a bundle of fundamentally different industries and occupations, from shop assistants and hairdressers to consultant surgeons and financial 'whizz-kids' in the City or Edinburgh's financial district. There is no clear consensus on defining or disaggregating the service sector, although such unpacking into more meaningful divisions is important to improve our understanding of relevant processes.[6] Using Office of National Statistics data[7] for more recent consistent information for Scotland, the number of employees in different sectors is shown in Table 4.1. The total number of employees in Scotland was 2,373,000 in September 2005, having risen by 17 per cent since 1985. However, of these, 1,943,000 were service employees, an overall increase of 44 per cent in the twenty years from 1985, while the number of non-services industry employees fell by 34 per cent in the same period.[8] The various services made up 82 per cent of jobs, if transport, storage and communication is included, as is more common, or 76 per cent if these are excluded, as is the case in the historic figures since 1841 given above. Each of the main parts of the service sector is now considered in turn.

Distribution services included 20.6 per cent of all Scottish employees in 2005. Within this, retail and wholesaling employed 15.1 per cent (358,100 jobs, a rise of 11 per cent from 1985 to 2005) and transport, storage and communication 5.5 per cent of jobs (130,000, a rise of 5 per cent since 1985).

Table 4.1 Scottish employment by sector, 2005 (September)

	Employees 2005	*Percentage change 1985–2005*
Agriculture, Forestry & Fishing	30,200	–22 per cent
Mining & Quarrying Industries	24,600	–4 per cent*
Manufacturing	232,800	–43 per cent
Electricity, Gas & Water Supply	12,800	–4 per cent*
Construction	129,900	–11 per cent
Retail & wholesale	358,100	11 per cent
Hotels & Catering	178,500	63 per cent
Transport, Storage & Communication	130,000	5 per cent
Financial Services	112,100	73 per cent
Real Estate & Business Services	323,700	89 per cent
Public Administration	161,400	18 per cent
Education	189,600	29 per cent
Health & Social Work	360,700	82 per cent
Other Services	128,600	35 per cent
Total	2,373,000	17 per cent

* Mining & Quarrying Industries and Electricity, Gas & Water Supply were grouped together until 1995.
Source: Office for National Statistics, Quarterly Employee Jobs Series (16/1/08), http://www.statistics.gov.uk/Statbase/Product.asp?vlnk=9765

Producer services employed 18.3 per cent of jobs in 2005: 4.7 per cent, 112,100 jobs, in financial services; and 13.6 per cent, 323,700 jobs, in real estate and business services. These two sectors, particularly important for Edinburgh and Glasgow, rose by a huge 73 per cent and 89 per cent respectively in the two decades from 1985.

Personal services made up nearly 13 per cent of jobs: 7.5 per cent, 178,500 jobs, in hotels and catering; and 5.4 per cent in other services, 128,600, some of which could be included with other sectors. These grew by a large 63 per cent and 35 per cent respectively from 1985. Finally, the largely public sector social services made up 30 per cent of jobs. Public administration accounted for 6.8 per cent, or 161,400; education 8.0 per cent, or 189,600; and health and social work 15.2 per cent, or 360,700. The growth of health and social work since 1985 was particularly great – 82 per cent – while education grew 29 per cent, and public administration by 18 per cent.[9]

Agriculture made up only 1.3 per cent of employees in 2005, a decrease of 22 per cent since 1985. Meanwhile, manufacturing covered less than a tenth of employees (9.8 per cent), nearly halving (a fall of 43 per cent) between these years. This partly reflected the continued collapse of manufacturing employment during and after the deep recession of the early 1980s. This decline was also due to increased capital intensity and issues such as the globalisation of manufacturing – especially competition from China and elsewhere – with employment in this sector reducing from 12.8 per cent as recently as 2001. These are huge changes, the implications of which have been notable for everyone, especially many of the older unemployed people. Construction made up 5.5 per cent of jobs, reducing by 11 per cent between 1985 and 2005. Mining and utilities, such as electricity, gas and water (excluding transport, storage and communication), together made up 1.5 per cent of employees (1 per cent and 0.5 per cent respectively), decreasing by a third (34 per cent), particularly in the 1980s and early 1990s.

So, overall, while agriculture, manufacturing, utilities and construction lost 223,000, or 34 per cent, of their jobs from 1985 to 2005, the service sector gained 591,500, or 30 per cent, new jobs. Of these new service sector jobs, 38 per cent were in the social services, 34 per cent in producer services, 20 per cent in personal services and 7 per cent in distributive services. So the rise of producer and personal services, in particular, can be seen to have played a considerable part in the later stages of the long-term, arguably over a century old, shift from a more manufacturing and heavy industry-dominated economy to a service-dominated one. It is important to see the range of social, class, cultural and even political changes in Scotland in recent decades and longer within this wider context of massive employment restructuring.

The large sector changes we have seen are linked closely with occupational changes and with gender. Clearly, factors such as: job characteristics; working conditions; flexible working; flexibility in the labour market; equality, demographic and cultural changes; technology; UK, Scottish government and EU policies; uncertainty over one's job; 'work–life balance'; and other factors all play a role in forming people's work opportunities and experiences. While each of these is influenced by the sector changes, they are also influenced by the changing gender and occupational balance within sectors, some of which are now considered.

GENDER

There has continuously been a considerable gender split in the sector of employment over history, with only 8 per cent of men but 44 per cent of women in the service industries in 1851. By 2005, just over half (51 per cent) of Scottish jobs were occupied by women (49 per cent by men), compared to 44 per cent (56 per cent by men) twenty years earlier. However, in 2005, 41 per cent of all jobs were occupied by men working full-time, compared to only 26 per cent by women working full-time. The remainder were taken up by men working part-time (8 per cent of all jobs) and women (24 per cent of all jobs).[10] From 1985, total jobs rose by 17 per cent, with male occupied jobs rising 5 per cent and female jobs 31 per cent (of these, full-time jobs for women rose by 15 per cent and part-time jobs rose 51 per cent). Although from a small base, part-time jobs for men rose by 107 per cent between 1995 and 2005 (data are not available for 1985), compared to a rise of 5 per cent for men in full-time jobs and a total rise of 6 per cent in male jobs; the equivalent 1995–2005 figures for women were an increase of 8 per cent for full-time and 31 per cent for part-time jobs and 17 per cent in total women in jobs.

Some of these differences can be partly explained by the greater growth in those services – especially personal and the generally public sector, social services – that are relatively dominated, in employment terms, by women. The burden of childcare also rests more on women than men, resulting in more women working part-time; part-time male workers tend to be those near retirement age. Part-timers often have poorer pay and promotion prospects than those in full-time posts, although legislation in recent years has sought to reduce these differences. Hence it is women with children who have generally been particularly disadvantaged in terms of pay and conditions, not only in Scotland but across most countries.[11]

OCCUPATIONS

While the sector in which someone works is important, there may be some commonalities in occupations that span industries. For instance, to what degree is a typist in an insurance firm different from one in a car factory, or a financial manager in a hamburger chain different from one in a window manufacturing company?

Over the last twenty years of the twentieth century in Scotland there was a sharp fall in the number of certain types of skilled people required in manufacturing with 'Skilled Trades Occupations', 'Processing, Plant and Machine Operatives' and 'Elementary Occupations' falling by around 50–80,000 each.[12] Meanwhile, 'Administrative, Clerical and Secretarial Occupations' had risen to become the second most common occupational group after 'Elementary Occupations'. So even within manufacturing, it can be argued that there has been a move towards 'service' sector occupations – which has been ongoing for over half a century.[13] This relative and absolute growth in service occupations is further reinforced by the rise of employment in the service sector. In total there was an increase of 80,000 in 'Professional Occupations', 70,000 in 'Managers and Senior Officials' and 64,000 'Associated and Technical Occupations'. 'Personnel Service Occupations' rose from a relatively small base, while 'Sales and Customer Service Occupations' only rose slightly, despite the increase in retailing and call centres.

This has led to considerable change in the structure of employment in Scotland. By 2005 the Annual Population Survey found that over a quarter (25.9 per cent) of the workforce were in managerial or professional occupations, with a further 13.8 per cent in Associate Professional and Technical occupations (Table 4.2). Only 12.2 per cent were in Elementary Occupations and 7.6 per cent were Process, Plant and Machine Operatives (largely traditionally 'blue-collar' manufacturing jobs).

Table 4.2 Occupations in Scotland, 2005

	Percentage
Managers and Senior Officials	12.6 per cent
Professional occupations	12.4 per cent
Associate Professional and Technical	14.3 per cent
Administrative and Secretarial	12.5 per cent
Skilled Trades Occupations	10.8 per cent
Personal Service Occupations	8.5 per cent
Sales and Customer Service Occupations	8.9 per cent
Process, Plant and Machine Operatives	7.9 per cent
Elementary Occupations	12.2 per cent

Source: Annual Population Survey, Scotland (2007).

Particular occupations within the service sector have not been immune to major shocks. Domestic service jobs declined by about a million in the UK, between the start of the great depression and World War II. Their share of employment fell from 8.2 per cent in 1931 to a mere 2.4 per cent twenty years later, before more than halving again to 1 per cent of the workforce by 1971. This resulted in Personal Service Occupations (including domestic service, hotels, eating places, entertainment) as a whole falling from 12.9 per cent to 9 per cent of employment. We can suggest some of the reasons for this rapid decline, such as changes in technology – the humble vacuum cleaner and washing machine – and the relative rise in wage rates, etc,[14] although a character in Gibbon's *A Scots Quair* controversially blames taxes and reduced incomes of the wealthy: 'Mr Mowat came suddenly home to Segget and sacked every servant he met in the House; he said that he Jahly well must, he'd no choice, he was taxed to death by those Labour chaps.'[15]

So we have a picture of major long-term changes in the place of jobs in the economy. In particular, there was a decline of agriculture – and also mining and fishing – the rise and then decline of manufacturing, and the rise of (some) services. Hence the employability skills and attitudes of employees that are desired by employers and the educational needs of the economy have changed dramatically. Also the expectations of workers have changed, partly due to wider and improved education, media influences and their own work experiences and those of their peers and those around them, as has their bargaining power as secured by legislation and, more generally, representative democracy. We have moved from the 'agrarian age' to the 'machine age' and on to the 'information' (or perhaps 'knowledge') age. This led some writers such as Daniel Bell to suggest that we have entered the post-industrial era.[16]

WHY THE GROWTH IN SERVICE INDUSTRIES?

It is worth considering some of the reasons why we have experienced such a growth in employment in certain service industries and a decline in others, such as agriculture and manufacturing.

First, a main argument suggested is that as national income rises, a greater proportion of it is spent upon services. Examples of such spending that spring to mind include health and education; as we become richer, we spend more on health, eating out and financial services such as pensions. Higher income may also be associated with greater urbanisation and lower self-sufficiency, particularly for developing countries. Indeed, the rise of health and education employment has been noted above, as has its differential effect by gender.

These have been long-standing trends, with early commentators on the transformation towards services including Clark, Fisher and Fuchs – although at the time of such commentary manufacturing had not yet then entered into long-term absolute employment decline in many developed countries. The Fisher–Clark hypothesis of the 1930s and 1940s suggests that countries move through three stages of development – agriculturally dominated, industrially dominated and services dominated – due to increased urbanisation and complexity, industrial expansion and labour productivity. Consumption patterns also change with income, as the number of products proliferates and there is a shift from essential to convenience and leisure products and services, and as societal tastes and desires change. However, there was some controversy about the model, and questions about whether it applied to developing and some other countries, and whether growth was always associated with service employment rises. The famous political economist John Kenneth Galbraith also stressed the importance of marketing by firms in stimulating consumer demand for certain products.

Second, demographic changes in society, such as an ageing population, influence demand for various products and services, such as health and some financial services. In the 1950s and 1960s the existence of the 'baby boomers' led to the growth in employment in school education, and their influence is moving through society rather like a meal moving through a python. Scotland's population rose to a peak of 5.24 million in 1974 (having exceeded 5 million for the first time in 1939) and then slowly fell to an estimated 5,116,900 by mid-2006.[17]

However, the total population figure disguises large changes in the age structure of the population. An important group of the population is the 15–29 year olds, who have often taken up the majority of entry-level and early career jobs. In 1927 this group was 1,369,915 (or 28 per cent of the population), although a large proportion of females would have left the labour market for childcare reasons. Gradually their population share fell to 25 per cent in 1952, 23 per cent in 1977 and 18 per cent in 2002, when they numbered 938,223. In the 25 years from 1977 to 2002 the number of individuals in this age range fell by 21 per cent, and projections by the General

Register Office for Scotland for the following quarter century are similar, with a predicted fall of 18 per cent. This is likely to change the experience of all workers and operations within work, as fewer young people enter the labour market. However, the situation in the future is perhaps more extreme, as during the period 1977–2002, the total population aged 15–64 years old rose by 2 per cent, while in the next period, 2002–2027, it is projected to fall by 12.5 per cent. So the last quarter century saw a decline in the number of young people, but in a context of generally rising numbers of people of working age. In the following twenty-five years a continued decline in the number of young people will be accompanied by a fall in the number of people of working age – even accounting for the rise in the state pension age for women from 60 to 65 by 2020 – and changing relationships at work.

An ageing workforce may potentially, but not necessarily, reduce national productivity, as, traditionally, older workers have often been considered less productive or entrepreneurial than younger workers on average; there will be a shift in expenditure from more to less productive sectors; and the relative size of the employed workforce is expected to significantly reduce under current trends, unless the length of working lives is extended. On the other hand, older workers may have greater life and work experience, longer tenures, lower turnover, lower absenteeism and require less supervision than younger workers.[18] Also, some loss of productivity among older workers may be due to skills obsolescence, rather than age, so training for older workers should become more important. However, there is less empirical evidence on the links between productivity and age in the skills needed in modern businesses using, for example, high levels of new technologies. The various 'active ageing' agendas of governments and others seek to change organisational and job structures and to work with health services to improve the ability of older workers to remain and be productive in the workplace. Again, an important aspect is that external forces, such as demographic change, have a profound influence on the changing employment experiences of all workers. Clearly, these changing experiences are influenced by the responses of key actors, such as government and other agencies as well as by employers, and by the way in which they are portrayed in society, including by the media.

A third factor in the growth of services focuses upon productivity growth differentials, which suggests that even if demand for goods and service products rise by the same amount, there will be relatively more growth in service employment. In other words, as productivity growth is generally lower in services than in manufacturing (and agriculture), so for any given growth in demand relatively more employment will be in the service sector. A similar line of argument is that the ratio between the price of unskilled labour in the sector producing goods and the productivity of that labour increased more than did the ratio between capital costs and its productivity, hence leading to a substitution of capital for (especially unskilled) labour. In the services the change in price/productivity was less evident, resulting in less substitution.[19] Within manufacturing (and agriculture), then, the growth in productivity, mainly through the introduction of new technology and technology processes, but also factors such as the growth in the size of firms allowing economies of scale and scope, have led to a large decline in the number of people working for any level of output.

A fourth reason for the growth of service employment, according to the American sociologist Herbert Gans, concerns the links to wider societal changes. Refering to the situation in the USA, though broadly applicable to Scotland, Gains identified such change as being the 'New Deal' era of the 1930s, the 'equity revolution' of the 1960s, the consumer revolution of the 1980s and 1990s, and the government support for a 'knowledge economy', with the need for greater education. All of these were powerful forces promoting the growth of spending on services and associated employment, particularly government-funded.

A fifth factor is globalisation and the international division of labour, with much manufacturing moved to lower labour-cost countries, such as in the Far East. This has put constant pressure on firms, and employees, to alter production processes and products to try to meet the challenges of competition from these much lower-cost countries, resulting in the large-scale movement of production abroad, particularly of physical goods, but increasingly services and capital intensification.

The sixth factor in the growth of the service sector is increased specialisation. Many specialist functions, where workers were employed within a manufacturing firm, say in advertising or finance, may now be carried out by 'external' specialist firms in the service sector. So although the work may be similar, and the labour demand comes from the manufacturing firm, the worker is classified as a service-sector worker. This is linked to the effects of technological innovations that result in new products, economic activities and methods of producing existing products. For instance, due to technological innovation, production processes may become more specialised, with more intermediate producers – most of whom may be classified as services – and there may also be a growth in smaller firms.[20] Greater complexity and sophistication of production also creates greater demands for education and training.

Other aspects of specialisation are considered by researchers such as Gershuny, who argued that the growth in service industries and occupations was largely a manifestation of the division of labour as 'planning, forecasting and organisational functions were removed from individual artisans and passed to other workers whose functions lie entirely within these fields and who are not directly involved in the physical manipulation of materials; hence the growth of 'white collar' clerical, administrative and management occupations'.[21] It could be argued that similar divisions of labour occur within service industries. Of course the introduction of the near-ubiquitous PC has transferred many administration functions from administrative support staff to the so-called 'artisans'.

Gershuny also noted that we do not have so much a service economy as a 'self-service' economy.[22] Instead of going to a concert, we can listen to the Berlin Philharmonic by downloading their performance onto our iPod. Instead of seeing a play or visiting a cinema, we can watch a DVD on our widescreen television. Instead of going out to a restaurant with friends, some of us go to Marks and Spencer to buy ready-prepared food for our guests. So what has been happening is a shift from directly buying services to buying manufactured goods with which we serve ourselves. Of course, most of the employment linked to these transactions involves service workers – from the actors and director of the film to the marketer and the internet music website or shop selling the DVD, with the manufacturing costs (and associated employment) being almost negligible.

Each of these factors has varying implications for both the level of employment in different sectors, and also the types of jobs and occupations within each sector and the content of each job.[23] Hence changes to craft skills over time are affected by these.

UNEMPLOYMENT AND EMPLOYABILITY

Unemployment affects the individual directly involved, but it also affects the perceptions and attitudes of those in employment – for instance, through their fear of redundancy – as well as wider society. The economic and employment changes discussed previously cast a shadow over unemployment across Scotland, partly leading to different unemployment experiences across time, space, industrial sectors and different groups within the population. Of course, many other factors are pertinent, but there is not space to discuss them fully here.

When considering unemployment across time, there are major problems with continuously changing definitions of who is, or is not, unemployed. The use of the International Labour Organisation (ILO) definition[24] has made comparisons easier and more consistent than using the number of people claiming unemployment-related benefits. Labour Force Survey figures show that from June 2004 to May 2005 there were 144,000 unemployed people, or 5.8 per cent of the working-age population, a steady decrease from ten years earlier (1994–5) when there were 219,000 (9 per cent).[25] By comparison, in May 2005 there were far fewer people (85,840) claiming Job Seekers Allowance (similar to what was previously called 'claiming unemployment benefits'), making up only 2.7 per cent of the working population. However, there were slightly more people claiming Incapacity Benefit – and not counted as being unemployed – at 92,350 (2.9 per cent of the working population), showing that the claimant count of unemployed people misses out on large sections of those not working but are of working age.

Over this time period, the number of inactive people (i.e., those neither employed nor unemployed) also fell from 701,000 (22.6 per cent of those of working age) in the 1990s to 635,000 (20.3 per cent) in 2004–5. Overall, during the last decade or more there has been a steady decline in unemployment and inactivity.

Spatially, various parts of Scotland have had markedly different unemployment experiences. The former heavy industrial areas have generally had high unemployment and inactivity rates, especially for those over fifty years old. As Table 4.3 indicates, the former heavy industrial, shipbuilding and mining areas generally had relatively high unemployment rates, such as Glasgow City (8.4 per cent), East Ayrshire (7.3 per cent), West Dunbartonshire (7.1 per cent), North Lanarkshire (6.9 per cent), North Ayrshire (6.7 per cent) and Inverclyde (6.4 per cent), well above the Scottish average (5.3 per cent). Local authority areas such as the more rural Perth and Kinross (3.0 per cent), Dumfries and Galloway (3.6 per cent), Moray (3.9 per cent), and more affluent 'suburbs' of the cities, such as East Dunbartonshire (3.5 per cent), and Edinburgh (4.8 per cent), all had unemployment rates below the national average.

Unemployment is heavily concentrated in the more deprived areas of Scotland. The 15 per cent Most Deprived Areas (according to the Scottish Index of Multiple Deprivation) have an unemployment rate of 13.6 per cent, compared to 4.2 per cent for the rest of Scotland. Large Urban Areas have slightly higher unemployment rates also (6.2 per cent), partly due to the concentration of heavy industry around them, but also due to the concentration of people with low skills, low qualifications and employability problems.[26] Other Urban Areas (5.5 per cent) and Accessible Small Towns (5.8 per cent) have slightly lower rates, while more rural areas are still lower (partly due to higher rates of out-migration), with Remote Small Towns and Accessible Rural each having a 3.6 per cent rate of unemployment and Remote Rural areas having a rate of only 2.9 per cent.

Table 4.3 Unemployment (residence-based) by local authority area, 2005

Aberdeen City	5.5 per cent
Aberdeenshire	4.4 per cent
Angus	4.9 per cent
Argyll & Bute	4.6 per cent
Clackmannanshire	5.4 per cent
Dumfries & Galloway	3.6 per cent
Dundee City	6.1 per cent
East Ayrshire	7.3 per cent
East Dunbartonshire	3.5 per cent
East Lothian	3.9 per cent
East Renfrewshire	4.0 per cent
Edinburgh, City of	4.8 per cent
Eilean Siar (Western Isles)	*
Falkirk	5.0 per cent

Fife	5.0 per cent
Glasgow City	8.4 per cent
Highland	3.3 per cent
Inverclyde	6.4 per cent
Midlothian	4.9 per cent
Moray	3.9 per cent
North Ayrshire	6.7 per cent
North Lanarkshire	6.9 per cent
Orkney Islands	*
Perth & Kinross	3.0 per cent
Renfrewshire	5.3 per cent
Scottish Borders	4.5 per cent
Shetland Islands	*
South Ayrshire	5.7 per cent
South Lanarkshire	5.0 per cent
Stirling	4.1 per cent
West Dunbartonshire	7.1 per cent
West Lothian	4.5 per cent
Scotland	5.3 per cent

* Numbers are too small to publish.
Source: Annual Population Survey (January to December) 2005

At different points in time, different groups of people are affected by unemployment or inactivity. An important factor is a person's employability, for both those in work and out of work. There are many definitions of employability, and no single organisation or author can claim that their definition is the only correct one, with the usefulness of differing concepts depending on what they are being used for. Some perspectives on employability focus 'narrowly' upon labour supply-side factors, such as an individual's skills or job readiness. However, under some circumstances (e.g., in a tight labour market) an employer may employ someone whom they would not accept if there were more job applicants (e.g., than in a looser labour market). In addition, some job seekers with the necessary skills may not accept a job due to factors such as the location of the job or problems with childcare. Many writers on employability stress the need to consider both demand and supply sides, and if employability is considered to be a person's ability to get (or change) employment, i.e. to make a successful job match, then such a broader perspective of employability is necessary.[27] The Scottish government used a wide definition for the purposes of their Workforce Plus (2006) policy in which they defined employability as 'the combination of factors and processes which enable people to progress towards or get into employment, to stay in employment and to move on in the workplace'.[28]

Even within buoyant labour markets characterised by low overall unemployment, there often remain pockets of high long-term unemployment. McQuaid and Lindsay (2002) found that even in Edinburgh, where general unemployment was low, long-term unemployed people tended to be selective regarding the sectors and occupations included within their job search, even though many of these jobs might be compatible with their general low level of skills, few or no qualifications, and lack of work experience over a number of years. In particular, many of them showed a lack of awareness of 'new', service-based opportunities within the labour market, and were reluctant to seek work outside traditional job roles. This, combined with their concerns about the instability and low disposable income provided by certain occupations (especially in the service sector), resulted in them adopting selective job search strategies that excluded a range of entry-level positions that would otherwise be relatively obtainable, particularly within high-demand, service-oriented labour markets where job-specific skills are relatively less important.

Lindsay and McQuaid (2004) focused on the attitudes of 300 registered unemployed people in Scotland towards specific service-sector jobs, examining whether there was a reluctance amongst job seekers to pursue service work, and whether it differed between job-seeker groups. When comparing job seekers' attitudes towards entry-level work in retail, hospitality and teleservicing or 'call centre work', they found a substantial minority of respondents ruled out entry-level service work in retail and hospitality under any circumstances. Older men, those seeking relatively high weekly wages and those without experience of service work (and who perceived themselves to lack the necessary skills) were particularly reluctant to consider these jobs. The differences between unemployed job seekers were much less apparent in relation to attitudes to call centre work, which was more unpopular than other service occupations across almost all groups. Hence they found that there do appear to be differences between various groups, e.g. age and gender.

The competition for unskilled jobs is intense, putting those without qualifications or who are in unemployment, at a great disadvantage. So even in the context of a local economy characterised by low unemployment and generally high labour demand, the selectivity that characterised the job-search strategies employed by many of the respondents may act as a significant and previously little acknowledged barrier to work.

CONCLUSIONS

This chapter has sought to consider some of the broad trends that influence and have influenced employment and parts of people's experience of employment. It has presented a picture of major long-term changes in employment, covering aspects such as sectors, gender and occupations. In particular, there was the well-known decline of agriculture, the rise and decline of manufacturing employment, and the rise of employment in some services. Hence the skills, attitudes and other attributes of employees that are desired by employers have changed markedly over time. One thing is clear: the broad demand for 'traditional', blue-collar and unskilled jobs has been contracting significantly for decades and is likely to continue to do so, although there will remain many jobs for the unskilled and semi-skilled, especially in various parts of the service sector. There is a continued need for good multi-disciplinary research to provide an in-depth and realistic understanding of complex change in the economy and society.

An important aspect of understanding peoples' experiences and attitudes towards their own jobs and careers are external forces, such as long-term industrial and demographic changes. Clearly these changing experiences are influenced by the responses of key actors, such as government and other agencies, as well as by employers, and by the way in which these changes are portrayed in society by the media and others. In addition, a key to an understanding of people's experiences and attitudes towards their own jobs and careers is an account of the responses of the people themselves to their own circumstances.

NOTES

1. The 1841 UK Census of Population was the first to include a detailed occupational classification, upon which official measures of social class have been based ever since. The industry of employment was included in the 1911 Census. The figures in this chapter are based upon the Vision of Britain Through Time project, http://www.visionofbritain.org.uk/index.jsp. They differ slightly from some other sources (e.g., Mitchell, 1962; Singlemann, 1978) and those of the Office of National Statistics, due to variations in the classifications of sectors into the broad groups of manufacturing, etc. The purpose in this chapter is simply to indicate the broad trends in employment.
2. Note that the early figures are based upon occupations rather than actual industry of work. As Visions of Britain Through Time notes, the early censuses used just occupations, mixing issues of social status, what the individual worker did and what their employer's business was. Care must be taken with the theoretical and practical purposes for which the information in the censuses was collected and subsequent interpretations.
3. Utilities are made up of 'Electricity, gas and water supply' and 'Transport storage and communication' using the 2001 census definitions.
4. Gibbon, 2006, 254.
5. Services include: 'Wholesale and retail trade, repair of motor vehicles', 'Hotels and catering', 'Public administration and defence', 'Education', 'Health and social work', 'Financial intermediation' and 'Real estate, renting and business activities', plus 'Other' ('other community [work]; social and personal service activities; private households with employed persons and extra-territorial organisations and bodies'). Utilities include: 'Electricity, gas and water supply' and 'Transport storage and communication', although elsewhere in the literature, transport is often included under services. Using Office for National

Statistics employer survey statistics, and a more common wider definition of services which includes transport and distribution, sets the level of services at 79 per cent in 2001.

6. Some writers simply separate public and private services, while others use more nuanced distinctions. One useful way was used by writers such as Singleman (1978), who split services into: Distributive Services (e.g., transport, retailing, wholesaling etc); Producer Services (FIRE [finance, insurance and real estate], legal services, etc); Social Services (medicine, education, welfare, government, etc); and Personal Services (domestic service, hotels, restaurants, barbers, entertainment, etc). In national accounting in the EU, services are grouped into two main branches: market services and non-market services. Market services are further broken down into: distributive trades and repair services, HORECA (hotels, restaurants and cafes), transport services, communication services, financial services (or services of credit and insurance institutions) and other market services.
7. Data are from the Office for National Statistics, on Workforce Jobs by Industry, although it should be noted that there are some relatively small differences from the historical data presented above due to changing definitions and the calculation of the industry composition, and so they are not directly comparable. Note that neither 1985 nor 2005 were at extreme stages of the business cycle, so this limits any cyclical effects on the trends. Another reason for choosing 2005 is that there was a discontinuity in the Office for National Statistics employee jobs series after September 2005.
8. Employee jobs exclude those in self-employment jobs (there were nearly 250,000 self-employed), those in HM Forces and government-supported trainees.
9. In Edinburgh the economy is heavily influenced by financial and professional services, and increasingly by government thanks to the presence of the Scottish parliament, although some attempts to disperse government jobs from the city have been made. The producer services of finance, insurance and real estate (called FIRE, now among the major private employers in Edinburgh), plus legal and other miscellaneous business services gradually rose from 2.6 per cent to 5.6 per cent between 1921 and 1971 (Singlemann, 1978). So they were still relatively insignificant in employment terms with, for instance, education employing more people, i.e., 5.8 per cent, in 1971. Significant growth in relative jobs came from social services such as medicine (3.9 per cent of employees in 1971), education, postal services (1.8 per cent) and government (6 per cent). Of course, in the UK most of the jobs in health and education are public sector, particularly after 1945.
10. The numbers of jobs taken by women working part-time do not translate directly into the number of women working part-time, as many have two jobs, so the percentage of women who declared themselves to be working part-time was 42 per cent.
11. OECD, 2007.
12. Green, 2001.
13. See McQuaid, 1986.
14. Similarly in the USA their share fell from 6.5 per cent of the American workforce in 1930 to 3.2 per cent twenty years later, although in the same period employment in agriculture fell from 22.9 per cent to 12.7 per cent, Singleman, 1978.
15. Gibbon, 2006, 455.

16. See Bell, 1973.
17. It is currently projected, by the General Register Office for Scotland, 2007, to rise to 5.13 million in 2019, before falling below 5 million in 2036.
18. Skirbekk, 2004.
19. There are issues concerning differing measures of productivity. For example, the trends in share of GNP in the US between 1929 and 1970 gives conflicting views of the relative growth of services output depending on whether you take current price or constant price measures. It is also dangerous to claim that higher output to employment ratios in market compared to non-market services indicates higher productivity, as it may also reflect low wages and an under-estimate of output if prices are limited by government policies, or if wages are used to partly determine output and the wages are kept low by monopsony power or otherwise (e.g., the NHS in the UK is by far the main employer of medical staff and so greatly influences medical pay levels and conditions). Also services are heterogenous with large productivity differentials within the sector and between occupations, suggesting the need to disaggregate the sector
20. After a sharp rise in the number of new businesses in Scotland in the 1980s and early 1990s, the number levelled off: Fraser of Allander, 2001; Glancey and McQuaid, 2000.
21. Gershuny, 1978, 92.
22. Gershuny 1978; 2000.
23. For instance the share of employment *within* manufacturing employment made up of non-production or ancillary workers, including accountants, office staff, etc has steadily risen for many decades: Delehanty, 1968; McQuaid 1986.
24. The ILO definition of unemployed used by the Labour Force Survey in the UK is of someone who is over sixteen years old and is: without a job, wants a job, has actively sought work in the last four weeks and is available to start work in the next two weeks; or is out of work, has found a job and is waiting to start it in the next two weeks. Other people are classified as employed or inactive.
25. NOMIS, March 2008. Nomis is a service provided by the Office for National Statistics giving access to the most detailed and up-to-date UK labour market statistics from official sources.
26. McQuaid and Lindsay, 2005.
27. McQuaid and Lindsay, 2005.
28. Scottish government website, http://www.scotland.gov.uk/Publications/2006/06/12094904/0.

BIBLIOGRAPHY

Bell, D. *The Coming of the Post-Industrial Society*, New York, 1973.

Buxton, N K. Economic growth in Scotland between the wars: The role of production structure and rationalization, *The Economic History Review*, New Series, 33, 4 (1980), 538–55.

Clark, C. *The Conditions of Economic Progress*, London, 1940.

Delehanty, G E. *Nonproduction Workers in U.S. Manufacturing*, Amsterdam, 1968.

Emi, K. *Essays on the Service Industry and Social Security in Japan*, Tokyo, 1978.

Fisher, A G B. *The Clash of Progress and Security*, London, 1935.

Fraser of Allander. *Promoting Business Start-ups: A New Strategic Formula*, Glasgow, 2001.

Freeman, C and Louca, F. *As Time Goes By: From the Industrial Revolutions to the Information Revolution*, Oxford, 2001.

Freeman, C and Perez, C. Structural crises of adjustment: Business cycles and investment behaviour. In Dosi, G, Freeman, C, Nelson, R, Silverberg, G and Soete, L, eds, *Technical Change and Economic Theory*, New York, 1989.

Fuchs, V R. *The Service Economy*, New York, 1968.

Gans, H J. *More Equality*, New York, 1973.

Gershuny, J I. The self service economy, *New Universities Quarterly*, 8 (Winter 1977), 1978, 50–66.

Gershuny, J I. *Changing Times: Work and Leisure in Post-industrial Societies*, Oxford, 2000.

Gibbon, L G. *A Scots Quair*, Edinburgh, 2006; first published 1932–4.

Glancey, K S and McQuaid, R W. *Entrepreneurial Economics*, Basingstoke, 2000.

Green, A E. *The Scottish Labour Market – Future Scenarios*, Glasgow, 2001.

General Register Office for Scotland, *Scotland's Population 2006 – The Registrar General's Annual Review of Demographic Trends*, Edinburgh, 2007.

Hall, P and Preston, P. *The Carrier Wave: New Information Technology and the Geography of Innovation 1846–2003*, London, 1988.

Haythornthwaite, J A. *Scotland in the Nineteenth Century: An Analytical Bibliography of Material Relating to Scotland in Parliamentary Papers, 1800–1900*, Aldershot, 1993.

Hood, N, Young, S and Peters, E. Multinational enterprises and regional economic development, *Regional Studies*, 28, 7 (1994), 657–77.

Lindsay, C and McQuaid, R W. Avoiding the 'McJobs': Unemployed job seekers and attitudes to service work, *Work, Employment and Society*, 18, 2 (2004), 296–319.

McQuaid, R W. Production functions and the disaggregation of labor inputs in manufacturing plants, *Journal of Regional Science*, 26 (1986), 595–603.

McQuaid, R W and Lindsay, C. The concept of employability, *Urban Studies*, 42, 2 (2005), 197–219.

McQuaid, R W and Lindsay, C. 'The employability gap': Long-term unemployment and barriers to work in buoyant labour markets, *Environment and Planning C-Government and Policy*, 20, 4 (2002), 613–28.

Mitchell, B R. *Abstract of British Historical Statistics*, Cambridge, 1962.

OECD. *Babies and Bosses: Reconciling Work and Family Life*, Paris, 2007.

ONS (Office for National Statistics). *Labour Force Projections 2006–2020*, London, 2006.

Royal Commission *Third Report on Depression of Trade and Industry, Minutes of Evidence, and Appendix*, Vol. XXIII, 1886, 496. [C. 4797], cited in Haythornthwaite, 1993.

Schumpeter, J A. *Capitalism, Socialism and Democracy*, 5th edition, London, 1976.

Scottish Government. *Workforce Plus – An Employability Framework for Scotland. Edinburgh*, Edinburgh, 2006.

Singlemann, J. *From Agriculture to Services: The Transformation of Industrial Employment*, Beverly Hills, 1978.

Skirbekk, V. *Age and Individual Productivity: A Literature Survey*, Vienna Yearbook of Population Research, 2004, 133–53.

Wright, R E. The impact of population aging on the Scottish labour market, *Quarterly Economic Commentary*, 27, 2 (2002), 38–43.

Scottish Government Website: http://www.scotland.gov.uk/Publications/2006/06/12094904/0.

Vision of Britain Through Time Project, available online at: http://www.visionofbritain.org.uk/index.jsp.

PART TWO

●

Food, Shelter and Movement

5 Farm Workers before and after the Agricultural Improvement Period

ALEXANDER FENTON

The conditions of farm workers and the shape of the labour force were very different before and after the Agricultural Improvement period, which was gaining momentum in the 1770s. There had been aspects of improvement long before this, however. Better marketing facilities were encouraged by innovations such as the use of lime as a manure, from the early 1600s, whence increased crop production, and higher levels of rent income for the landowners. In this, the Lothians were the pacemakers, with the bulk of the liming taking place on reclaimed outfield, on which oats were then sown.[1] However, though liming was shown to have economic advantages, nevertheless it required well-drained soil to be fully effective. Systematic underground drainage did not develop widely before the late 1820s. But this, in alliance with new plough types and cultivation techniques, the enclosing of farms and fields with stone dykes and hedges, and new crops such as turnips and potatoes requiring neat drills in squared-off fields, led to an almost total reorganisation of the farming landscape and indeed of the economics of farming. The modern farming landscape evolved at this time, and this is the background against which farm labour must be placed, with variations in skills according to the needs of the eight types of farming, which by the second half of the twentieth century were classified as hill sheep, upland, stock-rearing with arable, rearing with intensive livestock, arable rearing and feeding, cropping, dairy, and intensive (pigs, poultry, horticultural).[2]

PRE-IMPROVEMENT

A good deal of information about the composition and wages of the farm labour force in the mid-seventeenth century is to be found in an Assessment of Wages drawn up by the justices of the peace (JPs) for the Shire of Edinburgh (Midlothian) in 1656. Though it does not say so, it is likely that the farm servants in question were employed on single-tenancy or mains farms, rather than on multiple-tenancy farms. The categories specified were:

1. A 'whole hind, … [who] should not only perfectly know every thing belonging to Husbandry, but should also be able to perform all and every manner of Work relating thereunto: As to Plow, to

Sow, to Stack, to drive Carts, etc.' He was called a 'whole hind' because he had to maintain an 'able Fellow-Servant', enabling him to undertake the labour of a whole plough. A 'half hind' differed only in that he was not required to maintain a fellow-servant. However, the wives of the hinds:

> are to Shear dayly in Harvest, while their Masters corn be cut down. They are also to be assisting with their Husbands in winning their Masters Hay and Peats, setting of his Lime-kills, Gathering, Filling, Carting, and spreading their Masters Muck, and all other sort of Fuilzie, fit for Gooding and improving the Land. They are in like manner, to work all manner of Work, at Barns and Byres, to bear and carry th[e stac]ks from the Barn-yards to the Barns for Threshing, carry meat to the Goods (stock), from the Barnes to the Byres, Muck, Cleange, and Dight the Byres and Stables, and to help to winnow and dight the corn.

In return, a whole hind had a dwelling house (cot-house) and kail-yard, and for the year he received fifteen bolls (2,100 lbs) of oats, besides six firlots (6 bushels) of pease in summer. He was allocated sufficient ground to sow six firlots (8.7 bushels) of oats and one firlot of bere (0.998 bushels); and if living in the lower parts of the shire he had two 'soumes' of grass, i.e., grazing for two cows, or enough for three cows in the higher-lying districts. A half hind also had a house and kailyard, and half the amount got by the whole hind, with an additional two firlots (2.9 bushels) of oats. Food was provided at harvest time.[3]

2. A herd or shepherd, whose task was to keep, feed and herd his master's sheep, got a house and kailyard, eight bolls (1,120 lbs) of oats annually, a boll (140 lb) of pease in summer, an acre of land for sowing his seed, and two soums of grass in the lower districts or three in the higher.

3. A tasker (flail thresher) was responsible for threshing the grain, of whatever kind. If he was employed to thresh for some weeks, or even days, he was paid by getting a twenty-fifth part of what was threshed. This was called the 'lot' or 'proof''. If he worked on a mains farm he got constant threshing in winter, and worked at any farm jobs in summer and harvest. His wife was to work in the same way as the hinds' wives, and for this they had a house and kailyard, a boll of pease in summer, a soum's grass yearly and food supplied at harvest time.

4. A domestic or inservant, able to perform all the kinds of work relating to husbandry, had an annual wage of 40 merks Scots (1 merk = 1s. 1½d.), paid in equal proportions at Whitsunday and Martinmas.

5. Boys or lads who had their food in the house got 10 merks for a year's service.
6. A woman servant, able for work in the barns, byres, shearing, brewing, baking, washing and all other necessary tasks, outdoors or in, got 20 merks for a year's service.
7. A lass or young maid was to have the half of the woman's wage.[4]

Winnowing was an essential job after threshing and before grinding the grain or preparing it for sale. It demands a good deal of skill to get rid of chaff and foreign bodies with the aid of the wind blowing across the 'sheeling hill', and there is a hint that there were professional winnowers who could be hired. The *Accounts of the Treasurer of Scotland* for 1567 refer to eight taskers and 'winders'(winnowers) going out of Edinburgh to thresh the corn of Hepburne of Gilmertoun.[5] This looks like seasonally employed individuals, but it does not appear that winnowers were specifically included amongst the regular complement of farm servants.

By the beginning of the eighteenth century, little had changed, as indicated by the records of the justices of the peace for Lanarkshire. They appeared to keep a close control over wages. For example, in 1708, they set out a list of regulations, which included the following:

> A domestick servant man or inn servant who is able to perform all manner of work relating to husbandry, viz. to plow, sow, stack, drive carts and lay on loads, he is to have yearly for fee and bounty £24 Scots at Whitsunday and Martinmas by equall portions in full satisfaction of a years service and no more.
>
> Item, a manservant of younger years, commonly called a half lang, being a domestick servant is to have yearly for fee and bounty £16 Scots, to be payed as aforsaid and no more.
>
> Item, boyes or lads, haveing their meat in the house, are to have £8 Scots for a years service for fee and bounty, to be payed as aforsaid and no more.
>
> Item, a strong and sufficient woman servant for barns, byres, shearing, brewing, bakeing, washing and other necessary work within and without the house, is to have for fee and bounty, £14 Scots for a years service, to be payed as aforsaid and no more.
>
> Item, a lass or young maide is to have £8 Scots for a years service, to be payed as aforsaid and no more.

Special attention was paid to harvest shearers. The JPs ordained that they should not be paid a flat fee for the entire harvest but should have daily wages, which would include meat and drink. This amounted to 6s. Scots for a male shearer and 4s. Scots for a younger male shearer, called a 'half lang'. A female shearer was to get 5s. Scots, and a younger one 3s. Scots.

A tasker, if employed for a few weeks or days, was to get the twenty-fifth part of the grain he threshed; and if on a mains farm, where he got constant threshing in the wintertime, he did other work at other times, and his wife carried out the same range of tasks as the hind's wife. For this, he had a cot-house and a kailyard, with a boll of meal in summer and a 'soumes grass' (pasture for one cow or five sheep) yearly. He and his wife also got food during harvest.

But if a servant refused to accept the specified wages, he could be punished by imprisonment; if any left their master's service before the time had expired, for no cause acceptable to the JPs, they were to be treated as vagabonds and punished accordingly.

However, the masters did not get off lightly either. If any paid above the stipulated rate, they could be fined the amount paid, and a fourth part given to the informer, if there was one. Masters were also required to pay fees and wages at the time when they became due.

The JPs also agreed to put into practice certain laws and acts of the parliament against 'idle and solitary men and women and servants tyed to no certane service', namely the Act of James I, par. 3, cap. 66: Act that every man that hes nought of thir own shall labour for a living; and of James VI, par. 23, cap. 21, Act anent servants goeng louse and leaving ther master's service.[6]

On this evidence, farm servants and others were severely restricted in their ways of life, and firmly tied into a particular level of the social structure of the time.

In 1716, these points were repeated, in practically the same terms, and the level of wages remained the same.[7]

To judge by such sources, the main types of farm servant employed on mains and other single-tenancy farms were the ploughmen, the shepherd in appropriate hill areas, the tasker, the 'half-lang' or young farm-lad, and female workers, supplemented by the employment of casual labour when required. But much of the country was laid out in multiple-tenancy farms, which did not employ individuals to carry out such specific tasks and saw to the work themselves. Their organisation involved communal agreements and the sharing of draught animals and equipment, especially ploughs. They were not subject to any extent to acts of parliament or the dictates of JPs, and very little is published about their work organisation. One source of the late sixteenth century tells of the plough, drawn by a team of eight oxen, each of which is named. The 'hyndis' are also named, and it seems from one of the texts of the Plough-Song that there were twenty-five people present in the field, involving a man to control the plough, another to act as the 'gadsman' and keep the oxen moving with a pointed goad, some with wooden mallets or rakes to break up lumps in the ploughed furrow, some engaged in sowing grain single-handed from a sowing sheet, and others operating harrows to get the grain quickly out of the reach of foraging birds. Much of the community was involved:

> Higgin and Habken, Hankin and Rankin,
> Nicol and Collin, Hector and Aikin,
> Martin Mawer, Michel and Morice false lips,

Fergus, Rynaud and Guthra, Orphus and Arthur,
Morice, Davie, Richard, Philpie Foster and Macky Millar,
Ruffie Tasker and his marrows all,
Straboots, Tarboyes and Ganzel:
All that hes most domination and pastorie of your common ...

Some of these names are occupational: mawer is mower, a cutter of hay, tasker is a flail thresher. 'Straboots' may point to the practice, maintained into the twentieth century by women field workers, of winding straw ropes round the legs and ankles for protection during wet and muddy weather. The number of names pointing to classical legend and Arthurian romance seems to throw light on the nature of the oral traditions that circulated in such communities at the time.[8]

There is an increasing amount of evidence for a long-established settlement pattern that included numerous single-tenancy farms as well as multiple-tenancy farms, and a relatively small number of joint-tenancy farms. The multiple-tenancy farm appears to have been dominant in the pre-Improvement period, though the single-tenancy farm eventually took over.[9] There was more than one form of labour organisation, therefore, to match the differing circumstances of each form of land use.

POST-IMPROVEMENT

It appears that farm servants in the mid-seventeenth century who had cot-houses got their entire wages in kind. This was still the case for shepherds in some parts of the country in the mid-nineteenth century. At Strath in Skye, for example, a shepherd, instead of money wages, was allowed a house, 6½ bolls of meal, grazing for two cows and the keeping of forty to sixty sheep of his own.[10] Those living in the farmhouse, however, had money wages as well as their keep. The post-Improvement situation led to much more paid employment as farms were enclosed with sub-divided fields and single-tenancy or individual farming units became standard throughout at least the Lowland arable farming regions of Scotland. Although part-payment of wages in kind for the more essential full-time members of the employed farm staff continued for long, there was an increased use of money in the payment of wages, and more use was made of day-labourers who were paid by the day or sometimes by the hour for tasks not built into the seasonal round, such as dyking, ditching and draining.

The nature and size of the labour force on a farm is closely linked with the level of mechanisation. As a result of the technological advance that led to the widespread adoption of the threshing mill after Andrew Meikle produced the world's first fully successful model in East Lothian in 1786, threshing with the flail gradually became obsolete, though in the remoter parts of the country, especially in the crofting regions, flail threshing long remained in use. The tasker, therefore, was amongst the first of the older labour complement to vanish from the scene.

A fundamental change followed the replacement of the 'old Scotch plough' by the new type of plough, the shaping of which is credited to James Small in 1767. The old Scotch plough had been drawn by anything up to twelve oxen, or a mixture of horses and oxen. The animals were controlled by a 'gadsman' who put his pointed goad to good use, and the ploughman walked between the stilts and controlled the movements of the plough as it worked along the high-backed, corrugated ridges of the pre-Improvement farming landscape. A minimum of two people, therefore, was required to operate this plough. Ploughs of Small's type, however, were drawn by a pair of horses, or three in heavy ground, and the ploughman now controlled the horses himself, as well as the plough, with the use of reins. The manpower requirement was therefore cut from two to one, and the draught animals from up to twelve to two. This was a considerable saving for the farmer.

Overall, there was a greater degree of occupational specialisation than had existed previously. The hind, as the one responsible for ploughing, which was the first and most essential task in each new farming year, was the 'principal operative servant'.[11] This term, in the sense of a farm servant, goes back to the mid-fifteenth century, and is mainly used in the southern counties of Scotland and in England for a skilled man, married, who was provided with a house and played a responsible part in the running of the farm, ranking above the other farm servants. He was tied into the farm by the nature of his wages in kind, which in 1795 were recorded in Haddington as being occupation of a house, nine bolls of oats, two bolls of barley, two bolls of pease and a cow maintained summer and winter.[12] His status, then and much later, is indicated by a newspaper advertisement of 1955 from Dumfries: 'Married Man Wanted as Hind for Heads Farm for February 22nd, to occupy farm-house which has been recently modernised and redecorated; must be able to take full charge.'[13]

In the eighteenth centiury, the hind had carried similar high responsibilities. Improvement period attitudes had created a 'rage for great farms', which meant that a tenant could have in his possession several farms; in Roxburgh these could sometimes total 4,000 arable acres, though sheep walks could be even more extensive. A hind could be put in charge of such 'led-farms',[14] which were owned or rented by the possessor of another. The led-farm system seems to have been practised chiefly in the south of Scotland, but the creation of large farms from several dispossessed tenancies was widespread. In the parish of Logie Easter in Ross-shire, for example, such engrossment had taken place since the 1790s, leading to emigration to neighbouring towns, overseas to America, or a movement onto waste moors, which they tried to cultivate as best they could.[15] In Easter Ross, there were large farms where the farm servants were 'entire strangers'. There were few individuals who could trace their families back for two hundred years.[16]

The outworker that the hinds had to supply as a condition of their employment, normally a woman, came to be known as the 'bondager'. This term is largely confined to south-east Scotland and Northumberland. Elsewhere a woman or girl who worked in the fields was known as the 'out girl' or 'out woman'. Even by the time of the tractor the position had survived, as witness an advertisement in Fife: 'Foreman wtd. for pair horses with occasional tractor, with woman as out-worker.'[17]

Motivated attempts were being made in the 1840s to do away with the bondager system. A man called Thomson, from Tranent, tried to organise the ploughmen of East Lothian into refusing to engage with a farmer on feeing day unless the stipulation was waived.[18] This attempt failed, but the system gradually petered out in any case. Nevertheless, female outworkers continued to be employed, and they continued to wear the distinguishing dress of a bondager: a hat of black plaited straw with a wide brim trimmed with red ruching, a neat blouse and a drugget skirt, a striped apron, boots buttoned up the sides and, in muddy weather, straw ropes twisted round the ankles. The somewhat formal headgear was later replaced by a head covering of cloth, with a hood at the front and a flap at the back that gave protection from the sun. This was called an 'ugly' in East Lothian and a 'crazy' in Lanarkshire. Its use survived until well through the twentieth century.

Though outwomen were to be found on many farms, most of the female farm servants were employed as maids in the farmhouse. On a big unit, there would be specialisation, with a dairymaid and a kitchen- or housemaid. Smaller places had a single maid who combined these roles. She was known in north-east Scotland as the 'deem' or 'kitchie-deem'.

The shepherd was an important figure, who was reckoned to be £2 a year better off than the hind in the 1790s. The yearly keep of a number of sheep alongside the flocks of the master could give better returns than a fixed money wage. This was his 'pack', and the system was still surviving in the 1960s, when the allowance could be thirty-six breeding ewes and nine hogs (yearlings), though the numbers varied a little.[19] The system has now more or less died out.

In south-east Scotland, there was the 'grieve', overseer or steward, some of whom came from across the border, from Northumberland. Already by the 1750s, three farms in the parish of Swinton and Simprin were occupied by the Northumbrian stewards of three Northumbrian farmers. Possibly these were led-farms, but all the same they kept up a certain style and maintained a social distance at a time when the native farmers could scarcely be distinguished from their hinds in dress, attitudes and eating habits.[20]

The term 'grieve' is known in this sense from the fifteenth century. With the exception of south-east Scotland, where 'hind' was the norm, it is in general use for the foreman, as in an Inverness parish in the 1840s, when it was said that the 'superior servants or grieves get higher rates, and perhaps grass or foddering for a cow, according to the extent of the farms in their charge'. The ordinary wages specified for ploughmen and farm servants amounted to '£8 in money, 6 bolls of meal, and liberty to plant as much ground with potatoes as they could manure'.[21] In North Uist, farm servants got from £5 to £9, with victuals, and grieves or overseers £10–£15.[22]

Henry Stephens, in his *Book of the Farm,* provides a good deal of mid-nineteenth century information about the workforce on a farm. They consisted of the farmer himself, the steward or grieve, the ploughman, the hedger or labourer, the shepherd, the cattleman, the fieldworker and the dairymaid. The farmer was responsible for the entire system of management. He issued orders to the steward, where there was one, but otherwise he had to direct all operations. The steward or grieve (called a 'bailiff' in England) passed on instructions given by the master to the ploughmen and fieldworkers, and saw that they were carried out. He protected the farmer from problems created by any servant, 'yet it is not generally understood that he has control over the shepherd, the hedger, or the cattleman, who are stewards, in one sense, over their respective departments of labour'.

On a large farm, the steward did not always work with his hands, but he was still expected to deliver the daily allowance of corn to the horses (the horsemen were not trusted to do this, in case they overfed their charges), and to be the first up in the morning and the last to bed at night. On most farms, he did work, undertaking the most skilled jobs: sowing the seed-corn, superintending the fieldworkers in summer, tending the harvest field, building stacks and threshing corn. He might even work a pair of horses, but this was not considered appropriate, and on most farms it only happened rarely.

A steward was mainly active on arable farms but was less necessary for a pastoral farm. Where there was no steward, the farmer had to take direct charge himself, or appoint the hedger or cattleman to carry out his instructions.

The ploughman had charge of a pair of horses and worked them for all kinds of farm tasks, as well as grooming and feeding them. The usual hours of work (in 1844) were twelve hours a day for seven months of the year. 'Ploughmen are never placed in situations of trust; and thus, having no responsibility beyond the care of their horses, there is no class of servants more independent.' Stephens deplored the custom of precedence, according to which the foreman (first horseman) had to be first in every operation, with the second and third horsemen following in turn but never bypassing him.

The hedger, also called the 'spade-hind' or 'spadesman', was a high-grade labourer, who saw to the cutting and cleaning of the hedges and ditches on the farm. He was an experienced drainer. These basic tasks were mainly done in winter, and in spring he would sow corn and grass seeds, in summer shear sheep and mow hay, and in autumn build and thatch stacks. Again, hedgers were employed on arable and stock farms rather than on pastoral farms, and were less necessary where the farm had stone walls round the fields.

The shepherd undertook the entire management of the sheep: 'To inspect a large flock at least three times a day over extensive bounds, implies a walking to fatigue. Besides this daily exercise, he has to attend to the feeding of the young sheep on turnips in winter, the lambing of the ewes in spring, the washing and shearing of the fleece in summer, and the bathing of the flock in autumn.' Besides this, there is weaning, milking, drafting and marking at appropriate times, and action to be taken against insects.

The cattlemen were mostly engaged about the steading in winter, when the cattle were housed. They cleaned the byre and supplied the animals with fodder and bedding. Since these activities were a matter of routine, 'his qualifications are not of a high order'. In summer and autumn he brought in the cows to be milked, He saw to the serving of the cows by the bull, and kept a note of the time of calving.

Fieldworkers were mainly young women in Scotland, though more often men and boys in England. They carried out all manual field operations, including the use of small implements not worked by horses. They cut and planted sets of potatoes, gathered weeds, picked stones, gathered the potato crop, and filled drains with stones. The smaller implements were used in pulling turnips and preparing them for feeding stock or for winter storage, in barn work, carrying seed-corn, spreading manure, hoeing potatoes and turnips, and weeding and reaping corn crops. A big farm required several fieldworkers, who generally operated in bands.

The dairymaid lived in the farmhouse. Her main duties were to milk the cows, bring up calves, and make butter and cheese, sometimes also of ewe-milk. She also looked after the poultry.[23] The name for a milkmaid, recorded from the sixteenth century, is 'dey'.

For the same mid-nineteenth-century period, Stephens also provides data on wages. Wages in kind were paid in winter, after the crops were in and enough had been thrashed. Some farm servants were paid chiefly in kind, from the produce of the farm, with a small cash sum; others had a bigger amount in cash and a little in kind; and some had cash only. Wages in kind meant a strong dependence on the produce of the farm, and on its value, and should not be regarded simply as 'perquisites'. The wages of a ploughman or hind were taken as the standard against which the wages of others were gauged, since they were the main servants on a farm. Amounts provided were intended to support a ploughman and his family. Stephens, who farmed himself in Berwickshire, gives a breakdown of the elements and values of wages in kind for his county and for Northumberland:

Berwickshire

10 bolls (60 bushels) oats, at 12/10½ a boll	£6	8	9
3 bolls (18 bushels) barley, at 19/10½ a boll	£2	19	7½
1 boll (6 bushels) peas, at 23/3 a boll	£1	3	3
12 bolls (1,200 yards) potatoes, at 4/– a boll	£2	8	0
A cow's keep for the year	£8	0	0
Cottage and garden	£1	10	0
Carriage of coals	£2	0	0
Cash	£4	0	0
Total	£28	9	7½

Equal to 10/11¼ a week

Northumberland

6 bolls (36 bushels) oats, at 12/10½ a boll	£3	17	3
4 bolls (24 bushels) barley, at 19/10½ a boll	£3	19	6
2 bolls (12 bushels) peas, at 23/3 a boll	£2	6	6
3 bushels wheat, at 47/2 a quarter	£0	17	8½
3 bushels rye, at 29/4 a quarter	£0	11	0
40 bushels potatoes, at 1/– a quarter	£2	0	0
24 lb wool, at 1/–	£1	4	0
A cow's keep for the year	£9	0	0
Carriage of coals	£2	0	0
Cash	£4	0	0
Total	£29	15	11½

Equal to 11/5 a week

The Northumbrian farm servants were better off, therefore, than their neighbours to the north, and the produce supplied had a wider base, but the systems of payment were essentially the same.[24]

A key element at all periods was the occupation of a cot-house or cottage by the leading farm servants, but there was a change in the concept over time. In the sixteenth century, the cot-houses could have a piece of land attached to them, and a cottar occupying the house rented the land. There were also cot-houses without land, but with yards, and according to a sixteenth-century source, 'tuentie fuitis of leinth befoir thair durris'.[25] At this time, the cot-houses were an essential part of the farming system, in which the employees had a stake in the land as a major element in their livelihood. Later, the cottar was a married servant, a ploughman, who occupied a cottage along with his family as part of his contract, though the rural economy was primarily money-based. His cottage had no land attached.

Wages, more in cash than in kind, were paid mostly to single men, who were housed in the farm house or in a 'bothy'. This was a room in the steading or a separate building in which the men slept and ate, with milk and oatmeal being provided by the farmer. Each man got two pecks of meal a week, i.e., one stone of 17½ lbs, which at 1s. a peck was worth £5 4s. a year. Milk, usually two quarts a day, was estimated at £4 a year. The amount given in cash was £10–£14 a year. The men lived largely on 'brose', which was quick to make. A handful or two of meal went into a wooden bowl ('brose caap') and was sprinkled with salt. Boiling water was poured on, then it was stirred, milk was added, and it was ready to eat. 'I believe,' said Stephens, 'that no class of men can endure more bodily fatigue, for ten hours every day, than those ploughmen of Scotland who subsist on this brose thrice a-day.'

Stewards were paid in the same way as ploughmen but had a bigger cash wage. The shepherd was paid in the same way, but he had his pack of sheep in addition. The hedger was given a smaller proportion of his wages in kind than the first class of ploughmen, and more money, in toto £40 a year in value and often £1 a week, in view of his skills. The cattleman got the bulk of his wages in cash and a little in kind. He had no great status on the farm, and so had a low wage, about 9s. a week in kind. The fieldworker, usually a woman, was simply a day-labourer who earned 10d. a day. The first class of horsemen were also expected to support a female fieldworker during the year.[26]

Wages were higher in the arable areas than in the more pastoral counties. The *New Statistical Account*, for example, gives figures for the Western Isles at £6–£7 a year for men and £2 10s. to £3 for women (Harris); £5–£9 for men and £10–£15 a year for grieves or overseers, in each case with victuals (North Uist); £4–£7 without maintenance for men, and £2–£3 per annum with maintenance for women (Barra).[27] There were minor differences from island to island, though of course in the crofting districts the bulk of the work of the fields and byres was done by the families themselves, and the payment of regular wages even at this relatively low level, may have been somewhat speculative on the part of the ministers who compiled the parish accounts.

In the mainland parishes of the crofting counties, especially on the east side, farm service was more structured and wages were more tangible. In Pettie, a foreman had from £13 to £16 a year, and ploughmen £8–£10. The herd-boy and the women got £3–£4. It was said of this parish that most of the agricultural servants came from the interior Highlands, whereas the native lads preferred to acquire a trade where they could, presumably because there was the potential for making more money.[28] Thus in Cromdale, masons earned 15–18s. a week, and carpenters 12–15s., though it did not follow that there was full employment throughout the year. Male farm servants got £10–£12 a year, including maintenance, women £3 10s. to £4, and boys £4–£5.[29]

Wages in kind were probably more stable in value than the cash element, since they kept pace with the market. In Ross-shire, where there was a high standard of farming, a ploughman got a house, 7–10 barrels of coal, 6 bolls of oatmeal, 5 bolls of potatoes, a Scotch pint (a quart) of milk every day of summer and harvest, and £6–£7 in cash.[30] However, there is some evidence from the records of the Easter Ross Farmers' Club that farmers were not above taking a cartel-like approach to wage levels. In 1812, it was agreed that each member of the club should present 'a State of the Wages and every recompense or requisite given to his Plowmen & foremen or Grieve & Cottar'. In 1895, the working hours of farm servants were considered jointly by the Wester and Easter Ross Clubs. Nine, nine and a quarter, and nine and a half hours were reported from various farms, and it was agreed that this was not excessive and should remain unchanged. There was also consultation with the Black Isle Society, though this was done privately and no farm servants were invited. A weekly half-holiday was considered impractical. Ploughmen should get twelve days off a year, and were free from stable service after 8 p.m.[31]

Figure 5.1 Farm workers at Morphie, St Cyrus, Angus, in 1903. Source: SLA C4913.

Farm servants were hired at the fairs or feeing markets. Fairs were held in May and November for this purpose, single men being hired for half a year and married men for a year. The minister of the parish of Tarbat in Ross-shire noted that what was called in East Lothian the 'hind system' was used on almost all large farms and was considered the best in most improved districts of Scotland, both for the farmers and the servants. A grieve from East Lothian was even brought to Mr Macleod's farm of Little Tarrel (later called Rockfield) when he was improving the ground.[32]

By the mid-nineteenth century, the everyday work on improving farms had settled into a routine. Easter Ross farming, which came to equate with the best in the country, was at an in-between stage, with reclamation, reshaping of field boundaries, acquisition of iron ploughs and much else going on, in some cases complete and in others in progress. The daily activities on a big farm like Arrabella, in the 1830s, for example, required large numbers of servants, male and female. As part of farm modernisation, careful records were kept, including a *Weekly State of Labour on the Farm of Arrabella* for 1831–2.[33] Entries cover the six working days of the week, and Thanksgiving days (Thursdays, after communion) were observed, apart from byre and stable work. No work was done on 12 January 1832, as it was New Year's Day (old style). This source provides a usefully detailed analysis of the yearly range of tasks and the numbers of people employed at a period when improvement was proceeding rapidly.

In November–December 1831, under the eye of a 'superintendent', loads of marl were carted to manure the fields, and loads of stones, brushwood and thorns to make stone drains, 'gaus' (surface drainage channels) were cleared, several ploughmen were at work daily, harrowing went on, stacks were carted to the barn for threshing, turnips were pulled by women and brought in to feed the beasts in the byre, troughs for feeding were made, and carts made and repaired. Several women labourers were engaged at the threshing mill, dressing and measuring wheat, at 6d. a day, half the amount for male labourers. Cattle and young horses and pigs had also to be seen to. Collecting dung, scraping the roads, and whitewashing the houses kept servants busy, and there was a visit to the meal mill, kiln-drying and subsequent mending of the meal girnal by two carpenters. A load of wheat was taken to the trading port of Invergordon,[34] and barley also.

In January–March 1832, many of the same activities continued, but in addition there were the spring tasks of sowing wheat and rye grass and making drills for beans. Potato planting began. Animals let out from the byres had to be herded. Seed oats were prepared, and repairs of various kinds were made to barrows, carts and pigsties, and to grain sacks. There was a good deal of self-sufficiency: for example graip (fork) shafts were made, and wooden shovels. Dung was spread. Water furrowing was carried out to drain fields of surface water.

In April–May, palings (fences) were being mended, and dunghill work figured prominently, as well as the harrowing and sowing and rolling of seed. Ware (seaweed) was carted from the coastal village of Balintore, as manure. 'Diffats' (sods) were carted to the house, presumably as underlay for a thatched roof. Beans were hoed and more marl was brought home.

In April 1832, two men spent two days ploughing 'the crofts at Knocknakeen'. No doubt these crofts were occupied by people who worked on the farm from time to time. In May, a man went to Achany 'with the herd's furniture', and a day was spent in 'washing servants bed clothes'.

In May–June, activities included sowing turnips and putting on bone dust, carting stones, ploughing and spreading and trenching dung, liming, carting earth, stones and heather, raking grass, pulling thistles, hoeing potatoes and drawing straw. Oiling harness was also done. Scythes were got from Tain.

In July–August, herding went on, the Swedish turnip crop was hoed, and the cut hay was made up into 'coles' (small stacks). Tares were cut and carted, turnip seed was thrashed, and liming was done. A coffin was made for a cattleman who had died. The bothy was whitewashed, a house for the shearers was finished and beds were made for them. Herding continued. The threshing mill was completely renovated, the mill wheel was renewed and the mill lade was cleaned out. The cereal crops were stacked, and bosses were made to keep them ventilated. Ropes were made, presumably of straw.

In October–November 1832, there was herding, forking at the stacks, pitting potatoes, repairing the kiln, kiln-drying oats, sowing wheat, spreading dung and levelling new land.[35]

This is not a full catalogue of all the activities that went on, but it gives an impression of the seasonal round of work on a well-run cropping farm that was in the process of being improved. Numerous extra hands, many of them female, were employed. Timber for buildings and implements came out of the woods, and income that paid for the extra hands came from the cereals carted to Portgordon and then traded onwards by sea. Like other farms in Easter Ross, Arrabella was well on its way to achieving the standards of the farms of south-east Scotland.

As long as horses were the main source of draught power, a situation that gradually changed after the introduction of tractors during World War I, the workforce on farms remained fairly constant. In terms of status, horsemen took the lead over cattlemen, and a big farm would have several horsemen and 'cow-baillies', though of course on smaller units and on crofts the same individual would see to all the farm tasks. There were also young lads, in effect apprentice farm servants, known as 'halflins' (or half langs in the seventeenth century). A man who did odd jobs about the farm was the 'orra man', who also worked with the unpaired horse, the 'orra horse'. The organisation of the workforce on a fairly large farm in Aberdeenshire is shown in Figure 5.2.

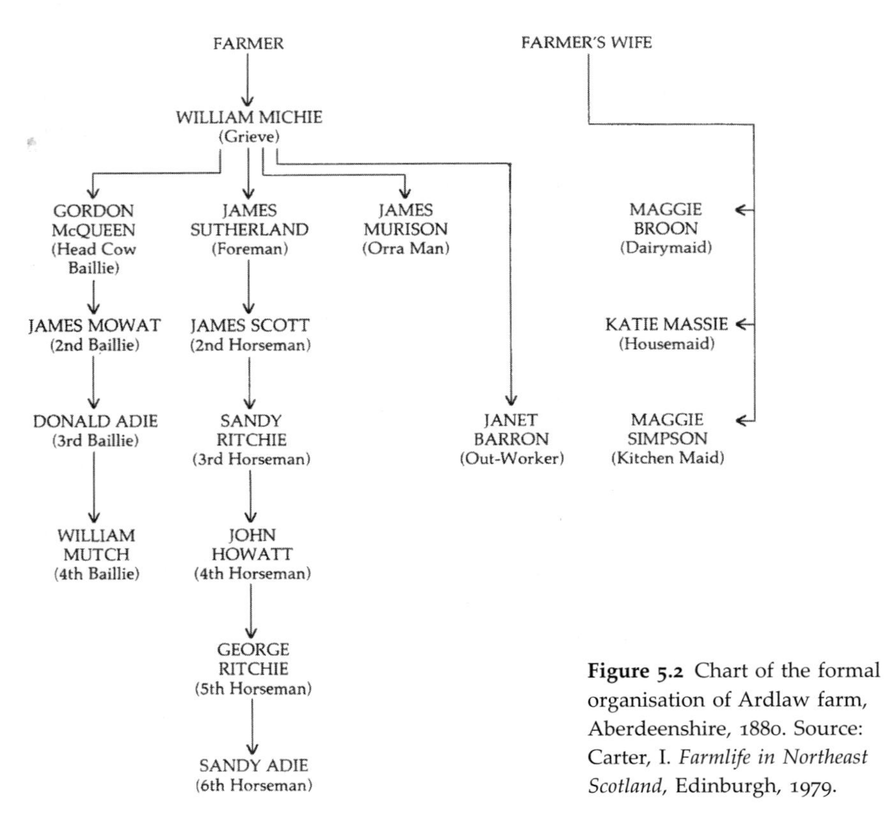

Figure 5.2 Chart of the formal organisation of Ardlaw farm, Aberdeenshire, 1880. Source: Carter, I. *Farmlife in Northeast Scotland*, Edinburgh, 1979.

Single men tended to move every six months, married cottar men once a year. This movement was not, however, as disruptive as might be thought, for it took place as a rule within a ten- or twenty-mile radius. This meant that the workers gained a detailed knowledge of their district, and of the farms and farmers within it, giving them a sense of brotherhood, even though it was difficult to persuade them to join unions. The elusive 'horsemen's word' was part of the means by which the men could exert pressure on farmers in seeking 'to obtain for Farm Servants Monthly Payments with Indefinite Engagements, Weekly Half-Holiday, except Six weeks in harvest, The Abolition of the Bothy System, the improvement of our Kitchen Dietary scale, and of our House and Sleeping accommodation'.[36]

The bothy system survived in Angus until the tractor era. The single men lived in the bothy and slept and ate there, the bulk of their food being oatmeal brose. An alternative in areas of smaller farms was the 'kitchie' system, when the single men slept in a 'chaumer' in the steading but had their food in the farm kitchen. There were also the cottar houses for married men.[37]

In the late nineteenth and twentieth centuries the major sources of discontent amongst farm servants, apart from wage levels, were poor housing or accommodation, monotonous or poor food, and time off. There was also some unease amongst them about the adoption of 'motor ploughing'.

The question of poor housing may be exemplified by the case of a farm servant at Adambrae, Midcalder, who entered a cottar house in May 1915. He complained to the sanitary inspector for the Calder district, but the response was slow, and so he got in touch with the secretary of the Scottish Farm Servants' Union, Joseph A Duncan, who went to see the house on 24 September. Here is what he saw:

> [The house] consisted of room and kitchen with a small mid-room which was not even suitable as a store room. The back wall of the house, divided the house from the byre, and the roofs of the house and byre drained into a gutter on this wall. One end of the house was built on to the milk house. The back wall was very damp, even although in the room it was lathed. A wooden floor had been laid down in the room on the top of the old floor which had the effect of giving a very low ceiling. There was one small window in the front wall of each room ... The grates were old and broken, that in the kitchen being quite useless. The flags in the doorway were worn and broken, and gave easy access to any water which naturally drained to the house door. In the room the walls behind the bed were damp, the back wall was damp, and in one corner was wet. Although it was a dry day and the family were living in the room with fire always on and the window was open, the room had a musty smell that proved that the dampness was of long standing. The kitchen was quite uninhabitable, and although Hughes had been four months in the house his kitchen furniture was unpacked, and the room was not used. The

family was living in a single end … Altogether the house struck me as being past repair.

Some work was done on the house in November, but the man had left it at the November term for another fee elsewhere. Duncan noted that he had stood it for six months, and then left: 'Most farm servants have got so used to the neglect of their complaints, that they suffer it for twelve months rather than wasting their efforts trying to get the matter remedied.'[38]

Not all cottar houses were of such poor quality, but a general matter that affected them was their status as 'tied houses', in the same way as bothies and chaumers were tied to individual farms. As long as the farm servants led a migratory existence, moving from farm to farm at the end of each term or year, there was little incitement for improvement on the part of either farmer or servant.[39]

Another source of unease in the early 1900s was the appearance of the motor plough and the double furrow plough. A *Scotsman* report on a demonstration at Ballencrieff, East Lothian, noted that it did not need a skilled mechanic to operate the machine, and with it one man could do the work of two men, 'no mean consideration in these times of shortage of labour and increase in the cost of horses'. It was feared that this would lead to the displacement of labour, as appeared in letters published in *The Scottish Farm Servant*, for example, but the secretary, Joe Duncan, was very much in favour of mechanisation:

> Let us have motor ploughs, motor reapers and binders, motor carts and lorries, milking machines, and every mechanical aid we can get So shall we lighten labour and quicken intelligence … More machinery on the farm and we should have shorter hours, keener and more alert men, and men more determined to have their share in life.[40]

CONCLUSIONS

By the time of World War II, which is the period at which this study ends, there had been a number of changes in the conditions of farm servants. In the pre-Improvement situation, farm servants were bound to their workplaces by a strict system that had the backing of law. At the same time, the widespread existence of multiple-tenancy farms involved communal activities in which the actors were the farmers themselves.

The long-continued tradition of providing a tied house for the married or main farm servants is a direct link with the cot-houses of the pre-Improvement period and marks the persistence of a feudal concept that was of practical importance for both farmers and servants. There was much discussion about tied houses throughout the twentieth century,[41] and the freeing of farm servants from the obligation to use them is one of the major changes.

Technological advances had a strong influence on labour. The spread

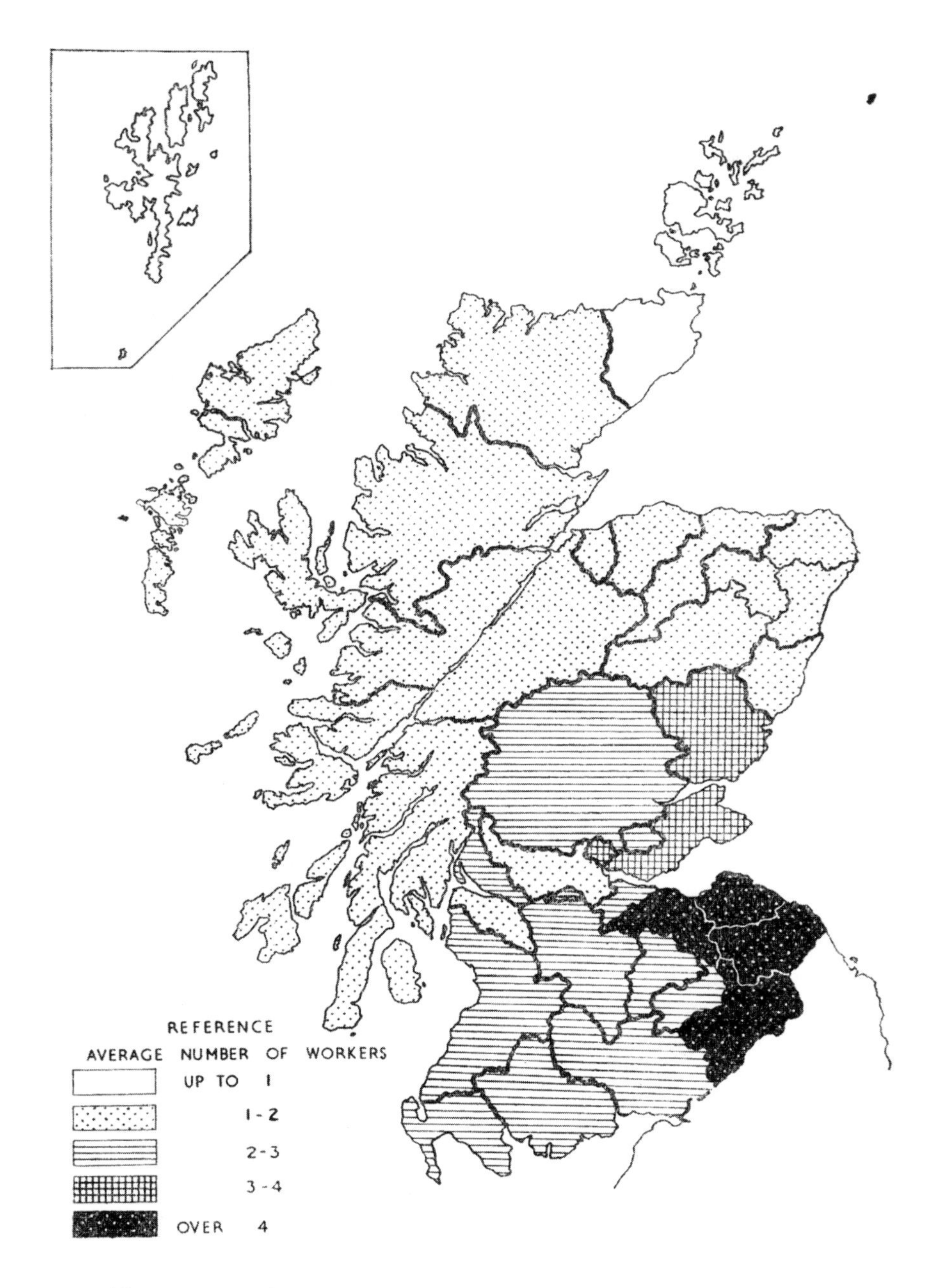

Figure 5.3 Regular hired labour force on full-time farms. Source: *Types of Farming in Scotland*, HMSO, 1952.

Figure 5.4 Working on the travelling threshing mill at Brownhill, Aberdeenshire, Easter 1966. Source: SLA C8107. Photograph: A Fenton.

of ploughs of James Small's type in the late eighteenth century led to a reduction in the manpower, and also animal power, required for cultivation. Oxen were largely replaced by horses as the main source of draught, and the 'horseman' eventually became the main farm servant in the hierarchy. About the same period, the barn threshing mill replaced flail threshing on the bigger farms, and the tasker disappeared. The reaping machine gradually reduced the need for large squads of harvesters with sickles or scythes in the first half of the nineteenth century.

With the building of new farmhouses on individual holdings came a greater degree of social separation between farm servants and farmers. Servants no longer had their meals together with the master and his family, but ate their brose in the bothy or, under the kitchie system, though they had their meals in the farm kitchen, it frequently happened that the farmer dined 'ben-the-hoose'.

The story of farm labour is a complicated one. It must take into account the smaller farms and crofts that did not employ farm servants and therefore did not pay formal wages. It must also evaluate the wide regional variations in the conditions of farm service according to geography and to the types of farming being pursued. One book that begins to tell this story is *Farm Servants and Labour in Lowland Scotland 1770–1914*, edited by Professor Tom Devine.[42] There is also Richard Anthony's book, *Herds and Hinds: Farm Labour in Lowland Scotland 1900–1939*.[43] This is a good beginning, but there remains the continuing story for the period following World War I in the Lowlands, including the effects of war, and of the areas outside the Lowlands, mainly the crofting counties and areas of small farms within the Lowlands.

NOTES

1. See Smout, T C and Fenton, A. Scottish agriculture before the improvers – An exploration, *Agricultural History Review*, XIII/II, 1965, 82–4 [73–93].
2. See Department of Agriculture for Scotland. *Types of Farming in Scotland*, HMSO: Edinburgh, 1952, 11–24; Coppock, 1976, 170–99.
3. For a discussion of harvest food, see Fenton, A. Special food at special periods. In Fenton, A. *The Food of the Scots* (Scottish Life and Society: A Compendium of Scottish Ethnology, vol 5), Edinburgh, 2007, 212–22.
4. Firth, 1899, 405–8.
5. *Accounts of the Treasurer of Scotland*, XIII (1567), 81.
6. Malcolm, 1931, 16–21.
7. Malcolm, 1931, 165–71.
8. Fenton, A. The plough-song. A Scottish source for medieval plough history, *Tools and Tillage*, I:3, 1970, 177, 181 [175–191].
9. See Whyte, 1979,137–45; Devine, 1994, 10.
10. *NSA*, XIV (1845), 310: Strath, Inverness-shire.
11. As he was called in an advertisement for a hind in the *Edinburgh Evening Courant*, 1 August, 1772, quoted in *SND* s v Hind.
12. *OSA*, III, 1795,197: Dirleton, Haddington.
13. *Dumfries and Galloway Standard*, 1 January 1955, quoted in *SND* s v Hind.
14. Ure, 1794, 15.
15. *NSA*, XIV, 1845, 53–4: Logie Easter, Ross-shire; *SND* s v Lead.
16. *NSA*, XIV, 1845, 27: Nigg, Ross-shire.
17. *People's Journal*, 20 October 1951.
18. Somerville, A. *Autobiography of a Working Man*, London, 1848, PAGE.
19. Littlejohn, G. *Westrig*, London, 1963, 53.
20. *OSA*, VI, 1793, 333: Swinton and Simprin, Berwick.
21. *NSA*, XIV, 1845, 19: Inverness.
22. *NSA*, XIV, 1845, 174: North Uist, Inverness-shire.
23. Stephens, 1844, I 220–9.
24. Stephens, 1844, II, 380–9.
25. *DOST* s v Cot-house.
26. Stephens, 1844, II, 380–9.
27. *NSA*, XIV, 1845, 157: Harris; 174: North Uist; 210: Barra, all Inverness-shire.
28. *NSA*, XIV, 1845, 401: Pettie, Inverness-shire.
29. *NSA*, XIV, 1845, Cromdale, Invernesss-shire
30. *NSA*, XIV, 1845, 33: Nigg, Ross-shire.
31. Fenton, A. The Easter Ross farmers' club, 1811–1898, *ROSC 16*, 2003–4, 121–2 [115–32].
32. *NSA*, XIV,1845, 465–6: Tarbat, Ross-shire.
33. The manuscript was kindly lent to me for examination by Mr W H M Gill, Invergordon.
34. Mowat, 1981, 72.
35. *Weekly Statement of Labour on the Farm of Arrabella 28 November 1831–24 November, 1832 Ms.*
36. From the revised Constitution of the Scottish Farm Servants' Union, Aberdeen, 1890
37. See Carter, 1979, 110–32.

38. Duncan, Joseph A. Housing of farm servants, *The Scottish Farm Servant*, III/35, February 1915, 6–7.
39. See, for example, Chapter 6, 'Rural Housing', in Smith, 1973, 121–45.
40. Motor Ploughing, *The Scottish Farm Servant*, III/26, May 1915, 7.
41. See ref 39.
42. Devine, T M, ed. *Farm Servants and Labour in Lowland Scotland 1770–1914*, Edinburgh, 1984.
43. Anthony, R. *Herds and Hinds. Farm Labour in Lowland Scotland, 1900–1939*, Scottish Historical Review Monographs Series, No. 3, East Linton, 1997.

BIBLIOGRAPHY

Coppock, J T. *An Agricultural Atlas of Scotland*, Edinburgh, 1976.

Devine, T M, ed. *Farm Servants and Labour in Lowland Scotland 1770–1914*, Edinburgh, 1984.

Devine, T M, ed. *The Transformation of Rural Scotland: Social Change and the Agrarian Economy 1660–1815*, Edinburgh, 1994.

Firth, C H. *Scotland and the Protectorate*, Edinburgh, 1899.

Malcolm, C A. *The Minutes of the Justices of the Peace for Lanarkshire, 1707–1723*, Edinburgh, 1931.

Mowat, I. *Easter Ross, 1750–1850*, Edinburgh, 1981.

Smith, J H. *The Scottish Farm Servants and British Agriculture*. Typescript printed by the RCSS, University of Edinburgh, in conjunction with the Scottish Labour Hisory Society, Edinburgh, 1973.

Stephens, H. *The Book of the Farm*, 3 Vols, Edinburgh and London, 1844.

Ure, D. *General View of the Agriculture of the County of Roxburgh*, London, 1794.

Whyte, I. *Agriculture and Society in Seventeenth-Century Scotland*, Edinburgh, 1979.

6 Managing and Working in the Food Retail Sector

ANNE M FINDLAY AND LEIGH SPARKS

INTRODUCTION

Food retailing in Scotland has changed radically from the market stalls, 'booths', producer–retailers and itinerant traders of the seventeenth and eighteenth centuries.[1] Superstores, hypermarkets, convenience and discount stores dominate the current retail landscape.[2] These transitions from producer–retailer to store or shop manager and from family business to multinational organisation have altered, perhaps forever, the commercial, social and cultural relationships in food retailing. The goat herder selling milk a hundred years ago in Springburn (Figure 6.1) has been replaced by the Asda Robroyston superstore (owned by the world's largest company, Wal-Mart).

Through snapshots of the personalities of Scottish food retailing entrepreneurs and histories of Scottish food retail companies and those employed in Scottish retailing, this chapter seeks to relate this transition from family to corporate business and from Scotland-based to transnational retailers.

This chapter considers the major organisational changes in retail business and how these have affected the people managing and working in the sector. The first section discusses the positions of independent, co-operative and multiple retailers. This highlights the role of Scottish retail entrepreneurs and the national trends which have 'revolutionised' food retailing. The second section focuses on the changing types of job, the introduction of self-service and the meaning of work for shop owners, managers and employees. Finally, we consider the changing meaning and identity of shopping for food, food retailing and consuming food in Scotland.

RETAIL STRUCTURE AND ENTREPRENEURS

The retailing sector in Scotland contains a number of organisational forms, which in turn operate a number of retail formats. The main organisational forms are independent, co-operative and multiple (or corporate) retailers. They operate a large range of shops in a variety of locational situations, ranging from corner shops to superstores in town centres and out-of-town

Figure 6.1 Upper: Goat herders selling milk in Springburn, 1895. Reproduced by permission of Springburn Community Museum Collection, Mitchell Library, Glasgow (http://gdl.cdlr.strath.ac.uk/springburn).
Lower: Asda Superstore, Robroyston, 2005. Source: Anne Findlay.

retail parks. Independent stores – where the retailer typically owns and manages the business – were the first fixed retail forms. These have been challenged, first by consumer co-operatives (essentially profit-sharing collective organisations equally owned by consumer–members; the consumer co-operative movement is discussed further in Chapter 24) and then by multiple or corporate retailers (who run chains of stores under the same fascia and are owned by institutional and other shareholders). The relationship between the management and staff of shops and these workers to the product and the location varies by organisational form and the retailing situation.

THE INDEPENDENT SECTOR

The independent retail sector in Scotland, as elsewhere, is the numerically predominant form. The market share of the local small shop run by the owner–manager has declined considerably, but their contemporary numerical significance should not be underestimated. In 1950 there were an estimated 23,500 food retailers in Scotland. Approximately 65 per cent of the 11,445 grocery stores (as opposed to specialised food stores) were independents, accounting for 37 per cent of retail grocery sales, the same market share as that held by the 1,698 consumer co-operative society stores.[3] As is the case throughout Western Europe, the number and share of independent shops has declined: by 1971 the number of food retailers in Scotland was 18,400, but by 1986 it was only 8,300.[4] In 2003 there were about 5,700 food stores in Scotland, of which 74 per cent were independents – including independents in associative organisations such as Spar and Mace – but they accounted for only 15.5 per cent of the turnover.[5]

The problems faced by independent retailers have been many and varied.[6] Amongst these are: inadequacies in the trading environment brought about through economic and social change; competition from other retail forms such as co-operatives and, latterly, multiples; and locational difficulties. These are compounded by inadequacies in the retail form itself which incur higher operating costs; have limited capital availability; and often suffer from product supply difficulties. Moreover, many independent shopkeepers have limited management expertise, limited knowledge of modern techniques, and may be growing older. Together these issues make independent shopkeeping a difficult vocation and often an economically marginal activity, although such shops can form the social and economic lifeblood of many areas in Scotland, particularly rural or urban deprived communities.

The Scottish Grocers' Federation (SGF), formed in 1918, has been a key Scottish organisation attempting to look after the needs of Scottish independent retailers. Scottish grocers began to feel that the many rules and regulations, particularly concerning pricing, that were being brought in by the UK government because of World War I did not take account of Scottish needs. The history of the SGF records:

> Difficulties such traders never dreamed possible were forced upon us by the war. At the outset everything looked black and it seemed almost impossible that we could carry on. It was then that the local association and the federation came into existence and that the organisation created an intangible something that has stimulated and assisted and strengthened the trade all over the country ... we are going to endeavour to do what we can for the good of our trade.[7]

The SGF has sought to establish buying groups, promote training and retail knowledge, provide legal advice, support the independent sector and liaise with Scottish policy makers. It continues these roles today, not least lobbying the Scottish Government and competition authorities.

Over almost 100 years, the SGF has represented the changing needs and ownership of independent food retailing in Scotland. The change in the sector over this period is encapsulated by the movement from an all-male executive committee in the early days of the SGF through the presence of women such as Mrs Mary Christie (Figure 6.2), who was on the federation executive from 1974 to 1978, to the recent appointment of Pete Cheema, owner of a Spar store in the Raploch in Stirling, who became the first Asian grocer to take over the presidency of the SGF in 2006.[8]

This significant development reflects particular cultural and ethnic dimensions of independent retailing in Scotland. Many independent stores and businesses have been owned and operated by immigrant families, during both recent and long-standing waves of in-migration, particularly in the urban conurbations. This is not a purely Scottish phenomenon, but it reflects the relative low entry barriers in small-shop independent retailing and the sheer hard work needed to compete in the retail industry. Long hours, hard work and often small returns are small compensations for the comparative low entry barriers.[9]

The pressures on independent retailers have mounted steadily, and

Figure 6.2 Mrs Mary Christie in her shop at Keppochill, Springburn, 1960s. Reproduced by permission of Springburn Community Museum Collection, Mitchell Library, Glasgow (http://gdl.cdlr.strath.ac.uk/springburn).

true independents are harder to find, often being located in specialist sectors. Examples include high-quality delicatessens such as Valvona & Crolla in Edinburgh or Clive Ramsey in Bridge of Allan. Both of these examples reflect another trend, that of diversification into the restaurant/bistro business. Many independents are now part of some larger voluntary or associative organisation such as Spar, Mace or Costcutter, or a buying group such as Nisa. These larger organisations to varying degrees provide support, expertise, buying power and branding, and help in meeting the challenges of competing with larger, branded multiple retailers in Scotland. Some operators such as Jim Botterill (a major Spar franchisee), Eddie Thompson (founder of Morning, Noon and Night, sold in 2004 to the Scotmid Co-operative Society) and David Sands have successfully developed strong local convenience-store businesses in Scotland. C J Lang & Son, the Dundee based wholesaler, is another business which recognised that in the independent and small shop sector there is interdependence between the retailing and the wholesaling operations. They have tried to bridge the gap between both sectors by supporting and running independent retail outlets.

CO-OPERATIVE RETAILING

The Industrial Revolution of the eighteenth and nineteenth centuries created new wealth-owning classes and new disparities in wealth. Additionally, in some industrial areas the development of 'truck' or company shops led to overpricing, product adulteration and profiteering. Together these created an environment in which co-operation, involving control of the supply of products and the sharing out of surplus amongst members of equal standing, was particularly attractive. The Fenwick Weavers Co-operative of 1769 and the Govan Old Victualling Society founded in 1777 are considered to be amongst the first co-operatives in Scotland. Co-operation was also an underpinning of Robert Owen's model community at New Lanark. In a similar vein, Forfarshire had 'soches' (bread societies following the principle of dividing profits amongst members) in the early nineteenth century. As co-operation and collectivism took hold, particularly in the mid-nineteenth century, a myriad of separate local consumer co-operative societies were set up in the neighbourhoods of cities and towns across Scotland.

St Cuthbert's Co-operative Society, which became one of the larger co-operative societies, opened its first store in Edinburgh's Ponton Street in 1859:

> We opened with considerable spirit and enthusiasm, and some of us found a new relish in our butter, ham and meal, in that it was turned over to us from our own shop, through our own committee. We were all yet working men, but we began to feel that we were something more, and would soon be business men reaping profits we had long been sowing for others.[10]

During the first years of the society there were many difficulties, including financial, staffing and management problems. Given that the local committee was made up of volunteers who were by profession joiners, foundry workers, blacksmiths and cabinetmakers, it was perhaps not surprising that sometimes they got into difficulties. Like other consumer co-operative societies of the time, St Cuthbert's gradually opened new branches in Edinburgh as well as diversifying into other retail sectors such as drapery, manufacturing activities such as bakery, service activities such as horseshoeing, and wholesaling activities, including a controversial meat-buying project which met with some resistance from other butchers.

As local consumer co-operative societies developed and expanded, the importance of establishing co-operative joint buying increased. In turn this led to the formation of the Scottish Co-operative Wholesale Society,[11] championed by James Borrowman, whose father, John, had been involved in the St Cuthbert's Society. Its aim was to organise such activities, based on 'traditional' co-operative values including surplus-sharing and one member, one vote. The Scottish Co-operative Wholesale Society, though, had to contend with personality conflicts, competing societies, traditional locational rivalries, unevenness in buying power, and different ideas about buying, production, distribution and finance. Getting co-operatives to co-operate proved hard work, something that has continued to the present time in areas such as branding and buying, for example. However, probably most problematic in the relations amongst local societies was the issue of whether it was in the co-operative spirit for co-operative society areas to overlap and thus for co-operative societies to compete directly with each other.

The peak of the consumer co-operative movement in the UK was in the late 1950s and early 1960s, when it dominated UK retail trade and boasted over 13 million members. However, even at its apogee, the tensions in the movement and the emerging threat of competition, particularly from ever larger and more aggressive multiple retailers, were apparent. Since the 1960s this pressure has taken its toll on individual societies and forced concentration and amalgamation. St Cuthbert's Co-operative Society proved more successful at independent survival than some other Edinburgh- and Scotland-based societies. In January 1981, St Cuthbert's Society merged with the Dalziel Society of Motherwell to form the Scottish Midland Co-operative Society (Scotmid), which continues today as the largest Scotland-based consumer co-operative society.

The 1957 Co-operative Union *Directory* records some 180 consumer co-operative societies in Scotland, most with one or two branches, but others with a large store network.[12] Today there remain three Scotland-based consumer co-operative societies: Lothian, Borders and Angus with forty-four food stores; Musselburgh and Fisherrow with one food store; and Scotmid with sixty-five food stores. All of these societies also run other non-food retail chains. Scotmid is now the largest Scotland-based food retailer. In addition, the Manchester-based Co-operative Group's Scottish Region operates 269 food stores in Scotland.[13] As might be inferred from this, in large parts of Scotland the very local dimension to co-operative structures and organisations has been diminished. The involvement of local people remains through committee structures, but the scale of the consumer co-operative organisation and the development of management structures similar to those of multiple retailers both mean that the particularly local dimension of local members managing and running the local shop has been somewhat lost.

MULTIPLE RETAILING

Multiple retailing has been the driver of success in modern retailing. Entrepreneurs developed their retail businesses, building chains of stores operating under a common fascia or trading name. In Scotland, familiar examples include Galbraith, Templeton, Lipton and William Low. Multiple retailers challenged the primacy of independent and co-operative stores. This challenge came mainly from the rapid speed of store development, business and store scale, and modern operating techniques. In particular, the centralisation of management and operations such as buying and distribution provided economies of scale, scope and replication. As consumer choices and patterns of shopping changed, so multiple retailers became successful common names across many Scottish towns and cities. One consequence of this was the development of a professional, retail management cadre with skills in store and head-office functions. This transition to a sector dominated in economic (if not numerical store number) terms by multiple retailing is a significant change to the structure of food retailing in Scotland.

Scotland has had its share of food retail entrepreneurs. Of these, Thomas Lipton and William Low are probably the best known. Multiple food or grocery retailing in Scotland began to emerge in the second half of the nineteenth century. This was a period when food product ranges were widening, with newly imported products such as tea and other exotic products coming from abroad. Increasingly, food production and sales were segregated, as producers no longer directly sold to consumers the products that they made or produced, but rather sold them through wholesalers, distributors or retailers. Entrepreneurs such as Thomas Lipton saw the possibility of organising the supply of food products and thus building a chain of retail outlets.

Sir Thomas Lipton is perhaps the 'archetypal figure of multiple retailing'.[14] The Lipton family were Protestant Irish smallholders who had fled Ireland during the mid-nineteenth-century potato famine. They established a shop in Crown Street in Glasgow, where Thomas Lipton learned the trade from his father. Thomas did not immediately follow his father into the grocery business, but at the age of fourteen went to North America. Returning to Glasgow, he opened his first shop in Stobcross in 1871. The family retained strong links with Irish farmers and were able to source food products from there. He opened new retail branches in the High Street and Jamaica Street in Glasgow, and by 1889, in addition to these and other Scottish stores (Figure 6.3), he had opened a store in London.

Lipton was an entrepreneur and he extended his trade, diversifying into new areas, including adding food processing to his retail business. The family links to supply sources gave him an understanding of the importance of supplier–retailer relationships. Lipton's purchase of tea plantations was probably the most important illustration of this. It is, of course, tea for which Lipton's (now a part of Unilever) remains best known.

Lipton shops and other businesses spread across both England and Scotland. His empire had already reached 200 stores by 1900, a couple of years after he was knighted. This number had doubled through organic growth by 1920.[15] This breadth marked him out from most of his contemporaries and gave him a business basis which they struggled to emulate. Lipton's was not just a Glasgow, or even a British, company, but a multinational business, although it remained within Thomas Lipton's control. Just prior to Lipton's death in 1931, the Scottish identity of the company was lost, as it was taken over by Home and Colonial Stores, another large retail business which had grown at a similar pace to Lipton's. Home and Colonial, together with Maypole Dairies and other firms including Scottish companies such as Galbraith and Templeton, together became Allied Suppliers. The Lipton store fascia, however, remained a feature of high streets until the 1980s.

Thomas Lipton had been successful in turning a Scotland-based company into a multinational one. Other Glasgow multiple retailers included William Galbraith, Andrew Cochrane, Robert Templeton and Alexander Massey, but their reach never rivalled Thomas Lipton's. Like Lipton, William Galbraith's father had a grocery shop; Andrew Cochrane had also worked in a shop before setting up on his own. These Glasgow-based grocers concentrated on particular lines and developed their own particular specialities. Galbraith's, for example, blended their own tea and had 'own label' products. Massey became an important supplier in the Highlands and Western Isles, and also had stores in towns such as Dundee. In 1910 Templeton's had fifty branches extending from Paisley to Blantyre. In 1918, Galbraith's is recorded as having had twenty shops in Paisley.[16] By 1920 Andrew Cochrane had 110 shops in Glasgow.

These firms were Scottish, operating in Scotland, and most did not seek to establish themselves across the UK. Similar 'regionally' focused retailers were also emerging across England at this time. Scotland-based retailers perhaps shared a common problem in that they were not well known in London. When a need came for them to raise capital – for example, to buy previously rented properties or to expand or to undertake new ventures – this could prove more difficult than for some other (English) firms. Their Scottish identity was to some extent an impediment to furthering business expansion. Additionally, England-based retailers such as Maypole Dairies were already trading in Scotland. All of these retailers were targeted for takeover, particularly by Home and Colonial Stores (which became Allied Suppliers, and eventually in the 1970s Argyll Group, and then Safeway). Food retailing in Scotland was becoming less connected with the Scottish

Figure 6.3 Lipton's Shop, Perth, 1880s. Reproduced by permission of Perth Museum and Art Gallery ©.

land and producers and the commercial infrastructure of Scotland's towns and cities. This is a pattern that has continued to this day.

Perhaps the most striking example of this is William Low. Like others, James Low, William's father, learned the grocery trade as an apprentice, in his case in Kirriemuir. He opened his own grocery business in 1868 (the same period as Thomas Lipton) in Hunter Street in Dundee. By comparison with contemporaries, Low's grew more slowly, but unlike the Glasgow firms the stores were more widely dispersed across Scotland. In 1900 there were some sixty-four stores across the country.[17] William Low followed his father into the firm, and the company became known as Wm Low & Co. The family interest in the firm and influence on the board was sustained until 1988. In the 1980s Wm Low positioned itself as a Scottish multiple retailer developing new store formats. In 1985 Wm Low bid successfully for Laws, extending the business into the north of England However, despite this incursion south of the border and efforts to build new superstores, Wm Low could not achieve the speed of growth or the economies of scale being produced by the larger more aggressive multiple retailers such as Tesco, Asda, Sainsbury and Safeway. In Scotland, Wm Low was also 'squeezed' by new discount retailers such as Shoprite. By 1994 Wm Low had fifty-seven stores and controlled 15 per cent of the Scottish grocery market.[18] However, just as Wm Low was seeking to establish itself in England, so Tesco and Sainsbury amongst others were seeking to establish themselves in Scotland. Wm Low became the subject in 1994 of a takeover battle, a battle won by Tesco. The Wm Low board recommended acceptance of the takeover, as they recognised that they could not compete successfully with larger more powerful rivals and that a takeover was the best way of realising value for the shareholders. Tesco, which won the battle for Wm Low from Sainsbury, has gone from strength to strength, whereas Sainsbury has struggled to fully establish itself in Scotland. As one commentator put it:

> Wm Low had tried to reconfigure itself since the 1960s on several occasions as retailing restructured. It recognised a need to become more 'national' in its approach, and its search for a British rather than a Scottish identity reflected wider considerations of space in British retailing and contemporary Britain. Tesco, on the other hand, realised the importance of a Scottish space as much as space 'down south'.[19]

This pattern of takeover and reorganisation of Scottish food retailing has been a recurring one. For example, Safeway plc was built by takeover from the 1960s onwards, with particular growth coming when the business was led by Alistair Grant, James Gulliver and David Webster, three Scots. For many, despite having headquarters in England, Safeway, with its Scottish directors and Scottish roots (through the Argyll Group and back to Lipton Stores), was seen as a Scottish food retail business. The takeover of Safeway by Yorkshire-based Morrisons in 2004 thus represented a further loss of what was left of Scottish retail identity. Morrisons in turn has sold off many of the Safeway northern and island Scottish stores and the smaller shops to another England-based multiple, Somerfield.

The recent pattern of food retailing in Scotland has therefore been the increasing domination by England-based retailers. There is another change, however. Until the 1960s the stores, regardless of whom they were owned and run by, operated out of small premises on high streets, or as local corner shops embedded in housing areas. The wave of development from the 1960s, particularly by English multiples, though copied by Wm Low and some of the Scottish consumer co-operative societies, was spatially different. Major retailers such as Tesco, Asda and Sainsbury initially expanded into Scotland by building larger more modern stores and then superstores on out-of-town sites or away from the traditional high streets. Asda was the first UK multiple to build a large superstore in Scotland. Coatbridge – chosen for the site of their first Scottish superstore in the 1960s – was an area where the Coatbridge Co-operative Society had been very active in promoting supermarkets. Other out-of-town stores followed, albeit later and at a slower pace than in England. The competitive effects of these new superstores and larger businesses began to change the face and place of Scottish food retailing. The growth of major retail businesses in Scotland (and the UK) has been a function of the need for companies to expand their businesses to achieve economies of scale, increase their buying power, penetrate new markets, increase their product range and establish new and larger formats in new locations. The out-of-town superstore, and latterly the hypermarket, are cost-effective ways of retailing food and non-food goods. Today the top four UK food retail multiples (Tesco, Asda, Sainsbury and Morrisons) are operating portfolios of stores including hypermarkets and superstores throughout Scotland.

There are now over 160 hypermarkets or superstores across Scotland, reflecting the changing consumer needs and shopping habits.[20] Superstores on out-of-town sites are cost-effective for retailers and for consumers. As consumers have gained mobility through the use of the car and the capacity to purchase and store larger volumes and quantities of products, so the shopping trip has altered. A once-a-week, fortnightly or monthly bulk-buy has become a familiar pattern. More recently, however, there has also been a resurgence of interest in smaller stores, although often operated by larger chains such as Tesco and the Co-operative Group. The convenience-store market has grown rapidly, both in stand-alone and other locations, for example, associated with petrol stations. Hard-line discount stores operated by foreign retailers such as Aldi and Lidl have also sprung up across Scotland. High streets have become the sites for new smaller food stores, such as the Sainsbury Central or Tesco Metro formats. Travel termini including airports, motorway service stations and railway stations also often have small supermarkets, convenience stores, delicatessens or a Marks and Spencer Simply Food store. Tesco, the UK's largest food retailer, now has hypermarket (Extra), superstore, high-street (Metro) and convenience-store (Express) formats to capture as much of the Scottish (and UK) trade as possible, as well as a very successful internet sales and home delivery operation (Tesco.com). One implication of this plethora of differing locations and shop sizes is that food retailing is now not so much a common experience carried out at the same times of day at broadly the same types of place and retailer and for a limited range of products, but instead has migrated from a solely daytime town centre or local corner-shop activity carried out by housewives to a multi-locational, multi-format, 24-hour activity undertaken by anyone.

The food retailing sector has thus witnessed a continual struggle for market share and domination, at local and national levels. Where we shop, how we shop and with whom we shop have all been transformed. What we buy and when we buy it are now very different to the past. People shop in different ways, at different times and buy different products from what they did a century ago, fifty years ago or perhaps even ten years ago. Changes in the scale of businesses and shops and the repositioning of the different parts of the food retail sector are responses to cultural and societal drivers. Retailers have had to strive to ensure that they maintain contact with consumers and markets to continually meet their changing needs. Understanding Scottish consumers and then meeting their needs remains the key purpose of the sector, despite managers having fewer ties to Scotland or the store location and the products being as likely to come from Peru as Portree, Thailand as Tillicoultry and Argentina as Alloa.

WORKING IN THE RETAIL SECTOR

The changes in Scottish retailing outlined thus far can clearly be seen to have had implications for managing and working in food retailing in Scotland. In this changed Scottish retail world, the nature of work in the store or the retail business has also been transformed. Nevertheless, there have also been other changes. While the organisational structure and the locations and shop types through which food retailing is carried out have altered, so too have many of the practices of food retailing changed. The most visible changes include the introduction of self-service retailing and the extension of trading hours, including widespread Sunday trading. Both practices demand an increased flexibility from the retail labour force. The nature of retail work has thus been significantly restructured. But the same is also true of the management task: 'A retailer of the 1950s would be totally confused by (today's) accounting procedures, labour relations, buying procedures and a host of other managerial activities.'[21]

Food retailing remains a significant employer of people in Scotland. Some 13.3 per cent (or *c*300,000) of the Scottish workforce were involved in the wholesale and retail trades in 2001.[22] Of these, 51 per cent were female employees and 38 per cent were working part-time. While in the past many food stores would have been small, local businesses, today many employees work in food superstores where employment numbers run into the hundreds. The scale and nature of the retail workplace has been transformed. The links to local products and local consumers are now for many retailers far weaker than they have ever been.

We can illustrate this changed Scottish retail world, with its very different work patterns and identities, through the following quotations, which provide the voices of two independent retailers from the 1920s–30s and the 1950s–60s respectively:

(a) Hurlford, Ayrshire

My mother had proved herself to be a most efficient wages clerkess at Hurlford Coop, then after at Lewis Isaac's Emporium of Kilmarnock. The business acumen and financial flair were to stand her in good stead in managing the new partnership in my father's business. They kept a property at Riccarton Road in Hurlford. It was here that my father stabled 'Nellie' his pony. There too he built storage sheds, and could use the warehouse sinks for the pre-cleaning and preparation of fish – a cold glutty, gutty affair in the bitter winter mornings. His main supplies of cod and haddock were ordered weekly from John Mowatt and Son of Aberdeen, delivered daily at 7am to

Hurlford Station. Early season herring came from Murray of Ardrossan. Sausages and bacon from Gold and Hardie. Tomatoes and summer soft fruit from DL Andrews of Mayfield and Jimmy Neill of Shawhill. Bananas, apples, oranges and pears were bought in the old Glasgow fruit market off the Candleriggs, where he went most Mondays to bid for barrels and crates of fruit.

The Aberdeen fish had to be collected daily from the station at 7am, gutted, filleted and loaded onto the float by 8am, so that he could be on the road before 9am. We carried a flask and sandwiches so that he wouldn't need to return to base before 6 or 7pm. Each evening the unsold goods had to be packed away, the pony stabled, groomed and fed. Only then could he go 'down the brae' to Riverbank Cottage for his own evening meal.

These were busy, happy days because, although the hours were long and hard, he was his own manager. He loved the open air life and he revelled in the afternoon rounds when he did his country runs. On Tuesdays he would visit every farm and cothouse from Templeton Burn to Moscow. On Wednesdays, it might be all the hamlets and farming coteries from Crossroads to Crosshands, then he'd trot the horse up the side lanes through Sornhill and Meikleyard to the outskirts of Mauchline. Thursday was reserved for the 'Craigie Run' and this became my favourite trip. On Fridays, he crossed the Haining ford and covered the area between Riccarton Moss and Craigie Mains.

In the days before supermarkets, the farms came to rely on his regular weekly appearance with a good range of fresh provisions, offered to them with his ever cheery crack and lively bantering tones. He knew the importance of presenting a smart, tidy impression, so his standard rigout was in freshly laundered khaki twill service coat with polished brass buttons and brightly burnished black leggings and highly dubbined boots. Rain, hail or shine saw him cantering along his regular routes whistling or singing one of the popular ditties of the day. No matter how severe the weather, his motto was to 'traik on regardless' until all of the day's perishable goods had been sold.[23]

(b) Geddes Fine Foods, St Andrews

Geddes for Fine Foods was a delicatessen in St Andrews with a reputation all across southern Scotland. It was a family business. George Geddes described how his father had first had a job with Gordon Baxter in Fochabers, later working for Maypole Dairies in St Andrews and then opening his own confectioner/fruiterer business in Market Street, St Andrews. The war years were hard years when many products were not available. However, even during this period Geddes was gaining a reputation for having unusual products not available elsewhere. This became the hallmark of the shop in the postwar years when George and his brother were running the shop. It became a speciality shop, so that even although there were many other grocery shops in St Andrews, including the Co-op, Liptons and William Low, as well as independent grocers, Geddes did not compete with them but established its own market niche. By the 1970s the shop stocked a vast range of products, including some 200 types of cheese. George tells of how he came to stock some alligator soup. He was not hoping for a quick sale but to use the product to intrigue his customers and spread the word back to Edinburgh and Glasgow with the summer visitors that there was alligator soup to be had in St Andrews and that Geddes for Fine Foods, as the shop was now known, had everything exotic and quixotic that you could think of. It worked, and the shop gained a reputation in many parts of Scotland, a reputation in which he took great pride. It was a labour-intensive job as George had to do all the sourcing himself, making contacts with suppliers, going to trade fairs, managing the shop and ensuring that the very large range of perishable products was correctly stored. He would go to Soho and see for himself the delicatessens in London. His shop was not a specialist in French or Greek foods like them, but rather stocked products from a wide range of countries. Suppliers were intrigued that any such business could survive so far north in the UK, and some even made a special visit to the store. At this time supermarkets had no interest in the delicatessen business making sourcing and ordering small quantities easier and of course posing no competition. George really liked the contact with people and the hands-on approach to the business, leaving his office on Saturdays to serve customers. There were six employees, and extras on Saturdays.

The shop had come a long way from the early open-doored shop with the dust blowing in and the bicycle boy, but with the retiral of his brother George found the burden of work too much. The hours were long. The market was changing and consumers were beginning to see the exotic and quixotic not as a special privilege but as something they had to have. He thought back to his navy experiences in the war and the poverty of many in the Third World and decided to close up shop in 1982. His interest had never been in setting up a business empire but in establishing an interesting business which gave him an adequate standard of living. It is not surprising therefore to learn that he went to work in the charity sector, with the Sue Ryder foundation, using his retail experience to promote their shops. He is also a champion of FairTrade.[24]

These extracts demonstrate the meaning that these retailers found in their chosen profession. In the first example, the linkages amongst local providers and the role of the travelling shop in sustaining farms is clear. In the second example, the ethos of specialist food retailers comes through, but even then the encroachment of the supermarkets eventually proved too strong. Family retail businesses, such as these, established on limited capital, encouraged a strong local and family bond, with often a generational continuity in the business. The satisfaction in being one's own boss and in meeting customers is evident, and for some retailing was a way of life or even a 'calling'. Business ambition was not driven necessarily by making money or creating a large business, but rather by earning an adequate living. Even Thomas Lipton had seen that his parents had no big goals for their retail business.

It was his American experiences of business that inspired his ambitions to expand.

> They had only a few pounds saved up and failure would have meant disaster. But they had faith and energy and determination; all they desired was a bare living and the thought of 'success' – in the ordinary sense of the word – and of making money never entered into their calculations.[25]

The grocery trade was a respected local trade. It represented upward mobility but required little capital investment to get started. In Scotland there was a strong association between work and religion amongst the middle classes.[26] This is reflected in the business ethic of many independent retailers, including those outlined in the extensive quotations above, both of whom had strong religious beliefs. Local food retailers were thus culturally, religiously and locationally bound.

The first St Cuthbert's Society co-operative store employed a single store manager. When the manager had a day off or was ill the local committee had to step in to operate the store.[27] The 1879 Liptons store in Paisley, by contrast, had a large horseshoe counter with twelve assistants, while the Dundee branch had fifteen assistants.[28] The assistants almost invariably had starched white uniforms, and indeed old shop photographs often show the staff posed outside the shop in these smart white aprons. Figure 6.3 illustrates this for the Lipton store in Perth in the 1880s, while Figure 6.4 shows a very similar, if less visually dramatic, situation at a St Cuthbert's co-operative store in Edinburgh in 1900. Comparing these illustrations, one can see how advanced Thomas Lipton was in his promotion, merchandising and display techniques.

Wm Low stores in the early part of the twentieth century had as many

Figure 6.4 St Cuthbert's co-operative store, Edinburgh, 1900. Reproduced by permission of the Scottish Midland Co-operative Society Ltd ©.

as seventeen employees at some branches, although others had only four.[29] These were typical local situations that persisted for many years.

In the nineteenth century, female participation in the workforce was limited, and in retailing the employment of men was more common. By the early twentieth century working in shops such as Jenners in Edinburgh became a respectable type of employment for young women.[30] About half of all female shop workers were under twenty in the late 1930s. They had little job security, were poorly paid and worked long hours. Although the Shop Assistants' Union tried to enforce better conditions, it was only in the co-operative sector that unions were influential in retailing. St Cuthbert's Co-operative Society prided itself on how well it looked after its workers. St Cuthbert's employed their first woman manager in the 1880s, although this was largely to keep costs down.[31] The main alterations in the male–female ratio in such stores came particularly during times of war. After World War II in particular, retailing, including food retailing, became more of a 'female' occupation, a contrast with the make-up of the workforces demonstrated in Figures 6.3 and 6.4.

Self-service in food retailing has been one of the main store-based operational changes to have occurred in Scottish retailing. Self-service has become the norm partly as a result of the increased availability of labour – particularly women – who required fewer skills to work in this environment, but it has been primarily driven by potential cost-saving and efficiencies. The stores in Figures 6.3 and 6.4 and even Mary Christie's store in the 1960s (Figure 6.2) offered counter service, an approach also seen clearly in the Mace store in Edinburgh in 1958 (Figure 6.5).

Customers waited at the counter to be served by staff, who selected items from the shelves for them, probably on the basis of an order book or list. In some situations the order was later delivered to the customer's house. Self-service, whereby customers picked products by themselves directly from the shelves and paid at central checkouts, was a dramatic shift in operations and power, both for customers and shop workers. The first self-service food stores opened in the UK in London in the late 1940s and early 1950s, taking off particularly as rationing and postwar austerity eased and consumers became more prosperous and adventurous. The location of the first self-service food store in Scotland is contested, but St Cuthbert's certainly had self-service co-operative food stores by 1956, as Figure 6.6 shows.

Self-service was a response to many things. Rising affluence, the end of rationing, the expansion of housing and home appliances, the introduction of mass television and television advertising, and the change to production and packaging technologies all played a part. The abolition of Resale Price Maintenance (RPM) allowed retailers to compete on the basis of price. The old loyalties, developed by local association and by rationing during World War II, which linked consumers to stores, broke down.[32] The efficiencies of independent stores based on bulk commodity buying and production of exact amounts in store for individual consumers was replaced by prepackaged, branded products on shelves, at a lower price.

Figure 6.5 Counter Service at Mace Shop, Gilmour Place, Edinburgh, 1958. Reproduced by permission of Scotsman Publications Ltd ©.

However, self-service did not take over immediately. Stores switched to self-service but were not necessarily successful, and some switched back to counter service. Consumer resistance was partly responsible for this, but so too was the 'bond' between the grocer, his staff and amongst the regular customers, as these quotations from Dundee shoppers remembering food shopping in the 1950s show:

> The service was very good, in the corner shop you always got a very friendly reception because you knew the person because you were maybe in quite a lot … I don't think I ever intentionally went there to meet someone but you were pleased when you did meet someone. And you always met someone you knew there and chatted for a wee while.[33]

> Everybody met there one way or the other, going in to shop or coming out from shopping, or standing at the shop. It was just a gossip shop, I suppose, in a lot of ways. There was no queuing as such but it didn't really matter because you would speak to somebody as you were waiting to be served. You just went in and just stood around until you were served; somehow the grocer always knew who was next.[34]

> I remember the first was William Low's, I think, and it was quite frightening going in and having to lift things off the shelves and put them in a basket –just the thought that you shouldn't be taking or touching them – but we got used to it quick enough.[35]

Figure 6.6 Self-Service at St Cuthbert's co-operative store, Drum Street, Gilmerton, Edinburgh, 1956. Reproduced by permission of Scotsman Publications Ltd ©.

Figure 6.6 shows clearly the new arrangement for customers serving themselves. The contrasts of the range, scale, depth of product assortment and basket size with the self-service superstores of today is remarkable (Figure 6.7).

These contrasts visually demonstrate the changing demands and opportunities for customers over a period of fifty years. For employees, self-service altered their jobs considerably. In many cases product knowledge was no longer required. Shop assistants were mainly shelf-fillers and money-takers, losing the requirement to inform and sell, as well as the opportunity to chat and pass the time of day. As Figure 6.5 shows, some staff were still required to carry out tasks for customers. Additionally, service counters such as delicatessens and butchers were retained or introduced, requiring more specialist labour. Checkout staff could also interact with customers, so not all customer–staff contact was lost. However, the relationship was clearly more impersonal. Recently, some larger modern stores have introduced self-scanning checkouts, where customers effectively serve themselves through automated tills. These changes have been described in the following way:

> Exchanging news with other customers may have reinforced shopping patterns but shopkeepers, and knowledgeable assistants, were also useful intermediaries for product and usage information. Functional specialization in food commodities bestowed an aura of expertise, and

> direct accountability for the quality of what was sold provided a more personal style of retailing than is currently typical.[36]

Self-service also provided other business economies and led in part to larger stores and centralised operations. No longer did the same person order the goods, contact suppliers, visit trade fairs, serve customers, refill shelves, and hire and fire staff. Retailers began to employ merchandisers, buyers, managers, personnel officers, supervisors, checkout operators, property and estates specialists, warehouse staff, logistics staff, site researchers, online retailing experts, personal finance experts, a corporate affairs department, in-store pharmacists and IT staff, amongst others.[37] From a Scottish perspective, the takeover of Scottish retail companies by English companies and companies headquartered in England has meant that the head office and other support jobs are now frequently located outside Scotland. Tesco, for example, only operates personal finance and customer services headquarters functions from Scotland, with all other head office functions being located in England. Being part of a large company brings with it a range of employment benefits, however, such as job security, training opportunities, career advancement, and sometimes better pay and pension schemes. Large retailers promote corporate allegiance that in many ways has come to replace loyalties to a local area or a local shopkeeper for retail employees. Tesco employees join in Tesco social activities and charity events with a Tesco identity. Asda prides itself on being one of the 'ten best companies to work for in the UK',[38] while Tesco recognises that 'our people are our most important asset'[39] and thus critical to the success of their growth strategy.

Figure 6.7 Asda, Robroyston, Glasgow, 2002. Source: Leigh Sparks.

At the store level, the increasing scale of shop units, as in superstores and hypermarkets, has meant that the nature of retail work has changed. Some hypermarkets can employ over 700 people, and even standard superstores have several hundred employees. Many of these jobs will be part-time, due both to the operational benefits that flexibility of labour provides and to extended trading hours (even 24-hour opening), which require at least one member of staff to be in the store almost every hour of the week. Over time, as retailers have searched for operational efficiencies within such stores, so the number of management layers has been reduced, enabled and aided by the widespread use of technology. Staff have had to become used to multi-tasking, doing several jobs in the store, but many tasks remain mundane and often repetitive. Some hypermarket jobs, however, are not mundane and require considerable skill and product knowledge. Retailers have sought within larger stores to replicate the functional trades of the high street such as butchers, bakers and fishmongers. The extent of product knowledge and trade skill required varies, though. In some stores, for example, meat will be brought in pre-cut, and bread might be part-baked and frozen, with only reheating required.

Nonetheless, despite the nature of some of the jobs and the constant search for store efficiencies, superstores and hypermarkets are seen as good and large local employers. While many would say that their trading impact on small stores causes job losses there, the impact of a superstore employing hundreds of people in an area is directly significant to that location. As a consequence, retail-led regeneration has become one way in which deprived or run-down areas can be improved. By introducing a superstore, not only is local retailing modernised, but local people obtain local jobs (Figure 6.8). In some cases the employment and training of local long-term unemployed people is a condition of obtaining planning permission to develop the store. Tesco, for example, has been involved in a number of Scottish regeneration projects, including several in Glasgow.[40] These schemes are specifically designed to offer employment to the long-term unemployed and to link community and retailer in a new shared local identity.

Longer shop opening hours and Sunday trading have also necessitated a different work regime. As a consequence of these, flexibility has been required in the labour force, which has generally been met by an increase in the scope and amount of part-time employment. Today, superstores and even convenience stores are offering a wide range of employment opportunities, with many short shifts that are compatible with family commitments. This has led to a further increase in the number of women working in these shops. For many, the major retail companies offer a job opportunity which matches the contemporary needs of those seeking a secure, local, part-time job with defined tasks in a socially integrated workplace, and people are often proud to work for such a company. The larger companies also offer a structured career and development ladder for those on the shop floor, in store management and in support functions. The small, local, often family-run shop cannot necessarily offer the same advantages.

Figure 6.8 Urban regeneration and local retail employment: Tesco, Shettleston, Glasgow, 2004. Source: Anne Findlay.

CHANGING RETAIL IDENTITIES IN FOOD – COMING FULL CIRCLE?

The identity of the retail industry has changed. Individuals used to identify with their local business community and had a closer relationship with their customers. Indeed, it was customers' requests for the unusual that set George Geddes (see above) on the path of being a specialist in unusual foods. Standardised corporate culture has replaced local identities for most of those employed in retailing in Scotland. This has happened, however, at the wider UK level as well, and in many industries and services, and so is not unique to retailing.

When Wm Low was taken over by Tesco in 1994, many lamented the loss of the last great Scottish grocer. But many bemoaning the loss never shopped in Wm Low, and those that did soon got used to Tesco – which did go out of its way to make links to Scottish producers and demands. We have recently seen similar negative comments in the context of another takeover with implications for Scotland, when Bob Stott, managing director of Morrisons in 2004 when the company took over Safeway, was reported as saying: 'We don't know Scotland at all well.' The concern of the *Sunday Herald* was easily stated: 'But what will the denizens of Morningside, Comely Bank, Newlands and Partick who have become accustomed to their fresh pizzas, rotisserie chicken and pitted olives make of this Yorkshire intruder?'[41] The reality, however, is that for some time and for most Scots, food shopping has not been seen as an aspect of life which expresses their Scottish identity. Similarly, for many of those working in retailing, the ownership and identity of the store matters little.

Changing identities in Scottish retailing are seen in the role of the individual within the shop and in the company. These are articulated in the changes from small to large companies and from local to national companies. The meanings of work have changed, but these are responsive to a broader repositioning of work ethics rather than being a specifically Scottish phenomenon. These changes reflect the wider transformation of retailing and are replicated in many other sectors of the economy, where new technologies, new markets, more competitive business and larger-scale operations have transformed businesses in Scotland from Scottish to UK to global operators.

This sense of a standardised, homogenised, corporate retail food sector in Scotland has, however, to be tempered in a number of ways. Some of these suggest a new local and Scottish identity for food and possibly food retailing. For some parts of Scotland, and particularly in the remote rural areas, locally run shops tend to be more significant. The local links may in such situations be strong and have survived, whether with local producers or through, for example, local shopkeepers or the local co-operative store. There will still be a local sense to the producer–retailer network. At the same time, however, the large corporate retailers have recognised the advantages of tailoring the local store offer to the local catchment. Using their databases and knowledge of sales (for example, from the Tesco Clubcard), stores sell product ranges that more closely reflect local demand. Moreover, most of the large chains are keen to emphasise their local producer links. In Scotland, saltires are prominent on Scottish food products, and Scottish producers are broadly encouraged, even if the retailer comes from Bentonville, Arkansas and the manager is from Barking.

More interestingly, perhaps, there has been resurgence in concern over the origins of the food that we eat and what it means to us. Probably sparked by various food scares, publicity over the plight of small farms and farmers, and more positively through recognition of the high quality produce available in Scotland and the merits of healthy eating, independent structures such as farm shops and farmers' markets[42] across Scotland have gained reputation and sales. Such enterprises reconnect the consumer with the producer or even the animal and have shown that Scottish consumers can be won over to good quality local products. From being initially simply a small extra income stream for farmers and producers, farm shops and farmers' markets have successfully tapped into a stream of consumer desires to understand their food and its origins better, and to support local producers and products at the appropriate times of the year for those products.

In some cases, substantial businesses are being developed, trading on the idea of high-quality local Scottish produce distributed over a number of channels. For example, in addition to the internet, the Perthshire/Fife-based producer Jamesfield uses a stall at a farmers' market and its own shop to sell its goods (Figure 6.9).

Many producers have been able to translate their 'localness' into a wider market via the internet and express parcel delivery systems, thus not only bringing Scottish produce to the local market but also taking it to people elsewhere who value knowledge of the product, traceability and the

Figure 6.9 Upper: Stall, Cupar Farmers' market, 2004. Source: Anne Findlay; Lower: Organic Centre and Farm Shop, Abernethy, 2005. Source: Anne Findlay.

quality of what they are eating. In some senses, then, the identity and operations of the Scottish retail food market has come full circle. Producers and consumers once again meet directly at a market or a retail site, be it a shop, a stall or even a website, and shoppers can choose to take the product home themselves or have it delivered.

NOTES

1. Hand, 2003, 614.
2. Dawson, 2000.
3. Dawson and Brooks, 1985.
4. Dawson and Broadbridge, 1988, 21.
5. World Advertising Research Centre, 2005.
6. Smith and Sparks, 1997.
7. Meldrum and Alexander, n.d.,10.
8. Movers and shakers, *The Grocer*, 28 May 2005.
9. These themes emerge clearly in Beech, J. New Scots. In Beech, J, Hand, O, Mulhern, M A and Weston, J, eds, *The Individual and Community Life* (Scottish Life and Society: A Compendium of Scottish Ethnology, vol. 9), 2005, 461–619.
10. Maxwell, 1909.
11. Kinloch and Butt, 1981.
12. Co-operative Union, 1957.
13. Data obtained from various Co-operative websites on 30 September 2005: Lothian, Borders and Angus (http://www.co-opunion.coop/live/welcome.asp?id=682); Scotmid (http://www.scotmid.com/), Musselburgh and Fisherrow (http://www.mfco-op.co.uk/); The Co-operative Group (http://www.co-op.co.uk); and their Scottish region at: http://www.co-op.co.uk/membership/regional/SC/SC_fr.html.
14. Mathias, 1967, 41.
15. Jefferys, 1954, 136–44.
16. Mathias, 1967, 55–72.
17. Howe, 2000, 27.
18. Sparks, 1996, 1465–84.
19. Sparks, 1996, 1465.
20. Calculated from store lists in the 2005 annual reports and on the websites of Morrisons, http://www.morrisons.com; Asda, http://www.asda.com; Tesco, http://www.tesco.com; and Sainsburys, http://www.sainsburys.co.uk.
21. Dawson and Brooks, 1985, 1.
22. Scotland's Census Results Online: http://www.scrol.gov.uk
23. Extract describing the family business of Allan H Findlay and subsequently his son William Findlay of Hurlford, from J H Findlay Findelsaga, 1990 Private Collection.
24. Report of an interview with George Geddes, August 2005, St Andrews.
25. Morgan and Trainor, 1990, 109.
26. Morgan and Trainor, 1990, 122.
27. Maxwell, 1909, 37–8.
28. Mathias, 1967, 46.
29. Howe, 2000, 27–8.
30. McIvor, 1992, 148.
31. Maxwell, 1909, 134.
32. Alexander, 2002, 39–57.
33. Lyon et al., 2004, 34.
34. Colquhoun et al., 2000, 109.
35. Colquhoun et al., 2000, 113.
36. Lyon et al., 2004, 28.

37. See the Asda and Tesco websites, at: http://www.asda.com/ and http://www.tesco.com/.
38. The Asda website, http://www.asda.com
39. See the Tesco website, http://www.tesco.com.
40. Cummins, Findlay, Petticrew and Sparks, 2005, 289.
41. Kemp, *Sunday Herald*, 2003.
42. See the farmers' markets website: http://www.scottishfarmersmarkets.co.uk/.

BIBLIOGRAPHY

Alexander, A. Retailing and consumption: Evidence from war-time Britain, *International Review of Retail, Distribution and Consumer Research*, 12 (2002), 39–57.

Colquhoun, A, Lyon, P, Kinney, D and Cockburn, J. Forgotten shopping: Memories of 1950s' grocery stores, *Education and Ageing*, 15 (2005), 99–116.

Co-operative Union, *The Co-operative Directory*, Manchester, 1957.

Cummins S, Findlay A M, Petticrew M and Sparks, L. Healthy cities?: The impact of food retail led regeneration on food access, choice and retail structure, *Built Environment*, 31:4 (2005), 288–301.

Dawson, J A, Brooks, D M. *Food Retailing in Scotland*. Paper presented at the IGFA Meeting, Peebles, 10 September 1985.

Dawson, J A and Broadbridge, A M. *Retailing in Scotland 2005*, Stirling, 1988.

Dawson, J A. 'Future Patterns of Retailing in Scotland'. Scottish Executive Central Research Unit, Edinburgh, 2000, available online at: http://www.scotland.gov.uk/cru/kd01/blue/retail–00.htm.

Grocer, The, 2005.

Hand, O. Structures associated with the retail trade. In Stell, G, Shaw, J, Storrier, S, eds, *Scotland's Buildings* (Scottish Life and Society: A Compendium of Scottish Ethnology, vol. 3), East Linton, 2003, 612–23.

Howe, W S. *William Low & Co: A Family Business History*, Dundee, 2000.

Jefferys, J B. *Retail Trading in Britain 1850–1950*, Cambridge, 1954.

Kemp, K. Safeway gives way to Yorkshire bitter, *Sunday Herald*, 12 January 2003.

Kinloch, J and Butt, J. *History of the Scottish Co-operative Wholesale Society Ltd*, Glasgow, 1981.

Lyon, P, Colquhoun, A and Kinney, D. UK food shopping in the 1950s: The social context for customer loyalty, *International Journal of Consumer Studies*, 28 (2004), 28–39.

McIvor, A. Women and work in twentieth century Scotland. In Dickson, A and Treble, J H, eds, *People and Society in Scotland, 1914–1990*, vol. 3, Edinburgh, 1992, 138–73.

Mathias, P. *Retailing Revolution*, London, 1967.

Maxwell, W. *First Fifty Years of St Cuthbert's Co-operative Association Limited 1859–1909*, Edinburgh, 1909.

Meldrum, A J and Alexander, A F. *The Story of the Scottish Grocers' Federation 1918–1993*, Edinburgh: SGF, n.d. (probably 1993 or 1994), (ISBN 0950035661).

Morgan, N and Trainor, R. The dominant classes. In Fraser, W H and Morris, R J, eds, *People and Society in Scotland, 1830–1914*, vol. 2, Edinburgh, 1990, 103–37.

Smith, A P and Sparks, L. *Retailing and Small Shops*, Edinburgh, 1997.

Sparks, L. Space wars: Wm Low and the 'auld enemy', *Environment and Planning A*, 28 (1996), 1465–84.

World Advertising Research Centre, *Retail Pocket Book 2005*, Oxford, 2005.

7 House Construction

RICHARD RODGER

Enric Miralles' Scottish parliament building was opened in Edinburgh in 2004. It has been, and remains, the subject of intense debate. In part, the Scots' reputation for financial 'prudence' was at issue as building costs escalated, but debate centred on Miralles' design, with its architectural references to Scottish heritage. Controversy also surrounded the juxtaposition of an ancient abbey and the modernity of Miralles' building.

Public opinion was polarised. Which events were most significant in the construction of the nation's identity? Where did myth overturn fact? Was there an economic dividend in terms of tourism, international publicity and a reputation for architectural ambition? Could the plural identities of Scotland be captured, far less reconciled, in a building and, if so, would it not be a mongrel of a structure? Can the democratic aspirations associated with the Scottish parliament be reconciled with the corporate world of global capitalism, or indeed can Scottishness survive internationalism except through a dumbed-down version of 'Scotland the brand'?[1]

These controversies, in pub and club, in print and in architectural practices, represent a critique of Scotland's cloned cities.[2] The sheer monotony, drab design and unimaginative planning that dominated much of the twentieth century are a product of the coincidence of public and private interests, of local authority control and corporate profit, and have resulted in acres of repetitive buildings reproduced with minor variations from one Scottish town or city to another. Profit and urban development are the symbiotic features that have cloned tower blocks and council housing in the twentieth century, but which also reproduced tenements in the nineteenth century and town houses in the eighteenth. In this regard, Scottish architectural form has interpreted and reproduced a concordance of values with a high degree of rationality.

The repetitive external built form of Scottish cities was customised internally by individual taste and preference, although in reality the introduction successively of inside toilets, electricity, white goods and electrical appliances from Comet/Dixons/Currys, and IKEA self-assembly furniture also produced a high degree of homogeneity. System, turnover, throughput and profit margins have been the language of business, and consumers have increasingly bought into it. Thus scale, increasing scale, and the American system of manufacture[3] based on standardisation and limited product ranges have dominated design, and in this the building industry and house construction

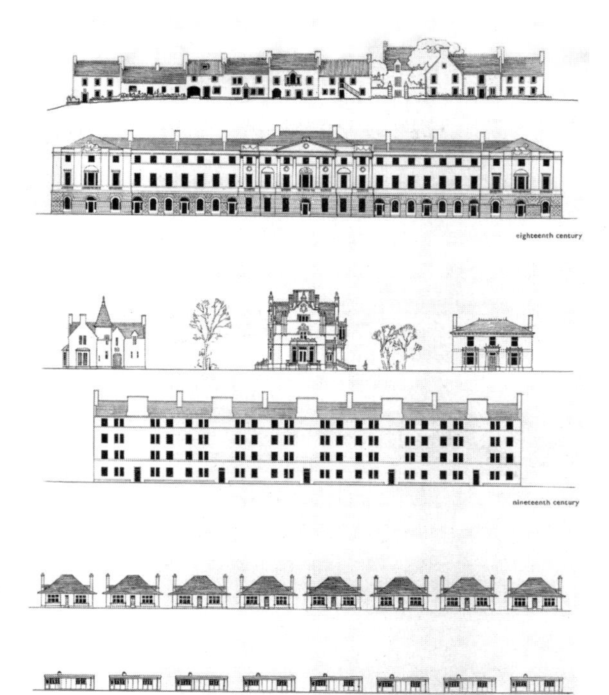

Figure 7.1 Scottish domestic architecture, 1700–2000. Source: Mears, F C. *Regional Plan for Central and South East Scotland*, 1948, 148.

sector are no different to other areas of consumption. In the last third of the twentieth century, Wimpey, Costain, Laing, Taylor Woodrow and McAlpine have replaced the housebuilding firms of Mactaggart and Mickel, Miller, and Ford and Torrie who dominated the middle third, and who in turn had displaced a multitude of smaller firms. Housebuilding was turned increasingly into civil engineering in the twentieth century, and only in specialist trades were skilled building workers able to go against the trend. In a world of new for old where replacement supersedes repair, the 'Do you know a good decorator/plumber/electrician?' plea of the middle classes echoes around the suburban dinner tables because skills in the building trades have been extinguished by system building, and because not everybody is willing or able to 'B&Q it' during their quality leisure and family time.

THE HISTORICAL LEGACY

After the Union of Parliaments of 1707 there was ultimately spectacular growth in overseas trade. The reorientation of trade from north-western Europe to the Americas as the colonies shipped increasing volumes of sugar, tobacco and cotton to West of Scotland ports redefined the Scottish urban system, with Glasgow and other West of Scotland burghs enjoying expansion.[4] The trade expansion was accompanied by an industrial one, and Scotland's manufacturers benefited from a late start, by making use of the technological advances developed previously in England. As Scotland entered the nineteenth century, significant productivity gains were derived from the application of steam power to industrial processes, from economies of scale obtained from larger-scale units of production, and from exploiting previously inaccessible coal and iron ore deposits. For a rising group of middle-class entrepreneurs and investors the expanding urban economies of west-central Scotland offered opportunities for profit. For labourers, employment in ports, mining villages and industrial towns provided a magnetic appeal in the wake of Hanoverian 'pacification', agricultural 'revolution', the Highland Clearances and, across the Irish Sea, harvest failure. In the course of the eighteenth century generally, production, trade and political relations were reshaped. New 'improved' towns, founded as part of the efforts to pacify the Highlands, and new estate villages, established in conjunction with changed agricultural practices and increased productivity in Lowland parishes, affected patterns of settlement as main streets were straightened and stone cottages were upgraded or extended.

After the 1745 Jacobite uprising, Scottish towns and cities enjoyed an urban renaissance.[5] New towns and villages were created, sponsored by

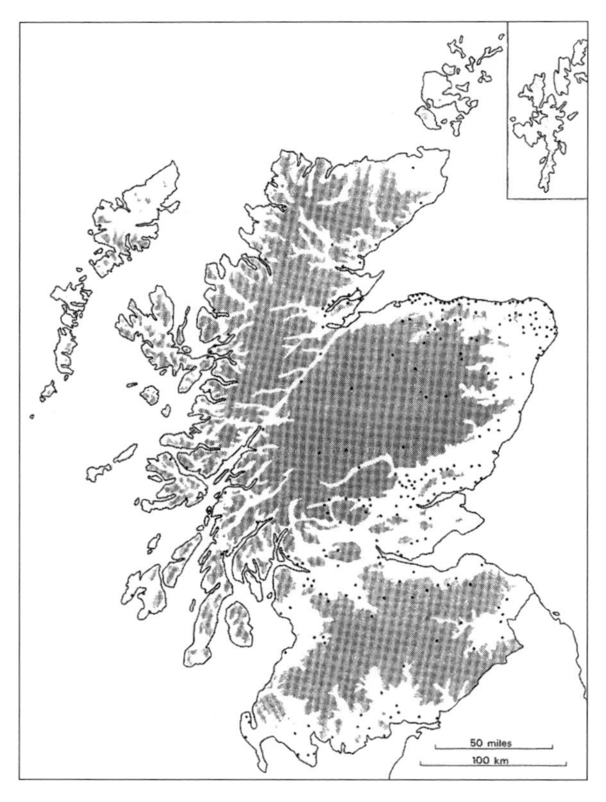

Figure 7.2a The planned villages of Scotland, 1720–1840.

landowners and the Forfeited Estates Commission.[6] New ideas and new fashions produced a vibrant urban culture, influenced more than a little by continental developments. The Scottish Enlightenment, with its centre in Edinburgh, sought to replace magic and belief with rational thought and scientific endeavour. Architectural developments were part of that process: the Edinburgh New Town, Gorbals and Blythswood estates in Glasgow, and the Bon Accord development in Aberdeen are the best known, but on a lesser scale other new housing projects flourished – in Leith and Dundee, for example. These high-status housing developments, the joint initiatives of the town councils and trustees of seventeenth-century institutions such as Heriot's Hospital in Edinburgh and Hutcheson's Hospital in Glasgow, provided extended opportunities for the building trades.[7] Feuing income accruing to Heriot's and the town council in Edinburgh as a result of the release of building land increased more than tenfold between 1770 and 1800, and by another tenfold by the late 1820s; in the 1790s and 1800s, Hutcheson's Hospital released 51 per cent of all its land for building.[8] By 1801, after England, Scotland was the most urbanised country in the world.[9]

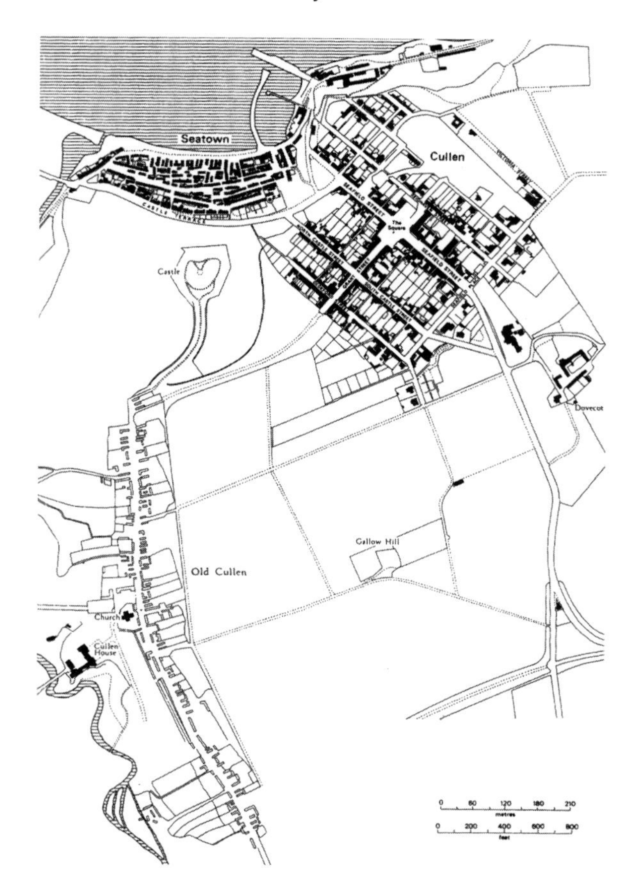

Figure 7.2b The new town of Cullen. Source: Adams, I H. *The Making of Urban Scotland*, London, 1978, 63, 70.

Builders showed considerable flair and independence of action in their Edinburgh New Town activities. They defied the town council's wishes concerning building height, and introduced storm windows, garret flats, chimney vents, bow fronts, and shops in opposition to the Building Acts that the council introduced. Twenty years after the Craig plan was adopted in 1767, the council had to concede that, in the face of concerted opposition from builders, they could not enforce sanctions for non-compliance with the plans they had approved. Partly this was because of the collapse of the Ayr Bank in 1772, which, like the City of Glasgow Bank failure a century later, resonated throughout Scotland and forced many builders either into bankruptcy or to suspend their business. The council's aspirations for the New Town design could not be achieved without builders and building tradesmen. This necessity, along with the economic recession of the 1770s, inclined the town council to soften its stance on conformity. A more relaxed political environment after 1781, combined with the adoption of fewer restrictions on workmen and their membership of a trade guild, and free trade principles, resulted in less control over the tradesmen of the town. The building industry made a recovery in the 1790s as a result of greater flexibility and wartime inflationary pressure.

By copying architects' plans and sharing practical solutions for construction, builders brought a degree of uniformity to James Craig's plan, while still maintaining the external appearance of individualised homes for customers. Based on fashion and financial constraints, builders responded to market forces and insinuated flatted tenements into the Edinburgh New Town where private houses or mansions had been intended in the quest for a socially exclusive suburb. The scale and pace of new building was unprecedented, and this placed builders in a previously unrecognised position of influence which affected almost every area of town council activity: administrative reform of the planning approvals system; taxation payments; relaxation of trade practices within the city; and even the political process itself. It was over his long-running legal and political feuds with property owners that the powerful figure of the financier and administrative leader of Edinburgh town council and promoter of the New Town, Sir Laurence Dundas, was eventually undermined in successive elections of 1774, 1777 and 1780.

Until recently, scholars have shown little interest in who built Scottish towns and cities in the era of expansion associated with the urban renaissance of the late eighteenth century. Recent research on the Edinburgh New Town shows how the expansion of Scotland's middle classes, and their wish to have town houses in the capital, was achieved by working in close partnership with builders. The New Town offered homes and work for merchants, lawyers and bankers and other professionals, as well as for noble families, and 'builders worked closely with the middle classes, and aspired to be recognised as respectable professionals'.[10] Builders functioned as consultants and middlemen, bringing all the elements of production together, either anticipating demand or working on a single project with clients often on a

Figure 7.3 Aerial view of the northern New Town, Edinburgh. Source: Reproduced by permission of RCAHMS, Crown Copyright.

partnership basis, or perhaps developing two or three feus simultaneously on a bespoke basis. In so doing they were highly influential in determining the form and internal arrangements, as well as the structural quality, of the houses on which they worked and, through custom and practice, were a force for the reproduction of the built environment.

These were issues that did not concern James Craig when he drew up his plan for the Edinburgh New Town.[11] Nor was Craig much concerned with elevations for entire streets or squares, and so the pragmatic, piecemeal approach adopted by builders towards developing Georgian Edinburgh was an important factor in the final appearance of the first new town. Indeed, the Edinburgh New Town plan lacked detailed elevations, and this provided sufficient flexibility to entice private individuals and builders to take up the feus (lots) offered precisely because building was unconstrained by design details. However, while this was advantageous in encouraging the building of new houses in the short run, it embedded longer-term difficulties over successors' wishes to change the use of a property in the future, and thus potentially to impair the capital value of a neighbour's property. This issue took several decades and many court cases to resolve before a House of Lords' decision in 1818 finally overruled the vagueness of James Craig's plan as a basis for building development.[12] Though it might appear as a mere legal technicality, this, and a later house of lords' decision in 1840 relating to the Bon Accord estate in Aberdeen, confirmed that commitments concerning house design and plot development were legally enforceable and binding on successive generations. Wilful changes of use and built form could not thereafter be introduced without due process. Stone-built Scots housing was intended to stand for centuries; legal decisions ensured that, after 1840, changes of use or appearance would be difficult. In addition, the legal cases affirmed the central importance of the system of land tenure, feuing, to Scottish building development which in turn ensured that intensively built plots of high-rise tenements would continue to be the dominant housing form.

The legal decisions confirmed the distinctively Scots form of tenure, feuing, by which landowners relinquished their land for an annual payment, a feu duty, payable in perpetuity. The legal decision of 1818 asserted that 'burdens' (charges) or restrictions on property could not be created retrospectively, and this meant that landowners (feuars) had a vested interest in creating the maximum burdens from the start of a proposed development on their land. Since the feu duty could be increased by sub-feuars in the construction chain – successively developers, builders and landlords – ultimately the tenant paid for this cumulative increase in feu duty through higher rents. As a consequence of sub-infeudation, land prices in Scotland were historically higher than elsewhere in Britain – central London excepted – and so plots

Figure 7.4 Interrupted building, Saxe Coburg Place, Edinburgh, 1826. Source: Richard Rodger.

were developed intensively in a distinctively Scottish form, the four-or five-storey tenement, so as to spread the cost of housing amongst a greater number of households. The right to levy and receive feu duties was a considerable financial asset. To institutions, widows and small savers, money invested in feu duties through a solicitor produced a predictable fixed income. To builders or developers, once they had acquired the feu duty and depending on the state of the market, it could be sold on for perhaps twenty to thirty times its annual value. The result was a handsome lump sum for the developer or builder, which was then used as working capital to build a tenement or house on the plot of land. These 'heritable securities' or rights to receive feu duties, and the compounding that resulted, guaranteed that intensively built plots would result in housing densities and environmental conditions that were considerably more hazardous to life than terraced housing in most English boroughs. Though accusations concerning living conditions and death rates were frequently directed at Scottish builders, it was Scots law and its market consequences that ensured that for the inhabitants of tenements, life expectancy was impaired and physical deformity was common.

Builders were not independent agents, and their activities were further constrained by the dean of guild court's jurisdictions over proposed building work that affected amenity, and by the town councils' expanding responsibilities for lighting, street cleaning and sewering under the Police Acts.[13] More significantly, it was the stop–go nature of the economy that influenced house construction. Prosperity and depression specifically affected a builder's ability to obtain both trade credit for materials and loans to pay for workmen's wages, and while this was also true of other sectors, the extended exposure to risk associated with the development period for a tenement to be completed meant boom and bust were more exaggerated. No other Scottish industry, shipbuilding included, experienced such dramatic fluctuations in output as the building industry. The abrupt downturn as a result of a national financial crisis in 1825–6 exposed the building industry to severe and sustained recession; for a generation – until the 1850s in fact – building projects remained incomplete, as in Saxe-Coburg Place, Edinburgh, and land bought to house the unprecedented urban population increase remained undeveloped. Much the same process applied to the booms of the mid-1870s and late 1890s when credit dried up.

THE CHARACTER OF THE BUILDING INDUSTRY

Classified by type of structure, there were four main elements in the building industry: residential; industrial and commercial; public; and alterations and additions. Predictably, there were variations between place and time, but during the second half of the nineteenth century Scottish building activity was divided roughly as follows: residential 46 per cent; industrial and commercial 25 per cent; public 15 per cent; and alterations and additions 13 per cent.[14] Occasionally builders would change track, becoming more involved with the 'minor works' of oriel windows or the installation of water closets, for example, as work in their specialist area receded. Reverting to employee status when work on their own account was scarce or risky was a strategy adopted by masons, who switched from residential to monumental work on gravestones and kerbstones, for which there was a more predictable demand. Public-sector building was the least volatile element, and the burgeoning civic responsibilities after 1870 created a considerable amount of employment for building workers on capital projects for schools, hospitals, asylums, abattoirs, wash-houses and municipal offices, as well as civil engineering contracts associated with trams and the gas and water departments.

Lines of credit and material supplies, and the location of the builder's yard, meant that the sphere of business operations was normally highly localised in Scottish burghs, with a workforce also drawn from a restricted orbit within easy walking distance of the site or yard. For these reasons, as well as because of the cost and logistics of moving heavy materials, most building firms, small and large, organised work from a local yard. Even larger firms rarely strayed far from their yard to build houses further afield, especially once they had injected sizable amounts of capital by installing steam-driven stone dressing machinery and woodworking lathes after the 1870s. The highly localised nature of house construction and building work was an important characteristic of medium and large-sized towns and cities, and had implications for the appearance, morphology and texture of the urban environment, since particular areas were strongly identified with particular builders. Thus John Pyper, W and D McGregor, Edward Calvert, and Galloway and Mackintosh each concentrated their activities almost exclusively on Marchmont, and James Steel, the largest Edinburgh builder in the late nineteenth century, never strayed into the southside suburbs or to east Edinburgh.[15]

Building was itself heavily polarised. The overwhelming majority of builders only ever put up one or two houses or tenements (see Table 7.1).

Table 7.1 The scale of building projects, 1873–1914

	Number of Houses in Building Project		
	1 house (%)	*2 houses (%)*	*3 or more houses (%)*
Cities (3)	39.6	39.3	21.1
Largest burghs (30)	53.9	25.8	20.4
Smaller burghs (69)	74.8	21.6	3.6
Scottish burghs (102)	52.9	30.9	17.2

Source: Rodger, 1979, 227.

Only 17 per cent of building projects were for three or more houses and, perhaps predictably, smaller burghs denuded in terms of natural population increase by migration to central Scotland relied almost exclusively on projects for just one or two homes at a time. Where there was new housebuilding in smaller burghs such as Oban, Peterhead, Lerwick, Stornoway, Alloa, Stonehaven, Grangemouth, Elgin, Lockerbie and Castle Douglas, and in the smaller towns of the central belt, it was retailers, merchants, salaried officials and widows who provided the backbone of finance for new housebuilding, with builders initiating no more than one in three houses.

More than half of all Scottish housebuilding (53 per cent) was undertaken on a one-off basis by builders (Table 7.1). In Aberdeen, 430 different individuals initiated 897 housebuilding projects in the decade 1885–94, so that 48 per cent of proposed housebuilding came from a person who had not previously built a house in Aberdeen, nor would seek to do so again. In Edinburgh, much the same pattern existed, with 42 per cent of planned housebuilding on a once-and-for-all basis. Though it might seem that the Scottish building industry rested on a highly fractured structure and was thus susceptible to speculative fluctuations, it is worth noting the diverse interests that, in addition to builders, financed housing construction. Of those non-building industry interests who initiated a house construction project in Aberdeen in the decade after 1885, the largest categories were represented by professional men (5.7 per cent), commercial interests (5.0 per cent), shopowners (4.3 per cent), manufacturers (4.2 per cent), artisans (2.0 per cent), and a diverse group of unclassified individuals, company and institutional interests, returning exiles, widows and spinsters that formulated plans to build new houses (totalling 18 per cent). The building trades initiated 41 per cent, and quasi-building interests such as house factors, architects and solicitors promoted a further 13 per cent. So in Aberdeen only a narrow majority (54 per cent) of finance for new housing originated through building interests, and if housing finance was more reliant in Edinburgh, where 73 per cent of planned housebuilding was initiated by building interests, there was still a 3:1 ratio of building interests to privately based financial backing that gave a measure of stability to the industry in Edinburgh in the decade of 1885–94.

Some impressively large operations also existed within this atomised structure of the Scottish house construction sector. Cooper and Son built 140 dwellings between 1902 and 1909 in Musselburgh; A Stewart and Co. built 152 tenements in Govan in the ten years from 1894; in Aberdeen, J Green and the partnership of Cameron and Matthew both completed building work in excess of £50,000 in ten years – twice the scale of the operations of their nearest competitors; and the five largest builders in Edinburgh themselves accounted for 19 per cent of all planned housebuilding during the 1885–94 decade.

It was not simply the structure of the industry, polarised as it was between numerous small firms and a few larger building concerns, that was the cause of instability in the house construction sector. It was the nature of building finance and the inherent difficulty posed by matching maturing loans with new sources of borrowing that was the real cause of building

Figure 7.5a Large-scale building development on the Merchiston estate, Edinburgh, 1897. Source: Edinburgh City Archives, dean of guild court registers.

Figure 7.5b Small building enterprise, 15–17 George Street, Whiteinch, Glasgow: P W and A Lightbody. Source: RCAHMS. Reproduced by permission of Mrs Esther Galbraith. The Lightbody firm survived the 1878 bank crash, produced houses in Partick in the 1890s, but succumbed to another crisis in 1905.

bankruptcies, and not simply how prevalent the small firms were in a town or city. With demand difficult to predict, and sales 'lumpy', since a building was only saleable once it had a roof, smaller producers found their cash flow problematical. With so many small independent firms, market trends were difficult to identify and so builders tended to build well beyond the point at which buyers were interested in making a purchase. Overshooting in the boom phase was mirrored by reluctance to recommence construction in the downswing.[16] Bankruptcy was never far away in the building industry. Poor standards of book-keeping meant builders' assessments of assets and liabilities were often very wide of the mark, as bankruptcy proceedings almost invariably show. Just one example gives a sense of this: John Duffes of Inverness assessed his assets:liabilities ratio in 1870 as 1:4; the trustee in bankruptcy calculated this as nearer to 1:12.[17]

Where funds were assured, as with private backers, institutions and companies, then the house construction sector enjoyed a measure of stability and fewer bankruptcies. Each burgh enjoyed such assured funding, but inevitably to different degrees. Few large-scale philanthropic settlements existed in Scotland, compared to the noted English examples at Saltaire, Bournville, and Port Sunlight. Where company towns did develop, as at Clydebank and Dalmuir, in mining settlements such as that built by the Arniston Coal Company, Summerlee Iron Company and many others throughout the central belt, the motives were far from philanthropic, since housing availability was instrumental in the management of the workforce.[18] Other tied cottages were scattered throughout Scotland, most conspicuously for agricultural workers and crofting communities, which were severely criticised in a Royal Commission Report in 1917, but they were also provided in towns and cities by railway companies, hospitals, golf clubs, and lodges and cottages for lock keepers and park keepers. In Aberdeen and Edinburgh, company and institutional housing constituted between 6 and 7 per cent of all accommodation; in many larger burghs it represented nearer to 10 per cent.

In most expanding nineteenth-century Scottish towns there were both a number of model dwelling blocks rented to artisans and building associations, rather like subscription clubs or terminating building societies, where a local group of people pooled resources and contracted with a builder to develop houses or a tenement. In Edinburgh, for example, there were ninety building associations active in the years 1869 to 1874. The subscribers drew lots for the homes, as they were completed. While this worked well for clerks and shopkeepers, it was normally beyond the means of the working classes. An important development amongst skilled building workers, though restricted to Edinburgh with more limited examples in Bonnyrigg and Dundee, was the direct building programme of the Edinburgh Co-operative Building Company (ECBC). Its mission was to provide sound housing for building workers at reasonable cost.[19] The ECBC initiative stemmed directly from a lock-out by building employers in 1861. With time on their hands and Free Church encouragement to improve their home life, building tradesmen formed a limited liability company of their own, becoming share-

Figure 7.6a Innovative designs for the Edinburgh Co-operative Building Company, 1862. Source: Richard Rodger. Note: Reid Terrace 'colony' houses with balcony access from one side and ground floor access to the garden flat on the opposite side of the terrace.

holders, setting up the Articles of Association to define the governance of the company, and eventually developing a mortgage and savings deposit scheme to assist potential buyers to garner the necessary deposit on a house. Dominated in its early stages by workers in the building trades, the ECBC was distinctive because it built its own houses – 2,000 of them between 1862 and 1914 – scattered across eleven different sites in the city. Initially these two-storey terraced designs relied on a first-floor flat constructed directly above the ground flat and accessed by an external staircase on the opposite side of the building to the ground floor flat (see Figure 7.6a). Within twenty years this distinctive design mutated to an internalised staircase, and as a result created the origins of the maisonette in the 1880s. Well over a century later none of these houses had been demolished, and the £150 price tag in the 1860s and 1870s now fetches £150,000 or more.

Building workers straddled a variety of trades, each with their apprenticeship system. Masons, bricklayers, builders, carpenters, joiners, plumbers, plasterers, slaters, painters and glaziers were each required in a fairly rigid sequence of work, and only the larger firms could afford to engage a fixed workforce, moving their different tradesmen around as each phase of the larger project was completed. For smaller firms, the availability of the preferred independent plumber or slater was difficult to dovetail with

Figure 7.6b Building trades insignias on Stockbridge housing. Source: Kenneth Veitch.

other elements in the work, and delay increased the costs of construction. In 1901, at its peak, the building trades (11 per cent), woodworking (3.5 per cent) and brick, glass and cement manufacture (0.8 per cent) constituted one in every seven (15.3 per cent) of male workers – there were virtually no women employed in the building industry – in those Scottish burghs with a population in excess of 30,000. Ten years later, in 1911, male employment in the building trades had slumped 32 per cent. This was the nature of the building industry: severe 20-year-long cycles of boom and slump, with frenetic periods of activity near the peak followed by deep and extended troughs. With its multiplier effects on local employment, the house construction sector, and building generally, had a profound impact on economic and social welfare.

LONG-RUN CONSEQUENCES

The unique Scottish land tenure system (feuing) and its financial implications for building development reinforced the traditional high-rise tenement form of house construction in Scotland. Land costs were higher in Scotland than in comparable English towns and cities, and this was reflected in rents that were up to 26 per cent higher in 1905 than in comparable industrial towns of the north of England; only London experienced higher rents. For food, Scots had to pay 5–9 per cent more than any other region of the UK, apart from London. Clydeside boilermakers and engine fitters obtained only 92–8 per cent of the wages of their Tyneside brethren in 1914; unskilled shipyard workers and miners were paid only 95 per cent of the UK average.[20] Indeed, cross-border wage differentials during the period 1860–90 were commonly 9–13 per cent adrift, to the disadvantage of Scots. Lower disposable incomes necessarily resulted in more basic accommodation for the Scottish working class, even though they spent a higher proportion of income on rent than in English towns and cities. The character of the Scottish building industry contributed to a highly unstable form of production, which resulted in severe shortages in housing accommodation. This is not to argue that all tenements were the same. Ground floor flats with private doors, bay windows, a boxroom in place of a box-bed, the exclusive use of a WC, plasterwork decoration in the public areas, and generous stairs were some of the features that internally and externally denoted predominantly middle-class tenement flats. A four-storey building could have between six and sixteen flats, another indication of the class status of the tenement itself, though where there were flats of different sizes within the block, a degree of social differentiation within the tenement occurred.

The result of the system of house construction was that urban Scots spent a higher percentage of their income on housing than elsewhere in the United Kingdom, London excepted. They bought less floor space, and consequently experienced more overcrowding, and more shared WCs and sinks. Tenements 60 ft high built on streets with less than 30 ft width were common, and so light, air and access were limited.[21] Of the Scottish population, 47.7 per cent in 1911 lived in one- and two-roomed houses; 7.1 per cent of the English population lived at such a density.[22] Had the registrar-general's criterion of overcrowding in England been applied to Scotland in 1911, then Wishaw, Coatbridge and Kilsyth would have been ten times more overcrowded than Salford or Liverpool, and residents in Clydebank, Cowdenbeath, Airdrie, Govan, Barrhead and Renfrew, with 60–9 per cent living more than two per room, were ten times more overcrowded than those in Hull or Manchester, and more in keeping, though still well above, the levels prevailing in Brussels and Budapest.[23]

If the rhythms of daily life – washing, cooking, bathing, sleeping, playing, courting, relaxing, visiting, socialising – were influenced by such basic accommodation and shared amenities, so too were those of sickness and death, where patient and corpse were laid out in scarce space and around which the household had to attempt to conduct normal daily routines. In such circumstances it was not surprising that mortality and life expectancy, as well as the incidence of tuberculosis, dental caries, scabies, skeletal and other infirmities, were notoriously high in Scottish burghs, as studies in Dundee, Glasgow and Edinburgh between 1905 and 1907 confirmed.[24] Physical condition was strongly and inversely correlated with the size of house; at all ages, both boys and girls from one- and two-roomed

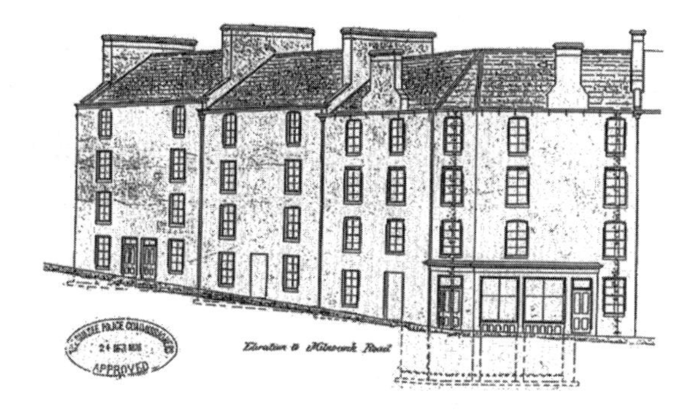

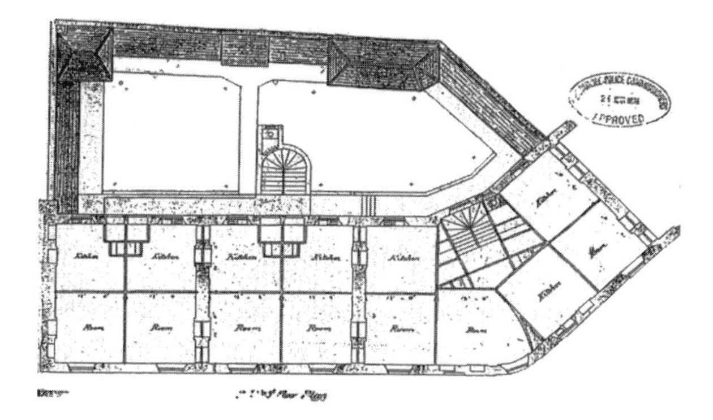

Figure 7.7 Drawing showing floor plan and section of tenements, Lawrence Street, Dundee, 1874. Reproduced by permission of RCAHMS, Crown Copyright.

houses were a few inches shorter and a few pounds lighter than those from three- and four-roomed houses. House construction and the proximity in which Scots lived to one another directly influenced their life chances and was emphatically an indicator of endemic poverty.

Inescapably, the conclusion was that a high percentage of nineteenth-century Scottish housing was dangerous to health. Residents were stunted, unhealthy and destined for an early grave if, as almost half of them did, they lived in a small, cramped flat in a tenement block. To address these conditions, action was taken on a number of fronts. By building a block of flats or terraced houses, philanthropists, churchmen and social reformers sought to improve housing by example. By demonstrating to tenants that their behaviour affected their welfare, and to landlords that a reasonable rate of return could be derived from working-class properties, reformers' model dwellings sought to alter the nature and character of house construction and management in Scotland. To these endeavours by influential individuals, a parallel and public approach was implemented based on the provision of minimum standards of construction. Consequently, burghs took on powers relating to improvements, nuisances and medical policing, and these were codified in 1862 and 1892 for smaller burghs to provide a more consistent approach to building byelaws and new house construction. Distinguished medical officers of health in the major cities and burghs encouraged civic authorities to intervene to close down unhealthy housing, and to introduce improvement schemes in the 1860s and early 1870s.[25] Public intervention in the sphere of private property involved compulsory purchase and compensation for land acquisitions and demolitions, powers which though controversial were grasped by the major cities with enthusiasm and provided a start on the learning curve towards more comprehensive town planning.[26]

Scottish burghs enjoyed a degree of optimism and expansion in the eighteenth century that continued into the mid-1820s. Development was arrested at that point due to financial crises and economic downturn, recurring periodically to produce booms and slumps on a cycle of about twenty years. Even in the years 1850 to 1870, as the fees derived by the Sasines Office for newly registered properties indicate, house construction was on little more than a hesitant upward trajectory. Of course, the wealthier elements in society could commission a suburban villa in spite of the general economic outlook, but as the first series of Ordnance Survey maps indicate, the real expansion of house construction in urban Scotland took place from the late 1860s, after several decades in which urban in-migration and natural population increase intensified the pressure on existing housing stocks. From the late 1860s, building byelaws also began to take effect, and new houses increasingly incorporated sinks, cold running water and WCs. A hierarchy of tenement housing developed. Before the nineteenth century, housing was vertically socially stratified within tenements; by the beginning of the twentieth century it was horizontally stratified. Thus the phase of new-town building during the Georgian period, described earlier as an urban renaissance, initiated a process that accelerated markedly from the 1870s, whereby Scottish towns and cities became increasingly socially and residentially segregated.

TWENTIETH-CENTURY RESPONSES

Between 1919 and 1939, 70 per cent of all houses in Scotland were built by the public sector. In England during the same period, 70 per cent of all houses were built outwith the public sector.[27] This inversion was perhaps the clearest single indictment of housing in nineteenth century Scotland. It showed both how deficient rural and urban housing was in quality and quantity at the end of the war, and how large an injection of new construction was required to upgrade the housing stock. Politically, with the rise of Labour in the West of Scotland particularly, it was convenient to support council housing initiatives and to indict the private sector's legacy of neglect and ill repair, of deficient amenities and dangerous structures, of health hazards and reduced life expectation, and of problems created by market forces that seemed, and probably were, irreversible by the private interventions of reformers and workers. It was a housing stock deemed not 'fit for purpose' in an influential Royal Commission report of 1917 (Ballantyne Report) because of its outdated, shared amenities, the difficulties of adapting tenements to running water, gas pipes and electricity cables, and the inability to guarantee the separation of the sleeping arrangements for boys and girls because of the preponderance of one- and two-roomed flats. Scottish housing presented a compelling case when treasury funding for 'Homes for Heroes' was available in the Housing, Town Planning, etc. (Scotland) Act, 1919. For sixty years thereafter, until the Tenants Rights (Scotland) Act 1980 enabled occupants to buy public sector housing, British government subsidies were central to efforts to improve house construction in Scotland.[28] They were also a frequent source of conflict between local and central government over construction standards, design details, deadlines, and bureaucratic and audit procedures, combined with the absence of sufficient specialist administrative staff in municipal offices to oversee major housing contracts.

Council house building was not new. Municipally owned and rented blocks built before 1913 represented 1 per cent of the total housing stock.[29] These 'deck-access' tenements were inserted here and there into existing congested central space, and rented not to the homeless or casually employed but to families with a steady income.[30] Government-sponsored housing was introduced during World War I for workers employed in manufacturing munitions at Gretna, Greenock and at the newly established naval base at Rosyth, where housing design and estate layout owed much to English-inspired concepts of community based on Ebenezer Howard's vision for

Figure 7.8 Pre-World War I deck-access municipal housing, High School Yards, Edinburgh, 1894. Source: Richard Rodger.

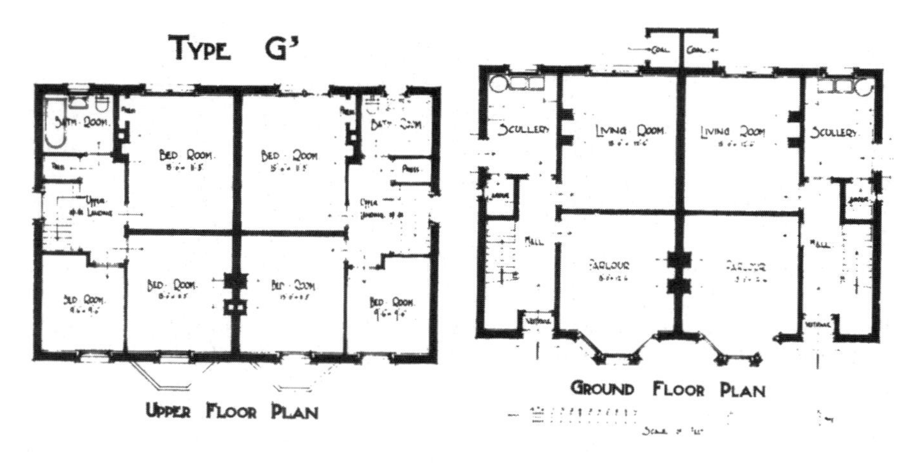

Figure 7.9a Type G3 semi-detached, concrete block, three-bedroom houses. Source: N J Morgan. £8 cottages for Glasgow citizens: Innovations in municipal house-building in Glasgow in the inter-war years. In Rodger, R, 1989, 134, fig. 5.4.

Letchworth and as implemented by Barry Parker and Raymond Unwin from 1903.[31] With the private rental market distorted by soaring wartime rentals and then frozen by the Rent and Mortgage Interest Act, 1915 – the first in a series of rent-control measures in force until the 1970s – the private house construction industry was never the same again until the last quarter of the twentieth century.

A number of pre-defined, ministry-approved house types were approved as the prospect of exponentially rising building costs was unacceptable to all parties, particularly during the 1920s when the Treasury imposed swingeing cuts on public expenditure. The result was council housing estates that throughout Scotland seemed indistinguishable one from

Figure 7.9b Oblique aerial view of Carntyne housing estate, Glasgow corporation, 1925–30. Source: Reproduced by permission of RCAHMS, Crown Copyright.

another, and an abrupt shift from a housing form that had characterised previous centuries of Scottish housing. Standard house types appeared with only minor variations, and drab uniformity resulted.

As a direct consequence of war-induced shortages in both traditional building materials and skilled building labour, there was some encouragement to move towards alternative house construction in the 1920s and 1930s. Glasgow City Council used clinker to make breeze blocks for some of its housing estates, and experimented with prefabrication; elsewhere, steel, timber and concrete were used as alternative materials in new steel-framed and poured-concrete construction techniques. It was a new organisational form of house construction production, though, that introduced the greatest change in Scottish housing during the interwar years. Large-scale projects involving thousands of houses were built simultaneously on the edges of towns and cities on land acquired from neighbouring county councils, often after considerable resistance. The logistics, finances and resources necessary to do so went far beyond the means of most builders of traditional pre-war Scottish tenements. Some, such as Mactaggart and Mickel, adapted and built medium-sized projects of a few hundred houses; others, such as McAlpine, developed poured granular concrete techniques. This ushered in a phase of system building, based on production engineering principles in which standards and tolerances were finely calculated by trained engineers in relation to the qualities of materials and their load-bearing capacity. Understandably, town councils were heavily involved and shouldered some costs in terms of site acquisition and development, and eventually in direct labour departments. For tradesmen there was ample work on the new estates, but system building deskilled house construction and casualised work. The effects were felt throughout Scotland, even in Orkney, where at Carter's Park, Kirkwall, terraces of cottages intended as a project to involve separate trades was built eventually by a single contractor in 1920–1.

By these means cloned estates were created. Mosspark, Blackhill, Hamiltonhill and Possil in Glasgow, Logie in Dundee and Lochend in Edinburgh were some of many examples of inter-war council house construction based on the Treasury subsidies of the 1920s. On appearances alone it was difficult to identify the city or burgh in which the estate had been built. Council housing was identifiable at a glance; stereotyping resulted. Scottish

Figure 7.10 Advertisement for Mactaggart and Mickel housing at Netherlee Park, Glasgow, from the *Bulletin*, 1 April 1933. Reproduced by permission of RCAHMS, Crown Copyright.

house construction had introduced a further layer of social segregation. Even though city architects Horsburgh Campbell and Macrae in Edinburgh sought to promote tenement-style designs and small-scale developments and worked in public–private partnerships with building firms to develop these, council housing was stigmatised. Homogeneity in form and design was characteristic of public housing ventures because government funding approved only a limited number of housing types and excluded the provision of amenities. So newly formed council estates lacked the facilities – meeting hall, surgery, recreation ground, library, pub, shops and frequent bus services to the town centre – essential to turn an estate into a community. Since council housing was so much more prevalent than in English towns, the homogenised built environment was more pervasive. This was the case in the Hilton, Ruthrieston, School Road, Torry and Pittodrie developments in Aberdeen, where 2,600 dwellings were built in four flatted blocks of two or three apartments each in the decade after 1925.[32]

No less visually repetitive were the private housing developments of the interwar years. Though numerically much less pervasive than council housing, monotonous lines of bungalows characterised many Scottish towns and cities, encouraged by the Housing Act, 1923. This enabled both local authorities and private enterprise to obtain a Treasury subsidy for twenty years for houses built to a minimum standard and which could be let or sold at any price. Local authorities could also advance money to individuals who wanted to become owners but lacked the necessary down-payment. So where town councils were not ideologically opposed to a contribution towards middle-class home ownership then serried rows and crescents of bungalows appeared, as in Mannofield, Ferryhill and Midstocket in Aberdeen, and Blinkbonny, Corstorphine Hill and Carricknowe in Edinburgh, the two places that embraced the private housing opportunity with greatest enthusiasm. Bungalow belts enjoyed further encouragement, as did housebuilding in general, when Britain left the gold standard in 1931 and the mortgage rate of interest was almost halved in 1932 and for the remainder of the decade.

The interwar years in Scotland were associated with severe depression in the shipbuilding, marine engineering, cotton, and iron and steel industries particularly. These were heavily concentrated in the West of Scotland, and regional policy as it developed under the Special Areas Acts sought to diffuse the blackspots of unemployment. One initiative was to build houses outside the worst affected areas in an effort to encourage workers to move to new labour markets. For this purpose the Scottish Special Housing Association (SSHA) was established in 1937, though little headway was made before World War II.[33] After the war, the policy imperatives were to encourage general economic development and specific initiatives, as well as to explore new materials and techniques. Between 1945 and 1950 the SSHA built 10,500 houses, equivalent to 30 per cent of its total in these years, for the mining industry under the newly reorganised National Coal Board. Though SSHA house construction for general purposes was concentrated in Kilmarnock, Clydebank, Coatbridge, Kirkcaldy, Greenock, Ayr and Aberdeen, it was scattered in 136 locations throughout Scotland between 1945 and 1950. Glasgow was largely avoided because the corporation was openly and ideologically hostile to the SSHA, and only in the late 1950s was there a degree of détente between it and the SSHA when tower blocks and overspill initiatives developed. House construction for specific economic regeneration at Erskine, in support of the Rootes car plant at Linwood, and at Tweedbank between Melrose and Galashiels in support of Scottish Forestry took place in the late 1960s and 1970s, and over 20,000 houses were built in Aberdeenshire, Morayshire and in Shetland to sustain economic developments there. These initiatives diffused the no-fines shuttered concrete method of construction promoted by the SSHA and based on Dutch experience. This was a technique that eliminated fine material, sand, from the normal concrete mix and used single-sized coarse aggregates surrounded and bonded by a thin layer of cement paste, giving the strength of concrete. The advantages of the method were lower density, lower cost due to lower cement content, lower thermal conductivity, less shrinkage, less capillary transfer of water, and better insulation than conventional concrete. No less important was the method of construction: the fineness of the mix allowed the concrete to be prepared in significant quantities and poured mechanically between large wooden shutters. The materials science, developed with SSHA assistance and technical resources, provided considerable benefits in a damp and cold Scottish climate both for builders and council and SSHA tenants. The techniques were made available to Wimpey, Costain and other construction

Figure 7.11a Weir quality steel flat roof, 1950. Four semi-detached and terraced houses: concrete foundations, prefabricated gauged steel sheet cladding, jointed and bolted. First floor steel joists on steel beams. Cast iron chimney, steel casement windows.

Figure 7.11b Orlit Bellrock, 1949. Load-bearing, pre-cast reinforced concrete structural frame, concrete outer wall, inner wall of pre-cast gypsum, hydrated calcium and sulphate concrete.

Figure 7.11c Miller O'Sullivan Construction, 1952. Three rooms, cavity brick walls, roughcast on a concrete strip base. O'Sullivan used patent 'on-site' aluminium alloy moulds to construct dense concrete walls.

Figure 7.12a Clone city: Livingston new town, 1960s. Source: Reproduced by permission of RCAHMS, Crown Copyright.

firms. Eventually, as policy priorities shifted from quantitative to qualitative provisions, rehabilitation, refurbishment, sheltered and special needs requirements formed the final phase of SSHA house construction in the late 1970s, and the organisation functioned in conjunction with dozens of agencies, such as the Glasgow Eastern Area Renewal project (GEAR), to achieve these objectives.

From the 1950s, the approach to house construction was redefined. There were two principal reasons for this. First, a change in the way local taxes were paid revealed a significant decline in landlords' rents, including those of council housing departments. Central government sought to recoup more of the housing costs from tenants, who, predictably, objected to increased rents and began a rent strike in 1957. On new construction projects, the government introduced larger council housing subsidies for overspill projects and high-rise tower blocks. So in the Scottish cities, where site development had historically been very intensive, building land was available on a sufficient scale only on the perimeter of the city. The inevitable result was the proliferation of council housing on the edge of the built environment – the peripheral estates inscribed in the annals of modern Glasgow in the form of Easterhouse, Castlemilk, Pollok and Drumchapel.[34] The absence of communal facilities and dedicated recreational space was justified by council housing departments on the grounds of the available

Figure 7.12b Tower blocks, Muirhouse, Edinburgh, 1960s. Source: Reproduced by permission of RCAHMS, Crown Copyright.

green space between blocks, although residents were inconvenienced by the distance from the city centre and the cost of travelling to work or for shopping that such remoteness involved. Glasgow City Council also continued its 'dry' policy, introduced in 1890, to ban pubs on its property. Whereas the council contracts of the 1920s and 1930s were a major undertaking, tower blocks involved construction on an altogether different scale, and like Birmingham and Sheffield, Glasgow City Council was quick to forge partnerships with major construction firms.[35] The climax of this phase of house construction was the Red Road scheme in Glasgow during the 1960s. The second fundamental change in Scottish house construction was also heavily associated with Glasgow housing policies. To address its housing needs, the city sought not just to build high-rise flats but to decant its population in an overspill policy that affected towns throughout central Scotland. Overspill projects accounted for 40 per cent of SSHA building in the 1960s as new towns, originally conceived as part of the Clyde Valley Regional Plan in 1946, came to fruition in East Kilbride, Cumbernauld, Livingstone and Erskine, and in significant additions to older towns such as Irvine and Dunbar. For every two houses built by new-town development agencies, the SSHA built three.[36]

As agent and double agent, insinuating itself behind the lines of elected councils and unelected quangos, the SSHA used its 'special' position to go where other agencies had no brief. It reached parts of Scotland where new housing construction had been almost unknown for decades. At its peak in 1980, the SSHA had built and managed over 111,000 houses; it was Scotland's second largest landlord after Glasgow City Council. It acted as a political counterweight to council housing, and set standards of construction and housing design to which many builders and councils aspired. With the tenants' right-to-buy legislation in 1980, the power, and identity, of the SSHA was dismantled in the rebadged quango, Scottish Homes.

OVERVIEW

House construction in the twentieth century redefined the Scottish townscape. The intense phase of council housing from 1926 to 1938 and again from 1948 to 1972 produced designs agreed between Westminster and the Scottish Office that resulted in an unprecedented expansion in the housing stock. Quantitative additions were paramount. Like factory production, house construction was highly standardised to reduce unit costs and maximise throughput. Much the same was true of the era of tower-block building from the late 1950s to the early 1970s. This built form, both cloned council estates and streets in the sky, altered the relationship of residents with their town or city. Centrifuge-like, they were spun out to the perimeter of the city. For some, in the harsh conditions of unemployment in interwar Scotland, the council house represented modernity and a fresh, clean start with a purpose-built kitchen, newly painted walls and electricity. For many others, the lines of credit, gossip and street banter, networks of association and sociability, as well as mutual concerns of stair and close or rotas and responsibilities, were deconstructed in that physical relocation. The social cement that helped stabilise communities in the nineteenth century was broken up, and in the city centres as well as on the peripheral housing estates, new social networks were developed that took time to root, if they ever did. This is not to gild Victorian and Edwardian tenement life, since many residents were keen to escape it and preferred their new council house, even though many subsequently regretted their move.[37] But in the drive to mass-produce house construction in the interwar years, and after World War II, the sheer scale and anonymity of the built environment damaged social structures and relationships that had been developed over many generations. By dismantling established communities, new house construction further undermined two other agencies that contributed to stable neighbourhoods: family ties and church attendance; and decanting population left a third agency, inner-city schools' with declining rolls. Thus the social fabric of the inner cities was also irretrievably damaged by the housing policy that prioritised new homes on peripheral estates. The price of quantitative additions to the housing stock in the twentieth century was a social fabric ill suited to the new challenges of the twenty-first century. In refocusing house construction since the 1980s to provide mixed tenures, small-scale and refurbished older properties, as well as some prestigious public projects with distinctive designs such as Miralles' Scottish parliament building, planners and architects have moved away from clone city.[38] Whether they can do so in regenerating the outer rings of housing remains to be seen.

NOTES

1. McCrone, Morris and Kiely, 1995, 1–12.
2. Glendinning and Page, 1999, 6–10.
3. The term dates from 1855 when a British delegation visiting America casually used this to describe the production process at the Springfield Armory. See Hounshell, D A. *From American System to Mass Production 1800–1932*, Baltimore, 1984. The term acquired currency in economic history with the publication of Rosenberg, N. *The American System of Manufactures: The Report of the Committee on the Machinery of the United States 1855, and the Special Reports of George Wallis and Joseph Whitworth 1854*, Edinburgh and Chicago, 1969.
4. Whatley, 2000, 48–95.
5. The phrase is from Borsay, 1989.
6. Adams, 1978, 59–71.
7. Rodger, 2001, 37–68; Kellett, 1961, 211–32.
8. Hill, 1881, 300; Rodger, 2001, 77, Fig. 3.4.
9. de Vries, 1984, 39–48.
10. Lewis, 2006, 16. Lewis shows how the assertions of Youngson and McKean, and others, are not founded on documentary evidence and rely on the repetition of limited and unrepresentative evidence.
11. Cruft and Fraser, 1995, 25–47.
12. Rodger, 2001, 59–76.
13. Rodger, 1983a, Police Acts applied to all royal burghs, and a few others, after 1833.
14. Rodger, 1979, 240, Table V.
15. Cant, 1984, 14–15; Rodger, 2001, 282, Fig. 9.1.
16. Cairncross, 1953, 12–36.
17. Rodger, 1979, 238.
18. Melling, 1981.
19. Rodger, 1999, 7.
20. Rodger, 1983b.
21. Rodger, 1986.
22. *Parliamentary Papers*, 1913 LXXX, Census of Scotland, Report, Table XLV. The report also showed (Table LXXX, 568) a steady decline in levels of overcrowding from 58.6 per cent in 1861.
23. Gyani, 1990, 169; van den Eeckhout, 1990, 99.
24. Rodger, 1985.
25. 31 & 32 Vict. c.108, Burgh Police (Scotland) Act 1862, and 55 & 56 Vict. c.55, Burgh Police (Scotland) Act 1892.
26. Allan, 1965; Smith, 1994; Rosenburg and Johnson, 2005.
27. *Parliamentary Papers*, 1943–44 IV, Cmnd. 6552, Scottish Housing Advisory Committee, Report on the Distribution of New Houses in Scotland, para 23.
28. For a summary of twentieth-century Scottish housing legislation see Rodger, 1989, Appendix B, 238–45.

29. *Parliamentary Papers*, 1917–18 XIV, Cd. 8731, Royal Commission on the Housing of the Industrial Population of Scotland, Rural and Urban, Report, 387. Glasgow, with 2,199 families housed in council property, was by far the largest provider of municipal housing (63 per cent), followed by Edinburgh (17 per cent), Greenock (6 per cent), Aberdeen (4 per cent) and seven other burghs.
30. Rosenburg and Johnson, 2005.
31. Whitham, 1989; Minett, 1989.
32. Williams, 2000, 312.
33. Rodger and Al-Qaddo, 1989.
34. Markus, 1999.
35. Dunleavy, 1981, 126.
36. The post-World War II new towns are discussed in Robertson, 2005, 372–402.
37. Miller, 2003, 242–69; Jephcott, 1971.
38. Keating, 1988, 138.

BIBLIOGRAPHY

Adams, I H. *The Making of Urban Scotland*, London, 1978.

Allan, C M. The genesis of British urban development with special reference to Glasgow, *Economic History Review*, 17 (1965), 598–613.

Borsay, P. *The English Urban Renaissance: Culture and Society in the Provincial Town 1660–1770*, Oxford, 1989.

Bowley, M. *Housing and the State, 1919–1944*, London, 1945.

Cairncross, A K. *Home and Foreign Investment 1870–1914*, Cambridge, 1953.

Cant, M. *Marchmont in Edinburgh*, Edinburgh, 1984.

Carruthers, A, ed. *The Scottish Home*, Edinburgh, 1996.

Cruft, K and Fraser A, eds. *James Craig 1744–1795*, Edinburgh, 1995.

Dunleavy, P. *The Politics of Mass Housing in Britain 1945–75: A Study of Corporate Power and Professional Influence in the Welfare State*, Oxford, 1981.

Van den Eeckhout and Brussels, P. In Daunton, M J, ed, *Housing the Workers: A Comparative History 1850–1914*, Leicester, 1990, 67–106.

Gaskin, M. Housing and the construction industry in Scotland. In Maclennan, D and Wood, G, eds, *Housing Policy and Research in Scotland*, Aberdeen, 1978, 92–109.

Gibb, K. Rebirth of the market: 1975 to the present day. In Glendinning and Watters, 1999, 259–70.

Gibb, K. Regional differentiation and the Scottish building industry, *Housing Studies*, 14 (1999), 43–56.

Glendinning, M and Muthesius, S. *Tower Block: Modern Public Housing in England, Scotland, Wales, and Northern Ireland*, New Haven, 1994.

Glendinning, M and Page, D. *Clone City: Crisis and Renewal in Contemporary Scottish Architecture*, Edinburgh, 1999.

Glendinning, M and Watters, D, eds. *Homebuilders: Mactaggart & Mickel and the Scottish Housebuilding Industry*, Edinburgh, 1999.

Gow, I. *The Scottish Interior: Georgian and Victorian Décor*, Edinburgh, 1992.

Gyani, G. Budapest. In Daunton, M J, ed, *Housing the Workers: A Comparative History, 1850–1914*, Leicester, 1990, 149–82.

Hill, W H. *History of Hutcheson's Hospital*, Glasgow, 1881.

Jephcott, P. *Homes in High Flats*, Edinburgh, 1971.

Keating, M. *The City that Refused to Die: Glasgow: the Politics of Urban Regeneration*, Aberdeen, 1988.

Kellett, J R. Property speculators and the building of Glasgow 1780–1830, *Scottish Journal of Political Economy*, 8 (1961), 211–32.

Lewis, A R. The builders of Edinburgh's New Town 1767–1795. Unpublished PhD thesis, University of Edinburgh, 2006.

McCrone, D, Morris, A and Kiely, R. *Scotland the Brand: The Making of Scottish Heritage*, Edinburgh, 1995.

Markus, T A. Comprehensive development and housing 1945–75. In Reed, 1993, 147–65.

Melling, J. Employers, industrial housing and the evolution of company welfare policies in Britain's heavy industry: West Scotland, 1870–1920, *International Review of Social History*, 26 (1981), 255–301.

Melling, J. *Rent Strikes: People's Struggle for Housing in West Scotland 1890–1916*, Edinburgh, 1983.

Miller, M J. *The Representation of Place: Urban Planning and Protest in France and Great Britain 1950–1980*, Aldershot, 2003, 189–298.

Minett, J. Government sponsorship of new towns: Gretna 1915–17 and its implications. In Rodger, 1989, 104–24

O'Carroll, A. Tenements to bungalows: Class and the growth of home ownership before the Second World War, *Urban History*, 24 (1997), 221–41.

O'Carroll, A. The influence of local authorities on the growth of owner occupation, 1914–1939, *Planning Perspectives*, 11 (1996), 66.

Reed, P, ed. *Glasgow: The Forming of the City*, Edinburgh, *c*1993.

Robertson, D. Scotland's new towns: A corporatism experiment. In Beech, J, Hand, O, Mulhern, M A and Weston, J, eds, *The Individual and Community Life* (Scottish Life and Society: A Compendium of Scottish Ethnology, vol. 9), Edinburgh, 2005, 372–402.

Rodger, R. Speculative builders and the structure of the Scottish building industry, *Business History*, 21 (1979), 226–46.

Rodger, R. The evolution of Scottish town planning. In Gordon, G and Dicks, B, eds, *Scottish Urban History*, Aberdeen, 1983a, 71–91.

Rodger, R. The 'Invisible Hand' – market forces, housing and the urban form in Victorian cities. In Fraser, D and Sutcliffe, A, eds, *The Pursuit of Urban History*, London, 1983b, 194–200.

Rodger, R. Wages, employment and poverty in the Scottish cities, 1841–1914. In Gordon, G, ed, *Perspectives of the Scottish City*, Aberdeen, 1985, 50–3.

Rodger, R. The Victorian building industry and the housing of the Scottish working class. In Doughty, M, ed, *Building the Industrial City*, Leicester, 1986, 151–206.

Rodger, R, ed. *Scottish Housing in the Twentieth Century*, Leicester, 1989.

Rodger, R. Urbanisation in twentieth century Scotland. In Devine, T M and Finlay, R J, eds, *Scotland in the Twentieth Century*, Edinburgh, 1997, 122–53.

Rodger, R. *Housing the People: The 'Colonies' of Edinburgh 1860–1950*, Edinburgh, 1999.

Rodger, R. *The Transformation of Edinburgh: Land, Property and Trust in the Nineteenth Century*, Cambridge, 2001.

Rodger, R and Al-Qaddo, H. The Scottish Special Housing Association and the implementation of housing policy 1937–87. In Rodger, 1989, 184–213.

Rosenburg, L and Johnson, J. 'Conservative surgery' in Old Edinburgh. In Edwards, B and Jenkins, P, eds, *Edinburgh: the Making of a Capital City*, Edinburgh, 2005, 131–49.

Sim, D. Glasgow's special housing initiatives. Ways of dealing with difficult estates,

Research Memorandum 3, Glasgow District Council Housing department, Glasgow, 1984.
Sim, D. The Scottish house factoring profession, *Urban History*, 23 (1996), 351–71.
Smout, T C and Wood, S. *Scottish Voices 1745–1960*, London, 1990, 13–35.
Smith, P J. Slum clearance as an instrument of sanitary reform: The flawed vision of Edinburgh's first slum clearance scheme, *Planning Perspectives*, 9 (1994), 1–27.
Smith, R and Wannop, U, eds. *Strategic Planning in Action: The Impact of the Clyde Valley Plan 1946–82*, Aldershot, 1985.
de Vries, J. *European Urbanization 1500–1800*, London, 1984.
Walker, F A. The Glasgow grid. In Markus, T A, ed, *Order in Space and Society*, Edinburgh, 1982, 155–99.
Whatley, C. *Scottish Society 1707–1830*, Manchester, 2000.
Whitham, D. National policies and local tensions: State housing and the Great War. In Rodger, 1989, 89–103.
Williams, N J. Housing. In Fraser, W H and Lee, C H, eds, *Aberdeen 1800–2000: A New History*, East Linton, 2000, 295–322.

8 A History of the Architectural Profession in Scotland

CHARLES MCKEAN

INTRODUCTION

In the technical sense of a person academically trained in the sciences of architecture, there were virtually no architects in Scotland until the end of the nineteenth century. By contrast, there were people practising the requisite skills and who were recognised by the term 'architect', or in those days, 'architector' (after the Latin) certainly by the early sixteenth century. It is therefore important to distinguish between the twentieth-century notion of an academically trained architect and those who were not, but certainly practised the craft, and were recognised as architects by their contemporaries.

So how might that craft be defined in the pre-academic days? It was, more or less, defined in a poetic panegyric to man celebrated as a royal master of works, Sir Robert Drummond of Carnock, who died in 1583. Drummond of Carnock was, by discipline, a wright, and his death was lamented by the poet Alexander Montgomerie:

> All buildings brave bids DRUMMOND nou adeu
> Quhais lyf furth sheu he lude thame by the laiv.
> Quhair sall we craiv sic policie to haiv?
> Quha with him straive to polish, build or plante?
> These giftis, I grant, God lent him by the laive[1]

An architect, therefore, had to be inspired by an inordinate love of buildings and have a primary skill in 'policie': that is, the ability to conceive of a plan or design. Further, he required skills in 'polish, build and plant', which may, in emulation of Sir Henry Wotton, be transcribed as 'delight' (i.e. cultural input), 'firmness' (i.e. technology and structure) and 'garden design'.

Inverted snobbery in the nineteenth century insisted upon regarding such skilled designers merely as artisans – masons or wrights – even where their contemporaries had regarded them as architects. Such is the case, for example, with Tobias Bauchop who practised from Alloa in the late seventeenth century.[2] Because he is known to have worked as a mason under Sir William Bruce on, for example, Kinross House from 1685 and the Stirling

Tolbooth in 1705, he is assumed to have been just a mason. Patterns were more complicated than that, and indeed remained more complicated until the formation of the architectural professional institutes in the 1830s–40s. Bauchop had been responsible for the design of his own rigorous ashlar-faced town house in Alloa, probably assisted the earl of Mar in his remodelling of the House of Alloa itself, and designed a mansion at Abercorn, Logie Church in 1684, and the very fashionable Town House in Dumfries, for which his clients there described him as the 'architector'.[3] Qualitatively, his designs far outshine some of the sublunary products of nineteenth-century provincial architects, for whom attribution of the title 'architect' appears to present no difficulty. So the examination of the Scottish Renaissance architectural world through the perspective of the European mediaeval pattern of 'a co-operative process between the patron … and the master mason, or other master tradesman'[4] can no longer be sustained as adequate.[5]

A further problem exists in that few drawings survive that pre-date the late seventeenth century. That, together with Scotland's strange refusal to adopt classicism as its architectural expression, has been taken as a further sign that architecture did not develop in the country until then. Whether there are any architectural plans that survive but remain undiscovered and bound within libraries, architectural plans there most certainly were. Sir James Murray was paid for one for the Parliament House, as was William Wallace for George Heriot's Hospital. The only way that client and builder of Gallery House, Angus, were to agree that the house had been built according to contract was that each was provided with the architect's draught against which to inspect the house upon completion.[6] When Sir Robert Innes decided to reformat his paternal seat of Innes in 1638, he obtained a draught from William Ayton – for which he paid two years later – and the earl of Lauderdale was forever exchanging designs for Brunstane and Thirlestane with Sir William Bruce, annotating them freely. Again and again, surviving building contracts refer to a 'draught'. Even the earl of Strathmore waxed vehement on the need for such drawings in his reformatting of Glamis.[7] Although drawings of decorative detail are recollected in the sole survivor of cusping details etched into the wall of the crypt of Rosslyn Chapel, other drawings must have existed to convey the three-dimensional complexities of even the medieval tower house. For example, when the masons began work on the hunting lodge of Hallbar Tower in the Clyde Valley, the thickness of the walls at ground-floor level varied substantially. That is because, three storeys above, there were variations of rooms and stairs contained within those walls. The implication is that three-dimensional, multi-storeyed thinking was required of the masons when laying the foundations at the very outset of the project; and some form of drawing is likely to have been one means of conveying the ideas to the team of workmen.

Our interpretation of the title 'architect' is based upon the twentieth-century conception of a distinct group of people who 'might be called upon to supply plans and drawings and to exercise professional control over building operations'.[8] It now implies a lengthy and specific training in design, whether through apprenticeship during the nineteenth century or attending an academic/vocational course during the twentieth century. It was different three centuries ago. Such training as there was – even on the continental model – was reserved for craftsmen, of whom masons were pre-eminent, followed closely by wrights. At the other end of the social scale, members of the élite took the expected Renaissance interest in the subject, usually by studying the ancients or the increasing number of interpretations of them in fashionable publications. It may be significant that it was in Scotland that, following William Schaw's drawing up of the first Statutes for Freemasons in 1598, masonic lodges first welcomed non-mason members, bringing the two together.[9] Schaw also laid down the expectation that the 'Art of Memory' should be central to masonic education.[10] So masons were expected to have knowledge and understanding of a highly charged symbolic language in which a building's significance was conveyed by far more than a set of efficient rooms. Such knowledge may always have been expected, but Schaw's 1598 statutes fixed it as a requirement. Too sharp a distinction, therefore, between architect, patron, mason, builder, developer or wright might be somewhat artificial before 1840.

The term architect, in its modern sense, was another rediscovery of the ways of the ancients newly arrived from Vitruvius through Alberti. It required a knowledge of architectural principles, the ability to design and the ability to instruct and to control building work. But Alberti required more: 'Him I consider the architect, who by sure and wonderful reason and method, knows both how to devise through his own mind and energy, and to realise by construction, whatever can be most beautifully fitted out for the noble needs of man.'[11] So the use of the term 'architect' or, in Scotland, 'architector' carried as much an implication of intellectual and cultural achievement as technical proficiency. Building was a principal means by which the crown and its magnates conveyed their cultural ambitions. Yet the historiography of the Scottish architectural profession has been bedevilled by the perspective of the 'ultimate triumph of classicism' which did not recognise non-conforming works during the sixteenth and seventeenth centuries as the product of architects.[12]

The first step, therefore, is to examine at what point the distinctive techniques of the architect became recognisably apparent in Scotland, and were accepted as such. The second is to consider how someone might reach the required level of skill, and when the role became exclusive. The critical issue in watching the profession develop is to observe how, in earlier centuries, the architectural function might be provided by a craftsman who had worked his way up into design, or sometimes by an interested and educated aristocrat. The Renaissance theorists held that they needed each other and, subject to proper skill, they were both architects – albeit of different kinds. The next step in the evolution is to observe how and when architects were also contractors, suppliers or developers, and when the professional managed to differentiate himself from the tradesman. Finally, whereas it is clear that regions developed their own architects quite early – some clearly teaching themselves from publications – a key feature in the development of the modern profession is the emergence of the architect who manages to create a market demand for his services outside his home territory.

THE MIDDLE AGES

No one has been identified as an architect in the Renaissance manner practising in Scotland during the Middle Ages, but there is at least one mason who was sufficiently confident to leave his mark. John Morvo worked between 1380 and 1420, and happily left his inscription upon Melrose Abbey. Ambiguously, he stated that he was born in Paris, but whether he was French or a Scot who had trained there is unclear. For a Frenchman, his Scots was good[13] and his delicate curvilinear work may also be seen at Lincluden, Paisley and possible Glasgow Cathedral. There are almost certainly others still awaiting rediscovery. However, the standing of masons was indicated by the founding of the Incorporation of Masons and Wrights in the capital in 1445, another which also included coopers, carvers, slaters and painters in Aberdeen in 1541, and one in Glasgow ten years later. It is not clear when there emerged an incorporation in Scotland's second city, Dundee, although a mason's lodge had been attached to the parish church of St Mary's for a long time before.[14] The seventeenth-century regulations of the Edinburgh Incorporation (the only one to survive) required applicants to make drawings and a pasteboard model of a house or a building, and there was an expectation of knowledge of European precedent, presumably to match that of their employers. The earl of Wemyss, in Fife, who had a fine library of architectural books, was accustomed to lend them to local craftsmen, possibly for this purpose.[15] From the evidence of the same initials carved on the dressed stonework of some country houses in Ross, we can surmise that estate masons would have been responsible, perhaps, for the main parts of construction, but that travelling quality masons might have been used for the all-important decorative and heraldic details.

THE RENAISSANCE

Evidence increases from the sixteenth century. There emerge virtual dynasties of master masons and wrights, exercising considerable responsibility, particularly in the royal works. The French family (father and son, both Thomas), for example, worked on Linlithgow Palace and other royal projects. The Mylne dynasty of Dundee and Perth contained members who designed, over the sixteenth and seventeenth centuries, bridges, forts, fortifications and alterations to churches, as well as working with architects in the construction of country houses and Holyrood Palace; their eighteenth-century successors became fashionable London architects. The accounts mention other masons and their patrons: Thomas Kedder (or Cadder), who built Boghouse of Crawfordjohn for the king was directed by Sir James Hamilton of Finnart; and Jean Roytel, French master mason of Dunbar who was made principal master mason by Mary of Guise, was directed by Sir William Makdowall.

However, the running was made, almost certainly, by the royal master of works, beginning with James V's cousin Sir James Hamilton of Finnart, an illegitimate nobleman who effectively exercised the role of royal architect from 1528 until his execution in August 1540. Recognised by contemporaries for his architectural skill, to Finnart may be ascribed the pioneering plan and defences of his villa at Craignethan, *c*1531, his seat at Avendale (Strathaven) after 1533, and the royal palaces of Linlithgow, 1535, and Stirling, from 1538.[16] All of these buildings displayed substantial innovation in planning and decorative programme, and in the year he died he was awarded a charter as master of works principal, leading a skilled construction team, with power to work on all royal buildings. By the 1550s, this role was taken by Sir William MakDowall, who was principally concerned with amendments and alterations to royal buildings – although possibly beginning at Drochil, by Peebles, and the earl of Mar's palatial town house, Mar's Wark, in Stirling, in 1569. He was succeeded by Sir Robert Drummond of Carnock, who was followed by William Schaw between 1583 and 1602. Shaw became, effectively, court architect to Queen Anne's court.

It is important to emphasise just how fluid this building world was. William Schaw, younger of Sauchie, was a member of James VI's court and occasional ambassador to Denmark. He was certainly the favoured architect of the 1590s, before his untimely death; but almost certainly acted conjointly with Sir Alexander Seton, Lord Urquhart and Fyvie, later earl of Dunfermline and chancellor of the realm, in the latter's house at Pinkie, and his ancient paternal seat at Fyvie. In true Renaissance fashion for aristocrats, Seton had trained, inter alia, as an architect in Rome,[17] and worked in tandem with another. James Murray was a wright, as was his son. His son became the king's master of works and principal designer of the later Jacobean court, bought a small country estate upon which he built a villa and, having been knighted in 1633, died Sir James Murray of Kilbaberton. To him may be attributed the 1617 palace at Edinburgh Castle, the contemporary wing at Linlithgow, the reworking of Holyrood, the 1634 Parliament House and, probably, the villas of Wrightishousis, Staneyhill, Binns and Winton. His colleague as joint master of works, Sir Anthony Alexander, designer of Pitreavie and the Argyll Lodging, Stirling, was son of the secretary of state for Scotland, Sir William Alexander of Menstrie. During the seventeenth century, it became customary for royal masters of works to be knighted and obtain a small estate.

That fluidity is not so apparent the next rank down. If both Seton and Schaw had occupied themselves with the design of Fyvie, the masterly execution of it almost certainly owed something to the Bel family of mason/architects of that region, to whom perhaps seventeen other great houses may be attributed through either similarity of plan or detail. Only later, at Castle Fraser, however, is there the signature of IB, taken to signal John Bel. Perhaps Fyvie was the project that made them fashionable.[18] Equally, it has been possible to identify a shadowy minor landowner, Alexander Con of Auchry, as the inspirer of a number of chunky houses of particular plan – Gight, Towie Barclay, Craig, Fedderate and perhaps Delgatie – deriving initially from the 1530s.[19] The equally shadowy Thomas Leiper left his initials on the Place of Tolquhon, which he built for William Forbes between 1584 and 1589, but his fondness for a variegated skyline and a particular form of heraldic gunloop implies that he was much involved in the counter-Reformation adaptations of many Auchry houses as well as, probably, the great garden at Edzell and the Wine Tower at Fraserburgh. In the 1620s, his son was responsible for the new gallery wing and arcaded courtyard at Castle Fraser before being dismissed and sued for breach of performance.

Closer to Edinburgh, William Wallace and William Ayton both trained as masons, and yet became architects. Wallace worked on the 1617 wing of Linlithgow, upon the houses of Peffermill and Winton, and produced a plan for Heriot's, whereas Ayton took over from Wallace at Heriot's (where his portrait remains) and accepted private work besides. Neither ended up a knighted landowner. Nor, indeed, did James Bain, the king's master wright and one of the principal entrepreneurs of the later seventeenth-century Scotland building industry. Bain was a significant timber supplier and, in the case of both Glamis and Brechin, acted in the role of main contractor, employing other trades directly. John Mylne certainly acted as architect when entrusted with the plan of Fort Charlotte in Lerwick and country houses such as Panmure, and his nephew Robert Mylne (1633–1710), was not only the master of works and undertaker of Holyrood, under Bruce, but also the architect/developer of the two private courtyards of mansion flats – Mylne's Court and Miln's Square – in Edinburgh's High Street. Mylne's contract for Holyrood expected him to know, design and plan the classical orders in proper sequence and proportion – and indeed leaves one wondering what exactly the contribution of Bruce might have been, particularly given the inexpert application of the classical orders. As their tombstones record, contemporaries regarded such men as architects, but they remained unknighted and were still classed by later generations as masons – building operatives[20] – with the implication that they had, perhaps, got above themselves.

The architectural profession, as such, has customarily been held to have begun with Sir William Bruce (*c*1630–1710), a minor landowner who became politically powerful in the 1670s before falling from favour. He was certainly involved in the design of the modernised Scottish house as exemplified in Dunkeld and Moncreiffe, as well as the more baroque extravagances of the 'ancient paternal seat' of the duke of Lauderdale at Thirlestane. However, his acceptance as architect has probably less to do with his substantial construction, his planning innovation, or his acts of patronage or directorship of works (for they were similar to that of many of his predecessors), than with the fact that in three of his most elevated buildings, Holyrood, Kinross and Hopetoun, he used the international language of contemporary continental architecture – even if with a very Dutch accent.

Somewhat erroneously, it has been suggested that Bruce's role within British architectural history was that he designed 'unfortified houses for the first generation of Scottish lairds to realise that the tower house was an anachronism, and to persuade them to abandon corbel and crow-step in favour of cornice and pediment'.[21] But Scottish domestic architecture had consisted of more than fortified tower houses centuries before Bruce, as the panegyric to Drummond had earlier demonstrated. Bruce's own two houses – first Balcaskie (a house achieved by infilling an existing small early U-plan Jacobean house) and then the majestic Kinross – exemplified his architectural approach. It is unclear just how much he owed, however, to the Italian-trained James Smith, Robert Mylne's son-in-law, who was certainly at work at Kinross whilst preparing a draught of the new 'castle' of Methven for Patrick Smyth of Braco.[22] Smith (*c*1645–1731), who trained as a Jesuit priest and then as a craftsman, died as the landed owner of the small Midlothian estate of Whitehill. He designed great houses like Melville in an updated version of Renaissance Scots, the powerful classical remodelling of the palaces of Hamilton and Dalkeith, and also a number of pert classical villas for the new businessmen, like the houses of Strathleven, Raith and his own house of Newhailes (Whitehill). He is the first Scottish architect for whom there remains what appears to be student *esquisses*.

The architectural consequences of the parliamentary union with England in 1707 were only to become striking after 1746. In the early decades of the eighteenth century, there is evidence of some convergence between Scottish and English architectures. New houses 'in the English manner', and a greater use of classical motifs and of symmetry, began to appear amidst the old patterns. The basic technologies of masonry construction, harled rubble and dressed stone, with imported timber, remained relatively constant. Likewise the élite, notably Sir John Clerk of Penicuik, continued the tradition of working as and with an architect to realise their designs – in Clerk's case with William Adam at Mavisbank. Their agenda was to recreate the ancient nobility of Scotland within the new British framework. Although we still know deplorably little about early eighteenth-century architectural patterns, the first clear emergence of a 'city architect' occurs with Alexander McGill (d.1734) in Edinburgh, which certainly did not preclude him from undertaking other commissions; and perhaps John Craig, the town's wright of Glasgow. However, Allan Dreghorn (1709–65) was identified with the principal civic monuments of that city, such as its adventurously classical town hall (1735) and baroque St Andrew's church with Mungo Nasmith (from 1737). As research deepens, the work of other architects is becoming better appreciated, notably that of John Douglas (d.1778)[23] who performed unlikely Scoto-baroque alterations upon a number of country seats including Archerfield, Taymouth and Lochnell, inter alia, and proposed some extraordinary transformations for Blair.

The undisputed leader was William Adam (1689–1748),[24] of very minor lairdly stock, who became industrialist, contractor, buildings materials supplier, and principal architect of Scotland, with none of these functions in contradiction with each other. Like Smith, Adam was as at ease with designs echoing old Scots (Duff House, with its corner towers may be read as homage to either Melville's south facade or to the English baroque of Vanbrugh) as in the most contemporary (the modernising and aggrandisement of Bruce's Hopetoun, or Edinburgh Royal Infirmary); and the many designs he produced to reformat existing country seats whilst retaining their ancient character (e.g. the House of Kelly, transformed into Haddo House, or Fasque).[25] Allowing for necessary rank difference, Adam was possibly more at ease with his clients than his predecessors, perhaps signalling a convergence between the craftsman and the professional, albeit the extent to which there had been a yawning gap in the earlier centuries is becoming increasingly doubtful.

After Culloden in 1746, Scots had to redefine themselves and their approach to architecture now that their recent Stewart history had become socially unacceptable. An architecture that paid homage to a noble ancestry was, for the time being, impossible. The path they chose was that of becoming the modernisers of Britain and the empire, and the next eighty years was the period for which it becomes most credible to refer to a 'British' architecture. The consequence was that the most striking architectural development of the later eighteenth century was the creation of 'new towns', some 500 of which were conceived – and most built to some degree – between 1750 and 1830. Most were grid-planned rural settlements intended to accelerate and capitalise upon agricultural improvement,[26] but after 1790 almost any community of substance (with the notable exception of Dundee)[27] eventually contemplated or built one or more grid-planned suburbs, exemplifying to some degree the virtues of rationalism, social segregation and classical architecture. Improvement brought increased wealth and a rapidly rising population, leading to what might be classed as architecture's first boom in Scotland. For the modernising of the country was expressed through new towns, new churches, new harbours, new farms and steadings, and new factories; and in the social buildings consequent upon an improving society, namely assembly rooms, concert halls, art galleries, museums and academies. Between 1750 and 1840 architects proliferated throughout the land. In 1752, only John Douglas and William Adam's son John had classed themselves as architects in the Edinburgh street directory, but by the end of the century there were many.

Gideon Gray was active in Stirling in mid-century, designing the Grammar School, the Red Lion Hotel and, probably, a new façade for Touch House. In Dundee, the town's wright, Samuel Bell (1739–1813), became the town's architect, designing its principal public buildings of St Andrew's Church, the Trades' Hall, and the English Chapel, as well as numerous new villas, eventually dying a wealthy property owner. George Jaffray may have performed a similar function in Aberdeen, and William Law, responsible for Marischal Street, certainly did. There were five Edinburgh architects who competed with James Craig for the design of Edinburgh's New Town in 1766, but we do not know who they were: possibly the developer James Brown, or

John Douglas, John Young, Robert Kay or John Fergus. In the hinterland, Sir James Clerk of Penicuik (1709–83) continued his family's practical interest in architecture but, lacking the collaboration of a practical architect, he did not quite achieve the distinction he might otherwise have gained in designs like Rossdhu, Hailes House or in Penicuik itself.

Bestriding them all was the Adam clan: eldest brother John (1721–92), practising from Edinburgh, whose houses followed a plainer form of his father's tradition; second-youngest brother James (1732–94), a somewhat flamboyant dilettante who also wrote on agricultural improvements and designed some houses (e.g., Hangingshaws) but may have been primarily responsible for the urban schemes in Glasgow such as Stirling's Square, the Tron Kirk, the Royal Infirmary, the Professors' Lodgings for the university, and unbuilt schemes for a Corn Exchange and George Square; the youngest brother, William, was a building contractor. The family's genius was Robert (1728–92), who practised from London. Robert, and to a lesser extent those who likewise were trained at St Luke's in Rome, brought to Britain four years of Italian cultural immersion[28] superimposed upon a Scottish training in both architecture and construction from his father and brother. The result was an architecture of strong geometry, invention and wilfulness in the deployment of architectural language, as well as spatially complex interiors realised in the

Figure 8.1 Drawing of Robert Adam by Pecheux, no date. Source: Reproduced by permission of RCAHMS, Crown Copyright.

most fashionable decorative language. Although most commissions followed normal architectural practice, and some were for drawings alone, he would occasionally include his own decorators and painters, and in the construction of the Adelphi development in the Strand in London, the Adam family – as in the tradition – were developers, clients, designers, builders and fitters-out, using largely Scottish workmen to whom they spoke in Scots. Most of Robert's country houses were in England, but from the 1780s his work predominantly lay in Scotland. It included an outstanding series of castellated country houses, with Culzean as its apotheosis, and the principal public buildings of the capital – namely Edinburgh University, proposals for the South Bridge and new buildings around Parliament Square, Register House, and the most splendid space in the New Town, the 1792 Charlotte Square.

The Adam office was a professionalised one, with pupils and apprentices (David Hamilton, Hugh Cairncross and Richard Crichton among them), and a particular type of drawing technique that was utterly recognisable. These drawings were sometimes intended to form part of a marketing brochure, as later was to be the case with similar portfolios by David Hamilton and David Neave.

London was the centre of British culture and fashion, and landed gentlemen tended to opt for a town house there rather than, as heretofore, in Edinburgh or Glasgow. Consequently, it became the focus of architectural patronage and attracted architects accordingly. Robert Mylne practised from it, as did the short-lived Angus-born James Playfair (1755–94) who promised to have become Scotland's finest neo-classical architect. Equally, the increasingly English-educated Scots élite were likely to select an English architect for their Scottish mansion and thus, in the early nineteenth century, did Sir Robert Smirke, Henry Holland, Thomas Harrison, Edward Blore, Thomas Rickman, Robert Lugar and William Atkinson dominate the country-seat market until the 1840s. Between 1800 and 1810 over half the best commissions in Scotland went to English architects. Perhaps inevitably, it was the new professional classes of Scotland who were the predominant commissioners of the smaller houses and villas of the period, and they – equally inevitably – selected Scots architects, usually from their own circle.

The architectural world was changing, in almost an existentialist manner. During the 1780s, the mason Alexander Reid purchased an individual site in Princes Street each year, building it and feuing it out, then accelerating by undertaking two sites a year, and later several. His son Robert joined him when developing Queen Street. In 1799, Robert set up as consultant architect and, four years later, Alexander likewise repositioned himself as architect. Robert, however, quit the house-developing world eventually to rebuild Parliament Square (for those who had occupied his father's houses), and Leith customs house. He was eventually appointed king's architect in 1818, from which lofty position he expressed some pioneering conservationist views about the treatment of ruined abbeys.

A more striking example is that of the brothers Archibald and James Elliott. Archibald, although exhibiting in the Royal Academy, remained in

London running the family's cabinet-making business, providing fittings for clients for whom James was providing the designs. At Taymouth between 1803 and 1810, James designed a new Taymouth to replace John Paterson's half-built unpopular version. He was considered architect, and paid 5 per cent. His brother was paid by the piece: the door, the wainscot, or whatever. After James died in 1810, Archibald became architect on a percentage commission but also continued to supply the house with timber fittings. The character he took depended upon what he was doing. Likewise, William Burn might produce a design as architect, but proffer a marble fireplace supplied by his father; or David Hamilton, acknowledged as the father of the architectural profession in the West of Scotland, might be a developer in Buchanan Street and have his architect's office fronting his marble works. David Neave, town's architect of Dundee between 1813 and 1831, likewise had his office by his own builder's yard.[29] Hamilton also had a business relationship with Glasgow's master of works, James Cleland. Yet he was regarded with the utmost respect by his peers, designed mansions for Glasgow provosts, villas in the manner of Soane for the new mercantile aristocracy, city blocks, churches, clubs and Royal Exchange Square, as well as great mansions. Of these, Hamilton Palace, undertaken in the 1820s, was one of the largest country houses north of London. Yet despite Hamilton's continuing links with trade, it was said of him that no one was more easy in the company of dukes and earls.

It was also a changed world so far as commissioning was concerned. Most of the principal buildings of the 'Athens of the North' were not commissioned by patrons but by committees or trustees, prefiguring what was to become customary for nineteenth-century civic building. As it happened, Athenian buildings virtually became the preserve of William Playfair (1790–1857) who, after an apprenticeship with William Stark (the man who broke the rigidities of Edinburgh's new towns with his emphasis upon the landscape, prospect and the picturesque), and then with Benjamin Wyatt and Sir Robert Smirke in London, came to design Edinburgh's principal monuments: the National Monument on Calton Hill (under Charles Cockerell), the Observatory, the Royal College of Surgeons, what is now the Royal Scottish Academy, and the National Gallery. His equivalents in Aberdeen were John Smith (1781–1852), who, like Playfair, combined stern neo-classicism in King Street and the North church with a romantic Tudor approach in his many country houses, and Archibald Simpson[30] (1790–1847), who was much more austerely classical throughout the north-east, whether in the Assembly Rooms (1820), a bank in Fraserburgh (*c*1835), or the masterful St Giles church, Elgin (1827).

Playfair's stern but romantic classicism, however, melted in the countryside, where his buildings became either Elizabethan (Floors) or 'Scottish revival' (Bonaly). For in 1814, Sir Walter Scott had made Scotland's past respectable once more with the subtitle to his first novel *Waverley*: ''Tis Sixty Years Since'. He had legitimised the Stewart dynasty once more. It took almost another fifty years for that recovery of memory to reach its zenith.

Figure 8.2 W H Playfair by Watson Gordon, no date. Source: Original oil by Watson Gordon at Donaldson's Hospital. Reproduced courtesy of RCAHMS.

In the intervening period, most of the classical architects devised some form of picturesqueness to match the new zeitgeist. Perhaps the most masterful was James Gillespie Graham (1766–1856), a sheriff's son whose over 100 commissions are to be found throughout the country, from the classical Moray Estate in Edinburgh to the romantic mansions of Brodick and Ayton. A self-taught architect who probably began as project manager at Lanrick *c*1798, Graham's proficiency at classical buildings was well demonstrated in the Moray Estate, Edinburgh (wherein errors earned him a lawsuit), but his genius was for a romantic external confection loosely based on Scots precedent, for which he devised the term 'baronial', a design intended to empathise with 'Caledonia stern and wild'. His interiors were likewise governed by a fondness for a picturesque romantic experience.

Of the great architects of the Athens of the North, Elliott, Playfair, Reid, Graham, Richard Crichton, William Sibbald, William Burn and Thomas Hamilton,[31] only Playfair came, as it were, to the manor born. Most of the others were related in some way or other to the building industry, normally as the son of a builder who may also have aspired to be an architect.[32] The world was existentialist, in that one's status and remuneration depended upon what was being provided to whom. The same man – Walter Newall is a good example[33] – might be a professional offering an architect's service, and a cabinet maker selling an escritoire or a marble fireplace to the same person on the same day, and therefore treated in two different ways.[34]

Yet it was not all fluid. A total outsider – carpenter George Meikle

Kemp, self-taught architect and antiquarian – won the second Scott Monument commission in 1838 and styled himself architect – and was allowed to be so by his friends. But the professional establishment led by William Burn generally froze out Kemp, who died relatively early in 1844.[35] The probability is that acceptance as architect had now become partially contingent on having undertaken an apprenticeship with a recognised office, and Kemp never had. The profession was undergoing change in the search for respectability. The Royal Institute of British Architects (RIBA) had been founded in 1834 on the presumption that an architect, as suggested by Sir John Soane, should never be tainted by any of the commercial transactions involved in building – contracting, subcontracting or supplying – in order to preserve the objectivity of his advice. However, the Limited Liabilities Act of 1855 and the Companies Act of 1862 together greatly reduced the risk of architects' speculative developments and encouraged the construction of commercial chambers for rental.[36] That might explain why certain notable mid-nineteenth-century Scottish architects demurred at becoming members of the RIBA with its prohibition on such activity.[37]

Following an abortive launch of the Institute of the Architects of Scotland in 1840,[38] the Architectural Institute of Scotland (AIS) was formed in 1841 and lasted for some fourteen years, its mantle then being carried forward by the regional architectural associations that had formed by then.

Figure 8.3 James Gillespie Graham and Thomas Hamilton, *c*1847. Source: Copied from Crombie's *Modern Athenians*, 1847. Reproduced courtesy of RCAHMS.

The AIS had several membership classes – for clients, for builders, for tradesmen and for architects – but they were kept discrete. A certain form of behaviour was expected of an architect. The AIS also sought to underpin the professional status of the architect, and lobbied – sometimes through its mouthpiece the short-lived *Building Chronicle*[39] – for a recognised fee level of 5 per cent (the construction of the Houses of Parliament and of Dunrobin Castle being two *causes célebres*). Although being a builder–developer was not absolutely banned, as had been the case in the RIBA, the notion of 'architect' had moved to the higher plane of wholly independent consultant who had learnt his craft not on a building site but as an assistant to an established member of the profession. So it was expected of the architect that he would take apprentices and, indeed, architects like David Bryce (1803–76), the outstanding architectural figure of mid-century, ran virtual academies. The title 'architect' was becoming exclusive to those following a particular career track.

That did not, however, preclude them from acting as developers. Glasgow had developed some five neo-classical 'new towns' up to 1840, with pleasant streets, terraces, crescents and squares designed by architects such as John Brash, John Baird, George Smith and, eventually, Charles Wilson and J. T. Rochead, which formed the context for the peculiar genius of Alexander 'Greek' Thomson (1817–75) and his brother George.[40] Impelled by religious conviction, Thomson refined a pure classical architecture which he applied to villas and showrooms alike, but in the case of his masterpiece, the St Vincent Street church, he bought the original site from the congregation, erected a warehouse/showroom upon it, and then provided the congregation with a new church on the proceeds.

Thomson's architecture is also interesting for exemplifying his response to the new materials then becoming available. He rejected iron as a façade material, as it had been used in the 1854 Gardner's warehouse in Glasgow's Jamaica Street, a rejection later shared by Charles Rennie Mackintosh, who bewailed iron and glass's 'want of mass'. Both architects had symbolic programmes for their buildings that demanded more from their façades than a mere curtain. Nonetheless, new materials, new technologies and new transportation were available, meaning that the essentially regionalised nature of building materials was no longer the case. Iron, later steel, would free buildings from the constraints of load-bearing walls, and hydraulic lifts (from the 1860s) and then Otis elevators (from 1888) would liberate architects from limitations of height once sufficient electric power became available in the following decade. During mid-century, however, the new technologies were restricted to leisure buildings, museums, greenhouses or railway stations, or were otherwise concealed behind a heavily expressive stone façade.

Perhaps not entirely coincidentally, the availability of new materials coincided with another flowering of 'Scottishness', the legacy of Scott being expressed in the 'baronial' advocated by Gillespie Graham, all catalysed by the publication of the *Baronial and Ecclesiastical Architecture of Scotland*.

Figure 8.4 Alexander Thomson, no date. Source: Reproduced by permission of RCAHMS (Waddell Collection).

These volumes had been prepared at the behest of Burn and Bryce between 1846 and 1852 by an English antiquarian, Robert William Billings, taking off from where Meikle Kemp had left a similar project on his death in 1844.[41] Baronial became the language of the modern Scottish house, filled with every contemporary gadget that could be invented, and it eventually spread to buildings symbolising authority, such as courthouses. It made little impact on the commercial world, which required a different symbolism. The Victorian belief lay in the 'appropriateness' of a style to represent the function for a building suitably. The whimsy of baronial was unsuitable for banks, clubs or mercantile premises. A cosmopolitan, usually Italianate, approach was preferred for those. Bryce was the outstanding baronial country-house architect, yet his banks were largely Italianate and his public buildings had a whiff of Scottish baroque. A similar, if slightly more sophisticated, approach was taken by David Rhind, architect of the Commercial Bank, who preferred the greater authority of a Renaissance *palazzo* as correct clothing for a bank. Most extensive was the work of the architectural 'factory' of Peddie and Kinnear,[42] whose enormous and efficient output varied from baronial in houses to Scottish in schools and an Italian villa style for their myriad banks.

Such firms were exceptional in their widespread clientele and frequent railway-travelling, for most architects remained regionally or locally based: Walter Newall in Dumfries, Thomas Mackenzie in Forres and the north-east, A & W Reid (who had inherited the William Robertson office) in Elgin, William Stirling in central Scotland, based in Dunblane, and even David Hamilton and Charles Wilson in the west (albeit the latter had a kinship clientele in Perthshire, Dundee and Fife). Local architects were often too dependent upon a single great estate or a small number of local patrons to become the independent consultant to which they aspired, or indeed to earn fees like them. The line between rising to being a true architect and remaining a factor or estate surveyor was perilously thin. The position of the estate architect does not appear to have been long-lasting, for estates moved toward preferring a salaried clerk of works. However, the offer of a house, salary and good standing in rural society was probably sufficient to attract architects into what was seemingly a lesser role. Conversely, independent architectural practices were more likely to be successful where estates were insufficiently large to justify having their own clerk of works and needed to share in obtaining relevant professional expertise.

For national projects, London architects loomed increasingly large. Either it took the form of London architects being appointed to major Scottish commissions – G G Scott to Glasgow University and Dundee's Albert Institute, for example – or their winning one of the burgeoning limited architectural competitions, such as that for Glasgow's Kelvingrove Art Gallery, attractive for its huge prize money. These caused resentment; but it was the brain drain that had the greater impact. Burn had gone south, eventually breaking his partnership with Bryce. An increasing number of architects followed him to London, since there they could find an architectural education in addition to apprenticeship which, until the 1880s, Scotland did not provide. The really ambitious travelled also to the École des Beaux Arts in Paris. When J J Stevenson transferred his office from Edinburgh to London in 1872, it acted like a metropolitan architectural finishing school. William Leiper and John Honeyman went south, and others went to G G Scott's office. Oddly, it was from a London base that James Maclaren (1853–90) designed houses in Fortingall that appear to have been particularly influential upon Charles Rennie Mackintosh (1868–1928). Maclaren was only one of a number of Scots architects who prospered in London and yet undertook the occasional commission back home. Not infrequently, their designs were even more distinctly ethnic for being half-remembered. An even larger number of Scots architects travelled abroad or emigrated, particularly to New Zealand and to Canada.

Most significantly, one of Scott's draughtsmen was Rowand (later Sir Rowand) Anderson, who was judged by contemporaries to be the 'premier architect' of Scotland. Anderson was to be the founder of the Royal Incorporation of Architects in Scotland (RIAS) in 1916. He was responsible for a fundamental shift in the architectural profession by seeking to establish a more formalised architectural education, and by insisting on students learning Scotland's architectural identity through studying ancient buildings in a manner comparable to doctors learning from the study of anatomy.[43] It was this, combined with Anderson's arts and crafts predilections, that led to yet another revival of Scottish architecture, evidenced principally in the country houses of his pupil (Sir) Robert Lorimer (1864–1929), such as Ardkinglas, Dunderave and Hill of Tarvit. In 1889, Anderson persuaded colleagues and sponsors to support the establishment of a school of applied art in Edinburgh, and in 1904 Glasgow followed suit. In that year, Eugene Bourdon arrived at the Glasgow School of Art, bringing to it both Parisian and American experience. With the educational skills now available in Edinburgh and Glasgow, a London or Paris education was no longer necessary. There appears, also, to have been a tendency amongst aspiring provincial architects to seek a job in Edinburgh or Glasgow, where they could also study the new courses there, possibly followed by a stint in London, leading to qualification for membership of RIBA. Nonetheless, in 1906, all of the leading architectural practices in Scotland joined RIBA for the first time.

For those who stayed at home, Scottishness was evolving as an architectural concept. Quite unlike Maclaren's, Mackintosh's designs – his houses, offices and the Glasgow School of Art – did not replicate the past so much as extract and abstract from it in the search of a contemporary Scottish identity.[44]

Figure 8.5 Sir Robert Rowand Anderson, no date. Source: Reproduced by permission of RCAHMS.

As had been the case in the mid-century wave of revivalism, the buildings deployed the most contemporary of technologies. Indeed, Mackintosh's close friend, James Salmon (1873–1924) used concrete only four-and-a-half inches thick for his ten-storeyed office block, Lion Chambers, in Glasgow's Hope Street, which he still hoped would convey a sense of Scottishness, whereas his 'Hatrack' building in St Vincent Street was virtually curtain-walled with only the thinnest veneer of Glasgow Style motifs combined with details from late seventeenth-century Glamis. There was something rather knowing about their designs, for Mackintosh, Salmon and their contemporaries and successors had all had some measure of academic training.

The growing number of local architects was reflected in the growth of local architectural societies or associations – Dundee in 1884 and Aberdeen in 1899 – and all Scottish cities had provided architectural opportunities with the urban restructuring that had followed the relavant Improvement Acts: Glasgow in 1866, Edinburgh in 1867 and Dundee in 1871. A particular result was the formalisation of the post of city architect and the rise of the salaried architect or architect–engineer: John Carrick in Glasgow, William Mackison in Dundee and John Lessels as, specifically, architect to the Edinburgh Improvement Trust. The government had already become accustomed to professional advice from an in-house chief architect relating to public works, ranging from advice on St Magnus cathedral and similar ancient monuments to the design of prisons, the Palm House in Edinburgh's Botanic Gardens, art galleries and other government property.

Rowand Anderson considered that the local proliferation of architectural societies and academies – and even of building contracts – was leading to confusion and a lack of critical effectiveness, and to that end he sought to recreate a unified body not unlike the original Architectural Institute. Thus in 1916, what became the Royal Incorporation of Architects in Scotland was formed, with the duty of 'fostering the national architecture of Scotland'. An incorporation of all the architectural societies, it regularised building contracts, prepared a code of conduct for those they called architects, and set about intensifying architectural education. It also, with RIBA, sought to restrict both the title of architect and the practice of architecture to those trained to new standards which it laid down. These, increasingly, required more and more academic input, in contradistinction to learning through the office. Although the two institutes failed to restrict the practice of architecture, Registration Acts in 1933 and 1938 succeeded in giving legal protection to the title 'architect'. As a consequence, the profession tightened its grip on architectural behaviour, forbidding architects from being developers or earning an income from their profession by anything but being an independent consultant. Yet it was indifferently policed. Although Paterson and Broom were expelled for refusing to relinquish house development, Tom Scott Sutherland of Aberdeen – who continued his role as house developer – remained a member. And no action was taken against William Beresford Inglis, cinema owner and architect, who sold six cinemas to build the 1938 Beresford Hotel in time for the Empire Exhibition.

The interwar period was one of significant change to architectural practice. In 1918, school boards were absorbed into the local authorities, and this led to a significant expansion in the numbers of their salaried architects. The emergence of subsidised housing, which was to become the predominant architectural preoccupation for the next eighty years, was associated with a further expansion in architects working for local authorities. The architectural consequence of technological development – in terms of the steel frame, ever-increasing new materials such as chrome, bakelite, rubber, plastics, etc., and the refinement of prefabrication and mass-production – was profound. Architecture had moved decisively away from being a craft-based technology. The stone skin of iron- and steel-framed buildings gradually became peeled back, as in T W Marwick's 1937 St Cuthbert's Co-operative building in Edinburgh's Bread Street, to reveal a pure curtain wall.

The third change was that the interwar period enjoyed perhaps Scotland's first consumer boom, which brought forward new building types – dance halls, cinemas, swimming pools, ice rinks, showrooms and the ubiquitous bungalow – in response to which some form of specialisation became normal. Architects emerged who, like McNair and Elder or John Fairweather and Son, might concentrate solely on designing a single building type like the cinema. Finally, the period began with an increasing proportion of the architect's training removing from the office and transferring to colleges: design training in an architectural college, and evening technical training in a technical college. That process was completed in 1958 when apprenticeship was finally replaced by a wholly academic five years of architectural training that became a prerequisite for registration as an architect. Generally, architects retained a regional base but some, as before, maintained a national client base founded upon a particular approach or building type. The majority, before World War I, remained in the private sector, and the opening of a new school of architecture in Aberdeen gave north-eastern architects a significant boost.

Commercial buildings in the west were dominated by the office of James Miller,[45] whose series of business chambers and banks in Glasgow took the form of American-influenced stripped classical designs of great quality, an approach shared by E G Wylie in the outstanding Scottish Legal Life building, and by Thomson McRea and Saunders in tall warehouses in Glasgow's Merchant City. Apart from some good banks in Edinburgh and Aberdeen, there was little to match Glasgow's transatlantic aspirations elsewhere in the country. Housing received its impetus from the First Scottish National Building Company, which employed Parker and Unwin's assistant A H Mottram to design their Admiralty suburb of Rosyth in 1917 in an English arts and crafts manner (see also volume 9 in this series, chapter 25). Despite experiments with steel houses and houses that could be constructed by unemployed miners, economics forced Scotland to return to the tenement in 1933, banned in 1918 in favour of the lower-density arts and crafts cottages. Some of the results were bad, but city architect Ebenezer MacRae of Edinburgh achieved an urban neo-Scots at Piershill,[46] and others attempted a

Figure 8.6 St Cuthbert's Co-operative building, Bread Street, Edinburgh. Source: Reproduced by permission of the department of architecture, Edinburgh College of Art.

more sweeping modernism in Glasgow and Aberdeen. Less dense housing, however, was much more likely to take a gentle neo-Scots form.[47]

The arts and crafts had died out quickly, reserved only to churches, church halls and certain high-quality charitable housing. In its place came the new Scottish architecture, or another revival. The Scottish literary renaissance had created a national self-consciousness which achieved political expression in the formation of the Scottish National Party in 1929, and architects followed suit by seeking to create a modern architecture that was truly Scottish. The initiative lay with the private sector – new, small or young practices attempting to survive the depths of the recession (so much worse than in England) – led by Robert Hurd, Basil Spence and Patrick Geddes' son-in-law Frank Mears. Harking back both to the Renaissance and to Mackintosh, they believed that Scottish modernism could be created from an abstraction from precedent to achieve a white geometric architecture which new technology had liberated from period details and confined rooms. Curiously, Scottish architects created far more white geometric 'modern movement' little boxes than would be expected pro rata with England. The fact is that they managed to sell the image of that type of modernism to a much wider range of clients – middle managers, professionals and the like – than their equivalents in England, on the grounds that Scotland must modernise but that it should remain recognisably Scottish. That was in total contradiction to the whole ethos of what became called 'the international style' and carried echoes of Renaissance Scots' rejection of classicism.

The apotheosis of the movement – that had to wait to flower until the Scottish economy revived from rearmament after 1935 – was the 1938 Empire Exhibition. Its master planner was Thomas Tait, Scotland's finest architect, leader of an *atelier* (studio) and designer of the Infectious Diseases Hospital in Paisley and St Andrew's House in Edinburgh. Tait had been the competition assessor for the Bexhill Pavilion, and his partner, Francis Lorne, had worked with Bertram Goodhue in America, so their firm was thoroughly au fait with contemporary thinking. The exhibition was intended to demonstrate Scottish prowess, and the speed of its erection required complete prefabrication. Tait assembled a team of young independent architectural practices which, through some form of symbiosis, managed to create an architectural language of mostly white (sometimes coloured) geometrically emphasised structures, that were intended to represent the Scottish architectural language of the twentieth century. Amongst these young men were Jack Coia (previously a designer of brick Byzantine churches) and Basil Spence, who had proved himself equally adept at white modernism, suburban, and arts and crafts.[48]

World War II intervened, and once it was over, architecture had become substantially more state controlled. Programme – that is to say, delivery of buildings at speed through fabrication – took priority over design. Scotland had ended the war short of an absolute 500,000 houses, and their provision became the priority. Yet although council housing became the predominant output, many other new building types emerged, principally hospitals, university buildings after the Robbins report of 1957, industrial estates and the rebuilding of city centres. Building conditions were often dire after the neglect or damage of the war years, so cities were required to prepare comprehensive development plans which, in essence, plotted the removal of most of their inner areas on the grounds of inefficiency, traffic obstruction and sheer outdatedness. Peripheral estates emerged on the boundaries of most cities, so as to reduce density in the inner cities and make it easier to clear the latter. Although there was an immense amount of worn-out urban infrastructure, much of the demolition had a political agenda: modernisation and the imagery of modernisation as a rejection of the past, for its own sake.

For the first time, public-sector architects became the dominant client, with the inevitable result that – since they were designing public-funded construction – architects were susceptible to even greater constraints on design and cost. As a consequence, speed of completion impelled a lesser use of architects and a greater use of 'package deals': that is, of contractors who proposed predesigned, prepackaged tower blocks, more often than not largely – if not entirely – lacking in architectural input. A similar attitude was taken to that other icon of modernity – the shopping centre. Indeed, in 1963 Dundee demolished one of the largest areas of medieval urban Scotland to survive – its Overgate – in order to build the first enclosed shopping centre in Scotland, designed by Ian Burke, Hugh Martin and Partners (which was sufficiently ugly for it to be demolished thirty-five years later).

Perhaps one consequence of a planning system that now had real sanction was the tendency of increasingly 'British' developers to appoint indifferent English architects who best understood how to make the system work. A number of the older Scots practices that had survived from the 1930s did not survive such loss of patronage. It was arguably only after the establishment of firms like Sir Robert Matthew's in the late 1950s that the position was reversed.

Craft-based design just managed to survive the mania for large-scale prefabrication, but principally on a diet of rare private houses, the patronage of a substantial client and, by 1969, the first stirrings of a growth in the repair of older buildings. Gillespie, Kidd and Coia, for example, the premier postwar practice, survived on commissions of churches and schools for the Roman Catholic diocese – notably the iconic St Peter's seminary, Cardross – with occasional university buildings in England. Peter Womersley and Alan Reiach principally designed private houses. William Kininmonth succeeded in going against the trend by winning a number of central Edinburgh commissions, such as Edinburgh University Union, and the Scottish Insurance building in St Andrew Square. All of these offices were run like *ateliers* welcoming students, but such commissions were becoming rarer.

The public sector became dominant, and by the mid 1970s, over 68 per cent of architects were working for new town commissions, local authorities or government agencies, to the extent that they formed their own distinct professional associations. London, moreover, had – if anything – increased its attraction for Scottish architects, four prominent members becoming presidents of RIBA.[49] It had taken the lead, in 1958, in the academicising of architectural education, and in 1962 it pioneered a study of 'the architect in his office'.[50] In 1968, it awarded its Royal Gold Medal to Jack Coia. RIAS, during the early postwar years, remained quiescent.

However, the construction wave 'broke', as Lord Esher put it, in 1969. An incipient recession combined with an increasing public outcry at the destruction of cities and the growth of amenity bodies established to prevent it. Architecture moved into yet another phase, in which legislation would finance the conversion of older properties into flats, as a result of which there emerged the 'community architect' pioneered by Strathclyde University's ASSIST office, which experimented with ways in which tenements in Govan, Glasgow, could be provided with modern sanitation in 1971. In the following twenty years, works to existing buildings – restoration, upgrading or reinhabiting ruins – became the predominant activity for Scottish architects. As a consequence, skilled conservation became a distinct discipline with its own courses and registration, and it reawakened an interest in traditional materials and techniques, reconnecting part of the profession to the craft side of building from which it had become severed several decades earlier. As in the AIS in the 1850s, architects, builders, craftsmen and clients found themselves members of the same organisations.

Perhaps coincidentally, it was during the 1970s that RIAS revived as an autonomous professional body to the extent that over the following twenty years it became the premier architectural institute in Scotland. In doing so, its agenda was as much public as professional, encouraging media coverage, exhibitions, architectural competitions and founding a publishing house. Its annual convention became the largest of its size in Europe. Balance was provided by its support for technical and legal service, marketing for architects overseas, and the founding of an insurance company. It was perhaps the high point of institutional penetration into the architects' professional life.

Again, perhaps as a consequence – but certainly related to the astonishing revival in the fortunes of Scottish cities – a distinct new wave of architects and architectural design became apparent by the late 1980s and 1990s, encouraged by competitions for lottery-funded buildings and the redevelopment of Glasgow's and Edinburgh's docklands. As had been the case during the Enlightenment, each city had to have its competitive jewels: its theatre, conference centre, museum or prestige tourist building. If RIAS had succeeded in creating a climate for architectural experimentation, it was limited to a certain type of client for certain types of buildings. For mass housing had moved back from the public to the private sector, with greatly reduced architectural input and quality, whereas commercial development – moving increasingly to out of town malls – largely eschewed any architecture whatever. New contracting procedures and the appointment of project managers – particularly consequent upon the Private Finance Initiative – have had the consequence of devaluing the architect's hard-fought-for role as provider of independent advice to that of a subordinate supplier of services. The reduced role for the profession was inevitably followed by a reduced income. Worse, when the putative 'competition' was held for the Scottish parliament, the government excluded all smaller architectural offices and permitted a few of the larger ones to enter only on condition that they were twinned with a non-Scottish firm. Such restrictions flew in the face of the success of the 1989 competition for the Museum of Scotland, which had been won by the little-known Benson and Forsyth. The belief grew that, despite the Scottish parliament's 'architectural policy', the profession did not enjoy the institutional support that it required.

So the portents for the profession, as it entered the new millennium, were mixed. On the one hand, there was a greatly enhanced public interest in architecture, government sponsorship for the Glasgow Lighthouse architecture centre, and the existence of a governmental architectural policy. On the other, the profession had failed to reposition itself in relation to the commissioning of buildings, or to retrieve anything like the position it had enjoyed in the 1960s, let alone the 1860s. Its income was diminishing, and its influence, in the broad mass of construction, was diminishing likewise. Its inherited position in society had slipped substantially, and it was not proving sufficiently fleet of foot to create a new role for a new world. The relevance of its purely academic training was now questioned (particularly by those involved in historic building conservation) and a new competitor in the form of computer-aided design packages had emerged.

However, the profession has always been resilient, and the remorselessness of the acreages of poorly designed houses and shopping malls may well create a backlash in favour of a higher quality of architectural design. If that challenge occurs, it is likely to be a smaller architectural profession that is ready to meet it.

ACKNOWLEDGEMENTS

I owe a particular debt of thanks to Professor David Walker for his endless support and help. Much of this paper was developed jointly with him for an essay on the Scottish architectural profession published in R Bailey's *Scottish Architectural Papers – A Source Book* (1996). Professor Walker's outstanding *Dictionary of Scottish Architects 1840–1940* is now online at www.scottish-architects.org.uk.

NOTES

1. Quoted in MacKechnie, 2000, 159.
2. Dunbar and Davies, 1990, 272–3.
3. MacGibbon and Ross, 1887–92, Vol. V, 127.
4. Dunbar, 1999, 236. Dunbar sees no change between the patterns in the reign of James I and James VI in this respect.
5. See also 'Design and Construction', Appendix 1 in McKean, 2001, and Howard, 1991.
6. Dunbar and Davies, 1990, 269–364.
7. Strathmore, first earl of, *Book of Record*, edited by A H Miller, Edinburgh, 1890.
8. Dunbar and Davies, 1990, 269–364.
9. For the significance of this see Stevenson, 1988.
10. Stevenson, 1988, chapter 5.
11. Cited in Kanerva, 1998, 142.
12. As was the belief of Stewart Cruden. See Cruden, S. *The Scottish Castle*, Edinburgh, 1963, 164–5.
13. See Fawcett, 1994; MacGregor Chalmers, P. *A Scots Mediaeval Architect*, Glasgow, 1895.
14. See Mill, M. Did freemasonry fill the role of Enlightenment clubs in Dundee? Unpublished MSc Dissertation, Dundee University Archives.
15. In the library catalogue at Wemyss. Information from Charles Wemyss.
16. For more about Finnart, see McKean, C. Sir James Hamilton of Finnart: A Renaissance courtier–architect, *Architectural History*, 42 (1999), 141–72.
17. Seton, G. *Memoir of Alexander Seton, Earl of Dunfermline*, Edinburgh, 1882, 63.
18. This is work currently under research by Matthew Davis.
19. McKean, 2001, 147.
20. Howard, 1995. Howard nails this neatly by commenting on the pleasing seventeenth-century paradox of someone being an architect, and capable of great aesthetic imagination, and having high craft skills at the same time. Sir Philip Dowson, founder of Arup Associates, regarded his training as a cabinet maker as invaluable.

21. Colvin, 1986, 173–4.
22. Wemyss, 2003.
23. Gow, I R. *John Douglas*, National Monuments Record of Scotland Exhibition catalogue, Edinburgh, 1989; Kinnear, 2001.
24. Gifford, Edinburgh, 1989. It is quite possible that Adam has been one of the beneficiaries of the Wemyss' habit of lending their architectural books, since one of the early borrowers was his Kirkcaldy father-in-law, William Robertson.
25. Adam, 1980.
26. Lockhart, 2000.
27. McKean, 1999, 15–37.
28. Robert Adam, Robert and William Mylne, John Baxter, John Craig, John Henderson and William Playfair were all educated at the Academy of St Luke in Rome.
29. There are such strong echoes of David Hamilton in David Neave's career as to imply that Neave may have trained for a time with Hamilton, although he first appears in Dundee in the guise of a mason.
30. Aberdeen Civic Society, *Archibald Simpson, Architect of Aberdeen*, Aberdeen, 1978.
31. Rock, 1984.
32. It remains the case in the twenty-first century that builders regard the elevation of their sons to the status of architect as something to be desired, implying an elevation in social rank.
33. Information from Dr David Jones via Prof. David Walker.
34. This is dealt with at more length in the *New Dictionary of National Biography* entries covering David Hamilton, Robert Reid, William Playfair, James Gillespie Graham and George Meikle Kemp.
35. Walker, 1976, 8–32.
36. Information from Dr D W Walker.
37. Sir John Soane's *Lectures on Architecture. As Delivered to the Students of the Royal Academy*, edited by A Bolton, London, 1928.
38. It aborted when the vice-president, William Burn, proposed his junior partner David Bryce for Fellowship; and Playfair, who disliked Burn as a vulgarian, objected on the grounds that Bryce was merely an assistant and that membership was restricted to partners.
39. *The Building Chronicle*, Edinburgh, 1855–8.
40. Stamp, 1999; Stamp, 2000.
41. Billings, 1852.
42. Walker, D W. Peddie and Kinnear. Unpublished PhD thesis, University of St. Andrews, 4 vols, 2002.
43. McKinstry, 1991.
44. Crawford, 1995; Robertson, 1990.
45. Sloan and Murray, 1993.
46. Frew, J. *Aspects of Scottish Classicism in the House and its Formal Setting*, 1690–1750, St. Andrews, 1998.
47. For a full discussion of the period, see McKean, 1987.
48. Edwards, 1995.
49. Professor Sir Robert Matthew, Graham Henderson, Sir Basil Spence and Larry Rolland.
50. RIBA, 1962.

BIBLIOGRAPHY

Adam, R and Adam, J. *The Works in Architecture*, London, 1931.

Adam, W. *Vitruvius Scoticus*, ed J Simpson, Edinburgh, 1980.

Adams, I. *The Making of Urban Scotland*, London, 1978.

Allen, N, ed. *Scottish Pioneers of the Greek Revival*, Edinburgh, 1984.

Beaton, E. *William Robertson 1786–1841*, Elgin, 1984.

Beaton, E. *Scotland's Traditional Houses*, Edinburgh, 1997.

Bailey, R, ed. *Scottish Architect's Papers – a Source-Book*, Edinburgh, 1996.

Billings, R W. *The Baronial and Ecclesiastical Architecture of Scotland*, Edinburgh, 1852.

Brogden, W A, ed. *The Neo-Classical Town: Scottish Contributions to Urban Design since 1750*, Edinburgh, 1996.

Brown, I G. *Scottish Architects at Home and Abroad*, Edinburgh, 1978.

Calder, A. *James Maclaren, Arts and Crafts Pioneer*, Donington, 2003.

Carruthers, A, ed. *The Scottish Home*, Edinburgh, 1996.

Chalmers, P MacGregor, *A Scots Mediaeval Architect*, Glasgow, 1895.

Colvin, H. The beginnings of the architectural profession in Scotland, *Architectural History*, 29 (1986), 168–82.

Colvin, H. *A Biographical Dictionary of British Architects: 1600–1840*, London, 1978.

Crawford, A. *Charles Rennie Mackintosh*, London, 1995.

Cruft, K and Fraser, A, eds. *James Craig 1744–1795*, Edinburgh, 1995.

Davis, M. *Scots Baronial*, Ardrishaig, 1996.

Dictionary of National Biography, Oxford, 2004.

Dunbar, J G. *The Historic Architecture of Scotland*, London, 1966.

Dunbar, J G. *Scottish Royal Palaces: The Architecture of the Royal Residences during the Late Mediaeval and Early Renaissance Periods*, East Linton, 1999.

Dunbar J G and Davies, K. Some late 17th century building contracts, *SHS Miscellany*, XI (1990), 269–364.

Edwards, B. *Basil Spence 1907–1976*, Edinburgh, 1995.

Emmerson, R. *Winners and Losers – Scotland and the Architectural Competition*, Edinburgh, 1991.

Fawcett, R. *Scottish Architecture: From the Accession of the Stewarts to the Reformation 1371–1560*, Edinburgh, 1994.

Fenton, A amd Walker, B. *The Rural Architecture of Scotland*, Edinburgh, 1984.

Fenwick, H. *Architect Royal – Sir William Bruce*, Kineton, 1970.

Fiddes, V and Rowan, A, eds. *David Bryce 1803–1876*, Edinburgh, 1976.

Fleming, J. *Robert Adam and his Circle*, London, 1962.

Frew, J and Jones, D, eds. *The New Town Phenomenon – The Second Generation*, St Andrews, 2000.

Frew, J. Concrete cosmopolitanism and low-cost housing: The architectural career of A H Campbell, *Architectural Heritage*, V (1995), 29–38.

Frew, J. Cottages, tenements and practical idealism: James Thomson and Logie, 1917–1923. In Mays, D, ed, *The Architecture of Scottish Cities*, Edinburgh, 1997, 171–80.

Frew, J. Towards a municipal housing blueprint, *Architectural Heritage*, XI (2000), 43–54.

Gifford, J. *William Adam 1689–1748 – A Life and Times of Scotland's Universal Architect*, Edinburgh, 1989.

Glendinning, M. National internationalist – Robert Matthew and the modern movement, *Architectural Heritage*, IX (1998), 51–62.

Glendinning, M, ed. *Rebuilding Scotland – The Post-War Vision 1945–1975*, East Linton, 1997.
Glendinning, M, MacInnes, R and MacKechnie, A. *A History of Scottish Architecture*, Edinburgh, 1996.
Gow, I and Rowan, A, eds. *Scottish Country Houses 1600–1914*, Edinburgh, 1995.
Gow, I. The problem of the Scottish baronial interior, *Architectural Heritage*, VI, (1995), 49–59.
Graham, C, Harrison, R and Henry, G. *Archibald Simpson*, Aberdeen, 1978.
Grose, F. *The Antiquities of Scotland*, London, 1797.
Howard, D. *Scottish Architecture from the Reformation to the Restoration, 1560–1600*, Edinburgh, 1995.
Howard, D. Scottish master masons of the Renaissance. In Chastel, A and Guillaume, J, *Chantiers de la Renaissance*, Paris, 1991.
Howard, J. *The Scottish Kremlin Builder*, Edinburgh, 1996.
Howarth, T. *Charles Rennie Mackintosh and the Modern Movement*, London, 1952.
Jones, D. A seventeenth century inventory of furnishings at Kinnaird Castle, Angus. In Frew and Jones, 1989, 56–64.
Hunterian Art Gallery. *David Hamilton Architectural Drawings*, Glasgow, 1995.
Imrie J and Dunbar J G, eds. *Accounts of the Masters of Works Vol. 11, 1616–1649*, Edinburgh, 1982.
Kanerva, L. *Defining the Architect in Fifteenth Century Italy*, Helsinki, 1998.
Kay, W. William Adam, new discoveries, *Architectural Heritage*, I (1990), 49–61.
King, D. *The Complete Works of Robert and James Adam*, Oxford, 1991.
Kinnear, H E B. John Douglas's country house designs, *Architectural Heritage*, XII (2001), 1–12.
Leith, C. *Alexander Ellis – A Fine Victorian Architect*, Aberdeen, 1999.
Lockhart, D. Planned villages in north-east Scotland, 1750–1850. In Frew and Jones, 2000, 25–40.
Macaulay, J. *The Gothic Revival 1745–1845*, London, 1975.
Macaulay, J. *The Classical Country House in Scotland 1660–1800*, London, 1987.
MacGibbon, D and Ross, T. *The Castellated and the Domestic Architecture of Scotland*, 5 vols, Edinburgh, 1887–92.
Markus, T A, ed. *Order in Space and Society – Architectural Form and its Context in the Scottish Enlightenment*, Edinburgh, 1982.
McAra, D. *Sir James Gowans*, Edinburgh, 1975.
McKean, C A. *The Scottish Thirties*, Edinburgh, 1987.
McKean, C A. *The Making of the Museum of Scotland*, Edinburgh, 2000.
McKean, C A. *The Scottish Chateau*, Stroud, 2001.
McKean, C A and Walker, D M. A history of the Scottish architectural profession. In Bailey, R, ed, *Scottish Architects' Papers – a Source Book*, Edinburgh, 1996, 18–61.
McKean, C A. Craignethan, the house of the Bastard of Arran, *Proceedings of the Society of Antquaries of Scotland*, 125 (1995), 1069–90.
McKean, C A. 'Not even the trivial grace of a straight line' – or why Dundee never built a new town. In Miskell, L, Whatley, C A and Harris, B. *Victorian Dundee: Image and Realities*, East Linton, 1999, 15–37.
McKean, C A. *Architectural Contributions to Scottish Society since 1840*, Edinburgh, 1990.
McKean, C. Sir James Hamilton of Finnart and the palace of Stirling, *Architectural History*, 42 (1999), 141–72.
McKean, J M. *Charles Rennie Mackintosh*, Edinburgh, 2000.

MacKechnie, A, ed. *David Hamilton, Architect: 1768–1843. Father of the Profession*, Glasgow, 1993.
MacKechnie, A. Evidence of a post-1603 court architecture in Scotland, *Architectural History*, 31 (1988), 107–19.
MacKechnie, A. Design in early post Reformation Scots houses. In Gow, I and Rowan, A, *Scottish Country Houses, 1600–1914*, 15–34.
MacKechnie, A. James VI's architects. In Goodare, J and Lynch, M, eds, *The Reign of James VI*, East Linton, 2000, 154–69.
McKinstry, S. *Rowand Anderson: The Premier Architect of Scotland*, Edinburgh, 1991, 31–43.
McKinstry, S. Rowand Anderson and 'Greek' Thomson: The Glasgow/Edinburgh divide, *Architectural Heritage*, IX (1998), 31–43.
Meller, H. *Patrick Geddes*, London, 1990.
Mylne, R S. *The Master Masons to the Crown of Scotland*, Edinburgh, 1893.
National Art Survey, 4 vols, Edinburgh, 1921–33.
Nuttgens, P. *Reginald Fairlie 1883–1952*, Edinburgh, 1959.
O'Donnell, R. *James Salmon 1873–1924*, Edinburgh, 2003.
Paton, H, ed. *Accounts of the Masters of Works, Vol. 1: 1529 – 1615*, Edinburgh, 1957.
Petersen, C. Robert Smith, Philadelphia builder–architect: From Dalkeith to Princeton. In Sher, R B and Smitten, J R, eds, *Scotland and America in the Age of Enlightenment*, Edinburgh, 1990, 277–99.
Reed, P, ed. *Glasgow: The Forming of the City*, Edinburgh, 1993.
RIBA. The architect in his office, *RIBA Journal*, 1962, 213–19.
Richardson, J R. *The Mediaeval Stone Carver in Scotland*, Edinburgh, 1964.
Robertson, P, ed. *Charles Rennie Mackintosh: The Architectural Papers*, Wendelbury, 1990.
Rock, J. *Thomas Hamilton*, Edinburgh, 1984.
Rogerson, R W K C. *Jack Coia*, Glasgow, 1986.
Sanderson, M. *Robert Adam and Scotland: Portrait of an Architect*, Edinburgh, 1992.
Savage, P. *Lorimer and the Edinburgh Craft Designers*, Edinburgh, 1980.
Sinclair, F. *Scotstyle*, Edinburgh, 1984.
Sinclair, F, ed. *Charles Wilson*, Glasgow, 1995.
Sloan, A and Murray, G. *James Miller*, Edinburgh, 1993.
Stamp, G and McKinstry, S, eds. *Greek Thomson*, Edinburgh, 1994.
Stamp, G. *Alexander Thomson: The Unknown Genius*, London, 1999.
Stamp, G, ed. *The Lamp of Truth and Beauty: The Lectures of Alexander 'Greek' Thomson*, Glasgow, 1999.
Stamp, G. *Robert Weir Schultz*, Bute, 1981.
Stell, G, The Scottish mediaeval castle. In Stringer, K J, ed, *Essays on the Nobility of Mediaeval Scotland*, Edinburgh 1985.
Stell, G and Riches, A, eds. *Materials and Tradition in Scottish Building*, Edinburgh, 1992.
Stewart, M. The earl of Mar and Scottish baroque, *Architectural Heritage*, IX (1998), 16–30.
Stevenson, D. *The Origin of Freemasonry: Scotland's Century, 1590–1710*, Cambridge, 1988.
Swan, A. William Kerr, *Architectural Heritage*, III (1992), 83–92.
Walker, D. The Architecture of MacGibbon and Ross: The background to the books. In Breeze, D, ed, *Studies in Scottish Antiquity*, Edinburgh, 1984, 391–449.
Walker, D. William Burn. In Fawcett, J. ed, *Seven Victorian Architects*, London, 1976, 8–32.

Walker, D, Sir John James Burnet. In Service, A, ed, *Edwardian Architecture and its Origins*, London, 1975.
Walker, F A. Tradition and discontinuity in architectural heritage, *Architectural Heritage*, X1 (2000), 80–90.
Walker, F A. Mackintosh the international nationalist, *Architectural Heritage*, IX (1998), 44–50.
Walker, F A. National romanticism. In Gordon, G, ed, *Perspectives on the Scottish City*, Aberdeen, 1989, 125–59.
Watters, D. *Cardross Seminary: Gillespie, Kidd and Coia and the Architecture of Postwar Catholicism*, Edinburgh, 1997.
Welter, V M. History, biology and city design: Patrick Geddes and Edinburgh, *Architectural Heritage*, VI (1996), 60–82.
Welter, V M. *Biopolis, Patrick Geddes and the City of Life*, Cambridge, 2002.
Wemyss, C. Paternal Seat or Classical Villa ? *Architectural History*, 46 (2003), 109–26.
Weston, R. *Richard Murphy Architects: Ten years of practice*, Edinburgh, 2001.
Youngson, A J. *The Making of Classical Edinburgh*, Edinburgh, 1966.

The *RIAS/Landmark Trust Illustrated Architectural Guides to Scotland* (29 volumes published to date) provide an unmatchable ready reckoner of information, with many hundreds of illustrations per volume. The Penguin volumes of Gifford, J, ed., *Buildings of Scotland* provide invaluable, much more detailed architectural/historical information.

9 The Forth and Clyde Canal

THOMAS J DOWDS

When proposals were laid before parliament in 1760 to construct a navigation 'betwixt the West and East Seas' the proponents of the bill were advocating the largest and most expensive project that Scotland had seen since the disastrous Company of Scotland seventy years earlier. By the time of its completion in 1790, it was regarded as a magnificent achievement that opened a new phase in the economy of the country. To many it combined the best of Enlightened ideas with commerce and industry, and it was at first a compromise between the two, although it was the hard-headed businessmen who ultimately triumphed. It took a generation to complete, largely as a result of unforeseen problems, administrative and financial due to the sheer scale of the project, and practical as difficulties arose with the construction. But these were overcome by ad hoc solutions which were later incorporated into standard business practice, and the canal not only proved that large-scale ventures were possible but provided a framework for organising such schemes and a skilled workforce capable of delivering them.

ORIGINS

Proposals for a waterway between the firths of Clyde and Forth were the first suggestions for a major canal in Britain, although the first Scottish canal was built by Sir Andrew Wood in 1495 to allow the retired admiral to sail to church on Sundays.[1] The earliest mention of the canal dates from the seventeenth century, when a safe passage for shipping was desirable during a period of military and commercial rivalry with the United Provinces. The opening of the Languedoc Canal in 1681 led to suggestions being made for a similar canal in central Scotland, but the cost of £100,000 prevented Charles II from emulating his cousin Louis XIV. Since government was unwilling to undertake the venture, private capital would be required, but the collapse of the Company of Scotland in 1699 with the loss of £200,000 had made Scots reluctant to risk further failure. Daniel Defoe noted the economic advantages of such a canal – 'Irish trade could pass through Glasgow to London via the east coast' – but he added that it was unlikely to materialise before the demands of trade persuaded people to invest.[2]

There was no will to undertake such a scheme that would have no return except in the long term. Prior to 1707, the Scots parliament was dominated by a landowning élite for whom trade tended to be directed towards the export of cattle to England and coarse linen and fish to the continent. With the Union, the security of the cattle trade was guaranteed, and Scots could share in trade with the British colonies. However, the loss of the Scots parliament left noblemen with a diminished role in political life, and they sought other ways to exercise power. Professor Phillipson has suggested that many saw participation in commerce and economic enterprise as a substitute for such political activity.[3]

With more time to devote to their estates, and for reading about developments elsewhere, the landowners started to improve their estates to make them more profitable. Encouraged by a number of societies, mostly based in Edinburgh, information on improvement was broadcast and reflected in the writings and thinking of the 'literati', who advocated involvement in commerce and business as suitable undertakings for members of 'polite society'.[4] This 'Enlightened' attitude applied the new rules of the world of commerce to the interests of the landed classes as a means of ensuring that they maintained their social status: 'Liberty and the interests of Scotland's most ancient families ... could only be preserved if the interests of landed society, like those of the commercial world, were regulated by market forces.'[5]

While this may account for the will to invest in commercial ventures in the middle of the eighteenth century, it has also been argued that a sense of duty had encouraged the gentry to participate in earlier schemes, most notably the Company of Scotland. What is certain is that the eighteenth century provided greater opportunity for such involvement as part of the belief in Improvement, the intellectual roots of which have been traced back to the seventeenth century.[6]

The fundamental obstacle to development was the lack of finance. As Glasgow began to enjoy the benefits of trade with the English colonies, particularly in tobacco, the idea of linking the east and west coasts through the Clyde-Forth valley was again raised. In 1723 a group of Glasgow merchants, anxious to encourage the transport of goods from the colonies to Europe, commissioned Alexander Gordon to examine the feasibility of a canal. His estimated cost of £200,000 – equivalent to the entire circulating capital of Edinburgh[7] – was in excess of what had been anticipated and the scheme was abandoned.

By 1750, however, the situation had changed. After the Jacobite Rising of 1745–6 the government was showing a greater interest in Scottish affairs, especially where schemes would lead to 'civilising and improving the Highlands',[8] and there was a suggestion that the Prime Minister, William Pitt the Elder, was prepared to make funds available as a means of securing colonial trade from French interference during the Seven Years' War (1756–63). However, Pitt's fall from power in 1761 ended hope of government finance. Since the 1660s Glasgow had experienced expansion as a result of the colonial trade. In 1750, Glasgow was a medium-sized town with a population of around 20,000. The main industry was textiles, but it had also cornered the market in tobacco and hides, producing a class of wealthy merchants – the tobacco barons – who wanted luxury goods that had to pass through the ports on the east coast. The tobacco industry created work opportunities in the city, and the growing labour force relied upon grain from the rich agricultural areas in the east. Since the chief market for the re-export of tobacco was Europe and Glasgow's imports traditionally came by sea from the west coast, being unloaded at Port Glasgow or Greenock and carried by lighters up to the city, the tobacco merchants sought a means of having their produce carried east to market more easily and cheaply. It was realised that a canal from the Clyde to the Forth would not only provide such a route, but would allow Glasgow to bring grain from the area through which the navigation passed, and coal from pits outside the city. Interest was such that Lord Napier of Merchiston paid for two engineers to survey a route from Abbotshaugh on the River Carron to Yoker Burn on the Clyde in 1762. The report by Robert Mackell and James Murray was favourable and a company was formed that raised subscriptions of £30,000 in four days. Although the report was not acted upon it began a period of intense activity.

Lord Kames approached the Board of Trustees for the Encouragement of Fisheries, Manufactures and Improvement with the idea of a canal, and John Smeaton was commissioned to conduct an 'exact' survey in 1764. His report suggested two routes: one from Carron, through Dullatur Bog to the Kelvin, thence to the Clyde at Yoker Burn, the other from the Forth north of Stirling to the Kelvin. The former was chosen as not requiring as many locks and hence being cheaper. In 1766 the *Scots Magazine* publicised the success of canals in England, and in the same year Robert Whitworth produced his

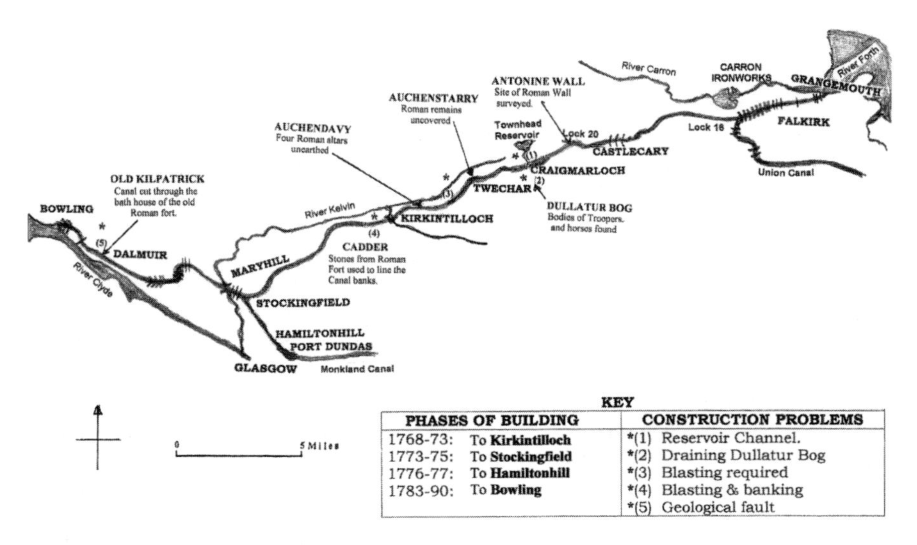

Figure 9.1 An overview of the Forth and Clyde Canal, showing phases of construction. Source: Thomas J Dowds.

book 'Advantages of Inland Navigation', extolling the virtues of this mode of transport. On 15 January 1767 the dean of guild informed the council in Glasgow that the Merchants' House had appointed a committee to promote 'a canal or cutt betwixt the firths of Forth & Clyde to come near the City at Broomielaw'.[9] Within days a petition was laid before parliament. Thus a combination of economic factors and a desire to 'further trade to the advantage of the city' led to the proposal for a four-feet deep canal to the Forth.[10] On 11 March, the MP for Glasgow, Lord Frederick Campbell, and Lord Mountstuart introduced the bill, and five days later it was given its first reading. At that point, Edinburgh interests intervened with suggestions of a larger canal.

CITIES IN DISPUTE

Edinburgh objected to the creation of a 'small canal' and the *Scots Magazine* in March 1767 questioned Glasgow's rush, commenting on Smeaton's original suggestion that the navigation should be 'the noblest work of the kind ever executed'.[11] It also pointed out that the Board of Trustees for Fisheries, Manufactures and Improvement had commissioned a second report from Smeaton and suggested suspending action until this was made public. The *Glasgow Journal* carried a detailed rebuttal over two weeks (28 March and 4 April), insisting that as the major user of the canal, Glasgow, had the right to determine the requirements of the canal and questioned whether a deeper canal would benefit agriculture in the surrounding districts. The readiness of Glasgow merchants to subscribe to the venture corroborates Professor Devine's claim that it was 'related to the tobacco men's need to link quickly with European markets'.[12] On 6 April the *Edinburgh Evening Courant* introduced the idea that the canal was not just an economic venture but was a matter of national pride. Arguing that commerce was of no concern to the capital, which was the seat of 'polite society', and that a 'grand canal' would attract the interest of noblemen and gentlemen, the Glasgow proposals were dismissed as 'a canal as narrow as their view, sordid as their dispositions, and shallow as their understanding – a canal that has been justly called a CUT between Carron and Glasgow, – a ditch – a gutter – a mere puddle – proper indeed for pedlars to puddle in'.[13]

This can be seen as merely an example of the rivalry that was becoming increasingly common between the two cities as Glasgow emerged as a competitor with the capital. National leadership was seen as being the role of Edinburgh, whose leading citizens had historically been the prime movers in matters of state.[14] It is also a criticism of the merchant class, who were seen to be aspiring to ideas above their station and usurping the role of the nobility by taking the lead in matters of national importance. This critique also served as a rallying call to the gentry to become involved in this scheme of Improvement. The Enlightenment had advocated that civic responsibility lay with the nobility as the natural leaders of society who had a duty to encourage schemes that were of importance to the community and nation. Writers such as Francis Hutcheson and David Hume[15] advanced the theory that civilisation passes through a number of phases, the last and most advanced being that of the commercial world where, put simply, the benefit of the greatest number was the purpose of society. Their thinking was to persuade the upper classes that involvement in the world of business was not only socially acceptable but indeed was the virtuous thing to do.

Argument continued when Sir Lawrence Dundas of Kerse, MP, persuaded the Commons to reject the Glasgow bill in favour of a canal twice as deep along the same route, but entering the Forth at Grangemouth. The change in the eastern terminus was seen as a means of developing the little village, owned by the Dundas family,[16] as a rival to Leith, angering Carron Company, whose shipping interests were threatened. Glasgow condemned the new plan as grandiose and unnecessary, while Edinburgh proclaimed the military advantages for safely transporting ships and troops to meet any threat from the continent, in addition to benefits that would accrue along the path of the navigation. In the midst of the controversy, on 11 April 1767, Smeaton produced his second report, advocating a £100,000 canal that would be seven feet deep and which could accommodate sloops of 40 tons. This convinced Edinburgh that the 'grand canal' was viable and that the resources were available for it, suggesting the capital represented the 'Wise men of the East' to whom the 'Fools of the West' should yield first place.[17]

Edinburgh was also using economic arguments to garner support for its plan. Grain and other agricultural goods, it was argued, would be the main cargo carried, and a deeper canal would be more suited to this trade. It also suggested that a deeper canal would open trade between Ireland and England, and illustrated this with a wide range of examples from a number of European centres.[18] The argument that the whole country could benefit by having a deeper canal won over the majority of the burghs, especially those producing grain. The Convention of Royal Burghs accepted a proposal by Midlothian that meetings be held throughout Scotland to rally opposition to the Glasgow scheme. As a result, parliament received petitions from Perth, Haddington, Selkirk, Linlithgow, Stirling, Elgin and Banff asking that the Glasgow bill be delayed and a new bill for a larger canal introduced in its place.[19] Smeaton was commissioned to examine a canal to take vessels of 60 tons and in October reported that a depth of seven to ten feet would be required. Edinburgh immediately pressed for a ten-feet deep canal and urged the Convention of Royal Burghs to seek financial assistance from the Commissioners of the Forfeited Estates. No funds were made available, but despite the estimated cost of £147,337, the new proposal won support from most of the burghs.

Faced with this united front, Glasgow withdrew the 'small canal' bill, in return for a cut being made to Glasgow and £1,200 as compensation for the costs of the earlier bill. A similar arrangement for a cut to Bo'ness and £300 compensation won over the merchants there, while a promised cut to Carronshore allayed the suspicions of the Carron Company, finally ending opposition. On 14 May 1767, Thomas Dundas, MP for Edinburgh, introduced the bill to the Commons, claiming it was 'of great advantage to the kingdom in general by reducing the price of land carriage'.[20]

FINANCE AND ADMINISTRATION

Canals required individual acts to legalise the construction, which necessitated the formation of a committee in London, to be close to Westminster to deal with members of both houses. Glasgow already had representatives in London, and when a new company was formed, the council sent Provost Murdoch to attend meetings to look after the city's interests. The participation of three cities meant the creation of three committees, each requiring copies of all the documentation, which in turn generated a great deal of inter-committee correspondence.

When subscription lists were opened, the shares were quickly taken up, and subscribers included most of the nobility of Scotland, together with the provosts of Edinburgh and Glasgow, many landowners and several of the Edinburgh literati, including James Hutton, the 'Father of Geology'.[21] Of the Glasgow tobacco barons only Alexander Spiers is named, although his £5,000 subscription was most likely on behalf of others.[22] The involvement of so many landowners, representing aristocratic and legal families, is an indication of the higher incomes that were arising from the improvements in agriculture. Increased wealth could be used for investment, and the canal offered such an opportunity. However, as it was realised as early as 1768 that there was little likelihood of dividends in the first ten years, it seems unlikely that those who subscribed were seeking immediate profit. In true Enlightened fashion, the majority of investors were content to delay returns in the knowledge that the enterprise was to be of benefit for all, and themselves in the longer term.

The financial and administrative problems faced during construction forced the company to experiment with new methods of organisation. Smeaton's original £150,000 estimate was quickly subscribed when the list opened, and a joint-stock company formed to administer the funds. It was agreed to make calls of up to ten per cent of the subscription at three-monthly or longer intervals to finance each stage of the operation.[23] However, not all calls were honoured, and Carron Company used its fifteen shares as a lever to coerce the canal company into making a cut to Carronshore to meet their shipping interest.[24] Even Glasgow City Council, one of the best payers, was frequently late in meeting the calls. As a consequence, the company's proprietors sought ways of enforcing payment when reminders and persuasion failed and gained greater powers, including through court action.[25]

The financial crisis of the 1770s resulted in an alteration in the power structure of the canal company. The collapse of Douglas, Heron and Company in 1772 caused the collapse of thirteen banks and severely damaged confidence in all financial institutions. The canal company felt threatened, particularly when it became clear that subscriptions would not cover the expenses already incurred. In an effort to solve the cash-flow problem, the canal company issued bonds guaranteeing five per cent against future income. This was psychologically unsound, as it was the same method the Ayr Bank had used unsuccessfully in trying to stave off failure. Equally damaging was the fact that several of the canal company's directors had been involved in the failed bank.[26]

At the same time, Glasgow was suffering an economic crisis. The dispute with the American colonies had reduced the trade in sugar and tobacco, undermining the financial position of the city and removing the main reason for advocating the canal.[27] Construction by 1773 had reached Kirkintilloch, with little prospect of further development as Glasgow refused to venture further funds without a guarantee that the western terminus would be in the city. No guarantee was given, and with the company owing £19,600 to the Royal Bank of Scotland and a further £70,000 needed to continue the operation, work ceased. In an effort to make the completed section pay and to give the benefit of the access the canal afforded, it was opened from Grangemouth to Kirkintilloch in 1773, but income failed to exceed £4,000, well below the cost of maintaining the waterway.

Appeals by Sir Lawrence Dundas and George Chalmers, the Edinburgh agent, failed to encourage the investment needed to complete the navigation. However, it seems money was not the main obstacle, as Glasgow merchants raised a private subscription to bring the canal to Stockingfield by November 1775, and to Hamiltonhill, nearer the centre of the city, in November of 1777. From 1777 to 1783 the canal languished and shares stood at 50 per cent below par. Professor Campbell has argued that the failure to move work forward was an example of the contemporary financial stringency that had led James Watt to suggest a cheaper route for the Monkland Canal.[28] However, the apparent caution of the Glasgow merchant community, as exemplified by their earlier reluctance to accept the 'large canal' plan, seems more likely the result of their desire to have the western terminus at the Broomielaw rather than a shortage of cash, which was readily available for the extensions into the city.

Repeated requests for government assistance were rejected as Westminster adhered to its strict policy of non-intervention. However, war with France from 1778 to 1783 served to highlight the usefulness of the navigation in providing a safe route between the east and west coasts, and the Canal Company was quick to exploit this advantage. The duke of Queensberry and Thomas Dundas, with others, met Lord Salisbury in 1783 to seek finance to complete the canal, and the following year the Prime Minister, William Pitt the Younger, agreed to consider their request. In an astute political move, the Canal Company donated £200 toward the government's scheme to resell the annexed estates of exiled Jacobites, and the Barons of Exchequer were later instructed to advance the company £50,000 from the Forfeited Estates Fund to enable the completion of the canal, in order to provide 'a safe transit for vessels across Scotland'. The exchequer took ownership of 2,999 shares and the Act of 1784[29] directed that dividends due from the navigation be used for road and bridge building. The Canal Company was also obliged to deliver an abstract of its books to the treasury each year.

Figure 9.2 Bowling basin. The canal entered the Clyde at Bowling, which became an important harbour where cargoes were unloaded from large seagoing ships to the canal vessels. This picture shows the famous puffers that plied the Firth of Clyde and to the Western Highlands and Islands. The one in the foreground is the *Texan*. Source: William Patrick Library, Kirkintilloch.

With these funds work resumed, and the navigation was cut to the west but by-passed Glasgow to join the Clyde at Bowling. Financial considerations also resulted in a change in attitude to the country's heritage. Where previously the finding of Roman remains had caused work to cease, in order to allow the recovery of artefacts, the need to progress with speed meant the destruction of later finds. At Cadder a Roman fort was discovered, but the company not only drove through the remains, it recycled some of the Roman stones to create the canal's towpath. On the discovery of a fort in the path of the navigation at Old Kilpatrick, there was no hesitation in driving the construction through the reputedly well-preserved bathhouse. Economic necessity had relegated the Enlightened ideas of the earlier stage to a poor second place, and no opportunity was given for more than the quickest description of the remains. The finds, however, sparked an interest in the Roman occupation, and John Anderson gave a series of lectures based on the finds.[30] One feature that brought pride to people in Glasgow was the magnificent and costly Kelvin aqueduct, designed by Robert Whitworth to carry the canal over the Kelvin.

By July 1790, the navigation was completed and the Forth and Clyde Canal joined the east and west seas. To mark the occasion, a group representing the Canal Company and Glasgow Council took part in ceremonially emptying a hogshead of Forth water into the Clyde on 28 July. A month later the sloop *Agnes* became the first vessel to sail from Leith to Greenock through the waterway, so formally opening the navigation. But work was not quite complete. Hamiltonhill was discovered to be an unsuitable terminus, as it was badly served by roads and was a mile from Glasgow. The city raised further funds to build a new basin, Port Dundas, half a mile closer to the city. A new road, Port Dundas Road, was constructed to take goods into the city, saving the cost of transport from Hamiltonhill.

Meanwhile, the Monkland Canal had been started in 1770 by James Watt to bring coal from the rich Lanarkshire deposits to Glasgow, and by 1790 the canal entered the city but coal had to be hauled over the ninety-feet hill at Blackhill, adding to the cost. The canal operated at a loss and its shares became worthless, until William Stirling bought the entire concern for £500 in 1782. He was a shareholder in the Forth and Clyde Canal and obtained an act to extend the canal to a new basin at St Rollox and to join it to the city by a new road – Stirling Road – before later building locks at Blackhill. In 1793 he obtained a further act to link the Monkland with the Forth and Clyde by the 'Cut of Junction'.[31] Thus Glasgow was finally linked to the Forth in the east and the coalfields of Lanarkshire to the south-east.

The canal proprietors had been accustomed to having to keep three centres informed of the company's activities. Glasgow was the prime mover and pivotal to the scheme, but Edinburgh, as the location of the Barons of Exchequer as well as many significant shareholders, had gained almost equal importance. From the beginning, however, London had a key role to play. Since acts of parliament were required to allow work to progress

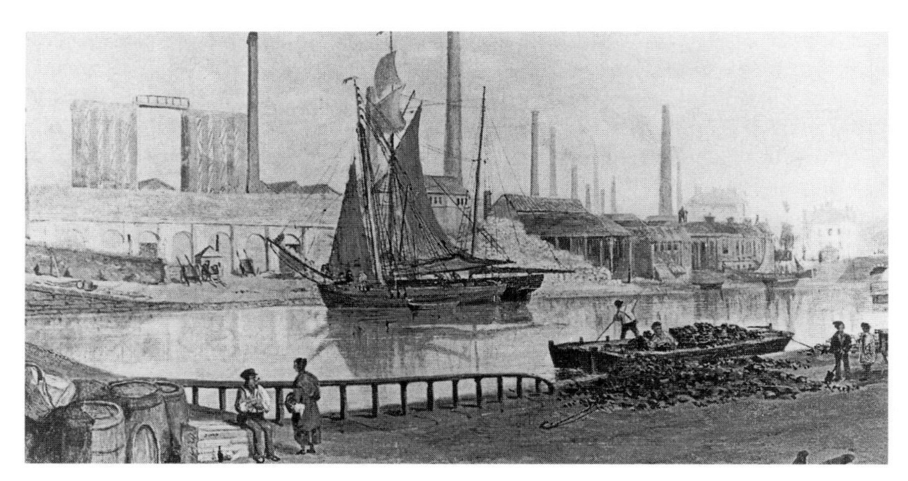

Figure 9.3 Port Dundas *c*1890. The meeting of the Forth and Clyde and Monkland canals at Port Dundas created one of the busiest harbours on the waterway. Large sailing vessels carried passengers as well as cargo, while the small scows and lighters handled bulk cargoes like coal. The harbour and canal attracted several industries, as is clear from this engraving. Source: William Patrick Library.

and to authorise any changes to the plans, a crucially important group of committee members had to be present in the city to maintain contact with the members of both houses. This meant that three copies of everything were required to allow each of the centres to be kept informed of developments. By the Act of 1783, the Canal Company was obliged to tender an abstract of its books to the treasury each year. Patrick Colquhoun, a former lord provost of Glasgow and the leading administrator of the company, copied the parliamentary system of monthly revenue analysis and end-of-year accounts, and introduced double-entry book-keeping and a general account current to improve the efficiency of the information,[32] effectively bypassing Edinburgh. So successful was the system that it came to be adopted as standard business practice by other large-scale undertakings and was incorporated into the acts granting charters to the banks in 1844.

The canal brought together the different classes, nobility and middle class, landowner and merchant, in a way that had not been possible half a century earlier. Involvement in the formation, organisation and running of the company conferred a greater social acceptability to the rising middle class and helped weaken class barriers. Among the Glasgow merchants actively involved in the canal venture were George and Adam Buchanan, Henry Glassford and Patrick Colquhoun, while one of the promoters of the 'small canal' and subscriber to the company, Alexander Spiers, married into the Dundases of Kerse.[33] True, the increasing wealth of the merchant and lawyer enabled him to purchase an estate and become a member of the landowning community and share in the political power that this conferred, but it did not always bring acceptance by the old-established landowning families. The opportunities for social intercourse provided by the Canal Company and similar ventures were important in giving both classes a common goal.

LABOUR

In the early stages of construction, the company relied on locally recruited labour, and this had been seen as one of the benefits to the local economy. However, the work was hard and dangerous and men tended to drift away, especially at harvest, leaving the construction at a standstill. Letters from landowners to the proprietors expressed concern that the building of the navigation might lead to an increase in the day-labour rates in the capital, and it was agreed that the labourers' wages would be set at 'not more than 10 pence a day' so as not to attract agricultural workers away from the land.[34] During the initial phase of cutting, there were 1,048 men employed by the company, and this later settled at around 700. In addition to the navvies, the company required to hire quarrymen, carpenters and masons as well as supervisors, and Smeaton expressed surprise that the last were paid more than their counterparts in England.[35]

Because of the unreliability of local labour and the need to train each new batch of workers, work progressed more slowly than had been anticipated. The hiring of non-local labour, from the Highlands at first and later from Ireland, was seen as a solution to this problem. Despite the first such recruits also being tempted to leave the construction to seek more familiar agricultural work,[36] the system succeeded, and by 1790 around half the workforce was Irish – the origin of the term 'Irish navvy'. The company realised that the creation of a full-time workforce was the most efficient way of operating, and by 1790 it had created a body of highly skilled workers who were to find employment in making roads, digging canals and building railways into the next century. As industrialisation progressed the need for experienced labour grew, and the Irish navvy found himself in demand throughout the country.

The canal passed through several different rock formations, and this made for slow progress at times as the whinstone required blasting to allow the navvies to progress. At Cadder Bridge, Inchbelly and Craigmarloch there was a need for particularly high banking, while Dullatur Bog required a team of fifty workmen to constantly rebuild the banks that sank into the bog. The bog also yielded the bodies of troopers from General Leslie's army who had drowned escaping from the battle at Kilsyth in 1645. Work stopped while the bodies were examined by experts and then removed for proper burial.

Work was slow and dirty, skills had to be learned quickly and the labour force had to obey strict labour discipline.[37] Having made the cut, men were required to level off the sides and bottom and then line these with a paste of sand and gravel mixed with clay, with a final layer of clay to make it waterproof. This was done by 'puddling', trampling down a layer ten inches deep and repeating the operation after each layer had dried out for two or three days, until a depth of up to three feet was obtained.[38] Accidents were common, and in an effort to reduce the cost of surgeon's fees the proprietors made a subscription of £200 to the newly created Glasgow Royal Infirmary in return for having its injured workers treated there free of charge.

The career of civil engineering was given impetus with the construction of canals, and the Forth and Clyde became a training ground for many of the leaders of the new profession. The habit of having a well-known consulting engineer led to several being associated with a number of canal projects. Robert Mackell was resident engineer on the Forth and Clyde, and James Watt, who had been involved with the Forth and Clyde, also surveyed the Monkland and the Crinan canals, Thomas Telford was involved in several canals, as well as being consulting engineer to the Exchequer Loans Commission, assessing transport projects. Robert Whitworth, the designer of the Kelvin aqueduct, together with Hugh Baird, resident engineer on the Forth and Clyde, both participated in the construction of the Edinburgh Union Canal. The Glasgow, Paisley and Johnstone Canal employed both William Jessop, who was to work on many English canals, and John Ainslie, who also engineered the Edinburgh Union Canal.[39] It is possible to trace the careers of the consulting engineers, the surveyors and the resident engineers whose names are recorded in later projects, but the nameless navvies who accompanied them have left only their work as a memorial to their presence.

IMPACT OF THE CANAL

From the opening of the Forth and Clyde to Kirkintilloch it was clear that a number of benefits would arise. Employment opportunities were created almost at once: lock keepers were required to operate the locks that allowed boats to pass through; stablehands to look after the horses that were the motive power of the barges; and maintenance crews to keep the waterway open.

Local building industry and quarrying received a boost, as houses were built for the lock keepers at all locks, as well as stables at regular intervals along the waterway, horses being changed every four miles on passenger boats.[40]

Kirkintilloch, which became the 'canal capital' in the late eighteenth century, experienced a doubling of its population from 2,640 to 5,888 in the half century after 1773, which the minister there attributed to the provision of employment.[41] It is interesting to note the correlating increasing increase in population in Glasgow with stages in the development of the waterway system:

28,000 in 1761, when Napier commissioned his survey
40,000 in 1777, when the canal reached Hamiltonhill
65,000 in 1790, when Port Dundas was opened
110,000 in 1811, when the Glasgow, Paisley and Johnstone Canal opened.

Figure 9.4 Horse-drawn barge. Horses were initially the chief source of power on the canals. The photograph shows a barge laden with timber, latterly one of the main cargoes, being drawn by two Clydesdale horses, which are having a well-earned snack. Source: History Research Centre, Callendar House, Falkirk.

Figure 9.5 Old Stables, Auchinstarry. The horses that pulled the boats along the canal required to be changed regularly to keep them fit, and stables were built at regular intervals to house the horses. Work for stablehands was provided by these stables until the canal fell into disuse. This one at Auchinstarry was demolished in the 1960s. Source: William Patrick Library.

Falkirk and Grangemouth were to experience similar population growth, the latter as the eastern port developed and the former when the Edinburgh Union Canal was opened.

When passenger boats were introduced on the canal, coach companies were quick to realise the opportunity this created for expanding their services. Coaches met passengers alighting at Falkirk and Kirkintilloch and conveyed them to more distant parts. By 1795, Perth, Aberdeen and

Figure 9.6 Aerial photograph of Kirkintilloch basin, 1930. Kirkintilloch was a major centre on the canal and attracted industry. This photograph shows the industrial and housing developments, as well as transport links, that straddled the banks of the canal. Source: William Patrick Library.

Figure 9.7 Canal passenger boat. The engraving shows one of the 'Swifts' used to carry passengers on the canal. The boat is pulled by a team of horses, and the rider wears a distinctive uniform. The curved prow was not decorative but was a sharp blade to cut the towrope of any barge that was too slow to let the passenger boat pass. Source: William Patrick Library.

Edinburgh were within reach, using canal and coach, and a new era of communication began.

The canal also spawned new industries, the most obvious being shipbuilding. The building of subsidiary canals, wharves and warehouses along the canal bank to stimulate industry was actively encouraged.

Timber imported from North America and northern Europe could be brought into the canal and stored in timber basins constructed at several points, particularly Firhill, Glasgow, and Kirkintilloch. It was natural for shipbuilders to set up business close to the source of their raw material, and Hays at Kirkintilloch was one of the earliest specialists in building vessels for canal traffic. An added advantage was that the ships could be launched directly into the canal. Likewise, timber-working firms were attracted and set up sawmills and finishing factories beside the basins. The presence of a plentiful supply of water and the convenience of bringing raw material in and sending the finished commodities out was particularly attractive to the brewing and distilling industry. As a result, a number of breweries and distilleries were sited along the banks of the Forth and Clyde, at Glasgow, Kirkintilloch, Wyndford and Falkirk, to slake the thirst of the growing industrial population of the central belt and beyond.

As had been predicted, agricultural goods were the chief commodities carried on the canal. With the growth of the centres of population the demand for foodstuffs increased, and the farming areas through which the canal passed were able to meet this demand by means of a cart boat, an early form of roll-on-roll-off boat that allowed laden carts to be conveyed to towns, especially Glasgow. One great benefit of this was that it made fresh food

Figure 9.8 Launch of SS *Nelson*, 1893. Shipyards developed along the canal to build the vessels needed by the transport companies. J & J Hays at Kirkintilloch was one of the earliest, and this photograph shows the launch of the *Nelson* from that yard in 1893. Note that the boat is being launched side-on to the waterway, and that the event has attracted a large crowd. Source: William Patrick Library.

available to town dwellers: 'Dairying gained substantially from transport improvements that allowed quick and safe traffic in perishable goods.'[42]

As well as vegetables and fresh fruit, bulk cargoes of grain were taken to distilleries and the towns by barges. This was of benefit not only to the consumers in the urban areas, but also brought financial reward to the landowners along the length of the navigation.

Two groups to whom the canal system was extremely useful were the owners of quarries and coal mines. Coal and stone were increasingly in demand for building and as domestic fuel, as well as for use by the industrial concerns that were developing in Glasgow, but the cost of transportation was prohibitive. Coal masters, like the Baird family, were prominent in the development of the Monklands link to the Forth and Clyde canal, and their interest continued into promoting railways in the following century. The canals gave the city access to the rich deposits in Lanarkshire and reduced transport costs, making coal much cheaper, while earning great rewards for the mine owners. To allow coal to reach the barges, owners copied the action of the quarry owners and laid iron tracks from the pits to the canal, along which wagons transported the cargo. The growth of ironworks was encouraged by the canals, and the early nineteenth century saw a rapid growth in this sector, coal masters like the Bairds being among the early iron masters.[43]

By 1793, the canals had proved their viability. The Forth and Clyde was realising an annual revenue of £22,000 and its shares stood at 325 per cent. The Monkland was even more impressive: the original £125 shares

Figure 9.9 Wyndford lock. Lying east of Castlecary the wooden lock gates are clearly seen in the foreground. On the left is the lock keeper's cottage, and through the trees in the centre Wyndford Distillery can be made out. Several distilleries were set up along the canal because it gave easy access for raw material deliveries and the export of casks. The distillery no longer exists, but the cottages have been turned into a restaurant. Source: History Research Centre, Callendar House.

had fallen as low as £25 in 1781–2, but by 1793 it had reached an incredible £1,550.[44] The optimism of the proponents of 1767 had been fully vindicated and investors were beginning to see great returns.

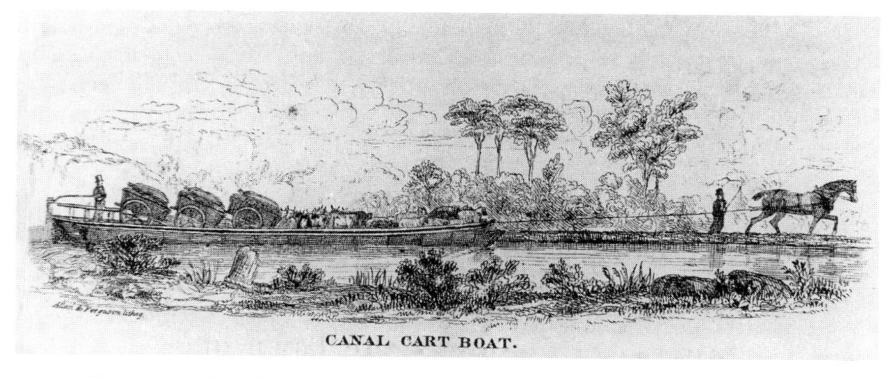

Figure 9.10 Canal cart boat. Farm produce was taken to towns from farms along the canal, and these boats were designed to allow laden carts to be easily wheeled on and off the vessel – an early type of roll-on-roll-off ferry. This enabled towns to have access to fresh food. The engraving shows cattle and produce being taken to market in one of these boats, pulled by one horse. Source: William Patrick Library.

Soon after the Forth and Clyde was started, a ship canal across the Kintyre peninsula, to avoid the dangerous voyage round the Mull, was proposed and James Watt surveyed a route in 1771, but no action was taken. The success of the Forth and Clyde in opening coast-to-coast transport prompted the duke of Argyll and others to resurrect the scheme in 1790. John Rennie was put in charge of the building of the Crinan canal, but the subscribed funds of £120,000 did not meet the cost of the fifteen locks required to allow ships passage. Government eventually funded the canal, and took over its running in 1848. Despite its lack of commercial success, the canal was popular with yachtsmen and was used by puffers for years.[45]

When the benefits of the Forth and Clyde Canal became clear, other schemes were launched to share in the anticipated economic boom that would accompany the making of a canal. During the 1790s plans were launched to make a number of canals – the Peterhead and Deer, Elgin and Lossiemouth, Montrose and Brechin, Cupar and Eden, Kilmarnock and Troon, Paisley and Troon, and Muirkirk and Ayr – none of which got beyond the survey stage.[46]

The canal linking Inverurie to Aberdeen harbour was rather more successful. John Rennie surveyed the route and work began in 1797, but he soon found that the eighteen-and-a-half miles of waterway required seventeen locks, making it very costly. Work went ahead over the next ten years, and the Aberdeenshire Canal was opened in 1807. While it provided an outlet for the export of agricultural produce and granite paving slabs from the quarries close to Inverurie to London and abroad,[47] it did not realise the anticipated economic benefits and was sold to the Great North of Scotland Railway. The railway company drained the canal and used it to lay the permanent way near Dyce.[48]

The prospect of exporting coal from mines near Glasgow to Ireland and importing Irish grain through the port at Ardrossan while developing local trade and industry brought Lord Eglinton, owner of Ardrossan, and mine owners together to promote a canal through Paisley and Johnstone to the coast. A route was surveyed by John Rennie in 1791 and later, in 1804, by Rennie and John Ainslie, before work began in 1807, financed by Lord Eglinton and the Govan Coal Company. Thomas Telford was appointed consulting engineer, and work began with plans to open branches to Clyde Iron Works and the Hurlet Coal and Alum Works. The deepening of the river Clyde, however, made the scheme redundant and the navigation never reached beyond Johnstone.[49]

Edinburgh also realised the potential of creating a navigation from the capital to the Monkland coalfields to compete with the Forth and Clyde, and from 1793 to 1807 a number of surveys were carried out by John Ainslie, John Whitworth and John Rennie to assess the viability of the project. Finally, Hugh Baird persuaded the magistrates of Edinburgh that the best interests of the capital would be served by linking Edinburgh to the Forth and Clyde

Figure 9.11 'Unique' bridge, Kirkintilloch, *c*1950. The picture shows a goods train passing under the aqueduct carrying the Forth and Clyde Canal and over the Luggie Water. Source: William Patrick Library.

canal at Falkirk, which had the advantage of 'bringing an ample supply of water … without any expense to the community'.[50] In 1817 an act was passed to create a five-foot deep canal from the city to lock 16 on the Forth and Clyde Canal, thereby creating a transport link between Scotland's two largest cities. The Edinburgh Union Canal opened in 1822 but was never a commercial success, and within a few years it faced competition from the Edinburgh and Glasgow Railway, whose line ran parallel to the waterway for much of its length.

Scotland's last completed navigation was the Caledonian Canal, which was first surveyed by James Watt in 1773 for the Commissioners of the Forfeited Estates, but his report was ignored until the French Wars of the 1790s. It was then realised that a ship canal through the Great Glen would improve the mobility of the British fleet and offer a route to oppose any threat of invasion. Government was aware that the marquis of Cornwallis had used the Grand Canal in 1798 to move his army from Dublin to the west coast of Ireland to defeat General Humbert's invasion force at Ballinamuck.[51] There was also a fear that the Highlands would become depopulated, and it was felt that the making of a canal would halt this process. Thus in 1803 Thomas Telford together with William Jessop began work on the great ship canal, although the two differed on the expected cost, Telford's estimate of £350,000 being considerably less than Jessop's more accurate figure of £434,000. Fear of invasion made government less cost-conscious, and the scheme progressed, giving employment to up to 3,000 navvies.[52] The canal was opened when a steamboat made the first passage on 23 October 1822, long after the threat from Napoleon had ended.

CONCLUSION

The Forth and Clyde Canal was the first Scottish attempt at a large-scale transport venture, and its ultimate success demonstrated that the country had the confidence and ability to undertake such schemes. In solving the administrative problems it encountered, it had set out a format for succeeding enterprises and promoted two occupations: that of the civil engineer and the navvy. The later generation of engineers who built the roads and bridges, the canals and railways, learned from those who had pioneered such work on the Forth and Clyde waterway. The labourers who were to transform the face of Scotland learned the hard way, making the cut between the Forth and the Clyde, digging, building bridges, aqueducts and retaining walls, and became a workforce with the skills needed to create the transport infrastructure of the later nineteenth century.

Yet the very success of the navigation was the cause of its decline. The coalmines and quarries that had laid trackways to carry bulk cargo to the canal bank were quick to appreciate the greater efficiency of the railways in reaching distant markets, and transferred their cargoes to the new transport system. As Professor Lenman has pointed out: 'Railways gradually developed from the pre-existing systems.'[53] With the advent of steam power, engineers like William Jessup built iron railways to replace the wooden trackways and turned to the skilled labour force from the canal-building era to undertake the work. Railways were easier to construct, did not require locks to climb inclines and could reach places that canals could not. By mid-century the country was in the grip of 'railway mania' and the age of the canal was drawing to a close as first passengers and then bulk cargo moved to the new form of transport.

When the Edinburgh and Glasgow Railway linked with the Central Railway at Greenhill, near Croy, it formed part of a transport network that ran from Aberdeen in the north to London in the south, and the Forth and Clyde Canal could not compete. The Forth and Clyde Canal, despite continuing to operate on a gradually declining basis for the rest of the century, was, in great measure, a victim of its own success.

NOTES

1. Dowds, 2003, xiv.
2. Defoe, 1962.
3. Phillipson, 1976, 109.
4. Skinner, A S. Introduction. In Campbell and Skinner, 1982, 2–3.
5. Phillipson, 1976, 116
6. Donaldson, 1981, 144–5.
7. Smiles, 1968, 92.
8. Hadfield, 1968, 45.
9. *Extracts from the Records of the Burgh of Glasgow* [1911], vol. vii, 1967, 237.
10. Campbell, 1981, 117; Devine, T M. Colonial commerce and the Scottish economy. In Cullen and Smout, 1977, 179; *Scots Magazine*, vol. xxix (1767), 131,

133; *Glasgow Journal*, 16–23 April 1767.

11. Smeaton, J. Report on the making of a navigation between the Clyde and the Forth, 1767, Introduction. Held in the William Patrick Library.
12. Devine, 1983, 179.
13. *Edinburgh Evening Courant*, 6 April 1767
14. Dowds, 2003, 25.
15. David Hume had published *Treatise on Human Nature* (1739–40) and the essays, *Of the First Principles of Government, Of the Study of History and Of Commerce* (1752). Frances Hutcheson had published *Inquiry into the Origins of our Ideas of Beauty and Virtue* (1725) and *A System of Moral Philosophy* (1755, posthumously).
16. The Dundas interest extended much further than Grangemouth, and when the canal was opened sloops carried alum from the family works in Yorkshire through the canal and returned with chemicals from the family's alkali works near Grangemouth. (Information supplied by Mr Peter Barton from his article: Shipping and the Alum Industry. In Miller, 2002, 82.)
17. *Edinburgh Evening Courant*, 6 April 1767.
18. Considerations on the intended navigation betwixt the Forth and Clyde from proceedings of the proprietors last meeting, October 1768, Letter to William Pultney, 25 November 1768. Held in the William Patrick Library.
19. Journal of the House of Commons, 31, 11 April 1767, 255–6.
20. Journal of the House of Commons, 31, 14 May 1767, 360.
21. Jones, 1982, 255.
22. Appendix to 8 George III, c 65, 1768, House of Lords Record Office.
23. Minutes of the General Meeting of the Forth and Clyde Canal Company, 3 May 1768, in Delvine Papers, NLS, MS. 1497, 123, 131–4.
24. Campbell, 1961, 122.
25. Letter of John Seton, 17 February 1773, Delvine Papers, 138.
26. Forrester, 1980a, 12, 14, 15.
27. Dowds, 2003, 55–6.
28. Campbell, 1965, 50, 89.
29. 24 George III c.55.
30. Anderson, J. Of the Roman wall between the Forth and Clyde and of some of the discoveries which have been lately made upon it. In the *Journal of John Anderson*, no date.
31. Lindsay, 1968, 56–64.
32. Forrester, 1980b, 117–8.
33. Forrester, 1980a, 10.
34. Allan, no date, 14.
35. Forrester, 1980a, 5; Handley, 1947, 36, 47.
36. Martin, 1971, 4.
37. Lythe and Butt, 1975, 176.
38. Handley, 1947, 43.
39. Lindsay, 1968, 70.
40. Lindsay, 1968, 93.
41. Parish of Kirkintilloch, County of Dumbarton. *New Statistical Account of Scotland*, vol. 8 (1834–5), 189.
42. Lythe and Butt, 1975, 124.
43. Corrins, 1994, 70–1.
44. Dowds, 2003, 63.

45. Butt, 1967, 168.
46. Butt, 1967, 169.
47. Donnelly, T. Tombstone territory: Granite manufacturing in Aberdeen 1830–1914. In Cumming and Devine, 1994, 84.
48. Butt, 1968, 169.
49. Lindsay, 1968, 88.
50. Massey, 1983, 6–7.
51. Dowds, 2000, 27, 34.
52. Butt, 1967, 170.
53. Lenman, 1977, 153.

BIBLIOGRAPHY

Allan, J K. *There is a Canal*, Falkirk, no date.

Anderson, J. Of the Roman wall between the Forth and Clyde and of some of the discoveries which have been lately made upon it. In the Journal of John Anderson, unpublished mansucript, no date.

Barton, P. Shipping and the alum industry. In Miller, 2002, 75–88.

Butt, J. *The Industrial Archaeology of Scotland*, Newton Abbot, 1967.

Butt, J. *Industrial History in Pictures*, London, 1968.

Campbell, R H. *Carron Company*, Edinburgh, 1961.

Campbell, R H. *Scotland Since 1707*, Oxford, 1965.

Campbell, R H. Stair's Scotland: The social and economic background, *Juridical Review*, (1981), 110–27.

Campbell, R H and Skinner, A S, eds. *The Origins and Nature of the Scottish Enlightenment*, Edinburgh, 1982.

Corrins, R D. The Scottish business elite in the nineteenth century – The case of William Baird. In Cumming and Devine, 1994, 58–83.

Cullen, L M and Smout, T C, eds. *Comparative aspects of Scottish and Irish Economic and Social History, 1600–1900*, Edinburgh, 1977.

Cumming A J G and Devine T M, eds. *Industry, Business and Society in Scotland since 1700*, Edinburgh, 1994.

Defoe, D. *A Tour thro' the Whole Island of Great Britain*, London, 1962, first published 1738.

Delvine Papers, NLS.

Devine, T M. Colonial commerce and the Scottish economy. In Cullen and Smout, 1977, 177–90.

Devine, T M and Dickson, D, eds. *Ireland and Scotland 1600–1850*, Edinburgh, 1983.

Devine, T M. *The Scottish Nation 1700–2000*, London, 2000.

Donaldson, G. Stair's Scotland: The intellectual inheritance, *Juridical Review* (1981), 128–45.

Donnelly, T. Tombstone territory: Granite manufacturing in Aberdeen, 1830–1914. In Cumming and Devine, 84–100.

Dowds, T J. *The Forth and Clyde Canal: A History*, East Linton, 2003.

Dowds, T J. *The French Invasion of Ireland, 1798*, Dublin, 2000.

Forrester, D A R. The great canal that linked Edinburgh, Glasgow and London. *Issues in Accountability*, 3 (1980a), 1–22.

Forrester, D A R. Early canal company cccounts: Financial and accounting aspects of the Forth and Clyde navigation, 1768–1816, *Accounting and Business Research*, 10/37a (1980b), 117–8.

Hadfield, C. *The Canal Age*, Newton Abbot, 1968.

Handley, J E. *The Irish in Modern Scotland*, Cork, 1947.

Hutton G. *Monkland: The Canal that Made Money*, Ochiltree, 1990.

Jones, J. James Hutton and the Forth and Clyde Canal, *Annals of Science*, 39 (1982), 255–64.

Knox, W W. *Industrial Nation: Work, Culture and Society in Scotland 1800–Present*, Edinburgh, 1999.

Lenman, B. *An Economic History of Modern Scotland, 1660–1976*, London, 1977.

Lindsay, J. *Canals of Scotland*, Newton Abbot, 1968.

Lythe, S G E and Butt, J. *Economic History of Scotland 1100–1939*, Glasgow, 1975.

Martin, D. *The Forth and Clyde Canal: A Kirkintilloch View*, Kirkintilloch, 1971.

Massey, A. *The Edinburgh and Glasgow Union Canal*, Falkirk, 1983.

Miller, I. *Steeped in History: The Alum Industry of North-East Yorkshire*, North York, 2002.

Phillipson, N T. Lawyers, landowners and the civic leadership of post-Union Scotland, *Juridical Review*, 21 (1976), 97–120.

Robertson, C J A. *The Origins of the Scottish Railway System 1722–1844*, Edinburgh, 2003.

Rolt, L T C. *Navigable Waterways*, London, 1969.

Skinner, A S. Introduction. In Campbell and Skinner, 1982, 1–5.

Smeaton, J. Report on the making of a navigation between the Clyde and the Forth. Unpublished document, 1767, held in the William Patrick Library.

Smiles, S. *Lives of the Engineers, with an Account of their Principal Works*, vol. 2, London, 1968, first published 1862.

Ward, J R. *Finance and Canal Building in Eighteenth Century England*, London, 1974.

10 The Hydro-Electric Schemes of Scotland

JAMES MILLER

In the thirty years immediately after World War II, the North of Scotland Hydro-Electric Board (NOSHEB) implemented a development programme in the Highlands and Islands on a scale never before seen in this part of Scotland. The board was created by the Hydro-Electric Development (Scotland) Act in August 1943, with David Ogilvy, the 12th Earl of Airlie, appointed as the first chairman and Edward MacColl as deputy chairman and chief executive. MacColl brought considerable experience as an engineer to the post; in the 1920s he had shown a flair for innovation in the development of the hydro-electric scheme in the Clyde Valley, and had followed this in the early 1930s with the development of the Galloway Scheme.[1]

During the interwar years many engineers had tramped and surveyed the Highlands, noting, often informally, the potential of the various river systems for hydro-electricity generation.[2] These explorations had been backed up by the investigations of official bodies, first by those of the Snell Committee[3] at the end of World War I and finally by the Cooper Committee in 1943.[4] Private firms such as the Grampian Electricity Supply Company had also been engaged in the construction and operation of hydro-electric generation plants. The earliest major hydro-electric development north of the Highland Line had been the work of the North British Aluminium Company, which built a plant at Foyers beside Loch Ness to generate power for the smelting of bauxite in the 1890s. Several local companies subsequently set up hydro-electric schemes serving particular localities. For example, the Grampian Power Company developed a large scheme using mainly Lochs Ericht and Rannoch in the early 1930s. This was the first major generating scheme for general public consumption. However, other proposals to develop hydro-electric schemes in the Highlands in the same decade failed to win parliamentary approval.[5]

The North of Scotland Hydro-Electric Board was conceived from the first as having a radical agenda. Its aim was not only to build hydro-electric schemes but, by means of such projects, to help revitalise the economic life of the Highlands and Islands as a whole. Electrification of this vast region was perceived as being capable of kick-starting new industries, thereby stemming the depopulation and emigration that had hitherto been such a feature of life in the region. These aspects of the board's work, very much the dream of Tom Johnston, the secretary of state who had initiated it and who served as its chairman from 1946 to 1959, became encapsulated in the term 'social clause'.

In 1944 the board produced its development scheme, an exhaustive list of 102 separate projects that seemed to cover every natural water system north of the Highland Line. In total the scheme predicted the potential generation of 6,247 million units of electricity for the national grid but recognised that, at least in the early years, many of the individual schemes would not be economically viable.[6]

The board published its first *Annual Report* in April 1945. At this time, with World War II still ongoing, its work was confined to planning; financed by borrowing from the Scottish banks, the total budget for its first year came to a modest £136,846.[7] The board appointed a panel of technical advisers – 'the best and most experienced engineers in the country' in their field – and showed its radical nature in also choosing three Scottish architects to cast a professional, aesthetic eye over construction.[8]

The proposals of the board created great excitement throughout the Highlands and Islands. Objections to the schemes were made on environmental and political grounds but, loud though some of these were, the majority feeling in the Highlands and Islands was that the board's proposals were a positive development of great promise. The first construc-

Figure 10.1 Monar dam under construction, winter. Source: Reproduced by permission of Northcliffe Newspapers.

tional scheme, involving Loch Sloy and two other small sites in Morar and Lochalsh, was subject to a public inquiry over the new year period of 1944–5 and was approved without much controversy for construction to begin in March 1945.

In contrast, constructional scheme no. 2, which involved a catchment area of 322 square miles and dams and power stations on the Tummel–Garry river system in central Perthshire, unleashed a storm of protest. After a stormy public inquiry, however, the board was given the green light and construction of the scheme was launched in a ceremony at Pitlochry on 25 April 1947.

The board survived two other serious threats to its existence and aims: nationalisation in 1948, and the threat of merger with the South of Scotland Electricity board in 1962, as recommended by the McKenzie Committee.[9] By the time the last major hydro-electric scheme was opened at Foyers in April 1975 the board had overseen the construction of over fifty major dams and power stations, almost 200 miles (322 km) of tunnel, 400 miles (644 km) of road and over 20,000 miles (32,000 km) of power line. In the process it not only gave employment to a significant proportion of the Highland workforce, but it also delivered the benefits of electrification to almost all of the north of Scotland (see Appendix).

Payne's account of the board has the nature of an official history and contains much political, administrative and technical detail.[10] It is complimented by two other works which give more informal accounts of 'the coming of the electric', as the board's work was often characterised in colloquial speech.[11]

HYDRO SCHEME WORKFORCE

All hydro schemes are based on the simple principle of harnessing a flow of water to power turbines to generate electricity. Natural water systems, such as in a river basin or loch catchment area, may be enlarged and channelled to ensure a dependable flow to the turbines. This usually involves the construction of dams, tunnels, aqueducts and weirs according to the natural disposition of the particular river system, and all of the Highland hydro-electric schemes show examples of these elements. The generated power has also, of course, to be carried to where it is required, and a very visible consequence of the Hydro Board's activities was the erection across the landscape of pylons and poles to carry power lines.

The construction of the large hydro schemes required a large and varied workforce. Large numbers of unskilled or semi-skilled labourers were needed for pick-and-shovel work and for the mixing and pouring of concrete by the ton. Skilled workers included joiners, electricians, crane drivers and fitters. The work was supervised by surveyors and engineers, and was administered by office staff. Schemes normally meant the presence of a temporary work camp to house and feed the workers, with attendant cooks and cleaners. The élite of the workforce were the tunnel diggers, the 'tunnel

Figure 10.2 Turbine hall, Ben Cruachan. Source: Perth Museum and Art Gallery, © Louis Flood Photographers, Perth.

tigers', who earned the highest wages but also at times faced great danger. Driving tunnels involved a number of specialist occupations, among them drillers, explosives experts, loco drivers, machine operators and engineers.

Work on a hydro-electric scheme went on around the clock – hard, physical labour often in conditions that were dangerous or at best uncomfortable and, at least in the early years of the Hydro Board, health and safety issues were largely ignored. Twelve-hour shifts were the norm, and the routine was for a man to work twelve hours, eat, sleep, then start on the next shift. It was common, however, for this pattern to be modifed, as Hugh McCorriston, a tunneller on Affric-Cannich scheme, made clear:

> We started at eight in the morning, until eight at night, and sometimes a lot more. I've seen me coming out the tunnel after Saturday night nightshift, and they had a batching plant outside, mixing the concrete, and half the boys hadn't turned up because they'd been in Inverness the night before. 'Will you come back up and work today on the outside?' Well, that wasn't too bad. We'd go down to the camp, get our breakfast and a packed lunch, and come back up, work twelve hours outside, go down, get our dinner, come back, and work the Sunday night as well – thirty-six hours straight through. We did that quite often.[12]

In the early years contractors had to resort to unusual measures to meet the demand for workers. The Loch Sloy scheme, which became operative on 28 March 1945, was dogged in its early months by an acute shortage of labour, a problem solved by drafting in German prisoners of war to work on the construction of access roads and other support facilities. In 1946, the Sloy workforce averaged 533 British and 521 PoWs, totals that changed to 1,156 British and 250 PoWs by the end of 1947.[13] PoWs were also employed on the early stages of the Tummel–Garry scheme: 320 in 1946, and 47 in 1947. The board stopped listing numbers of PoWs in its 1948 *Annual Report*, by which time this expedient was no longer significant. The lack of materials also delayed the schemes, a situation that did not begin to improve until 1948.

The contracting engineers recruited their workers from all over Britain, also drafting in many of the displaced persons (DPs) and refugees who had ended up in the country after the war. Wodek Majewski was among a number of Poles recruited by John Cochrane & Sons for the construction of part of the Affric–Cannich scheme in 1948: '[The agent] said he was looking for fifty men – strong men, and it's good pay and heavy work, but if you go it's too far to be sent back.'[14] This Polish contingent was taken by train to Inverness and then by bus to Cannich, where the work camp for the scheme was situated.

Many demobbed service personnel and itinerant labourers found their way to the Highlands, and in the long summer vacations students also found ready work. The schemes became noted for the cosmopolitan nature of their workforce. As well as Poles, there were men from most of the other nations of eastern Europe, especially from the countries under Soviet domination by the late 1940s. At Loch Sloy the DPs were taken on at half-pay but were highly motivated to prove their worth in their adopted country, fear of repatriation being strongly present in the early years. The workforce also included female DPs, some of them survivors of concentration camps, working as kitchen or cleaning staff.[15] As was common in civil engineering projects in Britain, the Irish were prominent. This was a tradition with a long pedigree,[16] and

Figure 10.3 Loch Sloy dam, Argyll. Source: SLA.

Irishmen from the poorer parts of Ireland such as Donegal found work in large numbers on the hydro-electric schemes. The multinational character of the workforce might have been expected to result in friction. In his account of his experiences on the schemes, Patrick Campbell from Donegal does give several instances of anti-Irish prejudice and fighting,[17] but the vast majority of workers maintain that everyone rubbed along very well and recall an easy, rough camaraderie as the prevailing situation.

The Hydro Board was called on from time to time to employ more Highlanders, but this was an unfair implied criticism as recruitment was the responsibility of the individual contractors; in any case, as time went on, the proportion of non-Scots in the workforce declined. The board argued that 'local men were the first to be employed and it was only when these had been absorbed that recruits were sought from further afield',[18] but there were still difficulties in finding men with certain skills. Exact figures for the composition of a fluid, large workforce are hard to come by but, for example, of the 1,739 employed on the Affric–Cannich scheme towards the end of 1949, 80 per cent were Scots, and half of these were from the Highlands; the others came from Europe (10 per cent), Ireland (6 per cent) and England (4 per cent).[19] On the Strathfarrar scheme in late 1959, 85 per cent of the workforce was local, and most of the rest came from other parts of Scotland.[20] The total number of workers employed on board schemes rose throughout the late 1940s and early 1950s: from 6,759 in 1948 to 8,500 at the end of 1955, a year in which its own staff numbered 2,645.[21] The workforce on the schemes fell steadily throughout the late 1950s and 1960s as the larger schemes were completed and as new construction techniques lessened the labour requirement.

THE WORK CAMPS

Some of the workers were new to camp life, but others, especially the Irish, were used to the lifestyle of the migrant construction worker, moving from job to job and usually sending money to help their families back home. Many of these men had hard-drinking, hard-playing ways, but they tolerated daunting conditions. On a cold evening in October 1947, Patrick McBride from Donegal arrived on the site of the Clunie dam in Perthshire. He and his colleagues were the advance squad on this scheme:

> We were introduced there to three buses on the side of the road, two for sleeping and one for the cookhouse. We had no running water, we had no light, and very little transport in them days ... The buses [had] three bunk beds on each side and a little stove in the middle. If that fire went out during the night you jumped to get some sticks on it. Our first job there was to build the camp.[22]

The camps consisted of ranks of rectangular wooden cabins, always called huts, and in some places ex-army Nissen huts were common. Facilities included the usual kitchen and canteen spaces, with ideally a cinema, a recreation hall, a sick bay and a post office and shop selling basic necessities.

The post office had a busy time at the end of the week, when many workers remitted sums of money to their relatives back home. At Cannich, some entrusted this task to Archie Chisholm, a young local apprentice, who recalled that a man might send home £7 as a typical sum, enough to do a family in Ireland for a week.[23]

Patrick McGinlay from Donegal who drove a crane on the Cannich scheme remembered pay day: 'You had to queue up on a Friday for your wages. At the pay office. They would shout your name and they had a big tray with all the paypackets lined out. It was all notes and cash.'[24]

The camps attracted peddlars. Patrick McGinlay again:

> A Pakistani came round with a suitcase. He could talk Irish Gaelic to some boys, he was that used to it. A lot of the men came from the west [of Ireland], all Gaelic speakers. He [the peddlar] would open up the case and you'd get shirts, shoes and underwear. He had the whole lot with him, more in a van.[25]

The board's *Annual Report* for 1947 described camp capacity as being 8,000, and said camps had libraries. It must be noted that the existence of the latter did not appear to be prominent in anyone's memory of their camp experience. The living accommodation was spartan but usually adequate by the standards of the time, and the comparative comforts of the camps were a common topic on the workers' grapevine. Wodek Majewski recalled:

> Some of them were quite good, you know. Clean, plenty to eat. Cannich – Cochrane's – was the best. There was everything there – concerts, cinema, and the food was quite good too. Balfour Beatty at Fannich was the same. I think the worst camp was [at Glascarnoch]. It was very rough. The food – if you go early in the morning, you get anything you want but if you go last you get nothing.[26]

Feeding the workforce was in itself a demanding task. Paddy Paterson helped in the family catering business that held the contract for several work camps in Ross-shire and Inverness-shire, and described the routine:

> We provided breakfasts, packed lunches and dinners, and there were cafeterias for the boys as well ... Breakfast started at six in the morning. In Fannich there were three eight-hour shifts – day shift, back shift and night shift – and in other places there were two twelve-hour shifts. Porridge was always a number one in the camps, and bacon and egg, scrambled egg, whatever. There was no choice, because of rationing, and these fellows were hungry. For the dinners, soups and stews were the main things – good, substantial homemade food. The Poles, Lithuanians, Irish – they all liked that sort of thing ... We had a thousand mouths to feed. You're talking about forty gallons of soup.[27]

Occasionally men would resort to catering for themselves, cooking on the stove in the hut or, as some times occurred, finding and sharing a private caravan close to the site. Donald Macleod from Harris and a pal from Fort William had another solution:

> We used to buy a dozen eggs and have two or three eggs in a big jug, topped up with lemonade, in the morning. Raw eggs. It was lovely. Give a good whisk and down that before we went to work in the morning. You got a piece, four slices of loaf and jam, provided in the middle of the day. And duck eggs. I've never seen so many duck eggs in my life. A lot of farmers were on a good number supplying duck eggs.[28]

The sleeping huts had a communal dormitory with perhaps twenty beds, with a few two-man and single cubicles at one end, for the more senior workers. Heating was provided by steam pipes around the wall, but these were not always up to coping with the winter cold. Hugh McCorriston recalled: 'But you were so tired when you came off shift you just lay down to sleep and that was it, cold or not. You just put an extra jacket on top of the bed.'[29]

Many found life in the camps reminiscent of what they had experienced in the services, and something of the wartime spirit of 'getting on with it' persisted. Health and safety were issues rarely attended to in the early years. As shown in many contemporary photographs, the men often worked in their own clothes – coats, bonnets, and rubber boots.

Figure 10.4 British and European tunnelling record, Loch Sloy, Argyll, 1951. Source: Perth Museum and Art Gallery, © Louis Flood Photographers, Perth.

Perhaps because they were foreign and could not be expected to have much, Wodek Majewski and his fellow Poles were issued with helmets, boots and oilskins for tunnelling, but this seems to have been exceptional in the 1940s and 1950s. The conditions improved as time went on, but all the construction schemes resulted in injuries and fatalities; the names of some of the workers who lost their lives are commemorated on plaques such as the ones on the Clunie Arch near Pitlochry and inside the power hall in Ben Cruachan. Trades unions were absent, and representatives who tried to organise labour within the camps were actively discouraged.[30]

Once the wartime shortages had been alleviated, the contractors found no difficulty in finding a workforce. Some of the workers, especially the itinerant Irish, might walk off the job if conditions or a foreman were not to their satisfaction, or if they felt slighted, and they did this in the knowledge that they would easily find another position on another scheme. A high turnover was characteristic of the 1950s' camp population.

Wages on the hydro-electric schemes were high in comparison to the alternatives on offer in the Highlands. Surface labourers might get 2s. 2d. an hour,[31] a rate still on a par with an estate worker,[32] but some trades and skills offered much higher rewards. As a teenage apprentice, Archie Chisholm felt himself to be well paid:

> I started off at £4.50 a week and rose to about £6. It was good money. The wage on the estate when I'd been working with my father had been £2 10s. You could pick up a pair of boots for 6s., you could go to the town on a Saturday with 5s. It was just after the war, and spirits were hardly available, and it would have been 9d. or 10d. for a pint.[33]

A shuttering joiner on dam-building might earn £16 a week in the 1950s, but the highest paid were the tunnellers, some of whom might receive up to £35 for their week's labour. Perhaps 30s. might be deducted for bed and board in the camp, but this left many feeling wealthy, with little outlet for spending beyond remittances home, the beer canteen and the occasional trip to town. The camps offered limited opportunities for recreation. Normally there was a 'wet' canteen that served beer and a 'dry' canteen offering soft drinks. Gambling schools flourished in some camps, despite the efforts of the authorities to stamp them out, with dice games such as 'crown and anchor' and card games. Much hard drinking went on. Roy Mackenzie, the local policeman at Invermoriston, estimated that 90 per cent of crime in the camp was fighting after too much drink had been taken (the remaining 10 per cent he ascribed to theft).[34] Mackenzie also recalled some discreet prostitution,[35] but encounters with the opposite sex were normally reserved for weekend excursions to the nearest town or village, where there would often be a dance to attend. Many of the workers met their future spouses while working on the hydro schemes. As the years went by, the proportion of Highlanders in the workforce rose and most men commuted daily or went home at weekends rather than staying on site.

Efforts were made to tend to the spiritual needs of the workers. A priest or a minister might visit to hold a service, and at Pitlochry some Catholic workers built a chapel in their spare time.[36] Workers from Butterbridge camp used to borrow a lorry to drive to chapel in Inveraray.[37]

Figure 10.5 Stages in dam and tunnel construction. Source: Perth Museum and Art Gallery, © Louis Flood Photographers, Perth.

WORKING CONDITIONS

Once surveying and design had been established, the construction of a hydro-electric scheme passed through a series of fairly well-defined phases, from piling of the river bed and the erection of a coffer dam, through the excavation and construction of the foundations, to the pouring of concrete for the main mass of the dam with attendant shuttering. At the same time, aqueducts and tunnels were constructed to feed the reservoir or conduct water to the power station.

Tunnelling was especially dangerous, working with heavy drilling machinery in a confined, dark space, amid the constant reek of gelignite, engine fumes and rock dust. The tunnellers regarded themselves as the élite of the workforce and revelled in the sobriquet of 'tunnel tigers', a reputation that often preceded them on their weekend trips to the villages and towns. The squads of tunnellers, perhaps thirty to forty men working under tunnel bosses, fostered a competitive spirit, at times cutting corners in an attempt to break a record or drill a few more feet before the end of their shift. A foreman on one scheme in Perthshire knew he could maintain good progress by having a squad of Irish tunnellers on the day shift and a competing squad of Poles and other Europeans on the night shift.[38] A 'holing through', when sections of tunnel driven from different directions finally met, was usually marked by a ceremony. To encourage productivity, contractors offered bonuses, an incentive that might lead to carelessness and the scorning of

Figure 10.6 Holing-through ceremony. Source: Perth Museum and Art Gallery, © Louis Flood Photographers, Perth.

risks, a state of affairs Wood has characterised as 'institutionalised haste'.[39] For all the risks and the discomfort, however, the workers on the board's hydro-electric schemes look back with pride on their achievement, as they confirm when asked.[40]

IMPACT ON THE HIGHLANDS

The impact of the sudden influx of several hundred men into a hitherto quiet Highland locality is often remembered more vividly by those who witnessed 'the coming of the electric' than the long-term changes wrought by the subsequent electrification. Gladys Farquharson, a young teacher in Killin when work began on the Breadalbane scheme, describes how life in the area was altered:

> Then this idyll disappeared in a welter of every kind of earth-moving equipment you could imagine ... On the hillside above and to the west of our house, rows and rows of hutted buildings appeared ... Buses, cars and lorries were on the move night and day. However, there was no hassle or real disturbance, and we just lived happily on the fringe, as it were.[41]

Away from the schemes themselves, lines of steel towers, or 'pylons', and wooden poles sprouted across the hills to bear the cables carrying the newly generated current. The squads who built these are also proud of their achievement, perhaps none more so than the men of the J L Eve Construction Company Ltd, who in 1954 put a line over the Corrieyairack Pass between the Great Glen and Upper Strathspey. This stretches for 25 miles (40 km) and rises to 2,600 feet (850 m) above sea level. A steel tower required a concrete foundation for each leg, and to accommodate these, holes 10–15 feet (3–4.5 m) deep had to be dug into the hillsides, a process that needed gelignite as well as pick and shovel. From the start, the great majority of the men recruited to erect towers and poles were local and tended to live at home, only resorting to periods in hostels or camps when commuting would have been too far for practicality. Jimmy Macdonald recalled some aspects of his experience:

> It was a fifty-hour week summer and winter. At the start [early 1950s], with Balfour [Balfour Beatty and Co., Ltd] we got about 1s. 10d. an hour ... When I started with the Hydro Board it was £9 a week – that was in the late 1950s. The men included a crowd from Inverness and on the Monday morning it was quite common to get the Beauly bobby up to summons them away to make an appearance in court. Our own lorries used to drive round to pick the men up and we left from Invergordon at half past seven, and went to Fort Augustus and back ... It was pretty tough up on the hill some days, up a pole on spikes, in a gale[42]

The board noted in its annual reports the progress of electrification of the Highlands. After ten years, it recorded proudly that 11,329 crofts out of a total of 21,800, and 9,223 farms out of 19,300 had been connected to the mains supply.[43] By 1959 it was estimated that 90 per cent of the region had been electrified.[44]

THE LATTER YEARS OF THE HYDRO BOARD

By 1966, the board had completed its plans for almost half of the water-power resources of the Highlands. By then fifty main power stations had been constructed, with a capacity in excess of one million kilowatts.[45] In August 1968, when the board celebrated its twenty-fifth anniversary, many power stations were opened to the public for a week, attracting over 25,000 visitors.[46] Despite this general popularity, however, opposition to the construction of more hydro-electric plants had grown, and there were fears even for the future of the board itself.[47] The Cruachan scheme, a massive construction project that involved the excavation of sufficient material to allow the construction of a power station deep inside the mountain, came into operation in late 1965 and demonstrated the advantages of pumped storage, where at times of low demand excess electricity is used to pump water from a lower to an upper reservoir. At the same time, four hydro-electric schemes – for Glen Nevis, Loch Laidon, Lochs Fionn and Fada beside Loch Maree, and Loch a'Bhraoin – lay before parliament for approval, but the green light was never given.[48] Plans were made for a pumped-storage scheme at Craigroyston on the shoulder of Ben Lomond, but these too were never realised. Describing the opening of the pumped-storage scheme at Foyers on Loch Ness on 3 July 1975, the *Inverness Courier* noted that it was 'a triumphant occasion, though also sad as it almost certainly marked the end of an era which has been vital to the Highlands'.[49]

APPENDIX

The major hydro-electric schemes built for the North of Scotland Hydro-Electric Board between 1945 and 1975. (The board also operated diesel and steam power stations in Dundee, the Hebrides, Orkney and Shetland.) Dates shown are dates of completion.

Loch Sloy
1950: Buttress dam on Loch Sloy, with tunnel and pipeline to a power station at Inveruglas Bay on Loch Lomond. Average annual output: 120 million units.

1959: Glen Shira phase: pre-stressed gravity Allt-na-Lairigie dam, and tunnel to power station on River Fyne. Buttress dam on Lochan Shira Mor, power station at Sronmor, dam on Lochan Sron Mor with tunnel to Clachan power station on Loch Fyne. Average annual output: 100 million units.

Tummel–Garry
1951: Phase 1: Clunie Dam at the east end of Loch Tummel, with tunnel to Clunie power station. Outfall was combined with the flow of the Garry to the dam

and power station at Pitlochry. Loch Faskally created behind the Pitlochry dam. Average annual output: 220 million units.

1955: Additional power stations: Cuaich 9 million units; Loch Ericht 11 million units.

1958: Phase 2: Enlargement of the Tummel catchment area by diversion of waters of the upper Garry to Loch Errochty; Trinafour dam built at the east end of Loch Errochty (1957) and tunnel from Loch Errochty to Loch Tummel via the Errochty power station. Dam and power station at Gaur (1958). Average annual output: 119 million units.

Conon Valley
1957: Phase 1: Dam on Loch Fannich with tunnel to Grudie Bridge power station at west end of Loch Luichart. Average annual output: 82 million units.

1957: Phase 2: Glascarnoch Dam, Vaich Dam. Tunnel from Vaich to Glascarnoch, and from Glascarnoch to Mossford power station on Loch Luichart. Barrage on River Bran and power station on Loch Achanalt. Dam on River Meig and tunnel to Loch Luichart. Dam on Loch Luichart and tunnel to Luichart power station. Dam and power station at Torr Achilty (Loch Achonachie). Average annual output: 280 million units.

1961: Phase 3: Twin dams on River Orrin and tunnel to Orrin power station on Loch Achonachie. Average annual output: 76 million units.

Affric–Cannich
1952: Dam on Loch Mullardoch, and tunnels via Loch Benevean to Fasnakyle power station. Average annual output: 231 million units.

Strathfarrar–Kilmorack
1963: Dam on Loch Monar, and tunnel to Deanie power station. Dams at Loichel (on Loch Monar) and on Loch Beannachran, and tunnel to Culligran power station. Mass gravity dams with power stations at Aigas and Kilmorack. Average annual output: 261 million units.

Moriston–Garry
1957: Garry: Dam on Loch Quoich and tunnel to Quoich power station on River Garry. Dam on Loch Garry and tunnel to Invergarry power station. Average annual output: 149 million units.

1957: Moriston: Dams on Lochs Loyne and Cluanie, with tunnels from Loch Loyne via Loch Cluanie to Ceannacroc power station. Dam at Dundreggan and Glenmoriston power station, with tailrace tunnel to Loch Ness. Aqueduct system to feed Livishie power station. Average annual ouput: 214 million units.

Loch Shin
1960: Dam and power station at Lairg. Tunnel from Little Loch Shin to Inveran power station. Cassley power station on Loch Shin. Average annual output: 137 million units.

Breadalbane
1961: Lawers section: Lawers dam on Lochan na Lairige, tunnel to Finlarig power station on Loch Tay. Average annual output: 64 million units.

1961: St Fillans section: Dam on Loch Breaclaich with tunnel to Lednock power station. Dam on Loch Lednock. Tunnel to St Fillans power station, Loch Earn. Tunnel from Loch Earn to Dalchonzie power station. Average annual output: 99 million units.

1961: Killin section: Tunnel, aqueducts, dam and power station at Lubreoch on Loch Lyon. Average annual output: 13 million units.

1961: Dam on Loch Giorra, tunnel to Cashlie power station. Dam on Stronuich reservoir and tunnel to Lochay power station. Average annual output: 185 million units.

Awe/ Ben Cruachan
1965: Pumped storage scheme: reservoir and underground power station in Ben Cruachan. Installed capacity 400 megawatts.

Associated conventional schemes: Dam on Loch Nant, with aqueducts and tunnel to Nant power station on Loch Awe. Dam on north-west arm of Loch Awe, with tunnel to Inverawe power station on Loch Etive. Average annual output: 127 million units.

Foyers
1975: Two dams on Loch Mhor, with tunnel to power stations of Foyers Falls and Foyers on Loch Ness. Pumped storage capacity 300 megawatts; average annual output: Foyers Falls, 6 million units.

Small schemes

Morar, 1948	Dam and power station on the River Morar. Average annual output: 3 million units
Lochalsh, 1948	Dam on the Allt Gleann Udalain and power station at Nostie Bridge. Average annual output: 6 million units
Mucomir Cut, 1951	Power station installed in a channel between the southern end of Loch Lochy, close to the Caledonian Canal, and the Spean River. Average annual output: 9 million units.
Striven, 1951	Dam on Loch Tarsan and power station at the head of Loch Striven, near Kyles of Bute. Average annual output: 22 million units
Kerry Falls, 1952	Reservoir at Loch Bad an Sgalaig and pipeline on the River Kerry, Gairloch, with power station. Average annual output: 5 million units
Storr Lochs, 1952	Dam and power station on the north-east coast of Skye. Average annual output 7 million units
Glen Lussa, 1952	Power station on Glen Lussa Water draining Loch Lussa, Kintyre peninsula. Average annual output: 10 million units
Loch Chliostair, 1954	Harris: dam, pipeline and power station on Eaval River. Average annual output: 3 million units

Loch Dubh, 1955	Dam, pipeline and power station in Strath Kanaird, Ullapool. Average annual output: 5 million units
Kilmelfort, 1956	Dams on Lochs Tralaig and na Sreinge, and on the Melfort Pass near the Ardrishaig-Oban road, feeding power station on the River Oude. Average annual output: 11 million units
Loch Gair, 1960	Dam on Loch Glashan, north-east of Lochgilphead, with pipeline and tunnel to power station on Loch Gair. Average annual output: 18 million units
Loch Gisla, 1960	Gisla River drainage system, with power station at Gisla, draining into Little Loch Roag, south-west Lewis. Average annual output: 2 million units

NOTES

1. Payne, 1988.
2. Interview with Hamish Mackinven.
3. *Parliamentary Papers* 1919.
4. *Parliamentary Papers* 1943.
5. Six proposed hydro-electric schemes failed to win parliamentary approval between 1928 and 1941.
6. MacColl, 2000. (MacColl's paper was originally presented in 1946.)
7. NOSHEB, 1945.
8. NOSHEB, 1945.
9. In an interview with the author, Hamish Mackinven, former secretary to Tom Johnston during his period as chairman of NOSHEB, identified the Tummel–Garry scheme, the threat of nationalisation in 1948 and the threat of merger in 1962 as the three main periods of danger in the board's life.
10. Payne, 1988. Peter Payne was professor of economic history at the University of Aberdeen.
11. Miller, 2002. Most of the quotes from interviews are taken from the work done for this book; Wood, 2002.
12. Interview with the author.
13. NOSHEB *Annual Reports*, 1946, 1947, 1948.
14. Interview with the author.
15. Wood, 2002.
16. See for example, McGill, 1999 which concerns the Irish labourers who worked on the Blackwater dam before World War I.
17. Campbell, 2000.
18. NOSHEB *Annual Report*, 1947.
19. Figures quoted by Sir Hugh Mackenzie in a newspaper interview, *Inverness Courier*, 27 December 1949.
20. *Inverness Courier*, 25 December 1959.
21. NOSHEB *Annual Report*, 1955.
22. Interview with the author.
23. Interview with the author.
24. Interview with the author.
25. Interview with the author.

26. Interview with the author.
27. Interview with the author.
28. Interview with the author.
29. Interview with the author.
30. Wood, 2002.
31. Interview with the author.
32. Wood, 2002.
33. Interview with the author.
34. Wood, 2002.
35. Wood, 2002.
36. Interview with the author.
37. Interview with the author.
38. Ford, 2000.
39. Wood, 2002.
40. Interview with the author.
41. Ford, 2000.
42. Interview with the author.
43. NOSHEB *Annual Report*, 1955.
44. *Inverness Courier*, 27 February 1959.
45. NOSHEB *Annual Report*, 1965–6.
46. NOSHEB *Annual Report*, 1968–9.
47. *Inverness Courier*, 9 April 1965.
48. A modified version of the Loch a'Bhraoin scheme was eventually built in 2000. Scottish and Southern Energy plc now owns Scottish Hydro-Electric, the privatised successor to NOSHEB.
49. *Inverness Courier*, 4 April 1975.

BIBLIOGRAPHY

Campbell, P. *Tunnel Tigers*, New Jersey, 2000.
Ford, G M ed. *Tunnellers, Tango Dancers and Team Mates*, Stirling, 2000.
MacColl, A E. Hydro-electric development in Scotland. In Leith T, Broadley I, Osborn H and Robb J, eds, *Mirror of History: A Millennium Commemorative Volume*, Glasgow, 2000.
McGill, P. *Children of the Dead End*, London, 1999, first published 1914.
Miller, J. *The Dambuilders: Power from the Glens*, Edinburgh, 2002.
North of Scotland Hydro-Electric Board (NOSHEB) *Annual Reports*.
Payne, P L. *The Hydro: A Study of the Development of the Major Hydro-Electric Schemes Undertaken by the North of Scotland Hydro-Electric Board*, Aberdeen, 1988.
Parliamentary Papers, Board of Trade, Interim report of the Water Power Resources Committee (Snell Committee), Cmd.79, 1919.
Parliamentary Papers, Scottish Office, Report of the Committee on Hydro-Electric Development in Scotland (Cooper Committee), Cmd. 6406, 1943.
Wood, E. *The Hydro Boys: Pioneers of Renewable Energy*, Edinburgh, 2002.

PART THREE

●

Security and Health

11 Scottish Soldiering: A Fading Tradition

TREVOR ROYLE

At the beginning of the twenty-first century, for the first time in over 100 years, the number of Scots joining the British Army fell below the national UK average. Having supplied 13 per cent of the British Army's infantry manpower requirements throughout most of the twentieth century, the figures plummeted to worrying new levels. A summary of enlistments in Scotland published in January 2005 showed that in 1999–2000 the number of recruits joining Scotland's six infantry regiments was 528, but by 2004–5 this had fallen to 301. Only in 2002–3 had there been a rally, when the figure rose to 514. At the same time the numbers being recruited into regiments or corps which offered trade training remained fairly constant. For example, 168 Scottish recruits joined the Royal Logistic Corps in 1999–2000; five years later this had only fallen to 159.[1] There were several reasons for this change in recruiting patterns across Scotland. In Edinburgh and the Lothians, unemployment had been low for a long period – as low as 2 per cent in 2004. The low-density population of the Highlands has always provided a smaller pool for regiments such as The Highlanders and the Argyll and Sutherland Highlanders; competing work opportunities in the region, especially the tourism industry, means that the pool from which recruits can be drawn is now more limited. Even high-density areas such as Glasgow and Ayrshire did not help The Royal Highland Fusiliers, which recruits from those areas. In 1999–2000 it recruited 89 soldiers, but this dropped to 56 in 2004–5. Pay was and remains a factor in recruitment. Two advertisements carried on the sides of buses in Lothian in the first months of 2005 help to explain why young people were turning their backs on the army. The first advertisement was produced by the army to encourage potential recruits to join Scotland's infantry, while the second, smaller advertisement offered jobs as bus drivers with a starting pay of £20,000, almost twice the amount paid to a trained infantry soldier. Even rates of pay for unskilled workers are relatively attractive, especially in the tourism and service sectors where tips and bonuses add to workers' incomes.

Education, job satisfaction and the falling birth rate also played a part in dissuading young Scots from considering a career in the army. With some 40 per cent of school-leavers going into tertiary education, the army is not seen as an attractive alternative, and in the period 1985–2005 there was a

Figure 11.1 Three infantrymen of the 1st Battalion, Royal Scots on duty on the Euphrates river, Iraq, December 2003. Source: Reproduced by permission of The Royal Scots Museum.

42 per cent reduction in the size of the 16–24 male age group in Scotland. As this is the army's prime client group, it had an obvious impact on the drop in recruitment levels. There is evidence to suggest that parents, guardians and partners have also been influenced by adverse publicity surrounding the army. Between 1995 and 2002 there were alleged abuses at the training centre at Deepcut in Surrey which resulted in the deaths of four soldiers, and in the same period there was widespread disquiet about the army's involvement in Iraq. Not only was this an unpopular and perhaps illegal deployment which had cost the lives of 92 British soldiers by August 2005, but British soldiers were also involved in allegations of brutality and abuses of human rights. Some of the cases were brought under the aegis of the International Criminal Court, which would make the soldiers war criminals if they were found guilty. Although the army in Scotland argues that it is hard to find clear evidence that adverse publicity has affected recruiting, there is agreement that high-profile opponents of the war, such as Mrs Rose Gentle, whose son was killed serving in Iraq, do not make life any easier for army recruiters.[2]

The army agreed to make radical changes to the organisation of its infantry regiments as a result of the Future Army Structure reforms in July 2004. The size of the infantry was cut from 40 battalions to 36, and the remaining 19 single battalion regiments were re-formed into larger 'super-regiments' consisting of several operational battalions. For Scotland this meant that the six existing regiments would be reduced to five through the amalgamation of The Royal Scots with The King's Own Scottish Borderers, and each new component would form a battalion within the new Royal Regiment of Scotland. It is made up as follows:

The Royal Scots Borderers
(1st Battalion The Royal Regiment of Scotland)
The Royal Highland Fusiliers
(2nd Battalion The Royal Regiment of Scotland)
The Black Watch
(3rd Battalion The Royal Regiment of Scotland)
The Highlanders
(4th Battalion The Royal Regiment of Scotland)
The Argyll & Sutherland Highlanders
(5th Battalion The Royal Regiment of Scotland)

The new Scottish regiment has bases in Edinburgh, Fort George, Northern Ireland, Canterbury and Fallingbostel in Germany and operates in a number of roles from light infantry to heavy armoured infantry. By creating a larger number of career opportunities within the regimental structure, the reformers hope to create a modern and versatile organisation offering a wide variety of career opportunities. One of the main instigators for change was the adjutant-general, Lieutenant-General Sir Alistair Irwin, the army's senior officer responsible for personnel matters, who is also colonel of The Black Watch. At the time of the reforms he was also a member of the executive committee of the Army Board. Writing in the *Journal of the Royal United Services Institute*, Irwin put the case for the new large regiments when he claimed that the existing system was not only inefficient but was bad for soldiers' careers: 'Does it make sense, in a small and highly active army, to select an officer who has proved himself to be a brilliant armoured company commander to command his battalion, now in the light role, simply because that is the regiment to which he belongs? Probably not.'[3]

Because Irwin was so deeply involved with The Black Watch his support of the new system angered regimental supporters, but there was also a growing awareness in the military community that change was on the way and that it had to be embraced for the good of the army. Like many other former soldiers, Lieutenant-Colonel W P C Macnair, formerly Queen's Own Highlanders, regretted the changes but recognised their inevitability:

> There are a number of different applications about what infantry means – it covers the whole spectrum from high intensity operations with Warrior fighting vehicles to providing aid to the civil community such as the recent help with foot and mouth. It is immensely important that senior personnel have a variety of things to do to make them employable. They will have to get that right to go forward.[4]

Given the country's attachment to the infantry regiments, the reforms created a good deal of opposition with the creation of a Save the Scottish Regiments campaign, which was set up to fight the changes and quickly created a prominent public profile. Not only do the Scottish regiments enjoy close links with their communities, but the deployment of the 1st Black Watch to support US forces in the 'triangle of death' in Iraq during December 2004 was deeply unpopular. To regimental supporters it seemed that the regiment was being stabbed in the back while it was taking part in a dangerous and unpopular mission. The deaths in action of five Black Watch soldiers only added to the sense of grievance felt by many, and there was further indignation when The Black Watch's commanding officer, Lieutenant-Colonel James Cowan, advised campaigners that the best way forward was to support the reforms and not fight them.[5]

One of the strongest and most emotive arguments advanced by the campaigners centred on the army's need to maintain its links with the regiments' traditional recruiting areas. During the late nineteenth century, following the Cardwell and Childers reforms which introduced 'localisation' by which regiments were given fixed, permanent homes in the hope that local pride in the regiment would stimulate recruitment, Scottish regiments had distinctive recruiting areas, and it was not unusual for sons to follow fathers into the same regiment. For example The Black Watch (comprising the 42nd and 73rd Highlanders) recruited from Angus, Fife and Perthshire, and the system continued throughout the twentieth century to the present day. This allowed the senior soldier in Scotland, Lieutenant-General Sir Peter Graham, to say in 1992: 'We have a great deal to be proud about. It goes right the way through the junior soldier. People are proud that you have a son in the army, serving in a Scottish regiment. It is an honourable profession.'[6] Graham, a Gordon Highlander, displayed many of the criteria that the civilian public expect of a Scottish officer: he was brought up in Scotland, though educated mainly in England; he served in a kilted Highland regiment with a keen interest in piping and military history; and he is a firm supporter of the virtues which the Scottish soldier brings to the British Army, claiming that 'there is much pride in what we as a small nation have achieved in feats of arms, in the wars, serving the British Army and the Crown over the years.' Graham's immediate predecessor, Lieutenant-General Sir John Macmillan, reinforced the idea that there are strong emotional links between the Scots and their tartan-clad regiments when he remarked in 1989: 'I think the advantage we have in Scotland is that there is a Scottish identity, even if it's only manifested by a feeling that those so-and-sos down in England don't understand us. At least it helps to keep us together.'[7]

Scotland's close relationship with the army and its attachment to the infantry regiments are rooted in the country's history, which is bloody with battles, some fought against the nearest neighbour, England. Many more were fought amongst the Scots themselves: family against family, clan against clan, Lowlander against Highlander, Catholic against Protestant. It also includes long periods of fighting abroad: from the sixteenth century onwards Scots mercenaries fought in the service of the kings of France, Spain, Russia and Sweden. Scots made good fighters, and in common with many other minorities on Europe's fringes (the Croat cavalry in Wallenstein's army, for example) they exported their skills to the highest bidder, becoming soldiers of fortune who gave good value for money. At least 25,000 were in the service of Gustavus Adolphus of Sweden, and half as many fought for King Louis XIII of France, often confronting their fellow countrymen on the field of battle, neither giving quarter nor expecting to receive it.

A further reason for the Scots' interest in soldiering was provided by the Volunteer craze, the Victorian fancy for part-time amateur soldiering which involved some gentle shooting practice and drills and, best of all, dressing up in turkey-cock uniforms. In Scotland the recruitment figures for the Volunteer units were twice the British average, a figure which was

Figure 11.2 Sergeant Donald Mackay, of the 1/9th (Highland) Battalion, Royal Scots (Territorial Force), 1908. This was a Territorial battalion which had been formed by 'Highland gentlemen' living in Edinburgh in 1900. Because of their Highland connection they were the only Royal Scots battalion to wear the kilt instead of tartan trews, hence their nickname of 'The Dandy 9th'. Source: Kenneth Veitch.

undoubtedly assisted by the creation of units with Highland affiliations, most of them in the central belt. With their panoply of kilts, tartan trews, ostrich feathers and ornate sporrans they were an irresistible attraction, and everywhere men rushed to wear them. Most of these outlandish uniforms owed nothing to tradition but were invented by local colonels, and they came to represent one of the flowerings of a self-conscious nationalism – or what the military historian John Keegan has described as 'a force for resistance against the creeping anglicisation of Scottish urban life'.[8] However, there was more to soldiering than putting on fancy dress. Being a part-time soldier meant following an estimable calling: it was companionable, offered self-respect and produced steadiness of character, all important moral virtues in presbyterian Scotland.

The vogue for soldiering can also be traced back to the defeat of the Jacobite army at Culloden in 1746 and to the subsequent subjugation of the Highlands and the despoliation of its Gaelic culture. In the aftermath of the revolt, measures were taken for 'disarming and undressing those savages' through the Disarming Act of August 1746. Not only were traditional weapons and the plaid made illegal, but Highlanders were forced to swear an oath of allegiance which left the clansman in no doubt that his day was over. What to do with them was another matter. Either they could accept modernity and the Union or they could be moved elsewhere to make new lives, courtesy of landowners who regarded themselves not as destroyers but as liberal reformers. As for the soldierly instincts of their tenants, these could be offered to the British Army at a time when it was being used as an imperial gendarmerie to expand the country's growing colonial holdings. It was during the Seven Years' War that the Elder Pitt, acting on a suggestion made by King George II, opened the door for the creation of the Highland regiments. It was a simple concept. Here was a ready supply of soldiers who would do their duty, were known for their abilities as fighters, displayed hardihood in the field and gave unbending loyalty to their commanders. One other factor intruded: the traditional clan structure produced a sense of coherence and loyalty which would translate into good military practice. As the days of the clan system were numbered after Culloden and would soon disappear, other than as a sentimental entity based on chiefdoms, tartan and yearning for a lost past, the Highland regiments became handy substitutes for the old clan way of life.

Between 1714 and 1763 a quarter of the officers serving in the British Army were Scots, proportionatly more than the English. Of 208 officers who were also members of parliament from 1750 to 1794, 56 were Scots. At the same time one in four regimental officers were Scots, and Scottish officers were used to receiving high command while fighting in the European and colonial wars waged by Britain throughout the eighteenth century. Between 1725 and 1800 no fewer than 37 Highland regiments were raised to serve in the British Army, and by the end of the period the numbers involved are estimated at 70,000 men.[9] These are impressive figures, and the names of many of the units lived on in the later Victorian army and in the British

Army of more recent times: for example, the 22nd (Seaforth) Highlanders, raised in 1778; Cameron Highlanders and Ross-shire Buffs, raised in 1793; the 91st (Argyllshire) Highlanders and 92nd (Gordon) Highlanders raised a year later; and finally, the 93rd (Sutherland) Highlanders raised in 1800. All of these formations came into being at the hands of a small number of eminent Highland gentlemen using the influence of the clan system with its age-old interconnection of family loyalties to raise the requisite numbers of men.

From the very outset the territorial links of the regiments were vital, not just for recruiting but also for maintaining group cohesion and loyalty. The system had other benefits. Landowners who had supported the Jacobites were able to demonstrate their loyalty to the Hanoverian crown by raising regiments as a quid pro quo. Most considered themselves to be Highland gentlemen dependent on their land for their standing and income. If estates had been forfeited as a result of supporting the Jacobite cause, the raising of a regiment was a useful means of retrieving family honour and making good lost ground. That was an important consideration, as the creation of a regiment required social status and financial capacity, the going rate for raising and equipping a regiment being £15,000. A landowner wishing to raise a regiment had to have contacts at the highest social level, as it was the king who gave authority for the regiment to be raised in his name. The actual commission for the commanding officer was signed by the secretary for war, but the final arbiter was King George II. Once the order and warrant had been issued, the regiment came into being and the commanding officer set about recruiting. For the senior officers a regimental commander would look to his closest family, friends and associates, and they in turn helped to recruit the soldiers from the tenantry on their estates.

With the allure of their uniforms, Scottish soldiers became an instantly recognised and widely feared element of the British Army, and their service in Africa, India and North America helped to consolidate Britain's growing mercantile empire. The long tradition of soldiering was also a powerful factor in establishing the Scottish people's relationship to the army. While there had been a falling-off in the numbers and the quality of recruits in the second half of the nineteenth century, leading to higher numbers of English and Irish soldiers joining the Scottish regiments, the army was not in itself considered a dishonourable career. War Office figures showed that 26.9 per cent of the men aged 15–49 volunteering for the army in 1911 were Scots, from a total male population of 2,308,839, compared to 24.2 per cent from England, from a total male population of 17,445,608.[10] In her study of the composition of the Highland regiments between 1870 and 1920, Diana Henderson insists that, despite seasonal manpower problems, 'soldiering was widely looked upon as a respectable profession in Scotland'.[11]

There was, of course, another side to the question of recruitment. Young men also joined the army because they had no option. In the poorest families an unemployed boy was an extra mouth to feed, and in every Scottish regiment there were large numbers of young men who had escaped

Figure 11.3 Sergent Donald Mackay and family, 1918. Source: Kenneth Veitch.

grinding poverty by becoming soldiers. A return by the army medical department in 1903 showed that of 84,402 recruits examined on enlistment the majority, 52,022, came from the lower stratum of the working classes.[12] This was especially true in Lowland regiments, where recruits often came from backgrounds in which families lived below the poverty line and overcrowding and a lack of decent sanitation were commonplace. Conditions in Glasgow and many parts of Lanarkshire were particularly bad. Although the prevailing squalor and degradation of the previous century had been addressed by a combination of philanthropy and the introduction of radical improvements in public health, the physical conditions for most workers in Glasgow were not good, with the population density in 1911 being twice that of Edinburgh and Dundee. In 1917 a royal commission on housing revealed a situation which was bedevilled by 'gross overcrowding and huddling of the sexes together in the congested villages and towns, occupation of one-room houses by large families, groups of lightless and unventilated houses in the old burghs, clotted classes of slums in the great cities'.[13] Faced by those conditions and the low pay which caused them, the army was often seen as an escape route. The initial period of service was seven years, after which they could be discharged to the reserves for a five-year period or elect to spend that time with the Colours, provided that they were found to be efficient and of good conduct. At the end of the 12-year period they could opt for a discharge or sign on for another nine years to complete 21 years of service. The best soldiers were promoted and in time became senior non-commissioned officers or warrant officers, the backbone of any self-respecting infantry battalion.

The officers were a different breed. The purchase of commissions had ended in 1871, but Britain's military officer class was still one which Wellington or Marlborough would have recognised, as most of its members came from the aristocracy, the landed gentry, the clergy and the professions, and had been educated at one of Britain's great private schools. All were expected to have private incomes, as the annual pay of a second lieutenant, the entry rank, was just over £95 a year, and that was insufficient to cover his uniform costs and mess bills, which amounted to roughly £10 a month. In 1914 the War Office recommended that the minimum needed to survive was £160 a year, but even that amount meant that a young officer would have to lead an abstemious existence. Some regiments were extremely expensive. Officers in cavalry or foot guards regiments had to purchase a variety of uniforms to meet all the variations in service and mess dress, and they were expected to live well in the mess and to keep at least two hunters and three polo ponies. It was not considered unusual for a smart cavalry regiment to insist on a young officer being in possession of a private income of up to £1,000 a year, an enormous sum in 1914. For the Scottish regiments, this made the Royal Scots Greys and the Scots Guards the most expensive, but Highland regiments such as The Black Watch or The Argyll and Sutherland Highlanders required a private income close to £400 a year. Lowland regiments were a little less costly, at an average of £250 a year, and the implied differences in social scale were not only understood but accepted by the men involved. So too were the differences in rank within a battalion. Officers were meant to be a breed apart, but they were also supposed to put their men's best interests before their own and prided themselves on setting high personal standards of behaviour.

Notwithstanding the dangers of generalisations, it remains ture that the Scottish infantry regiments of the regular army in 1914 were very much the nation in uniform, and they formed the basis for the regiments as they existed until recently. In each battalion there were representatives of the different social classes, from well-connected and wealthy aristocrats or landed gentry in the officers' mess, to the rank and file who had known only the subjugation of extreme poverty. In between were small numbers of the professional classes, clerks, shopkeepers, trained artisans and farm labourers, but the bulk of Britain's Regular Army was drawn from what one historian has called 'the lower end of the working class'.[14] There was a fair share of men who drank too much or gambled away their modest pay; equally there were temperance men and others who used their spare time to improve their educations. There were officers who did not fit in with the battalion's way of life and behaved irresponsibly, whilst there were those who took a Jesuitical interest in their calling and put a high premium on personal honour and courage. There were battalion commanders who would fail the test of combat, and men who forgot to die like heroes in the hellish fear of battle. But there were also those, the majority, who were bound up in common cause to serve the regiment which had become their physical and spiritual home. They might have spent a great deal of time grumbling, and they probably thought they were better trained and more professional than they really were but, warts and all, they represented the first flowering of Scotland's contribution to the war effort in the late summer of 1914.

During World War I the size of the army was increased by volunteers, and after 1916, for the first time, by conscription. By the end of the war, the number of Scots in the armed forces amounted to 688,416, consisting of 71,707 in the Royal Navy, 584,098 in the army (Regular, New and Territorial) and 32,611 in the Royal Flying Corps and Royal Air Force.[15] Although the army was reduced in size again after the armistice in 1919, the context in which young Scots continued to join the Scottish regiments remained fairly constant throughout the twentieth century. Apart from World War II, when conscription was reintroduced to make up numbers, the Regular Army was small – 207,000 in 1920; 102,000 in 2005 – and its soldiers were all volunteers. For the rank and file, the chance to lead an adventurous life was appealing and was a regular theme in the army's recruiting posters throughout the period. Unemployment was also a spur, and until fairly recent times most private infantry soldiers came from working-class backgrounds. By the same token the social background of officers remained much the same as it had been in the pre-1914 army, and it was not until the 1990s that the army made greater efforts to recruit officers from a wider social background.

Currently, soldiers sign on for a maximum of 22 years' service and under normal circumstances must serve four years of reckonable service dependent on age at enlistment, including basic military training and subsequent specialist training. For junior entry, recruits are in the 16–17 age group, while the age limit for adult entry is 32. Once soldiers have completed the full term of their career, the opportunity to extend their service beyond 22 years is open to selected individuals who still have valuable experience and skills. There are two main forms of continued service available: the Long Service List, which offers further employment, normally in periods of five years at a time up to the age of 55; and Limited Continuance, which provides short-term continuation of service for soldiers with specific skills or to provide cover where there are deficits in personnel levels. In its recruiting literature the army emphasises the training and educational opportunities that are on offer: not just military training, but the chance to gain recognised qualifications such as NVQs (or SVQs) and BTECs, or practical qualifications such as a driving licence or HGV licence. Career development is designed to be integral to this structure, and skills acquisition is continuous. According to army recruiters the following basic skills will be acquired by a soldier during his/her service:

> Responsibility: take responsibility for yourself – and as you gain promotion, be responsible for the others in your team who'll look to you for direction.
> Teamwork skills: you are a vital part of the British Army, so you will learn to work and live with others as highly skilled and dedicated as you.
> Decision-making: learn to think on your feet and assessing the best course of action in any situation.
> Technical expertise: you'll apply the skills you gain to any situation, and you'll be able to give advice to others and identify the solutions to problems, all using your newly acquired knowledge.[16]

Pay is low by most civilian standards (a trained private earns £13,866) but the army points out that the sums are not always comparable with jobs elsewhere, as soldiers do not always have to pay for their living costs. In return, the army offers a different kind of lifestyle, which can be summarised in the following terms: job satisfaction; working as part of a team in situations that most other organisations cannot handle; seeing the world; getting the most out of life by working with people of different ages and backgrounds; and experiencing active service. Recruits joining Scottish regiments give a variety of reasons for joining up: the desire to enter a family regiment; a need to belong to an identifiable organisation; the chance to wear the kilt; the longing for a life of adventure; and the opportunity to display a sense of pride in one's country. Some will have served already in the Territorial Army, and according to one recruiting officer most recruits bring with them a native toughness which stands them in good stead: 'Quite a few of them have hard and rough upbringings which helps them to adapt to the rigours of army life, not in a roughie-toughie sort of way but the kind of hardness you might need on active service.'[17] However, he cautioned that few have any real comprehension of what it is like to be a soldier other than what they have seen on television or in films. With that in mind, the training period includes time devoted to basic military history and regimental traditions. For the Scottish soldier there is also pride in uniform, especially the kilt, and the opportunity to take part in high-profile ceremonial duties, such as standing guard at Edinburgh Castle or forming the royal guard at Balmoral. Pride in regiment is also strong, and the formation of the new Royal Regiment of Scotland will test the tribal loyalties which lie at the heart of the Scottish regimental system. A recent historian of the British Army has expressed that sense of belonging: 'Scottish regiments are notable for the way their character is defined not so much by the officers' or sergeants' mess but by the Jocks themselves.'[18]

In practice this meant that each regiment had its own distinctive personality, which was formed by the areas from which they recruited.

Although every regiment in recent years gained recruits from commonwealth countries – mainly Fiji and South Africa, or Gurkha riflemen from Nepal – the battalions of the Royal Regiment of Scotland all have their own recruiting areas, and this helps to colour their complexion. The Royal Scots recruited from Edinburgh, the Lothians and Tweeddale; The Royal Highland Fusiliers from Glasgow and Ayrshire; The King's Own Scottish Borderers from the Borders and Lanarkshire; The Black Watch from Angus, Fife and Perthshire; The Highlanders from the north-east, Highlands and the Western and Northern Isles; and The Argyll and Sutherland Highlanders from Argyll, Stirlingshire and central Scotland. The structure of the new regiment maintains those traditions, but there are fears that the character of the battalions will be diluted by the inevitable cross-posting which allows soldiers to pursue their careers within the different formations of the new regiment. A former soldier in The Gordon Highlanders argues that each Scottish regiment is part family and part fighting formation, and that the links which bind the soldiers will be lost in the new large regiment:

> Each Scottish Regiment is a 'Family' with distinct and independent traditions. To lose these traditions, is to lose your identity which affects both morale and pride, and in the long run recruiting. It was always said that 'once a Scottish soldier always a Scottish soldier', even after leaving the Army you 'belonged' to the Regiment. The very spirit of the Scottish soldier will be destroyed with the loss of 'the sense of belonging' and with that goes everything that made OUR Regiments admired by our allies and feared by our enemies.[19]

Scots have also volunteered to join two other Scottish regiments: the Scots Guards, the third oldest of the foot guards regiments which number public duties at the Royal households as part of their role; and the Royal Scots Dragoon Guards, an armoured regiment operating main battle tanks. They also join other units of the army, such as the Royal Artillery and Royal Engineers, but the opportunity to serve in a Scottish infantry regiment is still one of the main reasons for joining the army. Not only does this give the soldier the chance to wear a smart and distinctively Scottish uniform, but the regiment – and in the case of the new formation, the operational battalion – provides a sense of belonging and group cohesion, the extent of which is perhaps unique in the workplace. Regimental attachments run deep. The Royal Scots are proud of the fact that they are the oldest line infantry regiment in the British Army, the 'First of Foot', and rejoice in their nickname of 'Pontius Pilate's Bodyguard'. The Black Watch is the senior Highland regiment and wears the distinctive black-and-green government tartan kilt with a red hackle in the soldiers' bonnets, and in common with the other Highland regiments it takes a great deal of interest in its traditions and heritage. On ceremonial occasions the Royal Scots Dragoon Guards still ride exclusively on big grey horses similar to those that charged at Waterloo and Balaklava, and the Scots Guards can trace their history back to the British civil wars of the seventeenth century. Of course, English, Irish and Welsh regiments are equally proud of their histories and traditions, but within the British Army the kilts of the 'tartan curtain' are difficult to ignore, even by those outside Scotland who sometimes resent the public attention given to the Highland regiments.

After the intensity of military life and the sense of being a member of a close-knit community, the return to civilian life can be dislocating, especially for those who have seen active service. The different regimental associations and the local branches of the Royal British Legion Scotland provide a structure for soldiers to stay in touch and opportunities to socialise, but the intimacy of service life is often difficult to replicate outside the army. Although the army offers opportunities for retraining before soldiers leave, these are not always taken up, or there is insufficient time for them to be implemented.

A growing number of ex-service personnel also suffer from combat-related psychological injuries, such as post-traumatic stress disorder (PTSD), caused by the traumatic events they have experienced in service. They are cared for by the Ex-Services Mental Welfare Society, which operates as 'Combat Stress' and has its main treatment centre in Scotland at Hollybush House in Ayrshire. In 2004 it cared for 868 veterans who were admitted for short-stay remedial care to help them rebuild their lives, leaving the charity to admit that it 'has never been busier than it is today'.[20] Some of those seeking treatment had served in recent deployments of the British Army in Northern Ireland, the Balkans and Iraq. One mother whose son served in Northern Ireland realised that he was suffering from PTSD when he came home from a tour of duty and she found him 'screaming in the night wrestling with his nightmares ... curled up in the corner of his room, shaking violently, beads of sweat sticking to his forehead'. After he was diagnosed with PTSD he was admitted to Hollybush, where 'he visits three times a year and is beginning to make progress with the support and care he has been given'.[21] Health workers involved with Combat Stress agree that the problem of PTSD is increased when soldiers leave the army and are no longer protected by the group loyalty and cohesion of the formation with which they served.

Despite those problems which have only recently been recognised and fully understood, and for all that there has been a great deal of turbulence in the wake of poor recruiting figures and the reorganisation of the Scottish infantry regiments, soldiering as a career still carries a cachet in Scotland. A glance at the regimental rolls shows that sons – and now, in some cases, daughters – are following their fathers not just into the army but into the same regiment. Amongst senior commanders hopes are high that the same loyalty and desire to be a Scottish soldier will be attached to The Royal Regiment of Scotland. Responding to critics who complained that the reforms hit at the heart of the old regimental tradition and would not work, Major-General Euan Loudon, general officer commanding 2 Division and colonel-commandant of the Scottish Division, reminded the campaigners that the regiment was not an end but a new beginning:

First, it has been made quite clear since the start of our work to form a new regiment that if a young man wants to follow family tradition and join The Royal Highland Fusiliers (for example), he can do so now and in the future. A crucial part of the new regiment is that the five battalions will retain their names and links to regional communities throughout Scotland. Secondly, army recruiting has peaks and troughs throughout the year depending on a range of factors; further education and buoyant employment also affect our ability to recruit young people.

However, the Army restructuring is not having a significant effect on recruitment. In fact, research among 300 young men interested in joining the army showed that 94 per cent said the restructuring would not affect their decision to join; and 86 per cent said they wanted to join a Scottish regiment. There is a minority of people, mostly within the retired army community, who are resistant to change and modernisation. I would say to them that we understand your feelings, and are working hard to retain historic identities and community links within the new Royal Regiment of Scotland. But this restructuring is what the army wants and needs. We must modernise and move forward to tackle new challenges, while also providing our soldiers with better career opportunities and stability for their families.[22]

In a crowded job market and against the background of a declining population, like any other employer in Scotland the army will have to fight hard to make sure that it continues to attract recruits and then retain their services. This is the major challenge for The Royal Regiment of Scotland. History and tradition have helped to turn the Scottish soldier into an iconic figure, and the kilted infantryman is one of Scotland's best known 'brands', but all of this counts for nothing if young people fail to join the Scottish regiments in sufficient numbers to ensure that they retain their distinctive identities.

APPENDIX

The proposed new structure of the infantry:

An infantry regiment consists of a number of operational battalions of 650 soldiers. The infantry regiments of the British Army belong to administrative divisions with their own headquarters: Household, Scottish, Queen's, King's, Prince of Wales's and Light. Individual battalions serve in operational divisions and brigades (see below).

The Guards Division

1st Bn Grenadier Guards
1st Bn Coldstream Guards
1st Bn Scots Guards
1st Bn Irish Guards
1st Bn Welsh Guards

The Scottish Division
The Royal Regiment of Scotland
Regular Battalions
  The Royal Scots Borderers, 1st Bn The Royal Regiment of Scotland
  The Royal Highland Fusiliers, 2nd Bn The Royal Regiment of Scotland
  The Black Watch, 3rd Bn The Royal Regiment of Scotland
  The Highlanders, 4th Bn The Royal Regiment of Scotland
  The Argyll and Sutherland Highlanders, 5th Bn The Royal Regiment of Scotland

Territorial Army Battalions
  52nd Lowland, 6th Bn The Royal Regiment of Scotland
  51st Highland, 7th Bn The Royal Regiment of Scotland

The Queen's Division
The Princess of Wales's Royal Regiment (Queen's and Royal Hampshires)
Regular Battalions
  1st Bn The Princess of Wales's Royal Regiment (Queen's and Royal Hampshires)
  2nd Bn The Princess of Wales's Royal Regiment (Queen's and Royal Hampshires)

Territorial Army Battalion
  3rd Bn The Princess of Wales's Royal Regiment (Queen's and Royal Hampshires)

The Royal Regiment of Fusiliers
Regular Battalions
  1st Bn The Royal Regiment of Fusiliers
  2nd Bn The Royal Regiment of Fusiliers

Territorial Army Battalion
  5th Bn The Royal Regiment of Fusiliers

The Royal Anglian Regiment
Regular Battalions
  1st Bn The Royal Anglian Regiment
  2nd Bn The Royal Anglian Regiment

Territorial Army Battalion
  3rd Bn The Royal Anglian Regiment

The King's Division
The Duke of Lancaster's Regiment (King's, Lancashire and Border)
Regular Battalions
  1st Bn The Duke of Lancaster's Regiment (King's, Lancashire and Border)
  2nd Bn The Duke of Lancaster's Regiment (King's, Lancashire and Border)

Territorial Army Battalion
  4th Bn The Duke of Lancaster's Regiment (King's, Lancashire and Border)

The Yorkshire Regiment
Regular Battalions
1st Bn The Yorkshire Regiment (Prince of Wales's Own)
2nd Bn The Yorkshire Regiment (Green Howards)
3rd Bn The Yorkshire Regiment (Duke of Wellington's)

Territorial Army Battalion
4th Bn The Yorkshire Regiment

The Prince of Wales's Division
The Mercian Regiment
Regular Battalions
1st Bn The Mercian Regiment (Cheshire)
2nd Bn The Mercian Regiment (Worcesters and Foresters)
3rd Bn The Mercian Regiment (Staffords)

Territorial Army Battalion
4th Bn The Mercian Regiment

The Royal Welsh
Regular Battalions
1st Bn The Royal Welsh (The Royal Welch Fusiliers)
2nd Bn The Royal Welsh (The Royal Regiment of Wales)

Territorial Army Battalion
3rd Bn The Royal Welsh

The Light Division
The Rifles
Regular Battalions
1st Bn The Rifles
2nd Bn The Rifles
3rd Bn The Rifles
4th Bn The Rifles
5th Bn The Rifles

Territorial Army Battalions
6th Bn The Rifles
7th Bn The Rifles

The Royal Irish Regiment
Regular Battalion
1st Bn The Royal Irish Regiment

Territorial Army Battalion
The Royal Irish Rangers

The Parachute Regiment
Regular Battalions
1st Bn The Parachute Regiment
2nd Bn The Parachute Regiment
3rd Bn The Parachute Regiment

Territorial Army Battalion

4th Bn The Parachute Regiment

Note: 1st Bn The Parachute Regiment will form the core element of the new Special Forces Support Group and will be removed from the formal Infantry structure.

The Brigade of Gurkhas

1st Bn The Royal Gurkha Rifles
2nd Bn The Royal Gurkha Rifles

Under Director Special Forces

22nd Special Air Service Regiment
22 SAS Special Reconnaissance Regiment

The SAS can be classed as an infantry unit but the members of the regiment are found from all arms and services in the Army after exhaustive selection tests.

A brigade is a collection of different regiments and supporting units that have been grouped together for a specific purpose. A fighting brigade will traditionally contain infantry, cavalry and artillery regiments together with many supporting cap badges. The composition of each brigade will differ depending on its responsibility but could often contain 5,000 soldiers. A division would traditionally be made up of three or four brigades, depending on the specific role it is to undertake, and is configured in a similar fashion to a brigade but on a larger scale. 1 (UK) Division and 3 (UK) Division are fighting divisions, whereas 2, 4 and 5 divisions are responsible for administrative support of specific geographical areas.

1 (UK) Division

4th Armoured Brigade
7th Armoured Brigade
20th Armoured Brigade
102 Logistics Brigade

2 Division

15 Brigade, North East
42 Brigade, North West
52 Brigade, Scottish
Catterick Garrison

3 (UK) Division

1 Mechanised Brigade
12 Mechanised Brigade
19 Mechanised Brigade

4 Division

2 Infantry Brigade
16 Air Assault Brigade
49 East Brigade
145 Home Counties Brigade
Aldershot Garrison
Colchester Garrison

5 Division

42 Wessex Brigade
143 West Midland Brigade
160 Wales Brigade

HQ Northern Ireland

8 Infantry Brigade
39 Infantry Brigade
10 Ulster Brigade

NOTES

1. *Summary of Enlistments Scotland*, Headquarters 2 Division, South Queensferry, 8 January 2005.
2. Dodd, V and Jones, S. Opposition demands Iraq answers, *Guardian*, 28 April 2005.
3. Irwin, 2004 .
4. Royle, T. Beat the retreat, *Sunday Herald*, 12 December 2004.
5. Ferguson, C. Top soldier's fury at Black Watch 'insult', *Courier*, 1 December 2004.
6. Royle, 1992, 71.
7. Royle, 1992, 71.
8. Keegan, 1983, 168.
9. Devine, 1999, 184–5.
10. *Parliamentary Papers*, 1927.
11. Henderson, 1989, 44.
12. Baynes, 1967, 133–48.
13. *Parliamentary Papers*, 1917, para. 1052.
14. Strachan, 2001.
15. Corbett and Newbolt, 1931, appendix J; PP, 1927; Raleigh, and Jones, appendices 35–6.
16. *Army Jobs: Army Life*, British Army recruiting literature, 2005.
17. Royle, 1986.
18. Beevor, 1990, 250.
19. Mac, 2005.
20. Woods, 2004, 2.
21. Woods, 2004, 2.
22. Loudon, Major-General E. Letter to editor, *Herald*, 19 August 2005.

BIBLIOGRAPHY

Allan, S, Carswell, A. *The Thin Red Line: War, Empire and Visions of Scotland*, Edinburgh, 2004.

Baynes, J. *Morale: A Study of Men and Courage, the 2nd Scottish Rifles at the Battle of Neuve Chapelle, 1915*, London, 1967.

Baynes, J and Laffin, J. *Soldiers of Scotland*, London, 1988.

Beevor, A. *Inside the British Army*, London, 1990.

Corbett, J and Newbolt, H. *Naval Operations*, vol. V, London, 1931.

Devine, T M. *The Scottish Nation 1700–2000*, London, 2001.

Henderson, D M. *Highland Soldier 1820–1920*, Edinburgh, 1989

Henderson, D M. *The Scottish Regiments*, Glasgow, 1996.

Irwin, Lieutenant General Sir A. What is best in the regimental system?, *Journal of the Royal United Services Institute*, 149, 5 (October 2004), 32–7.

Keegan, J. *Six Armies in Normandy*, London, 1983.

McCorry, H, ed. *The Thistle at War: An Anthology of the Scottish Experience of War, in the Services and at Home*, Edinburgh, 1997.

Mac, A. I watched my regiment disappear. Save the Scottish regiments website, available at: http://www.savethescottishregiments.co.uk/feedback/61to90/doc.htm, accessed August 2005.

Mileham, P J R. *Scottish Regiments*, Tunbridge Wells, 1988.

Parliamemtary Papers, Report of the Royal Commission on the Housing of the Industrial Population of Scotland, Rural and Urban, Edinburgh, 1917 C8731.
Parliamemtary Papers, General Annual Reports of the British Army (including the Territorial Force from date of embodiment) 1 October 1914 to 30 September 1919, 1927, XX, C1193.
Raleigh, W and Jones, H A. *History of the Great War in the Air,* vol. VI, London 1937.
Royle, T. *Awa' for a sodger: A portrait of the Scottish soldier.* BBC Radio Scotland, July 1986.
Royle, T. Scotland and defence. In Linklater, M and Denniston, R, eds, *Anatomy of Scotland: How Scotland Works,* Edinburgh, 1992, 71.
Royle, T. *Flowers of the Forest: Scotland and the First World War,* Edinburgh, 2006.
Strachan, H. *First World War,* vol. I, Oxford, 2001.
Wood, S. *The Scottish Soldier,* Manchester, 1987.
Woods, B. A view from the ground, *The Hollybush House Appeal,* Ayr, 2004 [leaflet produced as part of a fundraising drive on the part of Combat Stress, the ex-services mental welfare society].
Young, D. *Forgotten Scottish Voices from the Great War,* Stroud, 2005.

12 Insurance Industry

ALAN WATSON

INTRODUCTION

At the outset it is necessary to state that discussion of the history of the insurance industry in Scotland is somewhat problematic as there is limited material available. Most analysis is to be found in textbooks relating to the history of the United Kingdom insurance industry or is contained in insurance company histories. For example, *A Sense of Security: 150 Years of Prudential* and *A Premium Business: A History of the General Accident*, to name but two of many similar works.[1]

Lack of detailed research into the insurance industry in Scotland per se means that there is a need for further general research as identified by Pearson:

> Notwithstanding the attempt to be comprehensive, it must be acknowledged that certain areas have not received the attention they deserve. Scotland and Ireland are largely, though not entirely, viewed through the records of the English fire offices which operated there. The development of the native insurance industries in both countries remains to be properly researched.[2]

OVERVIEW

Insurance is a technically diverse, complex and dynamic industry, the influence of which in society is often underestimated and understated. The image of insurance as represented and advertised is often poor.[3] However, this does not detract from the fact that insurance provides essential financial protection in times of uncertainty. It provides financial security, compensation and protection to individuals, groups, industrial and commercial organisations in the United Kingdom and throughout the world. The diverse nature of insurance is extensive, encompassing marine, aviation, property, life, liability, accident and engineering risk exposures and many other areas. The Association of British Insurers has stated that:

> Insurance is, therefore, essential to the smooth running of the economy. It allows capital to be freed up and put to productive use. It also enables companies and individuals to go about their business in the

knowledge that if something unforeseen did occur, then they would not suffer financially. If this confidence did not exist, then many firms and individuals would reduce the extent of their business activities, while some would choose not to engage in any such activity at all. Furthermore, a lack of confidence would also curtail innovation and many worthwhile investment projects would not be undertaken.[4]

Insurance reduces risk and losses both directly and indirectly. Directly it may be done, for example, by insurers educating consumers on measures that can be taken to reduce risks, by campaigns against crime and arson, or by encouraging accident prevention. Indirectly, insurance encourages individuals and firms to take measures to reduce their risk and losses so that they can benefit from lower insurance premiums.[5]

The UK non-life market is probably the most volatile of any commercial or industrial sector.[6] This is in part due to the fact that the industry not only offers compensation and protection but also invests in industry and commerce, finances research in medicine and engineering, provides fire and accident prevention and contributes significantly to the nation's invisible earnings.[7]

DEFINITIONS OF INSURANCE

Insurance is a method of risk transfer, which spreads the losses of the few over the contributions of the many. It is a system of money collection – termed premium – placed in a fund by insurers so that when losses arise payments, or claims, may be drawn from this fund. According to Dickson, 'Insurance is a risk transfer mechanism';[8] Birds defines insurance as 'a form of financial protection' and 'a legal contract';[9] but Pfeffer considers that insurance is 'a device for the reduction of uncertainty of one party called the insured, through the transfer of particular risks to another party, termed the insurer, who offers a restoration, at least in part, of the economic losses suffered by the insured'.[10]

INSURANCE HISTORY

In order to appreciate the nature, function and operation of the insurance industry and its recent past and present position, it will be of value to describe and examine the factors that led to the present market situation. The insurance industry has been the subject of major influences, for example, the Great Fire of London, the Industrial Revolution, both world wars, the introduction of the motor vehicle, aircraft and spacecraft, and the technological revolution. The technological revolution can be considered to be the recent past, where computer-based technology, from personal computers and personal data organisers to large mainframe computers (including hardware and associated software), has had a major impact on individuals, industry, commerce, society, global communication and, in no small part, the insurance and financial services industries. All of these factors and many others have shaped the insurance industry in Scotland and the UK over many centuries.

The earliest contracts of insurance were legal arrangements, developed before the Christian era, that embodied elements of what later became social and marine insurance. The major documentation for this view is the Code of Hammurabi (*c*1750 BC), which took the form of a primitive bottomry loan or bond or form of insurance. Bottomry is where a ship owner pledged the bottom of his vessel in order to borrow money to buy cargo. The bottomry contract consisted of the loan, an interest rate and a premium for the risk of loss. In exchange, the owner did not have to repay the loan if the vessel was lost at sea. This was not fully developed until the ninth–eighth centuries BC. These facts suggest that the insurance concept is not modern, but reverts back to the commencement of maritime trading around the Mediterranean Sea. Disagreement exists as to the commencement of marine insurance: 'Although marine insurance dates back to 600 BC, most types of non-life insurance are less than a century old.'[11] The development of insurance through the expansion of commercial activity around the Mediterranean resulted in what could be described as a primitive form of insurance as we know it today, with marine insurance evolving around the twelfth and thirteenth centuries. Marine insurance was followed by ordinary life assurance in the sixteenth century, and the Great Fire of London of 1666 prompted fire insurance, with the concern of fire insurers further raised after the Tooley Street Fire in 1861, both fires destroying considerable areas of the City of London.

The earliest insurance associations were formed for commercial purposes, i.e. to conduct and transact insurance. Marine insurers in London met in the vicinity of the Royal Exchange and, in the seventeenth century, at Edward Lloyds' coffee house. The association formed at this time transferred in 1774 to the Royal Exchange. In 1802, in Liverpool, marine underwriters formed the Liverpool Underwriters' Association. An association of fire insurance managers formed in 1829 in Edinburgh. The Faculty of Actuaries in Scotland was a splinter group from the English Institute of Actuaries and was formed as a result of a suggestion made at a meeting of the Scottish Managers' Association in 1848. Various associations in Scotland were formed as a result, such as the Insurance and Actuarial Society of Glasgow in 1881.[12]

An interesting fact associated with the formation of insurance institutes in the UK is that by 1897 there were ten insurance institutes, an unusual feature being the omission of certain institutes in London, Edinburgh and Liverpool. Edinburgh's absence may be accounted for by the development of life insurance and assurance in the capital, and by the fact that insurance influence was catered for by the Faculty of Actuaries and a subsidiary, the Edinburgh Actuarial Society. The failure of the London Insurance Institute is difficult to comprehend. In 1886 an attempt was made to create such an insurance institute, but it foundered due to apathy and indifference amongst the senior managers.

The next major impetus to the development of the insurance industry was the Industrial Revolution in the UK and the implications for the Scottish economy that flowed from this period of change. This resulted in the emergence of a number of forms of insurance created mainly to protect employers from the uncertainty of loss due to the explosion of plant and machinery, thereby resulting in death or injury to employees or members of the public, and/or damage to their property. The mid-nineteenth century saw the development of pecuniary forms of insurance, for example fidelity guarantee, credit and business interruption insurance. A number of these 'modern' insurances were offered in many of the 'new' branches and insurers located in Glasgow and Edinburgh during the nineteenth century. This period also saw the development of many forms of insurance relating to financial compensation, personal financial provision as a result of accident and/or disease, financial provision for dependants through personal accident insurance, and permanent health insurance. Many of these types of insurance were offered in the 'new' financial services sector in the central belt of Scotland at this time.

Motor insurance developed as a result of the introduction of the motor vehicle as a means of transport with its attendant problems of injury, death to individuals and/or possible damage to their property. The Law Accident and Guarantee Insurance Society was offering policies to motor vehicle owners in 1889 in the UK. Since the early 1900s insurance in all its facets has grown considerably and has at all times recognised and reacted to the increasing needs of industry and commerce. Insurance has provided detailed, innovative and complex contracts to meet industrial and technological growth, insuring oil rigs in the North Sea, wide-bodied aircraft, super tankers, space satellites and space shuttles, advanced computer systems (IBM Greenock) and nuclear installations throughout Scotland at Douneray, Hunterston A and Hunterston B, and Torness.

THE INSURANCE INDUSTRY IN SCOTLAND AND THE UNITED KINGDOM

A brief introduction to the insurance industry in the UK at this point will help put into context the contribution of insurance to the UK economy and the situation of the industry within the global insurance market. The industry is a major employer, with some 370,000 employees in insurance and auxiliary occupations.[13] UK premium income as a percentage of gross domestic product (GDP) is the second highest in Europe, and fourth in world rankings.[14] The financing role of insurance companies is extremely important; a change in their investment portfolios may influence not only the terms on which capital is raised but also the consequences of debt management and monetary policy. Insurance companies, unlike banks, cannot create money, but they can, like other institutional investors, affect monetary demand by buying or selling securities, or by changing the structure of their investments.

Dunning provides a historical perspective on the importance of international insurance markets;

> Besides the UK, there are five other leading insurance countries: the US, which accounts for about three-fifths of world premium income, West Germany, Japan, France and Canada ... Switzerland alone, however, is an active exporter of insurance, particularly reinsurance. In premium income, and including the operations of foreign subsidiaries, the UK insurance industry is the second largest in the world, and it is considerably bigger than its European counterparts.[15]

Scotland has a distinguished history in financial services that dates back over some 300 years. It is now one of Europe's leading financial centres and the second largest financial hub in the UK next to London. Scotland is recognised for its strengths in insurance, banking, life insurance and pension provision. It also has a vibrant general insurance, corporate finance and broking services sector and a supportive and dynamic ancillary services sub-structure. The international financial services industry located in Scotland has an enviable record for excellence and innovation providing diverse and numerous financial products to industry and commerce in a global workplace.

The financial services industry in Scotland accounts for £70 billion (7 per cent) of Scotland's GDP, including leading international Scottish companies and major companies who are investing in the Scottish economy. The financial services sector continues to be one of the fastest growing industry sectors in Scotland with a growth rate from 2000 to 2006 of 55 per cent, whilst the corresponding growth rate for Scotland as a whole was 13 per cent. The UK financial services sector grew by 44 per cent over the same period. The financial services sector accounts for one in ten Scottish jobs, with over 108,000 individuals directly employed in the industry and over 100,000 employed in ancillary services.

Scotland ranks second to London in the European league table of headquarters locations of the largest banks in Europe, measured by market value. Scotland is the headquarters location of four clearing banks, including the Royal Bank of Scotland Group and HBOS. Many other international banks also have operating bases in Scotland, and the sector continues to expand, with Scottish Widows establishing its bank in 1995 and the Standard Life Bank opening its doors for the first time in 1998. Scotland's banks and building societies undertake a wide range of business activities, including retail and corporate banking, treasury, insurance, actuarial and consultancy services, mortgage advice, investment management and corporate finance.

The Union of the Parliaments in 1707 opened up lucrative trade to Scotland with the colonies in America and the West Indies, and the merchants of Glasgow and Edinburgh were not slow to identify the potential of this trade. The ensuing trade activity led to the import of sugar, tobacco, rum, wood and cotton and the export of linen and leather goods. The tobacco trade laid the foundation for Glasgow and created a sound foundation for trade across the globe. This in part ceased with the American War of Independence.

Around this time on the outskirts of Glasgow, in the environs of Lanarkshire in particular, ironworks and coalfields developed. This activity ensured the future of Glasgow as an industrial centre together with industries such as shipbuilding, engineering and manufacturing. Glasgow as a centre grew rapidly from a population of less than 30,000 in the mid-eighteenth century to some 400,000 by the 1850s. This growth continued as individuals gravitated from rural districts and from Ireland, and the city grew further by absorbing surrounding burghs

It was not surprising that the insurance industry in other parts of the UK was taking an interest in Scotland due to its people, industry and wealth. By the 1880s it is recorded that Glasgow had branches representing eighty insurance companies, and four life offices established head offices in Glasgow. The emphasis of business tended to be on fire and life insurance until the latter part of the nineteenth century and the earlier part of the twentieth century, with the advent of the motor vehicle and motor insurance in the 1930s. A major influence in motor insurance in Scotland was General Accident, with its head office in Perth.

The developing import–export trade had created a demand for marine insurance, and the Glasgow Marine Association was formed in 1818 to represent and develop the interests of the hull and cargo underwriters operating in the city at this time. The importance of the Clyde is reflected in the fact that the Salvage Association in the UK had a base in Glasgow, in addition to the ports of London and Liverpool. The Insurance & Actuarial Society of Glasgow was formed in 1881, only the second Chartered Insurance Institute local institute in the UK after the formation of the Manchester branch.

Such was the commercial background to the formation and development of the insurance industry in the West of Scotland that it continues as a major financial sector in Scotland and the UK, despite considerable change in both the structure of the market and the demand for diverse financial products.

One of the earliest records of insurance activity in Scotland is that of the establishment of the Friendly Society of Edinburgh in 1720. Movement of offices from London due to increased competition, overpricing, and the use of cartels resulted in rate-cutting and the development of offices in towns such as Glasgow and Edinburgh, and with prominent businessmen and gentry heavily represented on their boards, the country offices insisted on their local identity. The common law of co-partners recognised the legal status and limited liability of partnerships, and this circumvention of the 'Bubble Act' placed the Caledonian Insurance Company, 1805 and the North British, 1809 in a strong and well-protected financial position. The Bubble Act of 1720 was repealed in 1824.

Life assurance in Scotland began in 1815 when Scottish Widows was created to meet the needs of women whose husbands had been killed in the Napoleonic wars. Later in the nineteenth century, Scots pioneered the investment trust, the world's first mutual fund. The Scottish Widows Fund and Life Assurance Society opened in 1815 in what is now Chambers Street in Edinburgh. The formation of the society had been discussed from March 1812 with the purpose of providing for widows, sisters and other female relatives forced into poverty on the death of the fund holder. Its most famous customer was Sir Walter Scott, who effected a policy in 1824. Scottish Widows uses this fact in advertising its products to the present day, but fails to mention that Sir Walter died in 1832.

The managers of the Edinburgh fire offices began regular meetings in 1829, under the auspices of the Association of Managers of Fire Insurance Offices. Fifteen offices had joined the Scottish companies in a tariff on Manchester drying stoves used to dry wool in the eighteenth century, with an associated fire hazard, and a similar tariff operated for Glasgow and Paisley warehouses after 1833. The National Life appointed its first agents and medical examiners in Edinburgh in 1830 and operated through sixty agencies by 1856 (including an agent in France). The Associated Scottish Life Offices formed in 1834 in Scotland and the Life Offices' Association formed in 1889, serving as a forum in the United Kingdom for industry-wide problems of the ordinary life offices. Their records are important for the evidence they provide as to the impact of insurance and tax legislation. The two professional bodies, the Institute of Actuaries, 1848 and the Faculty of Actuaries in Scotland, 1856, were concerned with the actuarial significance of new classes of business or changes in mortality experience, where previously rates and premiums were based and insured on the proposer's historical loss experience. The Scottish Minister's Widows' Fund of 1744 is the earliest recorded actuarial-based fund in the world. The widening of the market for life assurance brought with it problems of rating new types of risk, reaching territorial agreements with competing companies, and meeting the requirements of domestic and overseas governments. The life assurance companies did not adopt the forms of combination used by fire offices, but many of these problems were dealt with jointly by representatives of the major companies.

Cotton mills and woollen mills often caused fire insurers considerable concern in the 1800s, due to their susceptibility to fire and associated perils, and no more so than in Scotland where it was noted that, according to the Royal Exchange Assurance's Special Fire Committee in 1841, there was to be an increase in premium rates 'in consequence of the very numerous and destructive fires in cotton mills, particularly in Scotland'. Apparently, over the previous years the cost of claims had considerably exceeded the premium income. Insureds were of a mind often in those days of allowing the conflagration to continue unattended until the last minute, in the hope of making a profit from the insurer (not unlike today).[16]

Competition for overseas life business in the UK did not emerge until the 1840s. Apart from Pelican, early exporters of life assurance included Standard Life (with a Quebec agency in 1883) and Scottish Amicable (in the West Indies in 1845 and Montreal in 1846). As a group, the Scottish Life offices developed a large and influential overseas business in North America in the mid-nineteenth century. The majority of British insurance companies promoted in the earlier nineteenth century were based in London. In 1860, all but 70 of the 213 operational insurance companies worked from London head offices. Edinburgh was the home of fifteen companies, Norwich of seven and South Shields in Durham, with seven marine insurance associations, had the next largest number of companies. Towards the end of the century, London companies were outnumbered by provincial companies, and by 1900 only 169 out of a UK total of 424 companies were based in London.

The Tooley Street Fire of 1861, which involved claims of £1 million, resulted in fire offices accepting the need for a municipal fire brigade in 1865, to which they made a financial contribution. At the same time, the fire offices formed and financed the London Salvage Corps, whose prime duty was to rescue and conserve property threatened by fire. Similar salvage corps operated in Liverpool from 1845, and in Glasgow from 1873. The idea of burglary insurance was meanwhile being developed in Scotland by the Mercantile Accident and Guarantee Insurance Company, established 1885 and the General Accident Fire and Life Assurance Corporation, established 1885.

Railway publications between 1885 and 1887 offered coupon insurance schemes with the Railway Passengers and General Accident companies. General Accident (now part of the Aviva Group) was the last of the independent companies of the Victorian accident offices. General Accident had included motor car insurance in its prospectus of 1896; its diversification into other classes of insurance culminated with the purchase of the Yorkshire Insurance Company in 1967. Most amalgamations in the nineteenth century had ended in the complete absorption of weak local or specialist companies. In the twentieth century, the purchase of large or strategically important companies was more common. Commercial Union, and Royal and Alliance, which had been established as fire or fire and life companies, accounted for the largest number of amalgamations in the early twentieth century. Their acquisitions included fire, life and marine and accident companies. The modern growth of General Accident and Eagle Star began with amalgamations in the same period.

An insurance quotation exists for vehicles entering the London to Brighton motor run on 14 November 1896, following the passing of the Locomotives on Highways Act 1896. The quotation was issued by the Scottish Employers' Insurance Company Limited, which also quoted annual rates for the vehicles.

Shortage of capital forced the Crusader Insurance Company to begin its career by taking its head office on a tour of Britain. Established in Glasgow as Mutual Property Recovery and Accident in 1902, the company sought new shareholders by moving to London in 1906, Manchester in 1910, and back to London in 1916. It is interesting to note that the selection and registration of a title was normally straightforward, unless there was a risk of confusion with another company; the General Life Assurance Company successfully opposed General Accident's plan to change its name to the General Assurance Corporation around 1903.

The London and Lancashire acquired an accident income of £50,000 per annum when it bought the Equitable Fire and Accident Office of Manchester in 1901. The years between 1957 and 1968 were an era of amalgamations, commencing with the acquisition of the Caledonian by the Guardian in 1957.

The archives of the insurance industry are amongst the oldest and largest collections of business records in Great Britain. Full lists of many of the Scottish companies are held by the national Register of Archives (Scotland), National Archives of Scotland and in the Chartered Insurance Institute's brochure entitled 'Sources of Insurance History: A Guide to Historical Material'.

The business records of the United Kingdom Provident Institution and the Scottish Mutual Assurance Limited, for example – which originally insisted on abstention from alcohol by their policyholders – throw light on the composition of the temperance movement. Suicide and duelling were other excepted risks.

List of Companies Established:

Caledonian Insurance Company, established in Edinburgh, 1805.

Scottish Widows' Fund and Life Assurance Society, established 1815.

Scottish Provident (mutual office), established 1837.

Ayrshire Employers Mutual Insurance Association Ltd, established 1895, wound up 1957.

British Legal Life Assurance and Loan Company Ltd, established in Glasgow, 1863.

Bute Insurance Company Ltd, established Rothesay, Bute, 1873.

Century Insurance Company Ltd, established in Edinburgh as Sickness and Accident Assurance Association, 1885, renamed 1901.

Church of Scotland Insurance Company Ltd, established as Free Church of Scotland Fire Insurance trust in 1888, renamed United Free Church of Scotland Fire Insurance Trust in 1901 and renamed Church of Scotland Fire Insurance trust in 1930. Renamed Church of Scotland Trust in 1974, and again renamed 1982.

City of Glasgow Life Assurance Company, established in Glasgow, 1838.

Colonial Life Assurance Company, established in Edinburgh, 1846.

Crusader Insurance plc, established in Glasgow as Mutual Property Recovery and Accident in 1899.

Dufftown Plate Glass Association, established at Dufftown, Morayshire, *c*1895.

Edinburgh Assurance Company Ltd, established as Edinburgh Life Assurance Company in 1823, renamed 1919.

Edinburgh Friendly Insurance Society, established in Edinburgh, 1720.

Employers' Mutual Insurance Association, established in Edinburgh, 1898.

English and Scottish Law Assurance Association, established in London and Edinburgh, 1839.

General Accident Fire and Life Assurance Corporation plc, established in Perth as General Accident and Employers' Liability Assurance in 1885. Renamed General Accident Assurance in 1891, and renamed 1906.

Currently Glasgow continues as a major financial centre and makes a contribution of 8 per cent to the GDP of Scotland, and the insurance industry provides employment for some 7,500 within the city. Whilst mergers and rationalisation activity over the past twenty years have reduced the number of insurers and insurance brokers, Glasgow retains its stature as one of the leading financial services centres in Scotland.

CONCLUSION

Recently there has been a movement towards national and international downsizing of insurance companies, brokers and financial services organisations through mergers and acquisitions. For example, Commercial Union merging with General Accident, then Norwich Union, finally changing its name to Aviva. Other financial services institutions such as banks and building societies also began acquiring and establishing insurance subsidiaries. All of these mergers and acquisitions have had an impact on the insurance market/industry in Scotland, with a reduction in employment and the subsequent effect on the economy. However, recently Glasgow (Clydeside Development) and Edinburgh have seen a resurgence in insurance and financial services, with both cities creating financial services districts, thereby developing an infrastructure to promote financial services in the future.

The influence of DNA profiling and research into gene therapy is of interest to the life insurance market, with its implications on life insurance underwriting in the future, and with many life offices resident in Glasgow and Edinburgh the impact will affect all potential proposers and insured in Scotland in the future. Academic interest in insurance and financial services is offered by a number of universities in Scotland, including Napier University, Edinburgh University and Glasgow Caledonian University, which in the past offered courses in insurance at diploma level and associateship courses for the Chartered Insurance Institute for some 500 students each year in the West of Scotland.

It is hoped that that this brief consideration of the history of the insurance industry in Scotland has been of some interest and benefit. The industry itself is considerable, with life, accident, fire, engineering, accident, liability, marine and aviation each deserving detailed attention in relation to Scotland and the Scottish economy. It is hoped that this can be undertaken at some future date. It must be understood and appreciated that a historical study of the insurance industry and market is in fact a study into the foundation and background of the UK and world economy, due to its wealth and drive over many centuries, and Scotland can be proud to have contributed to the present global insurance industry and market.

NOTES

1. Dennet, 1998; Young, 1999.
2. Pearson, 2004, 9–10.
3. ABI. *Insurance, Facts, Figures and Trends*, ABI, October 1998, 19.
4. ABI, 1998, 3.
5. ABI, 1998, 3.
6. Keane and Diacon, 1996, 82–9.
7. ABI. *Insurance Statistics Yearbook*, 1987–1997, 72.
8. Dickson, 1984, 3.
9. Birds, 1988, 7.
10. Pfeffer, 1953, 53.
11. Dunning, 1971, 37.
12. Cockerell, 1957.
13. ABI. 'The contribution of insurance to the UK economy', available online at: http://www.abi.org.uk/Display/File/523/Economic_Values_02.05.pdf
14. ABI. 'The contribution of insurance to the UK economy', available online at: http://www.abi.org.uk/Display/File/523/Economic_Values_02.05.pdf
15. Dunning, 1971, 7.
16. Supple, 1970, 24.

BIBLIOGRAPHY

Association of British Insurers (ABI). *Insurance, Facts, Figures and Trends*, ABI, 1998.

Birds, J. *Modern Business Law*, London, 1988.

Clark, G. *Betting on Lives: The Culture of Life Insurance in England, 1695–1775*, Manchester and New York, 1999.

Cockerell, H A L. *Sixty Years of the Chartered Insurance Institute 1897–1957*, London, 1957.

Cockerell, H A L and Green, E. *The British Insurance Business: A Guide to its History and Records*, Sheffield, 1994.

Dennett, L. *A Sense of Security: 150 Years of Prudential*, London, 1998.

Dickson, G C A. *Introduction to Insurance*. The CII Tuition Service Study Course 010, London, 1984, Chapter 2, 1–7.

Dunlop, A I, ed. *The Scottish Ministers' Widows' Fund, 1743–1993*, Edinburgh, 1992.

Dunning, J H. *Insurance in the Economy*, London, 1971.

Keane, J and Diacon, S. The Cause of Insurance Cycles in the UK, *The Journal of the Society of Fellows*, 10:2 (January 1996), 82–9.

McArthur, A. Investment firms more than just managing, *Scotsman*, 21 June 2005.

Pearson, R. *Insuring the Industrial Revolution, Fire Insurance in Great Britain, 1700–1850*, Aldershot, 2004.

Pfeffer, I. *Insurance and Economic Theory*, Holmwood, 1953.
Raynes, H E. *A History of British Insurance*, London, 1964.
Supple, B. *The Royal Exchange Assurance, A History of British Insurance 1720–1970*, Cambridge, 1970.
Young, P. *A Premium Business, A History of the General Accident*, London, 1999.

Websites
Faculty and Institute of Actuaries, Library Services, Edinburgh: http://www.actuaries.org.uk/Display_Page.cgi?url=/index.html.
The Insurance and Actuarial Society of Glasgow:
http://www.iasg-cii.com.
Scottish Financial Enterprise, Overview of the Scottish Financial Industry:
http://www.sfe.org.uk.
Scottish Graduate Programme in Economics, Economics in Scotland: http://www.sgpe.ac.uk.

13 Working in Scottish Banking in the Twentieth Century

CHARLES W MUNN

Scottish banking changed greatly in the twentieth century, but most of that change took place in the final quarter; for much of the century little changed. That is not to say that the industry was static: for example, the number of branches increased by 50 per cent in the interwar years. But the practices of management and administration were very little different in 1970 from what they had been in 1900.[1] What followed was both dramatic and, for many, traumatic. It is possible to trace the beginnings of change from the 1950s, when there were some amalgamations and when banks, led by the Clydesdale and North of Scotland Bank, began to buy up hire purchase companies. In the 1960s the banking cartel came to an end, as a result of government pressure, but it was not mourned by the banks. The British Linen Bank, a wholly owned subsidiary of Barclays bank, introduced Barclaycard in 1966. It was the beginning of a period of innovation, but acceleration did not occur until the 1980s and did not really speed up until the 1990s.

In 1971 the government policy of Competition and Credit Control was introduced, and new banking organisations and other financial institutions were encouraged to compete with the established clearing banks, which were widely thought to be uncompetitive. The dire straits of the British economy in the 1970s rendered government competition policy ineffective and may even have been counterproductive.

It followed that the introduction of computing technology from the mid-1960s made an enormous difference. Not only did it affect the financial services market, but it enhanced demand for services and drove new supply prospects in the shape of new products and new delivery channels. These factors combined to create an industry which, between 1980 and 2000, changed very dramatically in its structure, products, management, people and in the public's perception of it. This was in sharp contrast to the first three-quarters of the century, where it might be argued that there had been very little change. What follows will, therefore, deal with the experience of working in Scottish banking in two separate periods.

WORKING IN BANKING

By 1900 the Scottish banking system was well established. It had reached a stage of maturity quite early in the nineteenth century and had become the model for banking in other parts of the United Kingdom, and in other parts of the world. Its reputation for stability and sound management practices had been somewhat dented by the failure of the City of Glasgow Bank in 1878, but memories were short and Scottish-trained bankers continued to be in high demand for many years thereafter.

In 1900 there were about 7,000 people working in the banking industry, all of them male. By 2000 employment in banking totalled *c*40,000 of whom about 70 per cent were female.

Recruitment into the industry was by personal recommendation. 'Likely lads' with good penmanship and a facility with figures were recommended to local branch managers or agents by their fathers, uncles, ministers or schoolteachers. They entered the bank at age fifteen or sixteen and served a three or four-year apprenticeship, during which they were paid a modest salary. There was no involvement in the process by any central department of the bank: it was all done at the local level.

Sir Robert Fairbairn rose to be chief executive, then chairman, of Clydesdale Bank. He had entered the service of the bank in Perth in 1927 following a visit to the branch by his father. This was a surprise to him, as he had thought he was heading for a career in science.[2]

The apprenticeship was a standardised affair throughout the Scottish

Figure 13.1 Banking hall of Union Bank of Scotland head office, 191 Ingram Street, Glasgow, *c*1909. Source: HBOS plc Group Archives.

banking community. It was largely comprised of 'learning by doing' in the branch; it was very unusual for anyone to commence their career at head office. Work began at the back desk where the apprentice would be given responsibility for the postages and the messages. He would gradually assume other duties, which would include writing entries in passbooks and in ledgers, all of which he learned from other members of the branch staff. Before the apprenticeship was finished the fledgling clerk would be familiar with all ledgers in the bank, including the general ledger and head office ledger. In essence he would understand and appreciate the simplicity and elegance of the system. He might equally be driven to distraction by the routine and by the amount of checking or 'calling off' that took place, for this was a system that depended upon repetition. But it was also a system that delivered accuracy, quality and efficiency.

Breaks in the routine were few and far between. Relief was sometimes provided by the requirement to attend the local note exchange, where apprentices from the local banks would gather in one of the banks' offices to exchange the notes of other banks that had been taken in the course of business. However, this was not just an exchange but another opportunity for checking. Staff in other banks could not be trusted to deliver the amount of notes that was written on the packet, so everyone sat down to count the notes given to them by their competitors. In the bigger exchanges, such as Glasgow, this could be a lengthy process and could take the better part of the morning, although, on less busy mornings, apprentices would adjourn to the local coffee shop. It was not 'done' to return to their branch too early lest the manager came to expect an early return on every note-exchange day.[3]

Bank staff worked a five and a half day week, but working days were not long and most staff could expect to arrive at the bank just before nine o'clock and be on their way home before five. This provided plenty of time for diversions. Volunteer and Territorial Army-type activities were very popular, as were the usual pursuits of football and golf, although curling proved to be extremely popular in a number of banks. Evening hours would also be spent in studying for the examinations of the Institute of Bankers in Scotland. Failure in these would be a big factor in determining whether or not a young man would be offered a permanent position at the end of his apprenticeship.

Towards the end of a young man's apprenticeship, discussions would take place which would determine whether or not he would be taken onto the bank's permanent staff. Mostly he would not. Scottish bank apprentices provided a ready supply of staff for banks in many other parts of the world and for British companies in many other sectors of the economy. In reality, a ready supply of relatively cheap apprentices enabled the banks to run their offices in a highly cost-efficient manner.

A large determining factor in the decision to offer a clerkship to an apprentice involved his success in the examinations of the Institute of Bankers in Scotland. The institute had been set up in 1875 and is the oldest professional body for bankers in the world. From its Edinburgh base it arranged syllabuses and examinations in subjects relevant to banking.[4]

An apprentice who struggled with the examinations would be unlikely to be offered a permanent place as a clerk. The unsuccessful apprentices would have little difficulty in securing alternative positions, as it was widely acknowledged that a Scottish banking apprenticeship prepared a young man for a career in many other sectors of the economy and, of course, a sizable number went to banks worldwide. It was not just the care and attention to detail that was important: young apprentices would also have secured an induction into the world of business and would have gained a good understanding of general business principles.

For those who remained in the service there was the expectation that they would rise through the ranks. Indeed it was sometimes said that a young man entered the service of a bank with a field-marshal's baton in his knapsack. The recruitment of graduates into an accelerated management-training programme did not begin until the 1960s; until then every entrant was seen as having the potential to rise to the top.

CLERKSHIP

In becoming a clerk in the bank very little changed. A modest increase in salary would be welcome, but otherwise the routine would continue. Advanced-level work might come his way, but this would involve nothing more complex than looking after the securities ledger or becoming a teller. Nevertheless, becoming a teller was regarded by many as a significant step forward in their career. The acceptance of responsibility for looking after the branch cash brought a sense of self-importance to the individual, and many were content with this role for the rest of their working lives – especially if they were first teller in the branch. This role also exposed them to dealing with customers on a daily basis, which was an important element in their personal development. Most, however, saw it as a staging post to higher levels.

Studying for the higher-level institute examinations would consume part of their leisure time, and it became part of the tradition for ambitious clerks who had completed their institute examinations to study for the qualifications of the Institute of Secretaries, who provided tuition for those seeking employment as company secretaries.

The management regime under which the clerks worked was very controlling. This is explained by the need for accuracy, but also by the fact that, in dealing with money belonging to other people, a high level of supervision was required. It has sometimes been suggested that working in a bank implied that a young man had a position of trust, but the reality was that the degree of supervision and checking ensured that the system, and its people, were kept honest.

This degree of control extended into the private lives of clerks, as there was always a concern that young men might get themselves into debt or, worse still, into bad company. Drinking and gambling were frowned upon and so was marriage, at least until a clerk was earning enough to support a

Figure 13.2 Union Bank of Scotland Beith branch staff, possibly 1930s. Source: HBOS plc Group Archives.

wife. Permission to marry had to be sought from the branch manager, and this would nearly always be withheld until a young man was about twenty-seven years old.

Similarly, young men who held strong political or religious views would be required to keep these concealed lest the articulation thereof offended customers with the opposite affiliations.

Nor did clerks have much say over where they would work. As the twentieth century progressed, head office began to exert more control over staff matters, and it became common for staff transfers to be made to meet the exigencies of running a large branch network. This was not done as a matter of consultation with staff, who would merely be handed a letter from head office that had been sent to their manager. Little advance notice was given of a transfer – often only a few days. It was virtually unheard of for someone to refuse a transfer; such a refusal would be tantamount to writing a letter of resignation.

All of the Scottish banks maintained extensive operations in London, which were manned largely by staff sent from Scotland. A transfer to a London branch was usually seen as an accolade, and three or four years spent in London were often the precursor of a highly successful career.

ACCOUNTANT

It was extremely rare for a clerk to be promoted to branch accountant before the age of thirty, and often this promotion came much later. The branch accountant was in a rather strange position. The accountant was a 'signing' official and thus could sign documents that were legally binding upon the bank. This would include deposit receipts and similar documents. But the accountant also had a special role which was seldom discussed and never made public: this role was to keep a check on the manager, and he could report direct to head office if he thought that the manager had done something wrong or was up to no good. Otherwise his main role was to make sure that the clerks and apprentices were able to do their jobs. Amongst his other duties, he would issue the work rotas and supervise the holiday schedules. In reality this was hardly hard work as these changed little from week to week and from year to year. His role was very largely to ensure that the branch routines flowed smoothly.

THE WATCHERS

There were others whose role it was to ensure the integrity of the system. These were the inspectors and the superintendents of branches. An appointment to either of these roles was a virtual guarantee of a starred career. Someone who served as an inspector was virtually assured of a plum manager's role in the years ahead.[5]

The role of the inspectors was to visit the branches and inspect all the books. The favourite time for the inspectors to arrive was just after close of business. They arrived unannounced and proceeded to check everything in the branch, beginning with the postages and proceeding all the way to the cash ledgers and the vouchers. They stayed until well into the evening and would be back early the next morning to continue their task.

One of their tasks was to ensure uniformity in the branch system and to discourage any practice or process that did not conform to head office instructions. No deviation from the Head Office Instruction Manual was tolerated.

The important thing for the branch staff during an inspection was to create a good impression, as all staff were interviewed and reports written. Consequently, Sunday suits would be worn for the duration of the inspection, which could last for three or four days. The tone in which inspections were conducted was quite severe but would mellow as it progressed – provided, of course, that no problems were uncovered.

The superintendents of branches, who were based in head office, had a less exciting role than that of the inspectors. Although their title implies that they were supervisors, that was only part of their role. Their main function was to ensure that weekly returns from the branches were in order and had been completed in a consistent manner. Any irregularity or unusual activity would be followed up. But they also acted as a ready source of advice and guidance for the signing officials in the branches and so had a support function.

An article in the London-based *Bankers' Magazine* in 1900 said that: 'An excellence of administrative routine is a special feature of Scottish banking.' It went on to claim that: 'As regards book-keeping, Branch supervision and inspection the Scottish banks were superlatively thorough.'[6]

MANAGER

The position of branch manager was relatively new in this period, as the tradition was to have agents in charge of the branches. These were usually local businessmen, often lawyers or accountants, who ran the branch as an adjunct to their other interests. Bank staff felt that this situation created a promotion blockage for them, and the practice of appointing agents began to die out in the interwar years. In 1947 the Scottish banks undertook to make no more such appointments, preferring instead to promote staff from within the branch network.

Becoming a branch manager was, for most people, the ultimate aspiration. This was a major achievement, although there was a definite pecking order amongst the branches. Branches varied substantially in size from about three staff to about twenty. Location was also a factor – Giffnock or Morningside being much preferable to Govanhill or Muirhouse. Clearly the larger branches in the better places were more desirable.

The manager's main task was to bring in new business. However, this was not as simple as it seems. Until 1966 the Scottish banks operated a cartel, which effectively prevented competition. Consequently there was no marketing of bank services and no sales-type activities, nor were managers allowed to approach customers of other banks. In 1965 the manager of one bank in East Kilbride was seen at lunch with a customer from another bank. This caused something of a scandal. Similarly the National Commercial Bank began advertising on television in 1965. This was seriously frowned upon by the other banks as being not in the spirit of the 'Agreements and Understandings', which was what the cartel was called.

Managers, therefore, had little external help upon which to rely in their attempts to get new business. They had to rely on their own resources. Consequently, 'networking' may not have been called such, but it was vital in the getting of new business. Managers of banks could therefore be found in every social gathering. They were active in churches, clubs (especially sporting clubs), charities and other voluntary groups. Getting around and being seen to be involved at a local level was a vital part of being a branch manager.

In addition to their income from the bank, the managers were also allowed to operate in other areas for their own interest. For most, this meant that they would operate as agents for a range of insurance companies, and it was not unknown for managers to have higher incomes from this source than from their bank. Banks, as organisations, did not move into this arena until the 1970s, at which time the managers' insurance interests were bought out.

The other perk enjoyed by managers was the bank house. In many places this was a flat above the bank branch, whilst in others it was a substantial stone-built house, with the branch being housed in one of the front rooms. This was enjoyed rent-free in most banks, but by the interwar years it was beginning to be seen as a liability, as more managers wanted to own their own house. Thenceforth the bank house was enjoyed by more junior members of staff.

In the recruitment process penmanship was a prized skill, as the banks relied heavily upon the handwritten ledger. The letterpress, a potentially messy affair, was introduced in the late nineteenth century, but all it achieved was to make life more difficult for bank clerks. The typewriter and the telephone gradually made their way into banking in the early years of the twentieth century.

It was not until the interwar years that technology began to make a serious impact on banking routines. The introduction of the comptometer in the 1930s was certainly a radical departure from past practice. This was an electromechanical device for keeping ledgers and was used primarily for current accounts.

Compared with the technology in use today, it was a very cumbersome device which required a ledger card to be inserted into the carriage. The operator would then have to enter the balance from the previous transaction, followed by the latest transaction, whereupon the machine would calculate and type the new balance.

There were those who viewed these machines as a retrograde step and who argued that they were not a good investment as they did not actually improve productivity. Its introduction was hardly universal and, by the 1960s, when computers were being introduced, the comptometer had still

Figure 13.3 British Linen Bank ledger department, Glasgow, 1949. Increasing mechanisation in banks seems to have corresponded with a rise in the number of female staff. Women tended to be taken on to perform the lowlier, routine clerical work. Source: HBOS plc Group Archives.

not been installed in the branches of all banks. Where it had made a major impact was to begin a shift in the gender balance of bank staff.[7]

WOMEN IN BANKING

The introduction of women to the staff of banks really began during World War I, when banks experienced real problems in staffing their offices. In the Clydesdale Bank there were about 900 staff and, of these, some 600 signed up for service in the forces. One part of the banks' answer to this was to retain older members of staff – a move which gave rise to the creation of the first superannuation scheme. The second part of the answer was to employ 'lady clerks'.

This was a radical move. It was made clear to the women that they were 'temporary', and that as soon as the war was over they would return to their ordinary lives. The banks felt it was necessary to say this, as they had promised their staff who were in the forces that they would have their old jobs back when they returned fit from the war. Indeed, they usually paid their staff their full salaries during their military service, only deducting what the staff were receiving from the military. In London offices it was more likely that women would be recruited to the permanent staff, but this was certainly not the case elsewhere.

Of course, when the war ended many of the men did not return – either because they had been killed or were injured. But many, having tasted

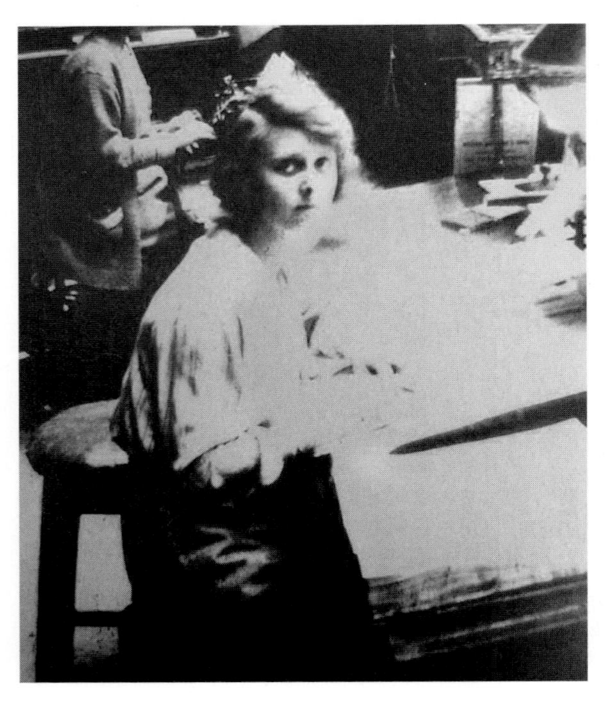

Figure 13.4 Lady clerk, Union Bank of Scotland, 191 Ingram Street, possibly 1920s. Source: HBOS plc Group Archives.

a life beyond banking and seen parts of the world beyond Scotland, decided to pursue careers elsewhere. Consequently there were jobs for women. Most banks maintained the illusion that women were not on the permanent staff until the 1930s, but the reality is that they were there to stay. It was discovered, of course, that women were particularly good at typewriting, and when the comptometers were introduced their use rapidly became an almost exclusively female activity.

In the post-World War II years the numbers of women clerks rose steadily until they formed a majority of the staff in most banks by 1965. New entrants were paid the same as the males, but the salary scales began to diverge after five or six years' service, and the impression was clear that females were not expected to progress into the management grades.[8]

ORGANISED LABOUR

Throughout most of this period the trade union movement was frowned upon by management. Young bankers were viewed as having the potential to become managers, and they were therefore seen as 'officer class'. It was a 'disappointment' when they were found to have joined a trade union.

The first sign of organised labour in banking came during World War I, but it was short-lived. Pressure had been building, and some banks began to introduce pension schemes to encourage retirement and thus release promotion blockages for younger staff.

In the 1920s it was clear that bank staff had lost ground in terms of real wages. The Scottish Bankers' Association was registered as a trade union in 1919 and quickly won concessions from the banks. This did little to assuage the thirst for higher wages, notwithstanding that the banks paid the income tax of their staff, and other benefits such as pensions and favourable house purchase loans were beginning to be introduced. Thereafter, membership of the association rose and fell, with bank staff having little to do with militancy of any kind.

Management's way of dealing with the rise of trade unionism varied between banks. Some tolerated the existence of the association but spoke to it grudgingly. Others set up their own staff associations, and banks gradually began to adopt a more paternalistic approach to their staff. Profit-sharing schemes and generous sickness pay became normal, if not universal. Matters deteriorated again in the early 1930s, but there was no appetite for militancy and the banks responded to their staff demands with pay increases.

At the same time the old controls over staff, such as limiting the age of marriage, were challenged at law by disgruntled members of staff. The banks lost. This was important in itself, but it also represented a watershed in bank–staff relations, as staff were increasingly likely to challenge management decisions.

World War II brought about a repeat of the problems of World War I, but thereafter it all settled down to a fairly balanced relationship. The Federation of Scottish Bank Employers was founded in 1945 and met with the Scottish Bankers' Association and its successor bodies on a regular basis until the 1980s. The banks acquired a reputation as paternalistic employers, and employment in banking was once again seen as a highly desirable pursuit for a young man and, increasingly, for young women. That is not to say that bank staff were universally contented: there continued to be a steady outflow of bank staff. Overseas banks continued to advertise for staff in the Scottish Bankers' Magazine – the Journal of the Institute of Bankers in Scotland – until the 1960s. There continued to be murmurings about pay in the Scottish banks, but it was rare for this to manifest itself in threats of industrial action.

THE MODERNISING PERIOD

In the last quarter of the century, change became a constant factor in the life of a bank worker. There were some signs in the late 1950s that change was afoot, especially following the publication of the Radcliffe report in 1959,[9] but when the Wilson report was published in 1981[10] it was clear that the industry was undergoing a period of growth. Wilson recorded that the banks offered 120 different services, but that was really just the beginning of a period which saw growth accelerate, with product development and differentiation leading the charge.

Despite sometimes being accused of indifference to change or even slothfulness, a long-term perspective shows that the banks embraced change with some enthusiasm.[11] Traditional boundaries between different types of financial institution disappeared as financial conglomerates took the place of financial specialists. And despite the doom-mongers who criticised them, it was the banks rather than the insurance companies that led the process. Critics in the 1980s were fond of saying that the insurance companies, or new arrivals on the financial services scene like Marks and Spencer plc, would 'see off' the banks. For a time it looked as if the critics might be right, especially when the world entered a recession in the early 1990s. Profits were extremely low, and in some cases losses were made.

But the banks had already learned about flexible management, the need to innovate, risk management and the need to cut costs. Some had learned better than others, and there is little doubt that the Scottish banks fared better than their English or European counterparts.

In short, the banking and financial system become more competitive, and in the process it changed the working environment and everything that went with it.

The cartel came to an end, and the committee of Scottish Bank General Managers transmogrified into the Committee of Scottish Clearing Bankers and became a very small trade body. The Federation of Scottish Bank Employers disappeared in 1983 as first one bank and then the others decided that they wanted to pursue their own staffing strategies.

New technologies were introduced, which took a lot of the drudgery out of branch work. The application of computers from the late 1960s made a big difference – first to the administration and then, from the late 1980s, to the delivery of services to the customers, as telephone banking and insurance, and then internet services, were introduced.

The implications of all this change for the staff were enormous. Many of the old sureties and business processes disappeared. Activities that had been in use for centuries were no longer needed or wanted, and the old psychological contract of a 'job for life' was no more.

It was clear, to anyone who looked, that the old ways were no longer appropriate to the changed business environment. If the old ways were no longer appropriate, then the old staff would have to change. This also raised a whole set of difficult questions about the strategies that would deliver change, growth and increased profits.

What ensued was an extremely difficult time for bank personnel. It was made worse by the recession, which ensured that an added impetus was given to the need for new ways of doing things. Such was the pace of change that planning became extremely difficult. This brought a rather short-term approach to much of the industry; for example, in 1988 the Scottish banks said that growth was so fast that by 1992 they would have to recruit every school-leaver in the country. When 1992 came, they recruited none.[12]

In an attempt to make management more professional, there was a period when senior people flirted with business consultants and management schools. Total quality management (TQM) and business process re-engineering were embraced with varying degrees of enthusiasm and

Figure 13.5 First Bank of Scotland computer centre, Robertson Avenue, Edinburgh, 1970s. Source: HBOS plc Group Archives.

Figure 13.6 First Bank of Scotland computer staff, 1962. Source: HBOS plc Group Archives.

varying degrees of success. Senior people understood the need for change, and it might be argued that the more successful were those who laid down longer-term strategies and took some time over it, as opposed to those who merely embraced the latest management fad.

The implications for staff were considerable. The job for life that many fancied as their right was unceremoniously taken away. In some banks it became fashionable to say that there was no such thing as a career. This particular canard did not last too long, as it soon became apparent how damaging it was to recruitment. Banks began to recruit more specialists for particular roles and, for a time, the banker became an endangered species. More and more specialised roles were created, and many people were recruited from outside the industry to work in the burgeoning head offices. There were roles for human resources specialists, finance gurus, marketing experts and salesmen.

In some places it became fashionable to decry the skills of the traditional banker. All of this was justified in the name of creating a 'new bank'. Anything that was associated with the 'old bank' was, by definition, redundant.

And so it happened that many bankers lost their jobs. Most of this happened in the early 1990s, and in some banks it was soon rare to find anyone working over the age of fifty. The old management theory that 'a manager is a manager' was applied in some banks, and people were brought in with no experience of banking. Some of them believed that selling financial services was just like selling any other commodity – with predictable consequences.

What was happening was that a sales culture was being introduced, and the traditional bankers were ill-prepared to meet the challenges associated with such a change. The idea of a job for life disappeared, and many of the old paternalistic patterns of a banking career also disappeared, to be replaced by a reward system that was a reward for performance rather than service.

This was also occurring in other sectors of the economy, part of a wider trend towards a higher level of competition in business, where shareholders expected higher returns and management teams were put under ever greater pressure to provide results. They could only do this by pushing down costs. What is perhaps surprising is that there was little militancy from staff resulting from these changes. Perhaps the generous early retirement packages that were provided helped dull the anger. It was also the case that fifty-something bankers generally found that their talents and skills were in demand in other sectors of the economy.

If the changes that were taking place in the management cadre were considerable, they were just as great at the other end of the spectrum. A vast expansion of the higher education system meant that more and more school-leavers went on to university. This removed from the recruitment spectrum a whole cohort of school-leavers who, traditionally, would have joined a bank. The prospect of a school-leaver entering a bank at seventeen or eighteen years of age and progressing through to become a senior manager or even chief executive largely became a thing of the past. The recruitment of school-leavers almost died out, as those who did not progress into higher education were thought unlikely to have the skills and abilities that were required to be an effective bank clerk.

What was evolving was a four-tier recruitment strategy which replaced the traditional single tier. Graduate recruitment, which had begun only in the 1960s, gathered pace. Graduates were recruited into accelerated management development programmes, and most were required to undertake the examinations of the institute which, in 1991, had become The Chartered Institute of Bankers in Scotland. Most would be sufficiently ambitious to pursue a career in one of the burgeoning head office departments. Running a branch had ceased to be an appropriate ambition for someone who wanted to get on in life.

Others, as has been mentioned, were recruited into specialised roles. As banks became ever larger they also became more complex. Additionally, changes in society resulted in banks employing people in roles that would never previously have been thought of. Public relations, parliamentary officers, community bankers and specialists in corporate social responsibility were all new areas in which banks had to develop their expertise.

Branch networks were, to an extent, rationalised in the 1990s, and this gave rise to adverse comment, but Scottish banks continued to have more branches per head of population than in most other countries. Within the branches, the nature of the work and the nature of the employee changed dramatically.

As the range of products expanded it became necessary to find sales outlets for them. It also became necessary to find people who would sell them. As the branches were still the primary source of customer contact, it was a natural step to use them more proactively as sales outlets. Consequently the branch staff, who were by this time mostly front-line staff, were increasingly expected to sell products to customers. Not all of them took kindly to this new responsibility and to the added pressures of having to meet sales targets. However, many took to the new tasks extremely well and clearly found in themselves skills that, until then, had lain dormant. Life also became more complex with counter staff, and others, having to become conversant with a wide range of regulatory requirements imposed upon them by a seemingly endless spectrum of organisations, and eventually by the singular regulator, the Financial Services Authority.

All of this coincided with a very substantial rise in demand for financial products, especially loans. Society was changing, and traditional habits of thrift rapidly gave way to a culture based on debt. This was a major sociological shift.

Moreover, the customers also had to adjust to the possibility that their weekly visit to the branch might involve having a discussion with bank staff about their need for other financial products. Yet at a time when banks were keen to cross-sell other products, their ability to do so was hampered by the fact that customers paid fewer and fewer visits to their branch. This was due to the increase in new channels and services such as ATMs, direct debits, telephone banking and internet banking.

The changing nature of branch banking brought with it additional complexity for staff departments in providing suitable staff. The school-leaver, who had been the traditional member of branch staff, was largely replaced by adult returners (mostly women) and by part-time staff who would work hours when the branches were busiest. These were the third and fourth tiers of the new recruitment strategy.

Similar types of staff were recruited into call centres. These were the new engines of financial, and many other, services. Again, these operations required a different type of staff from the traditional bank clerk. The ability to sit at a screen all day and deliver services over the telephone necessitated a different skillset and mindset.

CONCLUSION

The Scottish diaspora contained, and still contains, a large number of bankers and other financial specialists, for it is not just in the area of banking that Scottish financial skills are in demand: insurance, fund management, chartered accountancy and actuarial work are all well represented amongst the Scots in England and overseas. Scottish financial skills are in demand in many parts of the world – not least in Scotland.

Such has been the growth in demand for financial services, and Scottish financial services in particular, that employment numbers have grown to the extent that Scottish financial services is now the largest industrial sector in Scotland by employment and contribution to GDP.

It is still in the realms of speculation, but it may be that the reason for the success of Scottish bankers and financiers has something to do with the breadth of their education and development. It is, of course, part of a much wider sociological phenomenon in Scotland that education is considered important and is given a high priority. There were certainly signs in the maelstrom of recession and structural change in the mid-1990s that this had been lost sight of. However, there were healthy signs in the later years of the decade that education and training were back on the agenda of finance professionals.

The structural and other changes that took place in the 1980s and 1990s resulted in a series of takeover bids amongst providers of financial services. While the Scottish Life Assurance Offices mostly fell into foreign hands, the banks increased their power and status, with Royal Bank staging a successful bid for NatWest, and Bank of Scotland achieving a successful merger with Halifax. The implications for staff were considerable, but that is a story for the future.

NOTES

1. For a general history of Scottish banking see Checkland, 1975. Company histories include Tamaki, 1983; Munn, 1988; Saville, 1996; Cameron, 1995.
2. Munn, 1988, 312–4.
3. Personal reminiscence of the author.
4. Saville, 1996, 377–8.
5. Saville, 1996, 636–9.
6. *Bankers' Magazine*, 1900, 559–60.
7. Munn, 1988, 166–7.
8. Checkland, 1975, 683–4.
9. Committee on the Working of the Monetary System, 1959.
10. Committee to Review the Functioning of Financial Institutions, 1980.
11. Draper et al., 68–78.
12. Personal recollection of the author.

BIBLIOGRAPHY

Bankers' Magazine, 1900.

Cameron, A. *Bank of Scotland 1695–1995*, Edinburgh, 1995.

Checkland, S G. *Scottish Banking: A History 1695–1975*, Glasgow, 1975.

Committee on the Working of the Monetary System, 1959, Cmnd 827.

Committee to Review the Functioning of Financial Institutions, 1980, Cmnd 7937.

Draper, P, Smith, I, Stewart, W and Hood, N. *The Scottish Financial Sector*, Edinburgh, 1988.

Munn, C W. *Clydesdale Bank: The First 150 Years*, Glasgow, 1988.

Saville, R. *Bank of Scotland: A History 1696–1995*, Edinburgh, 1996.

Tamaki, N. *The Life Cycle of the Union Bank of Scotland, 1830–1954*, Aberdeen, 1983.

14 The Accountancy Profession in Scotland

STEPHEN P WALKER

INTRODUCTION

In the past, sociologists and historians of the professions tended to devote little attention to the 'new' vocation of accountancy. Accountancy has been perceived as less significant than older established occupations at the head of the middle class, such as the law, medicine and the church. During the last thirty years, however, there has been considerable interest in the accountancy profession, particularly within the accounting academy. This attention reflects the increasing significance of accounting and accountants in advanced capitalism and a number of compelling features of the occupation itself: its enormous growth relative to other professions over the last century; the interfaces between the accountancy profession and business; the complex organisation of the profession and its relations with the state; its enduring quest for social closure; its colonisation of new forms of work; and the shifting jurisdictional boundaries between accountancy and the neighbouring vocational fields of engineering and law.

A considerable amount of research on the history of the profession has focused on Scotland. There are two main reasons for this. First, it was in Edinburgh, Glasgow and Aberdeen that the first organisations of professional accountants in the modern era were established. The pattern of institutionalisation adopted in Scotland informed subsequent development in England and beyond. Second, the history of the accountancy profession in Scotland holds its own fascination. This centres on attempts to achieve control of the market for accountancy services, the dynamics of relationships with the larger London-based accountancy bodies and the diverse identities and values of accountants within Scotland.

A BRIEF HISTORY OF THE PROFESSION AND ITS ORGANISATION IN SCOTLAND

George Watson (1645–1723), 'accomptant' to the Bank of Scotland and founder of the Edinburgh college which bears his name, is usually identified as the first professional accountant in Scotland.[1] The first Scottish treatise on accounting was Robert Colinson's *Idea Rationaria* (1683). During the eighteenth century other important works on accounting were written by teachers of book-keeping and mathematics, establishing Scotland's reputation as 'a land of accountants'.[2] In addition to teachers, who often also practised accounting, some Edinburgh lawyers began to specialise in the craft, as 'writers and accountants'. A small number of Glasgow merchants also combined commercial activity with the provision of accountancy services.

During the late eighteenth century there was a marked expansion, especially in Edinburgh, in the number of teachers of book-keeping and of practising accountants.[3] Bankruptcy legislation in 1772, which transferred the management of bankrupt estates to trustees for creditors, encouraged an increase in the number of accountants. Separate bankruptcy laws resulted in a different pattern of development in England. Accountants in Scotland were also increasingly appointed as managers of – usually insolvent – landed estates, and received remits from the courts, municipal organisations and private firms to prepare and audit accounts. By the end of the eighteenth century there were almost forty accountants in Scotland, concentrated in Edinburgh and Glasgow. Half a century later there were over 300.

The earliest known organisation of accountants in Scotland was the Committee of Accountants Practising Before the Court of Session. This was formed by a group of Edinburgh accountants in 1834 to counter legislation which proposed the discontinuance of their appointment as auditors of the accounts of judicial factors. On successfully repelling the offending bill, the Committee of Accountants was disbanded. More permanent structures were established in 1853, with the formation of the Institute (later, Society) of Accountants in Edinburgh (SAE), which received a Royal Charter in 1854; the Institute of Accountants and Actuaries in Glasgow (IAAG) in 1853, chartered in 1855; and the Society of Accountants in Aberdeen (SAA) in 1866, chartered in 1867. In terms of membership and influence, the organisations in Edinburgh and Glasgow were by far the most important.

Explanations for the formation of the institutes in the two major cities in 1853 have been the subject of heated academic debate.[4] However, the evidence suggests that the SAE and IAAG were established as interest groups to protect practitioners from the threat posed by the appearance of the Bankruptcy and Insolvency (Scotland) Bill in 1852–3. This draft legislation, supported by powerful mercantile interests in England, was part of the movement to advance free trade by assimilating the commercial laws of Britain. The bill threatened to impose the cheaper 'officialist' system of English bankruptcy administration on Scotland, much to the detriment of accountants, who earned considerable fees as creditor-elected trustees on insolvent and bankrupt estates.[5] The SAE and IAAG successfully opposed the threatened legislation. In an ironic twist it was the imposition of the Scottish system of bankruptcy on England through the Bankruptcy Act, 1869 which inspired the formation of the first organisations of accountants south of the Tweed.[6]

The local societies in Edinburgh, Glasgow and Aberdeen remained the only organisations of accountants in Scotland until autumn 1880, when the Scottish Institute of Accountants was founded in Glasgow. In contrast to the city-based chartered bodies, this organisation was deliberately national in scope. Its appearance followed swiftly on the example set by the formation of the Institute of Chartered Accountants in England and Wales (ICAEW) in May 1880. The Scottish Institute of Accountants sought to recruit those excluded from the city-based organisations and made three unsuccessful attempts to secure equality of status through a royal charter during the 1880s and 1890s. Following this failure, the institute was disbanded in 1899 and its members were integrated into the Society of Accountants and Auditors, a London-based organisation of 'incorporated accountants'.

In 1891 the Corporation of Accountants, Ltd was formed in Glasgow by a small number of radical members of the Scottish Institute of Accountants who were dismayed by the failure to achieve a royal charter. The corporation was unashamedly established to challenge the exclusive use by the SAE, IAAG and SAA of the credentials 'chartered accountant', or 'CA', a right confirmed by the courts during the 1890s and 1900s.[7] Although it had limited success in Scotland, the corporation's membership increased in England. In 1939 its members merged with the London Association of Certified Accountants to form the Association of Certified and Corporate Accountants (now the Association of Chartered Certified Accountants) in London. The chartered societies in Edinburgh, Glasgow and Aberdeen therefore regained their organisational hegemony in Scotland. In 1951 the SAE, IAAG and SAA merged as the Institute of Chartered Accountants in Scotland (ICAS). The merger was provoked by the recognition that the existence of separate organisations in the three cities was cumbersome when it came to presenting Scottish opinion before government and in negotiations on the structure of the profession with other accountancy bodies.[8] ICAS is now one of the six major organisations of professional accountants in the British Isles represented on the Consultative Committee of Accountancy Bodies.

Although it is the only major organisation of UK accountants with its headquarters in Scotland, ICAS does not represent the whole professional community north of the border. At the time of writing, the following members of the six major accountancy bodies were resident in Scotland: ICAS, 9,200; ICAEW, 1,100; Chartered Institute of Public Finance and Accountancy (CIPFA), 1,000; Association of Chartered Certified Accountants (ACCA), 2,500; Chartered Institute of Management Accountants (CIMA), 2,400; Institute of Chartered Accountants in Ireland (ICAI), 50. As the accountancy profession has never achieved statutory registration, the offering of accountancy services by individuals who belong to minor organisations or no qualifying association at all has also been a persistent theme. Despite the incursions of members of other professional organisations into its territory, ICAS and its predecessor bodies have been, and continue to be, the dominant professional institutions in Scottish accountancy.

In common with the rest of the UK,[9] professional accountancy in Scotland has been gendered masculine. Although there were no explicit provisions in the rules to prevent women entering the institutes during the nineteenth and early twentieth centuries, the socio-cultural context and adherence to the Victorian ideal of 'gentlemanly' professionalism effectively prevented it. This was in stark contrast to the growing numbers of women who performed subordinate accounting functions as clerks and book-keepers from the late nineteenth century. The 1901 census recorded 38,511 male and 15,399 female commercial clerks in Scotland (compared to 180 in 1871), and 1,721 male but no female accountants.

When the issue of the entrance of women to the profession surfaced in Scotland at the opening of the twentieth century, chartered accountants responded by raising doubts about the capacity of women to perform the work of the profession: they emphasised the abilities of women as typists and clerks in accountant's offices; argued that women were better suited to domestic management; and contended that the entry of women to the profession would increase competition and reduce the capacity of men to earn an appropriate level of remuneration.[10] There is some evidence that in 1909–11 Scottish chartered accountants were willing to concede to the admission of women if it were attended by the greater goal of statutory registration of the profession, but this was not achieved.[11]

The question of the admission of women resurfaced during World War I, when the staffs of accountancy firms were heavily depleted as partners and apprentices enlisted for military service. Women were employed as clerks for the duration of the war to alleviate the shortage of male labour. In 1915, questions were raised about the possibility of taking women as apprentice accountants. Legal opinion indicated that references in the extant regulations to the entry of 'persons' effectively meant that membership was restricted to men. In 1916–17, consideration was given to the establishment of a separate Society for Lady Accountants. This was accepted by the IAAG but not the SAE and SAA. In 1918, the issue was reignited when other accountancy bodies decided to open their doors to women. In response to this, Scottish CAs devised a plan whereby women would be admitted as 'associates' rather than full members. However, the appearance of the Sex Discrimination (Removal) Bill in 1919 overtook these developments. An amendment to the bill proposed by the Faculty of Actuaries, Edinburgh, extended its provisions to incorporated professions and removed the obligation to secure amendments to royal charters. The passing of the act in 1919 allowed the admission of women as full members of the SAE, IAAG and SAA.

The first woman CA in Scotland was Isobel Clyne Guthrie, who was admitted to the Glasgow Institute in 1923.[12] The first women were admitted to the societies in Edinburgh in 1925 and Aberdeen in 1936. Despite this breakthrough, the number of women entering the accountancy profession in Scotland remained very small. While the statute of 1919 outlawed the

Figure 14.1 Isobel Clyne Guthrie, first female Scottish chartered accountant, admitted to the Institute of Accountants and Actuaries in Glasgow, 1923. Source: *The Accountant's Magazine*, 1923.

prevention of admission to the professional organisations on grounds of sex, the point of recruitment (as apprentices) was in the practising offices. Other impediments, such as the marriage bar, also remained. Hence, during the 1920s only 1.4 per cent of CA apprentices were women.[13] Several of these women were relatives of chartered accountants, and their recruitment was concentrated in a few firms.

The segregation of women accountants was epitomised at an ICAS function in 1960 attended by the Duke of Edinburgh. It is said that the duke noticed that lady members were seated at a corner table separate from the men, and he suggested that women form their own organisation to campaign for seats with their male colleagues.[14] A 'Lady Members' Group' was duly established in the same year. The group's objectives were not radical. It sought to encourage communication between lady members through social activities and by offering support to female apprentices. The group was reconstituted as 'Women in ICAS' in 1994. This organisation more actively promoted the position of women in the profession and sought the removal of obstacles to their progression. One year after its foundation, however, Women in ICAS was disbanded. This reflected the fact that since 1992 the primary focus for the advance of women's interests in the profession had become 'Women in Accountancy', a group drawing its members from the six major professional bodies in the UK.

As late as 1970 only 3 per cent of ICAS members and 7 per cent of apprentices were women. In 1980, the figures were 5 per cent and over 20 per cent respectively. The increasing number of women CA students, particularly from the 1980s, has begun to have an impact on the sex distribution of the membership. At the time of writing 21 per cent of members and 46 per cent of students are women. In 1994, a woman was elected president of ICAS. Despite these advances, studies of recent generations of female CAs in Scotland continue to reveal the segregation and marginalisation of women in the profession, and the obstacles to their occupation of senior positions in organisational hierarchies.[15]

LOCAL IDENTITIES: ACCOUNTANTS IN EDINBURGH AND GLASGOW

A distinctive feature of the history of the accountancy profession in Scotland is its organisation for almost a century on a local basis. In England, by contrast, the societies that formed during the 1870s in Liverpool, London, Manchester and Sheffield had merged as a national institute by 1880. This points to the distinctive identities of accountants in the east and west of Scotland.

Social network analyses of the founders of the institutes in 1853 indicate that the communities of professional accountants in Edinburgh and Glasgow were cohesive but based on quite different sets of relationships. Edinburgh accountants were associated with a landed–lawyer–actuarial nexus. Glasgow accountants were aligned to the mercantile, manufacturing and trading communities.[16] These differences reflected the socio-economic structures of the two cities and the composition of the clientele. In Edinburgh much of the work of accountants centred on the court of session, and there was a strong association with lawyers, especially solicitors who were the agents of landowners. Edinburgh accountants were accorded a lofty status on the basis of this association, one which distinguished them from accountants in Glasgow and elsewhere. Contemporaries considered accountants in the capital to be 'professional' gentlemen as early as the 1820s, well before their peers in the rest of Britain.[17] In his commentary on the social structure of mid-nineteenth-century Edinburgh, Heiton located accountants among the legal 'caste' – marginally below writers to the signet in esteem.[18]

One of the largest accountancy firms in Edinburgh was Lindsay, Jamieson & Haldane. Until the 1880s, when the audit of major financial institutions and companies assumed greater importance, the essential client base of this firm was the landed classes. Lindsay, Jamieson & Haldane was involved in the management of insolvent estates, the administration of heritable property and the management and audit of trusts, curatories, executries and judicial factories.[19] This work depended on close connections with lawyers. The firm constructed strong relationships with major firms of writers to the signet in Edinburgh and advocates in Aberdeen.[20] These networks were founded on kinship ties, business associations (such as directorships), and social contacts made through church and the membership of organisations such as the Royal Company of Archers.[21]

Their association with commerce as opposed to the landed–professional milieu meant that accountants in Glasgow were accorded a lower status, at least outside of their own domain. Although the extent of this can be overstated, it did receive some official recognition in the 1871 census. This classified three-quarters of Glasgow accountants as occupied in commerce. A similar proportion of Edinburgh accountants were enumerated as members of the legal profession.[22] In 1896 James Martin (not the most impartial of observers) concluded: 'I do not know whether it is the case that in Glasgow we have a lower type of CA than exists elsewhere. If I did say so I should not be the first to record the observation.'[23]

A major source of the differential status of accountants within Scotland and the 'professional' v. 'trade' tags accorded to practitioners in the east and west respectively, was the involvement of many Glasgow accountants in stockbroking. This association persisted in some degree until the 1940s. Twelve of the first sixteen chairmen of the Glasgow Stock Exchange were members of the IAAG.[24] Half of the founders of the IAAG were also members of the exchange.[25] For Edinburgh accountants (who, it should be noted, had strong connections with actuarial work and insurance), the receipt of brokerage commission was alien to notions of the independent public practitioner. Ethical issues also arose from the involvement of a number of Glasgow accountants (about 15 per cent in the 1890s) in broking. The existence of 'mixed' accountancy and stockbroking practices gave ammunition to those who challenged the pretensions of Glasgow CAs as professional gentlemen.[26] For example, the medium-sized practice of A and C T Sloan earned 38 per cent of its income from stockbroking between 1867 and 1878, 27 per cent from insurance agency and 35 per cent from accountancy work (involving the management and audit of trusts of public and private institutions, businesses and individuals). In the period 1879–98 the distribution was: accounting work 58 per cent, stockbroking 32 per cent and insurance 10 per cent.[27] During the late 1870s one partner's involvement in stockbroking almost ruined this Glasgow firm.[28]

Differences between accountants in the east and west of Scotland were also reflected in the sources of recruits. Recruitment to the SAE reflected the relationships of Edinburgh accountants with the professional classes and landed society. Fifty-five per cent of members admitted during 1853–74 were the sons of professional men, as were one third of entrants from 1875 to 1914.[29] A further 11 per cent in the former period derived from families of independent means. Of those SAE apprentices who commenced training before 1874, 27.5 per cent were the sons of lawyers.[30] The second sons of lawyers often became Edinburgh CAs, and the eldest sons of senior CAs frequently achieved upward mobility into the law.[31] By contrast, only 8 per cent of those admitted to the IAAG from 1853 to 1879 were the sons of lawyers. In Glasgow a higher proportion of CAs came from families in commerce, manufacturing and trade.[32]

There was also an ideological dimension to the different identities of Edinburgh and Glasgow accountants, at least as revealed by the structures and policies of the organisations which represented them. These divergences were manifested in tensions between the SAE and IAAG on a number of issues and were played out in the context of shifts in the distribution of power between east and west. The Edinburgh Society was numerically dominant until 1902 and had greater prestige. However, the membership of the Glasgow Institute expanded rapidly during the early twentieth century and was 2.5 times larger than the SAE and fourteen times larger than the SAA by 1950.[33] Disagreements arose, for example, over proposals for the amalgamation of the three societies during the late nineteenth and early twentieth centuries (these were opposed by the SAE). The Edinburgh Society was much more élitist in recruitment, imposing an apprenticeship fee of 100 guineas for a five-year indenture and an admission fee of the same amount during the 1880s. The IAAG considered itself more meritocratic and perceived apprenticeship fees as an impediment to the mobility of local youths. Further, the period of indentures in Glasgow was four years and the admission fee was half that charged in Edinburgh.[34]

Differences between Edinburgh and Glasgow accountants are much less apparent in recent times. The formation of a single national body in 1951 removed the institutional basis of separateness. There is also a stronger sense of primary allegiances to employers, such as major accountancy firms and corporations which are less parochial and increasingly global in outlook. The commercial taint of accountants in the west is insignificant in the context of advancing commercialisation and profit-seeking in the provision of accountancy services.[35]

The increasing employment of chartered accountants (in Scotland and beyond) in business represents an important departure from the originating concept of the profession and has also blurred the commerce–profession distinction.[36] During the nineteenth century the vast majority of professional accountants were partners or clerks in public practice or offered services to the public on their own account. After World War I, increasing numbers of CAs became employees as cost accountants, financial controllers, managers and internal auditors. A century after the formation of organisations for accountants in public practice, one half of the membership of ICAS was employed in industry and commerce.[37] Currently, 58 per cent of the employed members of ICAS resident in Scotland are engaged in industry and commerce, and 42 per cent are in practice. During the nineteenth century, sources of employment which compromised the notion of the independent fee-earner were deemed inconsistent with professionalism. The drift from public practice of members of the chartered institutes north and south of the border has necessitated the modification of the traditional professional ideal. The chartered bodies now emphasise the contribution of their members as financial advisers and managers, and offer students an 'education for business'.

NATIONAL AND INTERNATIONAL IDENTITIES: SCOTTISH CHARTERED ACCOUNTANTS

Despite the differences in the identities, statuses and ideologies of accountants in Edinburgh and Glasgow, commonality was found when collective interests were threatened. Divergences between east and west have been publicly subsumed in the pursuit of the cause of 'Scottish Chartered Accountants'. This construction emerged well before its formal institutionalisation as ICAS in 1951. Of particular importance to reinforcing the identity of the Scottish CA were the activities of the Scottish Institute of Accountants and the Corporation of Accountants, Ltd during the 1880s and 1890s. The attempts of these bodies to usurp the privileges of the SAE, IAAG and SAA seriously endangered the elevated socio-economic position of CAs and formed the catalyst for incremental steps towards the development of a 'national' chartered profession in Scotland. These included devising a common scheme of examinations (1892–1951), the appearance of a professional journal (*The Accountants' Magazine*, 1897), the invention of the collective title 'Scottish Chartered Accountants' (1896–7) and the formation of a Joint Committee of Councils in 1915 to discuss matters of mutual concern, such as formulating legislation to secure the registration of the profession in Scotland.[38]

Scottish CAs have been galvanised into responses which are overtly nationalistic, particularly when external threats have emanated from the metropolis. Indeed, it was demands from a London committee of merchants for the imposition of the English law of bankruptcy on Scotland and the appearance of draft legislation to secure that object which inspired the organisation of accountants in Edinburgh and Glasgow in 1853. At this time accountants featured as secretaries and members of the National Association for the Vindication of Scottish Rights. Other accountants wrote pamphlets defending Scotland's separate laws of bankruptcy and objected to the imposition of English practices and procedures north of the border.[39]

Subsequently, periodic proposals from London for structural change within the accountancy profession have aroused suspicion in Scotland. Although accountants in Edinburgh, Glasgow and Aberdeen were first to organise during the nineteenth century, the ICAEW soon became the dominant political force in the profession. It is currently the largest organisation of professional accountants in Europe. The structure of the profession in Britain has been problematic due to its representation by a large number of competing bodies. The city-based structures and closure practices initiated in Scotland and emulated elsewhere contributed to this. In 1930, there were at least seventeen organisations of accountants in the British Isles. Since then, there have been several attempts to integrate the accountancy bodies and secure the closure of the profession by registration. These proposals have invariably been driven by the ICAEW, which has faced greater competition from non-chartered accountants in its own backyard than has been the case in Scotland. Scottish CAs have frequently objected to the need for change and the way in which proposals from England for rationalisation have been insensitive to the character of the profession north of the border.[40] As recently as 1989, the members of ICAS rejected proposals for the creation of an Institute of Chartered Accountants of Great Britain by merging ICAEW and ICAS. While 87 per cent of English chartered accountants supported a unified institute, only 45 per cent of Scottish CAs did so. The Scots were concerned about surrendering their independence, losing their separate identity (as symbolised in exclusive use of the credential 'CA') and feared the creeping anglicisation of the profession.

In addition to the local and national identities of Scottish accountants, there is also an international dimension. The structures established in Edinburgh and Glasgow in 1853 informed professional organisation overseas; although in most nations, the problems of city-based bodies and the failure to close the profession by statute were avoided. In recent years, ICAS has been involved in establishing professional institutions in Uganda, Saudi Arabia, Greece, the Czech Republic and Romania. Moreover, many CAs have emigrated. Some Scottish chartered accountants were instrumental in founding accountancy firms and organisations which subsequently dominated the profession on the international scene.[41] A few departed from Scotland early in the history of the organised profession.[42] Most left from the 1890s, when anxiety was expressed about an oversupply of local practitioners. Later generations satisfied the increasing demand for accounting labour overseas, particularly in the USA.[43] Thereafter emigration became an acceptable career move for many young, newly qualified CAs.

The high status of CAs facilitated their ability to exploit employment opportunities outside Scotland. The city-based societies, anxious to increase their memberships and influence, appeared to encourage the diaspora.[44] Hence, by 1914 only one half of the members of the SAE were resident in Edinburgh. During the 1890s most emigrant CAs ventured to the metropolis, and in 1899 numbers were sufficient to form an Association of Scottish Chartered Accountants in London. During the early twentieth century the most popular destination for emigrant CAs was North America.[45] Subsequently, they ventured to South Africa, Australasia and continental Europe. After World War I, the proportion of CAs resident abroad began to stabilise, and from the 1930s there was a slow drift back to Scotland. Although members of ICAS can now be found in over 100 different countries, 61 per cent reside in Scotland. A further 25 per cent are located in other parts of the UK.

CREDENTIALISM AND THE LEARNED PROFESSIONAL

It will be apparent from the foregoing that Scottish chartered accountants have a particular attachment to their defining credential 'CA'; members of ICAEW use 'A[ssociate]CA'or 'F[ellow]CA'. Within the UK this nomenclature is applied exclusively to members of ICAS. The term 'chartered accountant' and initials 'CA' were adopted by the SAE in 1855 and by the IAAG a year later. In the competition for clients the title 'CA' soon proved advantageous, distinguishing the instructed practitioner from the non-professional accountant. Those entitled to append CA to their names were responsible for conducting over 90 per cent of accountants' business in Scotland by the 1880s and 1890s,[46] and the bodies that conferred it recognised (and continue to recognise) its value in attracting recruits. The use of the credential as a device for fee-seeking was lampooned by contemporaries:

> I'm a 'Chartered Accountant,' it sounds rather loud,
> As a 'Chartered Accountant' I feel rather proud;

For a 'Chartered Accountant's' a noble degree,
As a 'Chartered Accountant' confers £.s.d.[47]

Not surprisingly, the credential has been jealously guarded, and attempts by the Scottish Institute of Accountants and the Corporation of Accountants, Ltd to usurp it, especially from 1884 to 1905, met with fierce resistance.[48]

The elevated status of the CA has often been attributed to the superior professional education received by those who earned the right to use the designation. The chartered bodies in Scotland are strongly associated with erudition and have long been at the forefront of vocational education. The particular features of the system of training and examination for entry to the profession in Scotland have reinforced this source of learned respectability. The qualification systems north of the border, though not immune from criticism,[49] were often referred to as exemplars for new professional organisations. In addition to receiving practical instruction from their masters, a distinctive feature of the education of an apprentice CA in Scotland has been attendance at classes run by the institutes themselves.[50] This is recognised as a significant element in the socialisation of the CA.

The unique (among accountancy organisations in Britain) relationship between the profession and the universities has also contributed to the scholarly persona of the Scottish CA. Attendance at university law classes was required of Edinburgh apprentices from 1866.[51] In 1919 the SAE jointly endowed a chair of accounting and business method at Edinburgh University. The holders of the chair were nominees of the SAE, and CA apprentices attended classes at the university as part of their vocational preparation until the late 1960s.[52] In 1925 a chair of accountancy at Glasgow University was endowed by a former president of the IAAG.

PERCEPTIONS FROM OUTSIDE: THE ACCOUNTANT STEREOTYPE

Chartered accountants have attempted to portray themselves as reputable, altruistic professionals who contribute to economic progress. They favour the description 'doctor of business'. By contrast, representations in the Scottish media have emphasised four adverse, often compound images of the accountant: the rapacious fee-seeker; one whose career is built on the misfortune of others; the dispassionate expert insensitive to the human casualties of his decision-making; and the boring introvert obsessed with figures. Evidence of the latter image comes from the unlikely source of Sir Walter Scott. In 1820, Scott identified accountancy as a suitable occupation for his nephew, provided he was competent in arithmetic – 'steady, cautious, fond of a sedentary life and quiet pursuits' – but certainly not if he had 'a decided turn for active life and adventure, is high-spirited, and impatient of long and dry labour'.[53] Although some observed an earnestness peculiar to the Scottish CA, such perceptions of accountants were also common elsewhere.[54] Indeed, an earlier attachment to professional ideals and greater maturity in the institutions of the profession north of the border did insulate Scots from some of the vitriol directed towards their English counterparts during the nineteenth century. Accountants in Scotland were also largely spared the public wrath of other professionals, particularly lawyers. In England there were heated jurisdictional contests between lawyers and accountants over bankruptcy work from the 1850s, inciting the former to degrade the latter in the professional media as pettifogging interlopers.

The image of the callous accountant surfaced particularly during the mid-nineteenth century in relation to the appointment of some leading Edinburgh practitioners as trustees on insolvent estates in the Highlands and Islands. Accountants managed these estates in the interests of the creditors, their object usually being to sell part or all of the property to pay off debts. In their attempts to enhance the attractiveness of estates to potential buyers, tenants, cottars and crofters were sometimes cleared from the land.[55] Devine has commented that 'several of the most controversial clearances of the time [the 1840s and '50s] took place on estates which were managed by trustees'.[56] Many of these trustees were accountants. Senior Edinburgh practitioners were involved in clearances on the isles of Barra, Islay, Mull, Skye and at Ardnamurchan.[57]

When an estate fell under the control of trustees, the paternalist relationship between landlord and tenants was supplanted by a legalistic adherence to preserving the interests of the creditors. Accountant trustees were therefore vigorous in their pursuit of rent arrears and could be unhesitating in the sequestration of defaulting tenants. Their decisions were often made at a physical remove in Edinburgh, carried out by local factors, and infused with a Lowland–urban disdain for the Highland way of life. The accountant who was most renowned for this class of work during the mid-nineteenth century was the first president of the SAE, James Brown (1786–1864). From 1850 to 1856, Brown was trustee on the estates of Lord Macdonald in Skye and North Uist. Macdonald's properties had 15,000 inhabitants, many of whom were in a distressed condition. Brown ordered the notorious and highly publicised removal of tenants from Boreraig and Suishnish. He also incurred the wrath of Sir John McNeill, one of the managers of the Highland and Island Emigration Society, following his refusal to provide monies to allow the destitute to emigrate to Australia. McNeill considered Brown to be 'hard but keen' and accused him of 'trading on the misery' of the tenants who fell under his care. McNeill threatened to publicise Brown's unsympathetic stance to the poor on Macdonald's estate.[58]

Adverse representations of accountants have also surfaced in the wake of disclosures of wrongdoing, questionable practice, or when an accountant achieved notoriety in local society. James Wylie Guild (1826–94), a leading Glasgow accountant, was involved in a number of high-profile bankruptcies and liquidations.[59] He was also engaged in auditing and company formation. His questionable conduct in relation to the promotion of a chemical company on the Glasgow Stock Exchange resulted in his resignation as president of the IAAG in 1873. As a high-profile accountant, Wyllie Guild attracted the

Figure 14.2 James Brown (1786–1864), first president of the Institute of Accountants in Edinburgh, 1853–1864. Brown was also involved in the Highland clearances as a trustee on insolvent island estates. Source: Institute of Chartered Accountants of Scotland, Edinburgh.

attention of the press. In 1878, *The Baillie* described him as being at the head of 'the corbie species'; that is, Wyllie Guild and his fellow CAs were likened to carrion crows, reaping a rich harvest from the disastrous consequences of the City of Glasgow Bank failure.[60] In 1885, Wyllie Guild was described in *Clydeside Cameos* as: '*Oily Gammon* who … lives by misfortune … the misfortune of others. Like the ivy, he grows green upon ruins.'[61]

Although building a career on the back of insolvencies, liquidations and speculative ventures could result in expressions of popular distaste, these were nothing compared to the vilification heaped on accountants when one of their number crossed the boundary between sharp practice and illegal acts. The most notable case during the nineteenth century was that of Donald Smith Peddie, CA, in 1882. In public, Peddie appeared as a respectable, devout, professional gentleman of advancing years. He was an original member of the SAE, the son of Reverend James Peddie and the uncle of John Dick Peddie, MP, the classical architect. While in an impecunious state, Peddie became embroiled in property development. Threatened by his business partner with blackmail and revelations of sexual impropriety, Peddie generated funds by forging bills and defrauding the Dissenting Ministers' Friendly Society, of which he was acting treasurer. When these activities were discovered Peddie took flight, and the Edinburgh City Police offered a reward for information leading to his apprehension. Peddie was reviled in the press as 'a whited-sepulchre, the dishonourer of a good name,

Figure 14.3 James Wyllie Guild, leading Glasgow accountant. Source: Glasgow City Libraries.

the devourer of widows' houses, the reckless debauchee, the smooth-faced swindler and forger'.[62] The case, which damaged the altruistic façade of the Edinburgh chartered accountant, revealed to the SAE that it could not rely on admission procedures or the informalities of professional etiquette to ensure that its members were of enduring respectability and good character. A disciplinary code was therefore introduced.

Such revelations, though comparatively rare, reflected badly on the local accountancy profession. They appeared to confirm accusations made in the *Financial Record, Economist and Railway Review* in 1881 that the CA was a mercenary. The anonymous author of an article entitled 'Trickeries of Professions. The C.A.' wrote:

> This professional is associated with all who are in difficulties, and often helps them both into and out of their involvements. He is generally to be found (with, of course, certain notable exceptions) of limited education, of brusque and aggressive manners, insolent and irrepressible to secure an advantage.

According to this critic, although he cultivated a public persona of moral rectitude, philanthropy and altruism, the CA was in fact grasping and excessively competitive in his anxiety to 'contribute to his own profit'. He pursued connections through family and church to win clients and performed good works with the same object in mind: 'He engirds himself like the octopus with tentacles of continued motion and retention, and sets himself about to secure a large *clientele*.'[63] Other commentators objected to the way in which appointments were secured by CAs on the basis of nepotism. This was especially the case following the award of lucrative bank audits under the Companies Act, 1879.[64]

Rapacious behaviour could be displayed by accountants during competitions for bankruptcy trusteeships. These appointments were secured on the basis of election by the creditors. The need to obtain majorities encouraged the canvassing of creditors or their proxies and seeking recommendations founded on networks in related professions – especially the law – and business connections. Accusations of touting for trusteeships aroused widespread comment in England during the 1870s but also featured in Scotland,

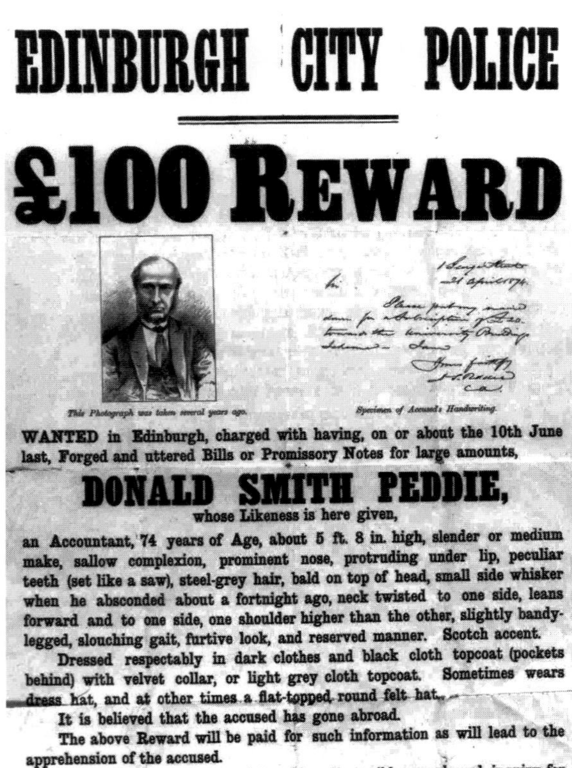

EDINBURGH CITY POLICE

£100 REWARD

This Photograph was taken several years ago.

Specimen of Accused's Handwriting.

WANTED in Edinburgh, charged with having, on or about the 10th June last, Forged and uttered Bills or Promissory Notes for large amounts,

DONALD SMITH PEDDIE,

whose Likeness is here given,

an Accountant, 74 years of Age, about 5 ft. 8 in. high, slender or medium make, sallow complexion, prominent nose, protruding under lip, peculiar teeth (set like a saw), steel-grey hair, bald on top of head, small side whisker when he absconded about a fortnight ago, neck twisted to one side, leans forward and to one side, one shoulder higher than the other, slightly bandy-legged, slouching gait, furtive look, and reserved manner. Scotch accent.

Dressed respectably in dark clothes and black cloth topcoat (pockets behind) with velvet collar, or light grey cloth topcoat. Sometimes wears dress hat, and at other times a flat-topped round felt hat.

It is believed that the accused has gone abroad.

The above Reward will be paid for such information as will lead to the apprehension of the accused.

Police-officers are requested to make all possible search and inquiry for the accused at Hotels, Railway Stations, Out-going Vessels, and other likely places; and, if found, arrest, and communicate with the Procurator-Fiscal, City Chambers, Edinburgh; or the Subscriber,

W. HENDERSON,
Chief Constable.

Chief Constable's Office,
Edinburgh, 7th December 1882.

Figure 14.4 'Wanted' notice for Donald Smith Peddie, CA, December 1882. Source: Royal Bank of Scotland Archives. Reproduced by kind permission of The Royal Bank of Scotland Group, © 2007.

as did claims of exorbitant fee-charging. James Martin, a leading critic of the chartered societies, claimed that in Glasgow:

> No sooner does an unfortunate individual call a meeting of his creditors, than a certain class of accountants are agog, and they swoop upon the unfortunate like vultures upon a carcase, and the whole city is raked up for claims upon the estate. The fact that an accountant has the matter already in hand seems to add zest to the desire in others to take it from him.[65]

In other forms of work, too, accountants found themselves under scrutiny for charging exorbitant fees. In 1868, the Law Courts (Scotland) Commission heard that when the courts remitted an investigation to an accountant the resultant report was invariably of excessive length – a device for generating higher remuneration.[66]

CONCLUSIONS

In common with the situation in the rest of the UK, the work of the professional accountant in Scotland has altered significantly since the first organisations were formed in Edinburgh and Glasgow in 1853. From the management of property under bankruptcies, trusts and factorships, accountants extended into the realm of corporate audit, taxation, business finance, corporate consultancy and IT advice. The occupation continues to grow at a rapid rate, sustained by its capacity to extend and modify the boundaries of professional activity. There remain, however, enduring issues confronting the profession. In the postwar era, accounting practice has been conducted within an increasingly pervasive regulatory framework that leaves less scope for the exercise of professional judgement. The profession has suffered from periodic crises of confidence, particularly in relation to high-profile cases of audit failure – though few of these have taken place in Scotland. The drift of accountants from public practice to employment in business represents a major departure from the founding ideals of the profession. The structure of the profession also continues to be the subject of debate, as the plethora of organisations which represent it appears outmoded in an era dominated by international accountancy firms, multinational corporations and global capital, and when unified representations of professional opinion are sought by governments in the UK and Europe.

As measured by membership statistics, ICAS has suffered decline in its position relative to other professional bodies in the UK. From the late nineteenth century to the late 1950s, the Scottish chartered societies represented between 20 and 25 per cent of the total members of the three institutes of chartered accountants in the British Isles. In 1958, the members of the Society of Incorporated Accountants and Auditors were integrated into the chartered institutes. Given the comparatively low number of 'incorporated' accountants in Scotland, this event did not have a great impact on the membership of ICAS. However, it substantially increased the size of the

ICAEW. Consequently, by 1961 ICAS represented 16 per cent of the members of institutes of chartered accountants in the British Isles.

The rate of growth of ICAS membership in recent decades has been less marked than the other chartered bodies, especially the Irish Institute (which encompasses the island of Ireland). ICAS currently accounts for 10 per cent of the members of bodies representing chartered accountants. There has also been enormous growth in the organisations representing non-chartered accountants, particularly in management accounting (CIMA) and public finance accounting (CIPFA). In 1961 ICAS represented 11 per cent of the members of the six major professional bodies in the British Isles. By 1991 the proportion was only 6 per cent. In recent decades, organisations such as CIPFA (1986) and ACCA (1998) have also established administrative structures in Scotland. These represent more than mere regional outposts of London-based organisations. Despite these indications of a relatively weaker position, ICAS appears determined to maintain its separate identity and to advance on the basis of its traditional strengths. As one recent president put it, ICAS intends to continue as 'an independent, vital professional body, not *in* Scotland but *of* Scotland', offering distinctive expressions of professional opinion to policy makers, building on the collegiality and cohesiveness of a smaller institute, extending its acknowledged excellence in education and furthering the elevated status of its members, who retain exclusive use of the coveted 'CA' credential.

NOTES

1. Brown, 1905, 183–5.
2. Mepham, 1988, 1.
3. Mepham, 1988, 19–38.
4. Macdonald, 1984, 1987; Briston and Kedslie, 1986.
5. Walker, 1995.
6. Walker, 1995, 2004.
7. Walker, 1991.
8. Shackleton and Walker, 1998, 99–101.
9. See Kirkham and Loft, 1993.
10. Lehman, 1992, 266–7.
11. Shackleton, 1999, 141.
12. The first female chartered accountant in Britain was Mary Harris Smith, who having been in practice as a public accountant since 1878, was admitted to the ICAEW in 1920 (*The History of the ICAEW*, 1966, 65).
13. Shackleton, 1999, 151.
14. ICAS, 1998, 3.
15. Paisey and Paisey, 1996.
16. Lee, 2000.
17. Walker, 1995.
18. Walker, 1988, 15.
19. Walker, 1993.
20. Walker, 1993.
21. Walker, 1996b, 70–2.

22. Walker, 1988, 17.
23. Martin, 1896, 18.
24. Walker, 1988, 16.
25. Lee, 2000, 27.
26. Shackleton and Milner, 1996, 116–21.
27. Shackleton and Milner, 1996, 103–15.
28. Shackleton and Milner, 1996, 121–5.
29. Walker, 1988, 256.
30. Walker, 1988, 85.
31. Walker, 1988, 241.
32. Kedslie, 1990, 86–96; Macdonald, 1984.
33. Shackleton, 1995.
34. Walker, 1988, 129–31; Shackleton, 1995.
35. See Hanlon, 1994.
36. Walker, 2000; Matthews et al., 1998.
37. ICAS, 1954, 97–110.
38. Shackleton, 1995; Walker, 1991.
39. Walker, 1995.
40. See Shackleton and Walker, 1998, 43–55; 2001, 73–7, 81–4, 213–34.
41. See Stewart, 1977, 171–6; Lee, 2002a.
42. See, for example, Carnegie et al., 2000.
43. Walker, 1988, 40–4; Lee, 2002b.
44. Walker, 1988, 44–51.
45. Lee, 1997.
46. Walker, 1991.
47. Anon., January 1881a, 22.
48. Walker, 1991.
49. Walker, 1991.
50. ICAS, 1954, 111–34.
51. Walker, 1988, 151–4.
52. Walker, 1994; Lee, 1996, 187–95.
53. ICAS, 1954, 14. For Scott's own account-keeping, see McKinstry and Fletcher, 2002.
54. The political career of the eminent Edinburgh accountant George Auldjo Jamieson was reputedly hindered during the late nineteenth century by an obsession with statistics: 'He would pile them up to an almost bewildering extent in his arguments for the cause he pleaded', Walker, 1996b, 76–7. For a historical study of the accountant stereotype, see Bougen (1994). There are of course exceptions to the characterisation of the 'boring' accountant. Most notable was James Balfour who died in 1795. As well as being successful in his profession, 'Singing Jamie Balfour' was noted as a blithesome and intemperate figure in the taverns of Edinburgh and was secretary of the Honourable Company of Edinburgh Golfers. Brown, 1905, 188–93.
55. Richards, 2000, 229.
56. Devine, 1988, 185.
57. Walker, 2003.
58. Walker, 2003.
59. Stewart, 1977, 87.
60. Kedslie, 1990, 10.
61. Quoted in Shackleton and Milner, 1996, 95.

62. Walker, 1996a.
63. Walker, 1996a; Anon., 1881b.
64. Walker, 1998, 44.
65. See Martin, 1896, 9–14.
66. Walker, 1993.

BIBLIOGRAPHY

Anon. The chartered accountant, *Scottish Banking and Insurance Magazine* (1881a), 22.

Anon. Trickeries of professions: The C.A., *Financial Record, Economist and Railway Review* (1881b), 281.

Bougen, P D. Joking apart: The serious side to the accounting stereotype, *Accounting, Organizations and Society*, 19 (1994), 319–35.

Briston, R J and Kedslie, M J M. Professional formation: The case of Scottish accountants – some corrections and some further thoughts, *British Journal of Sociology*, 37 (1986), 122–30.

Brown, R, ed. *A History of Accounting and Accountants*, Edinburgh, 1905.

Carnegie, G D, Parker, R H and Wigg, R. The life and career of John Spence Ogilvy (1805–71), the first chartered accountant to emigrate to Australia, *Accounting, Business & Financial History*, 10 (2000), 371–83.

Devine, T M. *The Great Highland Famine: Hunger, Emigration and the Scottish Highlands in the Nineteenth Century*, Edinburgh, 1988.

Hanlon, G. *The Commercialisation of Accountancy: Flexible Accumulation and the Transformation of the Service Class*, Basingstoke, 1994.

Institute of Chartered Accountants in England and Wales. *The History of the Institute of Chartered Accountants in England and Wales 1880–1965*, London, 1966.

ICAS. *Forward with Confidence*, Edinburgh, 1998.

ICAS. *A History of the Chartered Accountants of Scotland from the Earliest Times to 1954*, Edinburgh, 1954.

Kedslie, M J M. *Firm Foundations: The Development of Professional Accounting in Scotland 1850–1900*, Hull, 1990.

Kirkham, L M and Loft, A. Gender and the construction of the professional accountant, *Accounting, Organizations and Society*, 18 (1993), 507–58.

Lee, T A. Richard Brown, chartered accountant and Christian gentleman. In Lee, T A, ed, *Shaping the Accountancy Profession: The Story of Three Scottish Pioneers*, New York, 1996, 153–221.

Lee, T A. The influence of Scottish accountants in the United States: The early case of the Society of Accountants in Edinburgh, *Accounting Historians Journal*, 24 (1997), 117–41.

Lee, T A. A social network analysis of the founders of institutionalised public accountancy, *Accounting Historians Journal*, 27 (2000), 1–48.

Lee, T A. The contributions of Alexander Thomas Niven and John Ballantine Niven to the international history of modern public accountancy, *Accounting and Business Research*, 32 (2002a), 79–92.

Lee, T A. UK immigrants and the foundation of the US public accountancy profession, *Accounting, Business & Financial History*, 12 (2002b), 73–95.

Lehman, C R. 'Herstory' in accounting: The first eighty years, *Accounting, Organizations and Society*, 17 (1992), 261–85.

MacDonald, K M. Professional formation: The case of Scottish accountants, *British Journal of Sociology*, 35, (1984), 174–89.

MacDonald, K M. Professional formation: A reply to Briston and Kedslie, *British Journal of Sociology*, 38 (1987), 106–11.
McKinstry, S and Fletcher, M. The personal account books of Sir Walter Scott, *Accounting Historians Journal*, 29 (2002), 59–90.
Martin, J. *The Accounting Profession: A Public Danger*, Glasgow, 1896.
Matthews, D, Anderson, M and Edwards, J R. *The Priesthood of Industry: The Rise of the Professional Accountant in British Management*, Oxford, 1998.
Mepham, M J. *Accounting in Eighteenth Century Scotland*, New York, 1988.
Paisey, C and Paisey, N J. Marginalisation and segregation: The case of the female Scottish chartered accountant. In Masson, D R and Simmon, D, eds, *Women and Higher Education: Past, Present and Future*, Aberdeen, 1996, 72–83.
Richards, E. *The Highland Clearances. People, Landlords and Rural Turmoil*, Edinburgh, 2000.
Shackleton, K. Scottish chartered accountants: Internal and external political relationships 1853–1916, *Accounting, Auditing & Accountability Journal*, 8 (1995), 18–46.
Shackleton, K. Gender segregation in Scottish chartered accountancy: The deployment of male concerns about the admission of women, 1900–1925, *Accounting, Business & Financial History*, 9 (1999), 135–56.
Shackleton, K and Milner, M. Alexander Sloan: A Glasgow chartered accountant. In Lee, T A, ed, *Shaping the Accountancy Profession: The Story of Three Scottish Pioneers*, New York, 1996, 82–151.
Shackleton, K and Walker, S P. *Professional Reconstruction: The Co-ordination of the Accountancy Bodies 1930–1957*, Edinburgh, 1998.
Shackleton, K and Walker, S P. *A Future for the Accountancy Profession: The Quest for Closure and Integration, 1957–1970*, Edinburgh, 2001.
Stewart, J C. *Pioneers of a Profession: Chartered Accountants to 1879*, Edinburgh, 1977.
Walker, S P. *The Society of Accountants in Edinburgh, 1854–1914: A Study of Recruitment to a New Profession*, New York, 1988.
Walker, S P. The defence of professional monopoly: Scottish chartered accountants and 'satellites in the accountancy firmament', *Accounting, Organizations and Society*, 16 (1991), 257–83.
Walker, S P. Anatomy of a Scottish C.A. practice: Lindsay, Jamieson and Haldane, 1818–1918, *Accounting, Business & Financial History*, 3 (1993), 127–54.
Walker, S P. *Accountancy at the University of Edinburgh 1919–1994: The Emergence of a 'Viable Academic Department'*, Edinburgh, 2004.
Walker, S P. The genesis of professional organization in Scotland: A contextual analysis, *Accounting, Organizations and Society*, 20 (1995), 285–310.
Walker, S P. The criminal upperworld and the emergence of a disciplinary code in the early chartered accountancy profession, *Accounting History*, NS, 1 (1996a), 7–36.
Walker, S P. George Auldjo Jamieson: A Victorian 'man of affairs'. In Lee, T A, ed, *Shaping the Accountancy Profession: The Story of Three Scottish Pioneers*, New York, 1996b, 1–79.
Walker, S P. More sherry and sandwiches? Incrementalism and the regulation of late Victorian bank auditing, *Accounting History*, 3 (1998), 33–54.
Walker, S P. Benign sacerdotalist or pious assailant. The rise of the professional accountant in British management, *Accounting, Organizations and Society*, 25 (2000), 313–23.
Walker, S P. Agents of dispossession and acculturation. Edinburgh accountants and the Highland clearances, *Critical Perspectives on Accounting*, 14 (2003), 813–53.
Walker, S P. The genesis of professional organization in English accountancy, *Accounting, Organizations and Society*, 29 (2004), 127–56.

15 The Medical Profession in Scotland

DAVID HAMILTON

Characterising a 'profession' is no longer easy, but a simple working definition is that of a body of skilled persons offering expertise in a limited area closed to others, entry to which requires a period of nationally monitored, controlled-entry training after a high standard of previous scholastic achievement. After this training, national rolls of practitioners are maintained, adherence to an ethical code is expected, and a high-minded disinterest in financial rewards is encouraged. Performance in later practice is monitored, and discipline can be enforced by peer review, even leading to exclusion from practice.

For the medical profession, entry to university medical schools requires excellent school examination grades, and the licence to practise is by placement on the Medical Register, with monitoring throughout life by the General Medical Council.

The emergence of the various 'professions allied to medicine' has blurred this old simplicity, and it is of interest to look for the emergence historically of the medical profession. In the late 1700s there was a variegated medical 'marketplace' with a variety of healers offering a range of regimens to sufferers. From this disparate group, the university-trained cadre emerged as the most powerful player, successfully claiming a superior understanding of the body in health and disease. The medical profession gained its unique present status when parliament set up the Medical Register in 1858, which listed those emerging successfully from approved institutions.

THE PROFESSION

Doctors come in all varieties. In looking at their working lives, two kinds with direct contact with patients can for simplicity be studied here: the hospital doctor (for example, surgeons and physicians) and the general practitioner. There are other vital and important practitioners, who emerged in the late 1800s, including laboratory workers (research and diagnostic), public health doctors and administrators, whose rather different working lives complete the jigsaw of an increasingly team-based service.

In looking at these chosen doctors, it makes sense to reverse the usual historical sequence and examine the profession at present and then look backwards, a reverse chronology which moves from the familiar to unfamiliar and which may assist in showing the evolution of the working life. The changes will be examined at intervals of about 100 years. But useful historical records in Scotland are scarce since, though strong on scientific innovation and administrative change, they are remarkably deficient when looking for details of the doctor's daily routine. We know of Lister's achievements, but not what he did every day after breakfast. One explanation for this dearth is that traditional respect for patient confidentiality discourages the writing of medical memoirs.

AT PRESENT

As a career, medicine is favoured by highly qualified school-leavers, indicating that the profession is attractive to the young and that any perceived drawbacks are discounted. The medical profession continues to have high ratings in public opinion surveys.

University fees and living costs are still moderate in Scotland and are certainly less than in many other countries: hence it is hoped that few talented students are discouraged from medical training by costs. The proportion of women medical students has risen to well over 50 per cent, and the career restrictions that existed for them in earlier times have largely disappeared. In particular, prospects in hospital work have improved, aided by more flexible contracts for married women, and activists in this matter have been pleased at the appearance from the 1970s of the first women professors in the university medical schools and, a little later, a notable number of women surgeons emerged.

After graduation, further training schemes for the young doctor pave the way to established posts as a general practitioner or hospital consultant. The various national postgraduate training schemes have regularly shown unexpected difficulties in their well-intentioned attempts to match entrants to needs, but promotion to the grade of consultant is faster and less uncertain than formerly, when the grade was reached almost in middle life after much financial and professional hardship. Consultants are now well paid and semi-secret 'merit awards' can double an already excellent salary. All consultants can do some private work, consulting and operating in the new wave of well-equipped private clinics which emerged in the 1980s, adding in this way to the consultant's income by running a small business – though these rewards in Scotland have never been as high as elsewhere. En route to the top, the 'junior' doctors now have the luxury of a 48-hour week, imposed by legislation from Europe.

General practitioners now work from large well-equipped health centres assisted by multi-disciplinary team support, which ranges from physiotherapists to community nurses. Night-time and weekend work has lessened through sharing the task with colleagues, or handing over to out-of-hours clinical units and call centres. 'House calls' – a traditional and kindly part of the family doctor service – have largely disappeared, with their withdrawal justified on the grounds of the growth of car ownership among patients. General practitioners in the remoter parts of Scotland, many attracted there by lifestyle opportunities, use innovations in communication such as telemedicine, but they may retain the older style of house visits, often as a social call, and also act like the apothecary of old by dispensing their patient's medication.

Relative to the other ancient professions, doctors are now probably better rewarded than lawyers and are certainly well ahead of ministers of religion. Unlike other professions, doctor's salaries are monitored and supported by an independent review body. This arrangement has benefited the profession, particularly at times of inflation, and the doctors' negotiators are the British Medical Association, which reluctantly but successfully took on an overt trade union role in the 1970s. To add to this, the NHS final salary pension scheme is good – so good that it cannot last – and since the profession was the first group to shun smoking, these inflation-linked pensions have supported increasingly long periods of retirement.

During a working life, doctors often take part in the complex administration of the NHS or take office in professional bodies such as British Medical Association or the Royal Colleges. Some enjoy doing research, other do private practice and the larger hospitals have suitable salaried posts, usually university-linked. When off duty, aware of the need to preserve their health, doctors are often active in sport.

But while the working life is generally agreeable, some less comfortable dimensions have intruded recently. Hospital consultants are no longer generously appointed for life as before, but usually for only five years in the first instance. An additional constraint on their former freedom is the novelty of imposed periodic testing of fitness to practise and an analysis of 'outcomes', i.e. success rates in patients treated. Academic doctors now have to seek grant money to support their research, and tenure of post may require evidence of voluminous numbers of publications. Other matters are that patients are better informed, less deferential and have higher expectations than before, and they are not slow to take these concerns to lawyers or the new lay management of the NHS, who now have powers to rebuke or worse. In their NHS work, doctors were pleased and surprised that the government fairly recently decided to pay for any legal defence necessary, a huge personal cost in other countries.

Recourse by patients to 'alternative' medicine is paradoxically higher in this time of unquestioned medical progress, but the profession, securely salaried, has been increasingly tolerant of these other healers. By admitting that others have a place in the profession's traditional space, the doctors have allowed a little erosion at the margins of their turf, and permitted a modest return of the medical marketplace of old.

THE EARLY 1900S

Entry to medicine was less demanding a century ago, and largely middle-class students, mostly men, were attracted and could enter with little formality, with sons often following a medical father into medicine. A few bursaries in Scotland did enable the 'lad of pairts' to pay the modest fees, and their successes distracted attention from how few such lads did aspire to the professions. Women were beginning to be accepted as entrants, but after graduation they were largely excluded from mainstream hospital life and instead found places in less popular medical work. On marrying, they were expected to cease working. Other exclusions from the staff of the large city hospitals were on blatant, but unwritten, religious and ethnic grounds.

In hospital careers, the goal was to rise to 'ward chief', the senior member of the little internal 'units' of hospital doctors. In the large 'voluntary' hospitals of the cities and towns (so-named because their finances came from local subscribers and charitable fundraising) the senior staff were watched by the community, and their stature in the hospital led to a high local profile and ensured a flow of private work. The younger men in the units had poor salaries, and their uncertain wait for promotion might be assisted by having private means or making a 'good' marriage. Remarkably, the senior men (not called 'consultant' until later) had little or no pay from the major city hospitals, since the status of the hospital post and resultant private earnings were agreed to be reward enough. Payment was not to come to the senior staff until the advent of the NHS in 1948 when, to their surprise, they were paid for what had previously been done for nothing.

Private Practice

Prior to the NHS, the ward chief's routine was to make irregular, short, morning visits to oversee the work of the unit's younger assistants, and then the senior men were off to their private practice, their main source of income. Clinical research in the later style was largely unknown, but if scholarship was desired, then it found an outlet in textbook writing. Private work was done at their consulting rooms, and these 'Harley Streets' were found in the major Scottish cities. For serious illness, they did not admit their middle-class and well-off patients to the big voluntary hospitals, but instead they treated them at home, including doing serious surgical operations, with a private nurse hired for care during this time. An alternative was to admit the patient to one of the many small nursing homes often owned as a business by the same senior doctors. These small, badly equipped houses steadily fell out of favour from mid-century. Home treatment also faded as the century progressed, though the landed gentry might still demand the prolonged presence in a distant area of an eminent city practitioner.

For senior hospital doctors it was an entrepreneurial life with uncertainties, and subtle trawling for patients went on, most notably by courting the good opinion of general practitioners locally or those in country districts with well-off patients. Charitable work and free treatment of the poor similarly attracted favourable attention, as did the use of a carriage, or later a chauffeur, while on their rounds of home visits to paying patients. Lunching at good city restaurants and memberships of suitable gentlemen's clubs was essential. Treatment of other professionals and their families was often done gratis.

At this time litigation by patients against doctors was almost unknown. This immunity, regarded by the profession as natural at the time, was assisted by an informal code, found in other professions, which tried unsuccessfully to discourage gossipy denigration of colleagues but did succeed in precluding doctors from providing testimony in court to support legal action by a patient against a fellow practitioner. On the rare occasions when legal help was required, the medical defence unions proudly and reflexly defended their doctors, and membership of these unions was available to all doctors for a small annual fee.

General Practice

In retirement, personal pensions were self-arranged, though life expectancy was less than later. During working life, deaths from the work hazards encountered, notably infectious disease, were quite common. Taking out life insurance was a prudent move. But widows of medical men might be left in difficulties, and might be enabled to remain in comfortable respectability by an ancient arrangement in the Glasgow and Edinburgh colleges of physicians and surgeons which accumulated substantial Widows' Funds, contributed to by quite large levies on new fellows on entry, and these monies could be disbursed discreetly.

At this time, to start as a general practitioner a new medical graduate would add one year's hospital work and then go straight into practice without further training. He would start as an assistant, or by purchasing a practice, or even 'putting up a plate' in an under-doctored area and await the arrival of patients. Some law firms specialised in buying and selling practices, and if there were substantial numbers of established patients – the 'good will' – it was passed on at a price. The plates outside the 'surgery' were the only form of advertising allowed to medical men, and they were of fixed size, showing no more than the general practitioner's name and degrees, with the addition of 'physician and surgeon'. Some operative surgery was indeed done by general practitioners, particularly in remote areas, and many births were conducted in the home, applying forceps if necessary, with chloroform administered by a fellow general practitioner. Tonsil removal 'on the kitchen table' also existed, and life-saving unpleasant tracheotomy for diphtheria might be attempted. General practitioners usually worked single-handed from their small surgery in town or from their homes, with formal or informal partnerships. Urgent night-time consultations were increasingly expected as the urgent assessment of illness became accepted, and in response formal 'partnerships' were formed, usually with two doctors sharing the practice and its finances. Country doctors usually dispensed their own drugs, and this could be profitable. General practitioners, particularly in the country, were the first enthusiastic owners of motor cars.

Doctors' fees to patients were adjusted to social status and might be collected quarterly, but poorer patients could not pay and bad debts were common. Some salaried income might come from part-time work with Poor

Law infirmaries or charitable dispensaries, and compared with the later sharp divide between hospital and general practice, similar professional boundaries were not drawn at this time. Local businesses and friendly societies had a large network of arrangements of annual, fairly mean-spirited, contracts with local general practitioners.

The National Health Insurance Scheme of 1911 changed much and enabled working people to enrol on 'the panel' with a doctor for a fixed fee, and with government adding its share, this gave the doctor a reasonable and predictable annual income from this source. Though opposed by some sections of the profession as an intrusion on their freedom, the Scheme benefited the working lives of the inner-city practitioner. This was extended to give all of the population 'free' treatment with the advent of the NHS. But the service given by general practice stood still, and after World War II only a limited range of advice and treatment was available to patients waiting in long queues inside the doctors' spartan high street premises. Health centres were to arrive in the 1970s.

In this early part of the century, professional and academic meetings and societies were more widespread in town and country than later, and the evening meetings' agendas could range from scholarly lectures to discussions on maintaining the levels of professional fees and medical 'ethics' – then meaning the fraught matter of the gentlemanly code of relations with colleagues.

THE 1800S

When comparing the established professions in the early 1800s, lawyers, academics and ministers of religion outranked the medical men in status and financial rewards, and the medical profession suffered a setback with the Edinburgh grave-robbing scandals in the 1830s. But the status of the profession was to increase, and many other healers were marginalised, with the quacks of old almost gone. The repute of the profession was greatly assisted in mid-century by therapeutic advances, notably when anaesthesia and antisepsis made surgery safer, more humane and more adventurous. At that time also, the Medical Register sharply defined the profession.

Entry to medical practice was changing in the early 1800s. Formerly, those wishing to be surgeons were trained by formal apprenticeship to the surgeons of the town, but increasingly some time and instruction on 'day release' at university was desirable. This instruction was taken from the range of lecture courses offered by the professors. A final MD degree had not been taken even in the late 1700s, as it was not necessary for the practice of medicine. Instead, local approval of the colleges in Edinburgh and Glasgow, or a town guild, was required. But by 1800 it was thought desirable to have the MD, and the power of the colleges and status of their licences were waning. Increasingly, the separate teaching of anatomy, 'physic' and surgery were cobbled together in a loose form of medical curriculum which was modified by the needs of the pupil and led to the degree. Even the practice of obstetrics became a respectable part of the practitioner's working life, and the universities obliged by giving formal courses on the subject.

Huge numbers were trained at the successful Scottish medical schools, who took all comers without concern for the numbers entering the future profession; many who came from furth of Scotland returned whence they came, with market forces determining success or failure. Small, free-standing 'extramural' medical schools also existed, and humbler students could train and take the degrees offered by the Scottish colleges of physicians and surgeons, rather than the more expensive university degree. After qualification there was no formal training, even for surgery, in hospitals or elsewhere, and gaining experience on the job sufficed.

Medical Schools

By the 1800s the Scottish university medical schools were now dominant in Britain and were producing a multiskilled practitioner ready for all clinical challenges, unlike the lofty 'pure' physicians from the English universities. These products of the Scottish Enlightenment were prominent in founding many medical schools abroad. Those working with the armed forces, with the East India Company or elsewhere in the colonies, or working as ships' surgeons, had opportunities for natural history studies, and many publications and important insights resulted.

Since the students were now obtaining the MD or college equivalent regularly at the end of their studies, this meant a problem for established older practitioners. Disadvantaged by lacking the now-desirable degree, they were allowed to write back to their (or other Scottish) university for the MD, enclosing a fee and testimonials to their skills. The degree was then retrospectively awarded by post. Abuses followed, and ridicule of this 'Scotch MD' had to be endured for a while.

Hospital Practice

The hospital practitioner existed in very small numbers, though the university professors had their established position and, in addition to lecturing, often published extensively. For staffing the growing number of city hospitals erected by charitable subscription, the general practitioners of the area, who had broad skills in surgery and medicine outside the hospital, took on the work in rota. The Poor Law, now stretched almost to breaking point in the big cities, and charitable dispensaries gave free attention for the poor at clinics and occasionally within the poorhouses. Poor Law posts, notably the salaried 'district surgeon' jobs, gave young men a chance to start in practice, and perhaps if a practitioner caught the eye of the great and good on the various hospital boards of management, opportunities might open up.

In remote areas doctors were few and travel was difficult. Many deaths were 'uncertified' because a doctor had never been in attendance. There the Poor Law funds for care of the official poor of the parish might be crucial to the doctor's income, which otherwise came from the uncertain patronage of the landowners and their household.

General Practice

For newly emerging doctors, who did not settle to this broad-based general practice immediately or at all, there were alternatives, often in a career as one of the huge number of military surgeons, which also offered half-pay for those later becoming reservists, at this time of national nervousness. Taking a civilian ship's surgeon post could also give financial support and adventure before the uncertainties of starting on a settled medical working life. Medical mission work, which ranged from charitable work among the Scottish city poor to a full-scale distant missionary career funded by one of the Scottish churches, attracted some doctors later in this century. Colonial salaried medical work was seen as worthy, and offered early retirement, to which was added a pension, unlike in civilian practice – though doctors' life expectancy was low at this time. Some of the spas, increasingly emerging at this time for recourse to by invalids, had resident doctors.

THE 1700S

Early in the eighteenth century, three quite different types of doctor with different working lives existed. Learned physicians were the rarest, and only a handful attempted a career of this kind in Scotland, and to do so meant a preliminary education in Europe and taking the MD degree at an ancient university. The physicians claimed superior knowledge of disease mechanisms and the arcane complex drug therapy of the day, and their fees were high. One feature was that they could be consulted from a distance, and a letter describing the patient's problems would be answered by post with advice, prescription and prognosis. In their working lives the Scottish physicians had a leaning towards scholarship and often had a reputation as authors, naturalists and collectors of books, also taking the university professorial posts with distinction. In the troubled earlier times of the late 1600s they had been active in the struggles of Edinburgh's political and professional life and took chances with their careers, even risking exile. By the 1700s, this élite sector of practice in Scotland was at a low ebb, since the Scottish aristocrats were increasingly drawn to the court and parliament, now both in London.

A more numerous and available practitioner was the surgeon, a group of men now increasing in number and distancing themselves from the barbers with whom they had been joined in medieval times. There were also the apothecaries. They were not only dispensing chemists but had expanded to become practitioners, functioning as an economical form of physician. In the early 1700s training for both surgeon and apothecary was still by apprenticeship, but important changes were seen in Scotland at about this time. The surgeons and apothecaries put aside their rivalry and merged their organisations, thus inventing the surgeon–apothecary: an early general practitioner. No such harmony was achieved in England during this century.

The bulk of most aspirants to a medical career were apprenticed as young teenagers to the surgeons or apothecaries of the towns, usually for a five-year servitude, residing with and paying the established practitioner for his tuition. A school education and knowledge of Latin (later dropped) was probably required of these neophytes, and the apprentices' contracts of the time show that discipline was strict and necessary. Many dropped out and may have settled in unregulated country practice. Increasingly, as the century progressed the apprentices in Glasgow and Edinburgh added to their education the university lecture courses available locally.

Medical Practice

During training the apprentice learned the routine skills of a town surgeon, namely treating wounds and injury, at a time when civil violence was common. Added to this was the setting of fractures, dealing with skin disease, offering blood-letting and obstetrics, embalming and prescribing drugs for non-surgical disease. For the apprentices, it was a major career help to marry the boss' daughter, which might lead to inheriting the practice later on. No degree was taken at the end of these apprenticeships even if a period of study at the universities of Scotland was undertaken, since no degree was required to practise. What was required was permission either from the town's guild structure or, in Glasgow or Edinburgh, from the locally powerful corporations of physicians and surgeons, who carefully regulated the numbers of town practitioners. For the apprentices, an examination – with written, oral and practical parts – was necessary to be accredited in this way. Not all passed at the first attempt, and outsiders attempting to settle and practise in the main towns also found these tests to be a hurdle. The colleges and guilds in the cities and towns could revoke the licence to practise later, a form of 're-accreditation', if a licentiate was found unworthy of their support.

Established surgeons and apothecaries usually worked from a shop or pub. Often a contract to heal, with an agreed fee (and a deposit), was drawn up, and fees might occasionally come from other sources. The church-run Poor Law might make single payments for the care of deserving cases. The armies passing through Scotland, though less common than before, would offer temporary posts for surgeons. Young men might join the British armed forces for a while to fund later settlement in civilian life, and this gave not only considerable experience with surgery, but also increased skills in dealing with the many medical conditions which plagued and decimated the forces. Many broadly trained Scots doctors could still speak Latin, and were therefore doubly favoured as staff for the armies of Europe.

The surgeons of the town, as part of a town guild, were politically active. A surprising number of medical men were money lenders, as revealed by loans outstanding at the time of death. After death, the widow of an apothecary or surgeon might carry on in the business.

University medical education had started to grow in Scotland, and a few medical men could have careers in academia. This growth of medical teaching at the universities in Edinburgh, Glasgow and Aberdeen brought new income from student lecture fees to the professors. Publishing their lectures was an attractive idea, and many achieved income and wider fame through these textbooks, encouraged by the vibrant Edinburgh publishing sector. One other addition to the teachers' working lives was that they offered board and lodging to young students. To add to this was an occasional little earner when a personally devised pill or potion, with secret composition, caught public favour, and reward for its medical inventor and advocate followed.

An increasing number of vibrant societies sprang up at this time in Enlightenment Scotland, and during the working week they provided innovative proceedings and fellowship.

Doubtless it was easier to set up practice in remoter towns if there was no competition or the local practitioner was not popular. One factor in the working life of a country practitioner was the hope of the support from the local lairds or aristocrats, to which would be added the extra remunerative work of treating the household and estate staff. A surgeon–apothecary might only need the support of one landowner to make a good living, notably through the mark-up as an apothecary when providing the elaborate polypharmacy of the day. Country practice meant long journeys on horse or carriage and involved prolonged stays at a well-off patient's house while effecting a cure.

The Marketplace

But the medical men were far from supreme in the market place of the 1700s. Self-help was widespread, and traditional healers were well respected in their communities, often treating without fee. Herbal traditional remedies and herbalists were highly regarded, even by the learned physicians, who used the town's physic gardens to cultivate any Scottish plant favoured by folk tradition. The 'wise women' available ranged from alleged witches through to the wives of the gentry, who kept their remedies, sensible and otherwise, at the back of their cookbook, emphasising the overlap of cooking, flavouring and pharmacy. Less of a rival to medical men were the alleged magical powers of the healing wells and shrines, which had largely gone out of favour after the Reformation's strictures on such beliefs.

But itinerants of all kinds existed, with offerings ranging from cynical 'snake oil' preparations to sensible dentistry. Major surgery, in the later sense, was rare and was feared by the patient. Because of the high death rate, some serious surgery, notably on herniae or 'cutting for the stone' was left to these itinerants. When procedures like amputation were required of the town surgeons, a group of three might meet and certify that the procedure was necessary.

MEDIEVAL TIMES

A testimony to the different assumptions in this period, and lack of any organised profession, is that during times of epidemics – notably the visitations of the plague – medical men were not the only ones to give advice.

Even the diagnosis of leprosy, which meant exclusion of the sufferer from the town, was made by lay opinion.

Few full-time medical men can be found in Scotland in early medieval times, but the steady demand for healing at shrines and places of pilgrimage suggests that many sought a cure rather than stoically accepting disease. The earliest members of the nascent medical profession were the cleric–doctors, whose career was based in the church but who increasingly undertook medical practice on the side. Doubtless this arose in response to the needs of those seeking divine assistance at the churches and cathedrals, notably at the major pilgrimage sites such as St Andrews, where large numbers of sick arrived seeking help. Perhaps the clerics experimented with secular elements added to their spiritual healing. A relic of this bi-modal format was found in later medieval times, since secular surgeons were criticised if, while dealing with wounds or fractures, they failed to add in an incantation or two.

Some clerics who obtained a reputation as healers in this way appear as 'medicus' or 'physician' to the Scottish courts. But a working life at the gossipy court was uncertain and stressful. If a notable cure was obtained, then riches, or payment in kind, and preferment could follow. But if out of favour, doubtless church work was a useful fallback.

As secular practice of medicine outside the monasteries gradually developed and scholarship also moved out of the church's ambit to the four Scottish universities which emerged after 1400, the universities naturally set up posts for teaching medicine. But the few students who did study medicine rarely sought a medical career. The teachers did not seek to send out a cadre of professionals but simply aimed to broaden the education of learned men and future clerics, and no gentleman would be involved in the manual work of surgery or midwifery. Many of these university teachers had a period of study as physicians in Europe before settling back in Scotland. The salaries of these medical teachers were less than those for well-established university posts in other subjects.

Steadily, the church no longer wished to be part of medical practice. The universities could assist with providing a few élite physicians, but most medical manpower came from the emergence of secular surgeons and barbers in the towns – men skilled with use of sharp instruments – and they now united to make common cause. From 1500 onwards the barber–surgeons appear in some Scottish towns as a conjoint guild, though a lowly one. They were now organised and by the 1700s could take control of their working life, bring in formal apprenticeship, exclude others and raise their fees. A future profession was being shaped.

Highland and Islands

In the Western Isles of Scotland in the medieval period, and probably much earlier, there was a medical profession, of sorts. Entry was controlled by birth, and long lineages of hereditary healers, such as the Beaton dynasty, cared for the nobility and ordinary people in the area. Such was their reputation that they were occasionally summoned to consultations at the

Edinburgh court. Training was local, with experience elsewhere, notably in Ireland through the Gaelic-language connection, and these medical families held important libraries of manuscripts. This separate island tradition survived until the 1600s.

LOOKING BACK

At present, following a period of sustained progress in medical science and therapeutics, the medical profession has retained public esteem. The dominance in health care in Britain of the National Health Service and its mostly salaried posts has minimised the uncertainties of the marketplace experienced by those other professions with a more entrepreneurial structure. A puzzling resurgence of 'alternative' medicine has not threatened the profession with a return to the anarchic, diverse marketplace of old. In spite of traditional concerns about clinical freedom and new pressures from patients and government, the medical profession has been highly successful in obtaining a rewarding and reasonably secure working life for doctors, one expected to be followed by a comfortable retirement.

SUGGESTED FURTHER READING

Comrie, J D. *History of Scottish Medicine*, London, 1932.
Hamilton, D. *The Healers*, Edinburgh, 2003.
McCrae, M. *The National Health Service in Scotland*, East Linton, 2003.

16 Midwifery Practice in Scotland in the Twentieth Century

LINDSAY REID

The autonomous practice of midwives in Scotland for the care of women undergoing a normal childbearing episode has been approved by statute since the 1915 Midwives (Scotland) Act. Before that date, women who practised as midwives did so not illegally, but without state regulation. The majority were untrained and took the title 'midwife' by repute rather than by any specific education they might have received.

The term 'midwife' is very old and means 'the with-woman' – that is, the woman who is with a mother in childbirth. In Scotland other terms have been used, particularly for midwives with no formal training: 'howdie' or 'howdie wife' across the land, and the diminutive 'howdie wifie' in the north-east. Elsewhere there were other terms for the midwife, such as 'skilly' or 'skilful woman', 'handy woman', 'neighbour woman', or 'helping woman'. The Gaelic 'ban chuideachaidh', meaning 'aid woman', was used in many parts of the Highlands, while the people of the islands of St Kilda used the term 'bean-ghluine' or 'knee woman'. While legislation attempted to eradicate the practice of howdies or uncertified midwives in Scotland in the twentieth century, in some areas they were evident until the 1950s.

Midwifery in Britain moved towards regulation and increased professionalisation during the nineteenth century. First, a series of demographic changes due to disease, industrial change and war led to the existence of many widowed or unmarried women who needed to support themselves. This in turn brought changes in women's employment and occupations.[1] Second, dissatisfaction with the position of women produced the movement towards women's suffrage and increased employment rights.[2] Towards the end of the century, politically minded midwives in England, well-connected and belonging to the middle and upper classes, and using the influence of the newly formed Midwives Institute, moved towards the creation through legislation of midwifery as a respectable profession for middle-class women.[3] After twenty years of effort the first Midwives Act was passed in 1902. Arrangements were made for the registration of midwives in England and Wales, but not for Scotland and Ireland; legislation for midwives in Scotland came in 1915.[4]

This chapter examines aspects of midwifery practice in Scotland from 1915 to the present day, using archival sources and oral testimonies from midwives across Scotland. Oral history has considerable potential for the development of midwifery knowledge, being used to obtain information where little documentary evidence exists or is one-sided and may challenge an accepted, usually written, view of an issue. Midwives did not traditionally write down their experiences, and there is much that remains unknown about past midwifery practice. Many written archival sources for the history of midwifery in Scotland omit details of midwifery practice and the careers of midwives. It is therefore appropriate to use oral history where possible to examine and illuminate the work of midwives in Scotland in the comparatively recent past.[5]

The 1915 Midwives (Scotland) Act brought the first formal recognition of midwives as a group throughout Scotland and gave them a legal identity. However, at the same time its provisions restricted midwifery practice.[6] This chapter examines reasons behind midwifery regulation in Scotland and, briefly, the establishment of the Central Midwives Board for Scotland (CMB) and its work and duties. Throughout the period from 1916 to the end of the twentieth century, the nature of midwifery practice was subject to change both through institutional frameworks within which the women trained and practised, and alterations in practice before, during and after birth. Decades of societal change contributed to changing attitudes, both public and professional. At the beginning of the twenty-first century, midwives, more confident and better educated than those of 100 years ago, are acknowledged as full team members in the care of mothers during the full childbearing episode. Although interprofessional and political issues existed which impinged on midwives' identity and autonomy, constraints of space preclude their in-depth examination here.[7] Therefore, this overview of midwifery in Scotland in the twentieth century will consider midwives and their practice as it developed in the care of mothers antenatally, intranatally and postnatally.

REASONS BEHIND REGULATION

As late as the early twentieth century in rural Scotland most babies were delivered at home by untrained midwives or howdies.[8] Prior to the mid-eighteenth century, midwives cared for and delivered women in childbirth and called in a doctor in emergency situations. From the mid-eighteenth century, the presence of male practitioners in the delivery room became more common even for normal births.[9]

The 1902 Midwives Act preceded the Midwives (Scotland) Act by thirteen years.[10] The inclusion of Scotland within the act was discussed at the time in parliament but was ruled out. Some members said that extra provisions and amendments would be required to make the bill apply to Scotland, there were differences in the two legal systems, and one member commented that a Midwives Act in Scotland was unnecessary as 'these things are managed better in Scotland'.[11] Members of the Edinburgh Obstetrical Society (EOS) agreed that 'in Scotland at present [1895] there was no great need for the registration of midwives'. However, 'if anything was passed for England it would sooner or later cross the border'.[12] Registration of midwives was also opposed by members of the General Medical Council, particularly general practitioners, who feared that competition from midwives would increase if they were to be registered and their training regulated.[13]

Opposition to the EOS came from members of the Society of Medical Officers of Health for Scotland led by Dr A K Chalmers, who campaigned vigorously for legislation for midwives in Scotland in the early twentieth century.[14] The heart of their case was that levels of infant and maternal mortality in Scotland in general, but more specifically in Glasgow, were very high.[15] The implementation of the 1907 Notification of Births Act in some cities in Scotland disclosed a lack of records of midwifery qualifications, and further investigation indicated that most midwives held no certificate of proficiency.[16] Thus Chalmers identified a correlation between high mortality rates and a lack of formal midwifery training. Although persuasive at the time, Chalmers' data and views are contrary to other contemporary and recent studies, which show that maternal mortality figures in Britain for the late nineteenth and early twentieth centuries were, on average, better for those mothers who were delivered by midwives than by doctors.[17]

After years of campaigning and waiting, passage of the Midwives (Scotland) Act came quite quickly, influenced by the Memorial to the Right Honourable HM Secretary for Scotland and Right Honourable The Lord President of HM Privy Council pleading for 'the passing without delay of a Midwives Bill for Scotland'.[18] The Lord President of the Council agreed that the bill was urgent, all the more so because of World War I, which led to the call-up of many doctors for military service. Their absence created a void in maternity care, which was rapidly being filled by midwives, many of whom were unqualified.[19]

The Midwives (Scotland) Bill received royal assent on 23 December, 1915[20] and came into operation on 1 January 1916. However, the speedy enactment of the bill was due primarily to the shortage of doctors in Scotland as a result of the war, and not because of the need to recognise the importance of the profession of midwifery and its place in the healthcare of the people of Scotland.[21]

THE CENTRAL MIDWIVES BOARD FOR SCOTLAND

The 1915 Midwives (Scotland) Act made provision for the constitution of the Central Midwives' Board for Scotland, whose responsibility it was to implement measures to fulfil the aim of the act 'to secure the better training of Midwives in Scotland, and to regulate their practice'.[22] The 1915 act and similar subsequent acts provided the statutory framework pertaining to midwives and maternity care in Scotland until 1983, when the 1979 Nurses, Midwives, and Health Visitors Act took effect, superseding all previous acts.[23]

The 1915 Midwives (Scotland) Act laid down that the constitution of the Scottish CMB should consist of twelve members, 'two of whom shall be certified midwives practising in Scotland'.[24] Thus, within the board at its outset, there were two midwives, six medical practitioners and four 'lay' members. Medical practitioners remained the largest statutory group in the CMB, although the numbers of midwives grew over time both by statute and through the selection by appointing bodies of a midwife rather than a lay person as their board representative.[25] For many years male members of the medical profession continued to hold the positions of chairman and deputy chairman. Finally, in 1973 Sheelagh Bramley became the first midwife to hold the office of deputy chairman, and in 1977 she made the final breakthrough when she was elected chairman, with another midwife, Mary M Turner, as her deputy.[26]

Duties and Powers of the CMB

The 1915 Midwives (Scotland) Act was to go a long way towards improving maternity services in Scotland, supervising midwives and regulating midwifery as a whole. The duties and powers of the CMB included: regulating the issue of certificates; controlling conditions of admission to the roll of midwives; setting the course of training in midwifery; conducting examinations; and ensuring remuneration of examiners. The rules of the CMB were strict and detailed, with the board holding the power to investigate midwives who disobeyed its rules, and admonishment, suspension or even removal from the roll if it saw fit.

The CMB recognised three categories of midwife to begin with: there were the 'bona fides', who had to have been in bona fide practice for more than a year before the passing of the act; there were midwives who had obtained a certificate from one of a variety of institutions; and there were those who passed the CMB examination.[27] To begin with, many more midwives who had been 'bona fides' became enrolled than those with certificates. As shown below, this pattern changed dramatically over the first five years of the board's existence as it became the norm for midwives in Scotland to sit the CMB examination.

Table 16.1 Numbers of midwives enrolling with the CMB for Scotland 1916–21

Year	*By Certificate of an approved body*	*Bona fide*	*CMB Exam*	*Total*
1916–17	728	1229	69	2026
1917–18	624	465	195	1284
1918–19	45	20	216	281
1919–20	36	75	328	439
1920–1	15	6	470	518

(Figures extracted from annual reports of the CMB for Scotland.)

Under the act, each local authority became the Local Supervising Authority (LSA) over midwives within the district.[28] Local authorities had to pay midwives' expenses, medical fees arising in an emergency and contributions towards training. They also had wide powers of supervision and inspection of midwives practising within their district, and of investigating charges of malpractice, negligence or misconduct against a midwife, then reporting to the CMB. The LSA also had the power to suspend any midwife from practice to prevent the spread of infection. If a mother had a temperature of over 100 degrees Fahrenheit for over twenty-four hours, indicating a possible infection, the midwife had to report this to the local authority. Mary McCaskill, practising in Glasgow in the late 1940s as a 'Green Lady' (municipal midwife), recalled the rigours of the rules in this case and early use of antibiotics:

> Then you were suspended from practice [because of] the big danger in these years of puerperal fever. I had to go home and wash my hair, have a bath and change my uniform. I had to have a throat swab taken to make sure I wasn't carrying the infection and then the supervisor would arrange a day and a time when she would come to the house and inspect all my equipment and see that I had a clean lining in my bag. You changed your bag [lining] every week of course. [Antibiotics were] in their infancy. [We gave] sulphonamides then ... M&B, the great 963 ... the first one ... They were considered a great boon to controlling infection.[29]

MIDWIFERY PRACTICE: ANTENATAL CARE

Examination of the early days of antenatal care gives little evidence of the midwife's place in this field.[30] This slowly changed as the CMB included the use of antenatal clinics in midwifery training in the 1920s and made new rules to accommodate this.[31] However, the rules still decreed that when a mother booked a midwife for the birth, 'the midwife should advise the patient to avail herself of ... help' from, for instance, municipal clinics, rather than perform the antenatal care herself, even though the midwife was the professional most likely to be with the mother at the birth.[32] Many mothers did not attend the clinics, and either booked the midwife late in pregnancy or did not book her at all. The link has been made between the non-uptake of antenatal care and poverty and apathy.[33] However Margaret MacDonald, who practised in Glasgow in 1947, also said that mothers did not talk about their pregnancies and suggested this as a reason for non-attendance at clinics:

> Women in these days were different from women today ... people talk about everything. From sex, from conception right up to delivery ... In these days we didn't. It was all kept under wraps. Nothing was open and above board ... They didn't trust people to know about them. They didn't want to – my mother didn't do it [go to clinics] so I'll not do it. That was the attitude.[34]

The approach to antenatal care in Scotland in the 1920s and 1930s appeared to be undefined and patchy. In urban areas some midwives attended mothers in municipal antenatal clinics, urban hospital antenatal clinics and at home. But many midwives booked mothers for delivery, interviewed and advised them in accordance with the rules to attend the clinics, and then delivered the mother at home. Many of these mothers did not attend the clinics, and if they did there was no transference of notes from clinic to midwife. Margaret MacDonald recalled: 'if it was a booked case you knew all about her [through] … what the woman would tell me. How she'd got on at the clinic … If it was unbooked [we knew] nothing [about her history].'[35]

The 1947 NHS (Scotland) Act, implemented on 5 July 1948, had an impact on midwives and their practice. Antenatally, pregnant women could now go to their GP free of charge to 'book'. Midwives were thus missed as the first point of contact. GPs began to perform an increasing amount of antenatal care of pregnant women, which caused conflict between GPs and midwives.[36] In the cities, municipal clinics gradually gave way to hospital antenatal clinics. Ella Clelland, who was a pupil midwife in Glasgow in 1957, said: 'We saw the mother antenatally but mostly at the hospital at the clinics. Apart from visiting them at home to assess the situation and assess their needs and all that kind of thing, mostly it was hospital clinic visits.'[37]

In rural Scotland, where there was a midwife, she 'supervised' a mother in the antenatal period.[38] This was sometimes shared with the GP. A midwife in the Outer Hebrides recalled:

> I did the antenatal care all the time and sometimes on the first visit the mother would say, 'There's another one on the way but don't tell the doctor yet.' I would say, 'But I must tell the doctor.' The doctor went after you notified him. They didn't book early. After that we visited fortnightly, and weekly for the last month. We were on call when they came to term.[39]

If there was no midwife many mothers did without. In areas of Scotland where uncertificated midwives or howdies worked, they sometimes moved in with the family – not to provide formal antenatal care, but to help with the household work, as Doddie Davidson did in Aberdeenshire in the 1940s: 'I went an lived in the hoose afore the bairns were born. Usually they were needin some help especially fan there wis some little anes.'[40] Annie Kerr, working as a howdie in the Dumfries area also in the 1940s, said the same: 'I went a wee while before the baby wis born. Not as much as a couple o weeks … I wis with them and did everything that wis to be done ye know, to give her a rest. That workit in fine.'[41]

Occasionally rural mothers only saw the GP. One mother who had her first baby in 1963 in a small town in the north of Scotland said: 'The only professional I saw for my first pregnancy was my GP. There was no antenatal clinic. I just went to the surgery. I never saw a midwife at all until I went into hospital to have the baby.' Things were different when she decided to have her second baby at home: 'This time the midwife visited me at home the whole time. Once a month, then every two weeks and then every week. But not only that, a few days after the midwife had been, along came the doctor and did exactly the same.'[42] Such duplication of care was not cost-effective, emphasised the abnormal side of pregnancy by the constant medical presence, and deprived the midwife of an important part of her role while at the same time undermining her confidence in her ability to do her job properly.[43]

Midwives' opinions were often disregarded by GPs. One midwife working in the Outer Hebrides described what happened on such an occasion and how she questioned her own diagnosis when the GP did not agree with her:

> The first time I saw [this girl], I thought, 'She's got a breech' and notified the doctor. We both went and he said, 'Oh no, that's not a breech. There's the head down there'. Of course I was asking myself, is the doctor right or am I?
>
> I antenataled (sic) her twice over a fortnight and I was still convinced it was a breech ... Somebody ... called me at about two in the morning ... I found her very advanced in labour [with the breech presenting]. There was meconium staining and her uncle was dispatched for the doctor about eight to ten miles away. The baby was born and I had tidied up before the doctor arrived.[44]

At the beginning of the CMB's existence the midwife had no defined role in antenatal care. This gradually became clearer over the years. However, even as late as 1983, just before the CMB ceased to function, the Scottish Health Service Planning Council stated that the midwife's skills were largely underused and suggested that she should have much wider responsibility for the care of pregnant women in clinics and at home.[45] Eighteen years later, the Scottish Executive published *A Framework for Maternity Services in Scotland,* which said in its pregnancy section: 'Maternity services should provide a woman and family centred, locally accessible, midwife-managed, comprehensive and effective model of care during pregnancy with clear evidence of joint working between primary, secondary and tertiary services.'[46] Thus by the beginning of the twenty-first century the expertise of the midwife and her ability to be part of a team on an equal basis with other professionals in the field were at last acknowledged.

INTRANATAL CARE

During the twentieth century one aspect of midwifery practice which did not change was the mechanism of normal birth. Unlike the midwife's role in antenatal care, from 1915 her statutory role as a person allowed to care for a woman throughout normal labour has never been in doubt. Ann Lamb, who practised as a midwife in Scotland for many years from the late 1920s onwards, said: 'I delivered most of my babies at home without a doctor. I felt kind of safer with a doctor but I would still deliver the baby. Oh, yes, the doctor was just there to look on.'[47]

There was another group who had an important part to play in the field of childbirth. In rural Scotland in the early twentieth century, trained midwives were seldom available, and most of the babies were delivered at home by uncertificated midwives or howdies.[48] With the 1915 Midwives (Scotland) Act and subsequent legislation, the practice of howdies declined. Yet there is still evidence that howdies were practising in some areas of Scotland up until the 1950s.[49]

During the twentieth century there was a shift from an estimated 95 per cent hospital births at the beginning of the century to a peak of 99.5 per cent hospital births in 1981.[50] This, along with the increasing medicalisation of childbirth, removed the normality of many births and had a negative impact on normal midwifery. One midwife who practised from the 1950s to the 1980s said: 'I feel that midwifery now is all abnormal until it is proved normal which is not how it is intended to be and which is very sad.'[51]

The growing medicalisation of childbirth through the twentieth century[52] included the increasing use of medical forms of pain relief in labour, improved conditions in hospitals, and the development in the 1920s of maternity nursing homes.[53] Advances in medicine, instrumental in reducing maternal mortality in the 1930s and 1940s, also contributed to the increases in hospitalisation. These included the use of the first antibiotics for puerperal sepsis, particularly Prontosil,[54] the development of blood transfusions and better education of doctors and midwives. The 1937 Maternity Services (Scotland) Act enshrined in statute the requirement of local authorities in Scotland to provide the services of midwives at home, along with those of general practitioners, specialist obstetricians and anaesthetist help where necessary, thereby apparently approving home births for many women. However in 1939, before these arrangements were fully in place, World War II started bringing with it an acceleration of the trend towards hospital as a place for birth. The fundamental reason for this pattern was the lack of help in the house, as many women were replacing men in jobs that were usually male preserves. From the beginning of the war, pregnant women from urban areas were provided with emergency maternity hospitals, often adapted from suitable country houses.[55] By 1948 there were almost 3,000 maternity beds in Scotland, a rise of around 2,000 since 1934, most of which came as a result of the wartime maternity policy.[56] These were inherited and developed by the regional hospital boards under the National Health Service.[57] Thus the implementation of the NHS Acts in 1948 and the changes that arose as a result of this also contributed to the trend towards hospital births. The Queen's Institute of District Nursing gave other reasons in 1959:

> It was due to excessive propaganda from hospital specialists stressing greater safety, lack of suitable housing in certain areas, insufficiently developed or insufficiently flexible home help services, economy to the mother ... in spite of the increase in the home confinement grant

> and encouragement by GPs [to have a hospital birth], sometimes irrespective of medical, obstetric or social need.[58]

The change in place of birth also affected the intranatal practice of domiciliary midwives. In the years after World War II and the beginning of the NHS, when the number of hospital deliveries was rising, midwives on the district became anxious about what would become of them:

> Towards the end of that five years [1952] – the home deliveries had just sort of imperceptibly started to decline … older midwives … probably in their fifties were beginning to talk about – they didn't have so many bookings and they were wondering … what was the future and what would they be used for … would they be maybe diversified into some other duties?[59]

There was also the issue of lack of job satisfaction. Experienced district midwives expressed frustration at what they saw as a lack of domiciliary midwifery. Their remit had moved from giving full intranatal and postnatal care to some antenatal visits, and to postnatal visits to women who had been confined in hospital. In addition it became difficult to arrange intra-partum training for pupil midwives.[60] As the number of hospital births rose, very few student midwives delivered a baby at home, although much depended upon the area and the number of maternity beds there. For example, one midwife recalled:

> I was appointed as a district midwifery sister from 1957–1964. Govan was densely populated at that time and the birth rate extremely high. Because of the shortage of beds in the maternity unit of the Southern General Hospital home confinements were essential and during my seven years in the community, I delivered 1,322 babies.[61]

Any student midwives with that midwife would have had no problem in obtaining deliveries, and this situation in Glasgow prevailed for some years, with one student midwife as late as 1969 having plenty of home births.[62] Yet Alison Dale, who also trained in 1969 but in Aberdeen, said: 'Only one girl in our set went and saw a home delivery and there were very few by the time that we were training.'[63] Thus, sooner or later, home births were phased out as a compulsory part of the midwifery training syllabus. In addition, because of the increasing rate of hospital births, midwives who worked in the community 'very quickly lost their intra-partum skills'.[64] One solution to this issue was a suggestion in a DHS Report (the Montgomery Report), which said that 'domiciliary midwifery might become a hospital responsibility, which would allow some interchange of midwives between domiciliary and hospital services'.[65] This early suggestion on integrating the maternity services was to come to fruition in the future.

Another suggestion in an effort to provide more continuity of care for mothers, and at the same time help community midwives maintain their skills, was what became known as the domiciliary in and out (DOMINO) scheme. The first reported use of this scheme was in West Middlesex in 1971, although ideas for similar schemes were mooted in the early 1960s. Mothers having a DOMINO delivery were looked after antenatally by their community midwife and escorted into hospital in labour by the community midwife on call, who then delivered the baby and looked after the mother and baby for a few hours afterwards before escorting them home. Postnatal care continued as usual. DOMINO deliveries had the added bonus for the midwives that they could give intranatal care as well.[66] Jan Fenton, practising in Whitfield, Dundee, had many DOMINO deliveries over the years. She said:

> We started the DOMINOS in the 1970s. A lot of the girls ... were frightened of hospitals. [There was] a little lass ... and she wanted the baby at home and of course those were the days when ... it was very much frowned on. She was determined. She discovered that she was having twins ... But we went into hospital with her ... Dr Smith [the consultant] got out of his bed at two o'clock in the morning and sat beside the bed so that there was no interference from the staff. You see this is the rapport ... that we had. I delivered the twins ... it was marvellous. And then eventually I was delivering babies from all over the town ... They'd obviously had a bad experience the first time. I got up in the middle of the night, went in, delivered a baby and came home. But you see I was using my expertise in the hospital and this was one way of keeping it up ... These were all DOMINOS ... It was brilliant ... I've even done it on my day off because they were – if they were frightened they were frightened.[67]

Two issues which inevitably arise when discussing midwifery and intranatal care are those of equipment and transport. Midwives on the district carried most of their equipment with them. In the days before midwives had cars, this involved carrying what they could while walking, cycling or using trams and buses. Their loads were often heavy and awkward.

The standard piece of equipment was and is the midwife's bag. Myles indicated the need for two bags, particularly for a confinement: 'separate delivery and puerperal bags of metal or leather'.[68] Nearly every midwife I spoke to mentioned her bag. No midwife on the district then, or now, went without it. Working in the 1920s, Ann Lamb spoke about her 'brown bag', and Moira Michie from Aberdeen stressed the importance of her howdie grandmother's bag.[69] James Tweedie wrote about his grandmother, howdie at Douglas (later renamed Happendon), from 1877 to 1923. When she was called out to a tinkers' camp, the bag was the first thing she picked up:

> In case the lass was in distress,
> Gran took her black bag from the press
> And off on that frosty morn she went,
> To help the wife in the tinker's tent.
> And there in the moonlight's fleeting beam,
> The bairn was born and washed in the stream.[70]

Bags, used every day, had to be repacked every day. Agnes Morrison, working in Leith in the 1940s, said: '[When] we came in … it was up to us to have it all cleaned out, things sterilised [and] boiled up … we might have to do it mid-day [as well].'[71]

Pupil midwives on the district had to have their bags inspected before they went out. Linda Stamp, in training at Rottenrow in the 1940s, recalled this, and also how they made special cords for tying the umbilical cord, which they kept in the bag:

> To make the umbilical cords we had very fine thread just like string. It was very white and it had to be a certain length and we had so many taped together … The sister there was pretty old and supervised what we took out with us. She inspected our bags before we went out. She was very very fussy. We took all that was essential. We had our sterile things in little packs.[72]

But certified midwives also had their bags inspected by the supervisor. Mary McCaskill said: 'You had your bags inspected regularly … every one to three months.'[73]

The bag was looked upon as a passport, for instance helping midwives get to the top of the queue for the Glasgow trams, and Margaret Dearnley recalled:

> You wore your uniform, you're carrying your bag. And if you went to stand in the tramcar line … they would say 'Oh,' – great big queues of people of course – 'Here's the nurse, some puir sowel's waitin on her comin.' And you were pushed up the line and they let you on. And the driver would say … 'Where are you going nurse? … They would stop at the close that you wanted to go to. Wasn't that fantastic?[74]

Anne McFadden remembered how her midwife's bag gave her a clear passage in 1940s' Edinburgh:

> The whole culture was quite difficult. You could call them the drug addicts but it wasn't the kind of drugs we know about … They used the gas fittings [from lamps] and the methylated spirits and they were lying asleep down the Vennel Steps down into the Grassmarket. But, they were so respectful, whenever they saw a black bag, up they got … and said … 'Sorry, nurse, let you pass' … And you had your black bag with you.[75]

Another important piece of equipment was the delivery pack or box. One midwife at the beginning of her midwifery career in the late 1930s said:

> First of all … we made up these packs and I had a gown and masks and bonnet and cotton wool and swabs but then latterly the boxes came all ready and you just opened it in the house. But, you know, it was amazing, you had children there and we wrapped this up in

> brown paper and put it up on top of the cupboard. Nothing was ever touched, nothing.[76]

Sterile delivery packs were supplied free to mothers at a cost (in 1951) to the local authority of 18s. 6d. (92½ pence) each, on condition the pack was opened only by the midwife or doctor, and if not used it should be returned unopened.[77]

When the baby was born, the midwife delivered the placenta, which had to be disposed of. In most areas the midwife examined the placenta and, if it appeared complete, wrapped it in newspaper and burned it on the fire. However, Rottenrow pupil midwives were required to take the placenta back to the hospital for examination. As Ella Banks recalled, this involved an unusual piece of equipment:

> We had to have a sponge bag and in that sponge bag, you had to put the placenta and bring it back to the hospital to be checked to see that it was complete and healthy. The sponge bag had strings that you pulled across. And you put your placenta in there and took it back to the hospital.[78]

Midwives had another heavy piece of equipment to carry after 1946, when the CMB allowed midwives to administer gas and air on their own responsibility. This sometimes meant carrying the Minnitt apparatus with

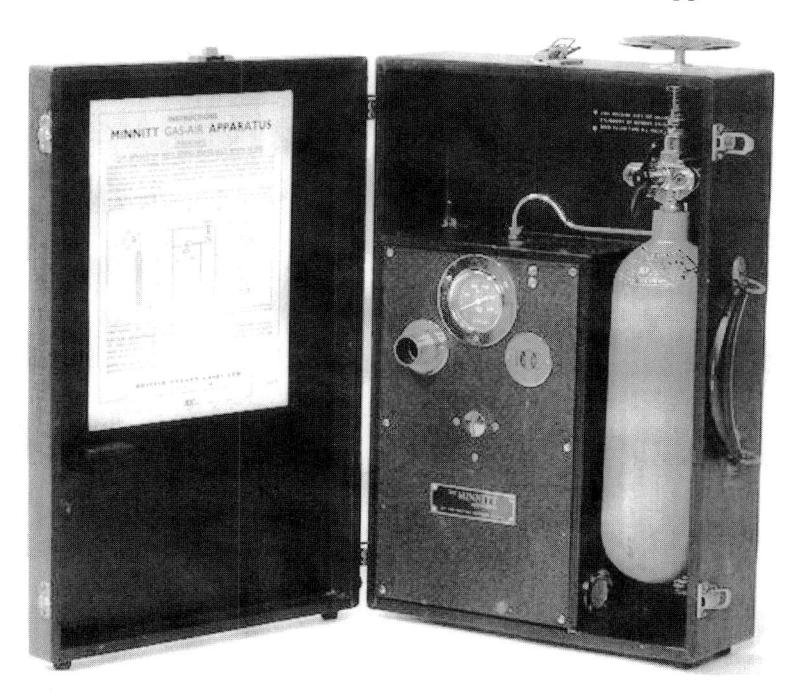

Figure 16.1 Minnit gas and air anaesthetic device.

them as well as everything else. Mary McCaskill recalled: 'We had a portable Minnitt machine. [The cylinder and the machine] were all in the box together. It was quite heavy even although it was portable … quite often we would ask the husband if he would carry [it].'[79]

However bulky, heavy and awkward the equipment, it was necessary for successful domiciliary midwifery. In 1959 the CMB responded to a complaint about the weight of equipment a midwife had to carry by justifying the need for all the articles in the bags but also saying that employing authorities should provide adequate transport.[80] Before this, in some areas taxis were used, and Mary McCaskill described what happened in Glasgow from the early 1950s:

> Somewhere between eight and nine at night till six o'clock in the morning, Glasgow Corporation provided a car and a driver … If it was during the hours of darkness he would come to my house and take me to where the confinement would take place and then he would come to the house and bring me [home].[81]

This seemed to be common practice, as Margaret Dearnley corroborated:

> If you were beyond the tram route or very far away or it was very early in the morning you got the green car, the town car … You phoned in to say you were leaving, where you were and it was two o'clock in the morning you had to go back to Rottenrow, so they sent a 'green caur' as they talked about which was the … town limousine.[82]

This did not always solve the problem. Many midwives walked, cycled or used public transport when it was available. In Tullibody in the 1950s Anne Bayne was given a Vespa scooter to use. On the negative side, she said that 'the wee ones would take the cap off and put stones in the tank', but on the positive side Anne found her petrol bill was always secretly paid for her.[83] On some memorable, very early-morning occasions some midwives hitched a lift on the back of a refuse lorry. Anne Chapman in Glasgow in the 1940s recalled:

> We had to walk everywhere … We used to stand up on the back of a refuse cart and get a lift … If the driver saw us he would stop and yell, 'Where are ye goin?' and he would take us … 'Jump on'. And there we were clutching on the back – the student and I. Rattling away and then he would shout, 'What street are ye goin to?' These men were great.[84]

The 1949 Working Party on Midwives highlighted the problems of inadequate transport endured by midwives on the district, and recommended the provision of a car for every midwife, financial help to run it and assistance with driving lessons.[85] Eventually it was a condition of the job that community midwives should be able to drive, and local authorities either supplied a car or assisted midwives to buy or lease cars, thus helping to solve both transport and equipment problems.

MIDWIVES AND THE POSTNATAL PERIOD

Since the time of the 1915 Midwives (Scotland) Act there has been a statutory requirement for midwives to attend upon and examine a mother and her infant for the ten-day postnatal period. For many years, it was twice a day for the first three days and at least daily thereafter for a minimum period of ten days, or longer if required.[86] The only exception to the ten-day rule was between 1939 and 1965, when the statutory period was fourteen days.[87] These rules did not mention other, less tangible, aspects of postnatal care. The three main areas of responsibility for midwives during the postnatal period are physical, educational and psychosocial.[88] The physical role was acknowledged from the outset. The CMB gradually fulfilled the educational part of these responsibilities from the 1930s until, by the 1980s, parentcraft became an important part of the midwife's role. The importance of the psychosocial and emotional aspects of postnatal care did not receive full acknowledgement until the later decades of the twentieth century.[89]

As the place of birth changed and the number of home births declined, most women (often in hospital for up to ten days postnatally) did not see the district midwife. More midwives were employed in maternity units and fewer on the district. The role of the midwife on the district, already eroded by having fewer home deliveries and little or no antenatal care to give, was diminished and very nearly extinguished by the late 1960s. The rising number of hospital deliveries and the corresponding requirement for antenatal beds contributed to a shortage of institutional maternity beds.[90] At the same time the postnatal care of mothers changed, with increasing early ambulation and self-care. To cope with these issues, 'early discharge home' became an occasional feature which started in the 1950s and developed into an accepted part of postnatal care over the next twenty years. The need for the district or community midwife gradually became apparent again, but not usually to give full midwifery care.

By 1980, 99.5 per cent of babies in Scotland were born in hospital, and community midwives were employed mainly to give postnatal care. These midwives, their numbers originally reduced because of the lowered home-birth rate, coped with increasing numbers of mothers discharged early from hospital. They had less time to give to each woman and, in addition, fragmentation of care increased. Much antenatal care was performed by GPs and obstetricians, most births were conducted by hospital midwives, and postnatal care was given by midwives on the community. Thus, mothers were not obtaining continuity of care throughout the childbearing episode; neither were community midwives maintaining their midwifery skills in areas other than postnatal.

Full postnatal care did not always happen. Ann Lamb remembered working (privately) as a midwife in Inverurie in the 1940s. She said: 'I just stayed in the house for a week … after the birth for the postnatal care but there wasn't really much postnatal care in my day – just a wee look.'[91] Care was also less defined when howdies were in charge. Occasionally in rural areas there was no midwife to give postnatal care. This happened in Aberdeenshire in the 1940s, where the howdie in this instance lived in the house postnatally. She said:

> There wis nae midwife ... I looked after the mother after the baby was born. Oh aye, I washed the Mam and see that she wis [all right], ye ken. And the bairn too – I bathed the bairns. An then ye stayed, sometimes a wik sometimes mair, sometimes ye didna hae time ti spare but ye aye hid aboot a wik wi them or ten days.'[92]

Yet one howdie in particular, Johnann Roberton, working in Aberdeen in the early decades of the twentieth century, did things differently, even though still probably not conforming to CMB rules. According to her granddaughter:

> Postnatally, she had a strict routine and according to my late mother, Grandma would clean up the patient first after expulsion of the afterbirth and immediately wrap her up in a supporting binder round the stomach to prevent sagging of the abdominal muscles. My mother felt the benefit of this in later years! The binders were usually made of flannelette sheets torn to the required strip for size. She also collected old sheets from neighbours and made the bandages herself. If the mother had insufficient milk she too was 'binded' to prevent drooping.[93]

Occasionally a mother would not allow the midwife in to do postnatal care. Anne Bayne said: 'Often when the Green Lady went [to do a postnatal visit] she did not get in.'[94] Thus, the ideal of postnatal care as laid out by the CMB was not always performed in the way the board expected, although it was easier to keep to the rules when mothers were in hospital and under a watchful eye.[95]

At the beginning of the twentieth century, medical practitioners considered that, post-delivery, a mother should be kept in bed for at least ten days, and in her bedroom for a fortnight.[96] In practice, this rule was difficult for a midwife to achieve and left a mother weak and often debilitated through lack of exercise. Mima Sutherland, practising in Unst, Shetland in the 1930s and 1940s said: 'After the birth I used to see them twice in the day. We were supposed to see them fourteen days. By the time they put their legs oot ower the side of the bed they would say, "Look at my legs. What are they like?" They were wasted you see.'[97]

Midwives accepted the reality that most mothers could not stay in bed for the stated length of time because they had so much to do. Therefore it was often up to the mother. Doddie Davidson, howdie, said:

> Keepin the mothers in bed depended on the mother a lot. Some o them wanted tae lie, some o them didna. But they were aye a few days in their bed. Sometimes aboot a wik. Sometimes if they hid ither little anes they wanted tae get up. But some Mams wanted tae lie. It aa dependit on them. Once they got up they were OK.[98]

May Norrie, midwife in Friockheim and Carmyllie in Angus in 1947 was realistic too. She said: 'Oh well, I just left [them] to their own devices. But I knew perfectly well some of them were up and doing things for their family, you know.'[99] Margaret Foggie, pupil midwife in the depressed Glasgow of 1934, added another aspect to the postnatal period. She recalled:

> The people ... were very dispirited ... It was very sad. It was a very bad time ... I never looked after anyone who had a husband working. You had to try and keep the mothers in bed but it was very difficult. You couldn't really insist on it. But you told the husbands – look after these children and remember the woman is not well ... Run the messages for her and look after the weans. They would say, 'yes', but I'm afraid they very often – well I don't think they did it. They used to stand around on the street, talking.[100]

Financial problems were echoed by Mary McCaskill, who explained why in the late 1940s she sometimes had to negotiate times of visits:

> [The mothers] sometimes weren't there. In some parts of Glasgow the mother was in receipt of dole money and they would hop off to sign on. That took precedence over everything else but usually you just had to work round it. Because you couldn't chain them to the bed ... for seven days.[101]

However, mothers who would not stay in bed probably did themselves a favour, apart from avoiding wasting of the muscles. A well-known hazard of staying in bed with no exercise was deep vein thrombosis. Annie Kerr, howdie in the Dumfries area in the 1940s, recalled:

> I wid mebbe be there a fortnight. It wis reckoned they shouldna be out o bed till the nine days were up. That wis quite a common thing. Ye had tae wait in yer bed intil a certain time and then get up ... I remember one woman gettin up ... she walked across to the baby, and picked it up ... She was never any the worse of it. They say they liked to get ye up on yer feet. It saved clottin.[102]

Keeping mothers in bed for long periods of time eventually became a thing of the past. Changes in postnatal care developed, with early ambulation and self-care of mothers, and the use of aseptic techniques and antibiotics where necessary. Increasingly, mothers were encouraged to tend to their babies themselves. In the postnatal wards, and increasingly at home as early discharge home became the norm, the role of the midwife changed from one who performed full physical care for mother and baby to one who taught mothers to look after themselves and their babies, with support and help where necessary.[103]

CONCLUSION

This chapter has explored some facets of the work of midwives in Scotland in the twentieth century. Through midwifery legislation, starting with the 1915 Midwives (Scotland) Act, this work was regulated by the CMB until

1983, and latterly through the United Kingdom Central Council for Nurses, Midwives and Health Visitors (UKCC) and the National Board for Scotland for Nurses, Midwives and Health Visitors (NBS).[104] Midwives were and are required to keep within the rules and Code of Professional Conduct. The act brought the first formal recognition of midwives as a group throughout Scotland and gave them a legal identity. At the same time, legislation and the rules of the CMB restricted midwifery practice.

Midwifery practice in all its aspects has evolved from the restrictions placed on it by the first act. As the events of the twentieth century unfolded, societal attitudes changed, matching a change in attitude to maternity care in general and midwifery in particular. This became particularly apparent from the late 1980s onwards. In 1993, the consultative document Maternity Services in Scotland: A Policy Review was published, heralding the change which became known as 'woman-centred care'. *A Framework for Maternity Services in Scotland*, published in 2001, continued this aim and challenged the NHS to provide an essentially community-based, midwife-managed service with easy access to specialist services wherever needed.

Acknowledgements to Professor Marguerite Dupree and Professor Edith Hillan.

NOTES

1. Mander and Reid, 2002, 2.
2. Reid, 1843, 22–32.
3. Mander and Reid, 2002, 4; Dingwall et al., 1988, 145–72.
4. Cowell and Wainwright, 1981, 33; Loudon, 1992, 207.
5. Hunter, 1999, 426.
6. Fleming, 1998, 46.
7. Reid, 2003 focuses on the extent to which the CMB, midwifery training, the changing location of births and the changing nature of midwifery practice, along with relationships of midwives with those in the medical profession, shaped their identity and limited or facilitated their autonomy.
8. Ferguson, 1958, 510.
9. Donnison,1988, 34.
10. Jenkinson, 1993, 83.
11. Hansard, Commons, vol. 109, 6–24 June, 1902, cols 58–9, quoted in Jenkinson, 1993, 83; Dow, 1984, 151.
12. *Transactions of the Edinburgh Obstetrical Society*, vol. 20, session 1894–5, Edinburgh, 1895, 164–82.
13. Donnison, 1988, 138.
14. Tait, 1987, 413–40; Chalmers, 1914, 251–4.
15. Tait, 1987, 413–40; see also Chalmers, 1930, 259.
16. Ferguson, 1958, 546; Chalmers, 1930, 261.
17. Tew, 1995, 273, 281 and 283; Loudon, 1992, 241, also cites a lack of improvement in MMR figures in the years 1910–1934 associated with increased [iatrogenic] interference and the 'employment of surgical measures in normal and moderately difficult labours'.

18. Memorial, 1915.
19. Hansard, 8 December 1915, col. 569–570; Hansard, 26 October 1915, col. 1167; Hansard, 1 December 1915, col. 814.
20. Hansard, Commons, vol. 27, 23 December 1915, col. 806.
21. Jenkinson, 1993, 84.
22. Midwives (Scotland) Act, 1915, 1 (3).
23. NAS, CMB, Index, 1.
24. Midwives (Scotland) Act, 1915, section 3, 1.
25. NAS, CMB 1/5, CMB Minutes, vol. 2, 27 November 1936, 25; NAS, CMB 1/7, CMB Minutes, 4 November 1951, 1; NAS, CMB 1/7, CMB Minutes, 14 August 1952, 1.
26. CMB Report, 31 March 1973, 1; CMB Report, 31 March 1977, 1.
27. Loudon, 1992, 208.
28. Midwives (Scotland) Act, 1915, 16.
29. Interview no. 27 as recorded by Lindsay Reid.
30. CMB for Scotland, rules framed under the Midwives (Scotland) Act, 1915, 5 (1), 26 August 1916.
31. CMB Minutes, 27 June 1924, vol. 9, 27; CMB Minutes, 28 May 1925, vol. 10, 21; CMB rules, 1926, 12; CMB rules, 1926, 10.
32. CMB rules, 1918, 16; CMB rules, 1931, 34.
33. See Reid, 2006.
34. Interview no. 125 as recorded by Lindsay Reid; for further oral testimonies concerning midwives and antenatal care, see Reid, 2006.
35. Interview no. 125 as recorded by Lindsay Reid. The term 'hospital emergency' was used when a mother wanted a hospital delivery but had not booked, and thus frequently did not obtain a hospital bed when she went into labour. These mothers often did not have antenatal care. Mothers who went to the clinics and who were booked to be delivered by choice on the district told the midwife verbally how they had fared at the clinic.
36. Ministry of health et al., 1949, paras 101 and 102; Towler and Bramall, 1986, 234; Robinson, 1990, 61–91.
37. Interview no. 9 as recorded by Lindsay Reid.
38. Interview no. 56 as recorded by Lindsay Reid. See also Reid, 2006.
39. Interview no. 99 as recorded by Lindsay Reid.
40. Interview no. 101 as recorded by Lindsay Reid.
41. Interview no. 110 as recorded by Lindsay Reid.
42. Interview no. 123 as recorded by Lindsay Reid.
43. Cumberlege, 1948, 47.
44. Interview no. 99 as recorded by Lindsay Reid.
45. Scottish Health Service Planning Council, 1983, para.1.6.
46. Scottish Executive Health Department, 2001.
47. Interview no. 47 as recorded by Lindsay Reid.
48. Ferguson, 1958, 510.
49. For further discussion on howdies, see Reid, 2006.
50. General Register Office for Scotland.
51. Interview no. 9 as recorded by Lindsay Reid.
52. van Teijlingen et al., 2000, 1; Here Edwin van Teijlingen et al. define medicalisation of childbirth and midwifery as: 'the increasing tendency of women to prefer a hospital delivery to a home delivery, the increasing trend toward the use of technology and clinical intervention in childbirth, and the determination

of medical practitioners to confine the role played by midwives in pregnancy and childbirth, if any, to a purely subordinate one'.

53. Moir, 1986, 1, 5. Twilight sleep was first used in Germany in 1902; Murray, 1930, 145–53; Johnstone, 1939, 1020–8; registration and inspection of maternity homes was made statutory in the Midwives and Maternity Homes (Scotland) Act, 1927 [17 &18 Geo. 5. ch. 17.], Part II, 5–9.
54. See Loudon, 1992, 258–61, for a discussion on the use of antibiotics, especially prontosil rubrum in puerperal fever, firstly in hospitals and then in general practice.
55. For further discussion on change to hospital birth, see Reid, 2006.
56. Sturrock, 1980, 173–87.
57. DHS, Montgomery Report, 1959, 7.
58. DHS, Montgomery Report, 14–16.
59. Interview no. 27 as recorded by Lindsay Reid.
60. DHS, Montgomery Report, 1959, 20.
61. Interview no. 34 as recorded by Lindsay Reid.
62. Interview no. 112 as recorded by Lindsay Reid.
63. Interview no. 94 as recorded by Lindsay Reid.
64. Interview no. 74 as recorded by Lindsay Reid.
65. DHS, Montgomery Report, 1959, 20.
66. Murphy-Black, 1993, 115; Campbell and Macfarlane, 1994, 93.
67. Interview no. 116 as recorded by Lindsay Reid.
68. Myles, 1962, 654; see also Watson, 1914, 139–41. With variations, a midwife's bag for delivery in the early twentieth century contained: an enema syringe, catheter, bath and clinical thermometers, disinfectant soap, nailbrush, biniodide of mercury, lubricant, safety-pins, scissors, tape-measure, cord-thread/tape, ergometrine, chloral hydrate, sal volatile, measuring glass, syringe and needles, cottonwool, lint, permanganate of potash, suturing materials, clean aprons, mackintosh sheeting and apron, notebook and copying-ink pencil, four-hourly charts, scales for infant weighing. While these requirements have modified over the years, the basics remain the same.
69. Interviews no. 3 and 47 as recorded by Lindsay Reid.
70. Interview no. 42 as recorded by Lindsay Reid.
71. Interview no. 35 as recorded by Lindsay Reid.
72. Interview no. 115 as recorded by Lindsay Reid.
73. Interview no. 27 as recorded by Lindsay Reid.
74. Interview no. 20 as recorded by Lindsay Reid.
75. Interview no. 108 as recorded by Lindsay Reid.
76. Interview no. 99 as recorded by Lindsay Reid.
77. DHS 1951; Myles, 1962, 653; see also Reid, 2006.
78. Interview no. 2 as recorded by Lindsay Reid.
79. Interview no. 27 as recorded by Lindsay Reid; Myles, 1962, 270. The Minnitt-minor apparatus for nitrous oxide–air analgesia in domiciliary midwifery practice weighed 12.5 lbs. See Reid, 2003, 128–33, for further discussion on midwives and the use of inhalational analgesia.
80. NAS, CMB 1/7, CMB Minutes, 12 March 1959, 1.
81. Interview no. 27 as recorded by Lindsay Reid.
82. Interview no. 20 as recorded by Lindsay Reid.
83. Interview no. 91 as recorded by Lindsay Reid.
84. Interview no. 11 as recorded by Lindsay Reid.

85. Ministry of Health et al., 1949, 57.
86. CMB rules, 1980, 10; Interview no. 117 as recorded by Lindsay Reid; Fay MacLeod, practising in the 1960s, said: 'Yes, ten days or you visited them till the cord came off. Or if there [were] any complications like breastfeeding and [problems] like that.'
87. In the 1930s there was much anxiety surrounding the maternal mortality rate (MMR). It is reasonable to suppose the extension of the postnatal period from ten to fourteen days could have been in response to this.
88. Murphy-Black, 1993, 120–46.
89. Raphael-Leff, 1991, ix, but see the whole book for insight into maternal and family psychology related to childbearing; and, for postnatal psychological complications, chapter 33, 477–97; Watson, 1914, 319–20. Here, one and a half pages are given to puerperal insanity, mania and melancholia.
90. Murphy-Black, 1993, 120–46.
91. Interview no. 47 as recorded by Lindsay Reid.
92. Interview no. 101 as recorded by Lindsay Reid.
93. Interview no. 3 as recorded by Lindsay Reid.
94. Interview no. 91 as recorded by Lindsay Reid.
95. Interview no. 9 as recorded by Lindsay Reid. Midwife, Haddington, 1950s: 'We kept the mothers ten days and with a first baby sometimes fourteen. We had them in bed for five days and we swabbed them by douching them with Dettol water. We used to have a big trolley and all these jugs of Dettol douche water and pans underneath it and you went along and put them on the bedpan, and you had the douche and swabbing equipment, turned them on to their side examined their stitches and made sure that that was well healing.'
96. Watson, 1914, 289. This was so that the mother could recover from the birth in a peaceful atmosphere with few visitors and establish breastfeeding; Towler and Bramall, 1986, 77.
97. Interview no. 56 as recorded by Lindsay Reid.
98. Interview no. 101 as recorded by Lindsay Reid.
99. Interview no. 70 as recorded by Lindsay Reid.
100. Interview no. 50 as recorded by Lindsay Reid.
101. Interview no. 27 as recorded by Lindsay Reid.
102. Interview no. 110 as recorded by Lindsay Reid.
103. Murphy-Black, 1993, 124.
104. These regulatory bodies came into being in 1983 on the implementation of the 1979 Nurses, Midwives and Health Visitors Act. Further legislation took effect in 2002.

BIBLIOGRAPHY

Brotherston, J. The National Health Service in Scotland: 1948–1984. In McLachlan, G. ed., *Improving the Common Weal: Aspects of Scottish Health Services 1900–1984*, Edinburgh, 1987.

Campbell, A and Macfarlane, A. *Where to be Born? The Debate and the Evidence*, 2nd edn, Oxford, 1994.

Chalmers, A K. On the need for a Midwives Act in Scotland: The Journal Of Midwifery, A Weekly Record for Midwives and Maternity Nurses, *The Nursing Times* (21 February 1914), 251–4.

Chalmers, A K. *The Health of Glasgow, 1818–1925*, Glasgow, 1930.

CMB Minutes, Edinburgh, 1916–1983 (National Archives of Scotland (NAS)).
CMB (NAS 4/2/10, Schedule): CMB for Scotland, Rules Framed under Section 5 (1) of the Midwives (Scotland) Act, 1915 (5 and 6 Geo.V. c.91), 26 August 1916.
CMB Rules, Edinburgh, 1918, Records of the Central Midwives Board, NAS.
CMB Rules, Edinburgh, 1926, Records of the Central Midwives Board, NAS.
CMB Rules, Edinburgh, 1931, Records of the Central Midwives Board, NAS
Cowell, B and Wainwright, D. *Behind the Blue Door: The History of the Royal College of Midwives, 1881–1981*, London, 1981.
Cumberlege, G. *Maternity in Great Britain*, London, 1948.
Department of Health for Scotland (DHS) Circular no. 84/1951. *Dressings for Confinement and Lying-in Period*, 8 August 1951.
Department of Health for Scotland and Scottish Health Services Council. *Maternity Services in Scotland* (Montgomery Report), Edinburgh, 1959.
Dingwall, R, Rafferty, A M and Webster, C. *An Introduction to the Social History of Nursing*, London, 1988.
Donnison, J. *Midwives and Medical Men*, New Barnet, 1988.
Dow, A D. *The Rottenrow: The History of the Glasgow Royal Maternity Hospital 1834–1984*, Carnforth, 1984.
Ferguson, T. *Scottish Social Welfare 1864–1914*, Edinburgh and London, 1958.
Fleming, V. Autonomous or Automatons? An exploration through history of the concept of autonomy in midwifery in Scotland and New Zealand, *Nursing Ethics*, vol. 5, no. 1 (1998), 43–51.
Galt, J. The howdie: An autobiography. In *The Howdie and Other Tales*, Edinburgh, 1923, 3–28, reproduced from the original manuscript.
Hansard, Commons, vol. 109, 6–24 June 1902, cols 58–9, quoted in Jenkinson, J, *Scottish Medical Societies, 1731–1939*, Edinburgh, 1993, 83.
Hansard, Commons, vol. 25, 26 October 1915, col. 1167.
Hansard, Commons, vol. 26, 1 December 1915, col. 814.
Hansard, Commons, vol. 27, 23 December 1915 col. 806.
Hansard, Lords, vol. 20, 8 December 1915, col. 569–70.
Hunter, B. Oral History and Research Part 1: Uses and Implications, *British Journal of Midwifery*, vol. 7, no. 7 (July 1999), 426–9.
Jenkinson, J. *Scottish Medical Societies, 1731–1939*, Edinburgh, 1993.
Johnstone, R W. *The Simpson Memorial Maternity Pavilion, Royal Infirmary, Edinburgh*, Manchester, 1939. Reprinted from the *Journal of Obstetrics and Gynaecology of the British Empire*, vol. 46, no. 6, 1020–8.
Leap, N and Hunter, B. *The Midwife's Tale: An Oral History from Handywoman to Professional Midwife*, London, 1993.
Loudon, I. *Death in Childbirth*, Oxford, 1992.
MacGregor, A. *Public Health in Glasgow, 1905–1946*, Edinburgh and London, 1967.
MacKenzie, L. *Report on the Physical Welfare of Mothers and Children, Vol. 3: Scotland*, Dunfermline, 1917.
Mander, R. Autonomy in midwifery and maternity care, *Midwives Chronicle* (October 1993), 369–74.
Mander, R and Reid, L. Midwifery power. In Mander, R and Fleming, V, eds, *Failure to Progress: The Contraction of the Midwifery Profession*, London, 2002, 1–19.
Maternity Services (Scotland) Act, 1937 [1 Edw 8 & 1 Geo 6].
Memorial anent a Midwives Bill for Scotland, of the Medical Faculties of the Universities, the Royal Medical Corporations, and the Medical Officers of the Maternity Hospitals in Scotland, to the Right Honourable HM Secretary for

Scotland and the Right Honourable The Lord President of HM Privy Council, 19 August 1915.
Midwives (Scotland) Act, 1915 [5 & 6 Geo. 5. Ch. 91].
Midwives and Maternity Homes (Scotland) Act, 1927 [17 & 18 Geo. 5. Ch. 17].
Ministry of Health, Department of Health for Scotland, Ministry of Labour and National Service. *Report of the Working Party on Midwives*, London, 1949.
Moir, D. *Pain Relief in Labour*, 5th edn, Edinburgh, 1986.
Murphy-Black, T. Care in the community during the postnatal period. In Robinson, S and Thomson, A M, eds, *Midwives, Research and Childbirth*, vol. 3, London and Glasgow, 1993.
Murray, E F. Some observations on puerperal sepsis with special reference to its occurrence in maternity hospitals. In Transactions of Edinburgh Obstetrical Society, Session LXXXIX, Read 14 May 1930, *Edinburgh Medical Journal*, new series, vol. XXXVII, no. 10, October (1930), 145–53.
Myles, M. *A Textbook for Midwives*, 4th edn, Edinburgh and London, 1962.
Oakley, A. *The Captured Womb*, Oxford, 1984.
Raphael-Leff, J. *Psychological Processes of Childbearing*, London, 1991.
Reid, L, ed. *Scottish Midwives: Twentieth-Century Voices* (Flashbacks no. 12, European Ethnological Research Centre), East Linton, 2000.
Reid, L. Scottish midwives 1916–1983: The Central Midwives Board for Scotland and practising midwives. Unpublished PhD thesis, University of Glasgow, 2003.
Reid, L. Childbirth. In Storrier, S, ed, *Scotland's Domestic Life* (Scottish Life and Society: A Compendium of Scottish Ethnology, vol. 6), Edinburgh, 2006, 440–57.
Reid, M. *A Plea for Women*, Edinburgh, 1843; reprinted Edinburgh, 1988.
Robinson, S. Maintaining the independence of the midwifery profession: A continuing struggle. In Garcia J, Kilpatrick, R and Richards, M, eds, *The Politics of Maternity Care*, Oxford, 1990, 61–91.
Scottish Health Service Planning Council. *Shared Care in Obstetrics: A Report by the National Medical Consultative Committee*, Edinburgh, 1983.
Scottish Home and Health Department. *Maternity Services: Integration of Maternity Work* (Tennent Report), Edinburgh, 1973.
Scottish Office Home and Health Department. *Provision of Maternity Services in Scotland: A Policy Review*, Edinburgh, 1993.
Scottish Executive Health Department. *A Framework for Maternity Services in Scotland*, 2001.
Should midwives be registered in Scotland? *Transactions of the Edinburgh Obstetrical Society*, vol. 20, session 1894–95, Edinburgh, 1895, 164–82.
Sturrock, J. Edinburgh Royal Maternity and Simpson Memorial Hospital. A Sir James Y. Simpson memorial lecture delivered on 11 July 1979. Reprinted from the *Journal of the Royal College of Surgeons of Edinburgh*, vol. 25 (May 1980), 173–87.
Tait, H. Maternity and child welfare. In McLachlan, G, ed, *Improving the Common Weal: Aspects of Scottish Health Services, 1900–1984*, Edinburgh, 1987, 411–40.
van Teijlingen, E, Lowis, G, McCaffery, P and Porter, M. *Midwifery and the Medicalization of Childbirth: Comparative Perspectives*, New York, 2000.
Tew, M. *Safer Childbirth? A Critical History of Maternity Care*, 2nd edn, London and Glasgow, 1995.
Towler, J and Bramall, J. *Midwives in History and Society*, London, 1986.
Watson, J. *A Complete Handbook of Midwifery for Midwives and Nurses*, London, 1914.
Williams, S. *Women and Childbirth in the Twentieth Century*, Stroud, 1997.

17 Being the Queen's Nurse: Work and Identity as a Queen's Nurse in Scotland

RONA DOUGALL

> Oh I felt very proud to be a Queen's nurse. People were so funny too because I remember one little man saying to me, 'Did you really nurse the Queen?' 'No, it doesn't mean that, it doesn't mean that at all.' But, yes, it was lovely. I treasure that.[1]

From the late nineteenth century until the 1970s the 'Queen's nurse' was synonymous with the district nurse in many Scottish communities. Although it was not obligatory, most district nurses working in Scotland from the 1940s until the 1970s had a specialist training for work on the district. Provided by the Queen's Nursing Institute of Scotland (QNIS), this training allowed acceptance onto their roll and the adoption of the informal title of Queen's nurse. Based on a collection of interviews with Scottish district nurses working from the 1940s onwards, this chapter will look at the culture in which the Scottish Queen's nurses trained and practised, and the ways in which they expressed the unique Queen's identity which remained with them throughout their working life.

THE ORGANISATION OF HOME NURSING

Originally a voluntary service, home nursing was first organised in the late nineteenth century by nursing associations controlled locally by committees comprising upstanding people of the community. The role of the association committees was to ensure the delivery of a home nursing service to the sick poor in their area. To this aim they raised money to fund such a service, appointed district nurses, administered their salaries, provided local accommodation for the nurse and organised off-duty periods and holidays.

Class difference was evident in the relationship between the nursing association committees and the working nurses they employed. Committees comprised the great and the good of the area and typically included medical doctors, ministers of the church, accountants, owners of local business and industry, and the lady wives of professional men. However, the district nurses employed by the nursing associations were usually women of a more modest background whose fathers were employed in a variety of occupations, including shipyard workers, crofters, blacksmiths or small business managers. Where class tension was keenly felt by some nurses with regard to the nursing associations, the QNIS managed to engender a greater respect amongst those who, even in retirement, retained a sense of belonging to the QNIS and a pride in having been a Queen's nurse. With the founding of the Queen's Nursing Institute (QNI) in England and Wales in 1887 and the QNI Scotland (QNIS) in 1889, the work of the nursing associations throughout the country was effectively underpinned by a central institution providing specialist training and clinical standards. Part of a national body with royal patronage, the QNIS quickly developed a high profile, and by the 1930s many more local associations in Scotland had formed and become affiliated to it. Affiliation brought the associations under an obligation to employ only Queen's trained nurses and as the number of affiliated associations grew, so too did the prominence of the Queen's nurse.

THE DISTRICT NURSE AND THE ASSOCIATION

A number of those interviewed in this study began their working life as district nurses employed by local nursing associations. By their accounts the relationship between the association and its nurses was not an easy one. Essentially, the committee was responsible for raising funds by encouraging people to contribute regularly to a subscription scheme in return for a nurse's services. Committee members had no input on nursing matters, and patient confidentiality was guarded by the nurse. Hence the committee's sphere of interest in the nurse, often revolving around her house:

> All the district nurses, every area had a district nursing association. Now this is a very bad thing, I would have said. In one way it was good, others it was bad. Now it was usually the lady of the manor, and the lady of the manor would have … gone around collecting money, you know, for the district nurse, and give the patients a card, 'now if you need the nurse you'll have her services free'. Now this was how it was done … You had to go for interview by this lady of the manor. And this lady of the manor and her committee thought they owned you, and they would have come to your house, you see, all nurses had a house supplied to them, they paid a rent, but, they would say, 'now what have you got nurse? Now I think you could be doing with a more comfortable chair. Have you got a chair in your house Mrs. So and So?' – 'Oh yes, I'll supply nurse.' You see, charity. And, 'Now, I noticed you were at Mrs. Smith's door, what's wrong with Mrs. Smith?' Now, that was none of their business.[2]

The nurse's input into the organisation of the association and its service was negligible and controlled by the committee. Minutes of the Leith Jubilee Nurses Association reveal that the nurse was not present at committee meetings, although she was regularly brought into the meeting near the close of business to give a brief report on her work, which was usually noted in the minutes as 'satisfactory'. At least in Leith the nurse was invited to appear in person, but in some districts the report was merely submitted to the committee and approved in the nurse's absence.[3] Without the basis of a shared professional interest and exacerbated by class difference, the relationship between the nurse and the committee was often viewed by nurses as personally patronising.

> I liked old Mr. L., was a lovely old chap ... he was the banker [on the committee] ... But the funny thing about his wife, and I was quite friendly with her, but this attitude of that anything goes, which really you just have to laugh about ... anything does for the nurse. Everybody isn't like that, but you know there's a wee bit of this attitude. I suppose, maybe it's upper or lower crust business.[4]

While home nursing remained a voluntary service, the question of fundraising was always uppermost. By and large, interest in the nursing association subscription schemes was keen, but the financial position of many associations, particularly the smaller ones, was insecure. For larger associations or those benefiting from affiliation to a larger county association, subscriptions and donations were often adequate for the needs of the service. However, this position had to be maintained and so committee members, their friends and families organised musical evenings and bazaars to raise funds, or took part in the Scotland's Gardens Scheme.[5] Each association engaged its own collectors to go round homes and workplaces collecting the subscriptions, which entitled people to free home nursing at a time of need. Sometimes general collections were organised directly by the QNIS, and nurses were asked to take part in special collections.

> Well, my sister once had to go in with a collecting bowl to a football match. For the Queen's, yes ... And I remember being sent to a theatre in the Tollcross area, or somewhere, with my bowl and, sitting at the back and soliciting cash from the customers as they were going out ... In my uniform, oh yes, in my uniform. And the actor or the entertainer, whatever way you describe him, he put out a message, there is a Queen's nurse at the back, and they do very valuable work, and we would be very pleased if you would find it in your means to make a little donation.[6]

People who were not subscribers were also allowed the services of a nurse on the payment of a fee. However, the collection of patients' fees by nurses was not always strictly adhered to.

> Patients were charged fees if they were able to pay ... Unless you were a member, each area in Glasgow, in Edinburgh ... they all had their local district nursing associations. And probably for perhaps half a crown a year you could become a member, and that entitled you to the services of a Queen's nurse free. If you weren't a member, then, and it

was apparent that you could afford to pay, then you were charged a fee. But in the district I was allocated to, they were neither members, nor could they afford to pay … So they weren't charged.[7]

In 1944, largely as a result of wartime conditions, the QNIS (to whom most nursing associations in Scotland were by this time affiliated) experienced a period of financial insecurity. Fundraising became more difficult, and the income from the Gardens Schemes also decreased. Furthermore, the immediate postwar period saw a restructuring of the nation's health system, signalling a radical change in district nursing organisation. In 1948 local nursing associations ceased to function as such, and home nursing became a part of the statutory health provision under the new National Health Service. However, given their expertise and experience, many local associations continued to act as agents for their local authority, and the QNIS continued to train nurses for district work. Although the obligation to employ Queen's nurses was removed as local authorities developed their own services and training schemes within the NHS, it was some twenty years before the QNIS lost its influence altogether, heralding the final demise of the Queen's nurse.

CASTLE TERRACE – TRAINING, TRADITION AND THRIFT

As we have seen, financial accountability was the principal concern of each nursing association within its own locality. Almost all eventually affiliated to the QNIS and thereby contributed to the cost of training and the establishment and maintenance of a pension fund for the Queen's nurses. The QNIS raised and managed the larger part of the funds for the nurse's training homes and benefits to nurses in sickness or retirement.

Although much of the work of the QNIS related to the clinical aspects of nursing (i.e., providing lectures and instruction, developing procedures and guidelines, as well as organising district placements and supervising nurses in post), financial constraints necessarily impinged on daily routines in the training homes. Conditions there conformed to common notions of thrift and economy. Hence the QNIS headquarters and principal training home in Castle Terrace, Edinburgh, earned the reputation of being a rather austere place with little concession to comfort and relaxation, a reputation heightened during the lean times of the 1940s. Interviewees who trained at Castle Terrace between the late 1930s and 1950s remembered its traditional boarding-school atmosphere with nostalgia:

> It was quite Spartan. It took a bit of getting used to, yes … I rather think it would have had, because our economy measures were quite stringent. I mean, the lift for example. You carried your luggage upstairs. You see … the lift was really meant for heavier stuff. Because of the electricity that it would consume on the way up. Our pay, we got paid once a month, whatever little it was. And I suppose, there would be a slip showing what our salary was. But we had to, it was,

> our pay came in an unsealed envelope, and we had to have, hand the envelope back, and it was used for us next month.[8]

In addition to the physical economies of Castle Terrace, regulations contributed towards a traditional and conservative atmosphere.

> When you went in to Castle Terrace, you, you couldn't make a cup of tea at night, in the kitchen without having full regalia on. You had to have your cuffs on, you had to have your white hat on, which you never wore, but you had to, all this nonsense … There was a bust of Queen Victoria on a window, halfway up these stairs. And we used to look at her and play like, you know, we used to talk to her and say, 'hasn't changed a bit since you were here'. Poor old thing had nothing to do with it, except that she formed the Queen's nurses.[9]

Other home-nursing services existed in Britain at this time, working on evangelical principles (notably the Ranyard mission nurses), but district nurses trained through the QNIS were discouraged from bringing their religious beliefs to the nurse–patient relationship. Records suggest that although Christian prayer was built in to the daily routine at Castle Terrace there was no sectarian bias in those accepted for training. However, prayer was not strictly observed by all nurses:

> What I found surprising, but because of my own upbringing, I found it very, almost pleasingly acceptable. We had, there was, family worship was held in the morning … we gathered about … five to ten minutes before breakfast. And, in the council room, or in the duty room where we were given our work, would it be, in the council room. And Miss Sinclair read a portion of scripture, a few verses, and then we all went on our knees and repeated the Lord's Prayer. And then off we went into the dining room for breakfast. Almost, most people came, most nurses came. I never heard of any compulsion, but most came to this worship session. But occasionally, if you were at the end of the queue leaving the room, a few stragglers coming down the stairs would join, just in case they missed their breakfast, we went to breakfast immediately after that. I think our breakfast had half a roll on our plate, when we went into the dining rooms, half a roll. There would be bread and there would be, I forget what else, but we made a joke of your half roll.[10]

Despite the stringent conditions, many interviewees recalled a prevailing sense of camaraderie, whilst others remembered a divisive hierarchy, extending down to the trainees:

> I think that, if it was a fault, [it] was made up for, to some considerable extent, by the feeling of camaraderie there was among the trainees, there was really … We were more, were more contact, more close, we came in closer contact, we were together as a group. In general

training, you're not together as a group, there's the ward group, the various wards. But there, in the morning, everybody was there to get their work allocation. And then again, in the evening, after work was done, we were all in the sitting room, everybody was resident. So we were taught then, and you found it, a tremendous feeling really, camaraderie.[11]

Now there was two grades of nurses. There was the midwives and there was those doing their district nurse training. Now, the midwives, I was a midwife before I went to Castle Terrace, but those who were doing their midwifery were, training, they had done, been up at Simpson's Maternity for six months, and they were back at Castle Terrace to do their district maternity. Now they were thought the world of. They, oh, they were just perfection, these midwives, as they called them. So the midwives sat at the main table, you see, and they were up at the top beside the milk. The underlings were along the bottom or round about. Well, you were lucky if you got any milk at all.[12]

The economies of the QNIS were reflected in the low pay received during training, and nurses were often grateful for gifts from relatives.

We did get a small salary and ... of course because we had our uniform and food and lodgings ... my mother ... she was quite good, she would give me something perhaps once a month or something ... I remember there was an occasion there was four of us in a group and it was the day before payday, none of us had any money left an I said oh I'll just go write to my mother I got a letter from her this morning. Well, inside her letter she had enclosed a ten shilling note so we all went out on the strength of this ten shilling note, you could have got a lot for ten shillings in those days, you know.[13]

PRACTICE: OUT ON THE DISTRICT

Many district nursing posts were, and still are, based in rural or remote areas, often subject to harsh weather conditions and sometimes with no resident doctor. Curnow provides a working definition of 'remoteness' in the context of health services, noting that 'geographical isolation is of course important but it is not the only consideration'.[14] He contrasts an offshore island in calm waters having a good weather factor with the same island set in the rougher North Sea. In this scenario, remoteness is not defined by distance but by transfer time to and accessibility of 'a clinical facility providing sufficient medical services'. Given that the weather and terrain of much of northern Scotland often makes travelling difficult and time-consuming, it is reasonable to define this region along with many mainland districts as remote. District nurses in these areas may have had to travel miles over difficult terrain to reach a patient, or deliver babies in isolated situations with no access to distant medical services. Even with improved communication and transport, nurses in some remote areas remained relatively isolated and as a result experienced a level of nursing responsibility which differed from that of their urban colleagues.

TRAVEL AND TRANSPORT

Given this geography and the weather, just getting around from one patient to another in rural or remote areas often required great physical endurance. Travelling tales became a feature of district nurses' stories, even in urban areas, where the car had also transformed the nurse's travelling experience. One rural nurse spoke of travelling to confinements during thick snow on a horse-drawn sledge.[15] Another city nurse recalled the brakes being removed from her motorbike while she attended a case.[16] But this kind of behaviour towards the nurse and/or her property was unusual. On the contrary, the public were generally aware of the amount of travelling the district nurse's job incurred and usually she would be given a seat on a bus or have her bag carried up tenement stairs for her.

> Mind you everybody looked up to the Queen's. If you got into a bus even an old buddy would offer you a seat and I would say, 'No thanks, but if you take my bag.' 'Oh what a weight,' they would say when you put the bag down on their knee.[17]

People often tried to help the nurse with transport. Local garage workers would rescue nurses whose cars had become stuck in mud or deep snow, boatmen remained on call to ferry nurses to and from patients, bicycles were designated by locals for the nurse's use, and in particularly inaccessible parts farmers offered the use of horses and carts. But nurses had their preferences, and not all offers of help were welcome:

> And then one day, at dinner time, the back door went, and here's the man with a motor bike. And I said, what are you doing with a motor bike? He says, 'It's for you.' 'Well,' I says, 'you can take it away,' I says, 'I'm not putting my leg over that.'[18]

Scottish winters did not make travel any easier, and in country areas the lie of the land could be lost under deep snow. In the following story about travelling to a patient despite heavy snow which had obscured the roads and paths deep below, the attitude of the nurse echoes a commonly expressed nursing ethos where commitment, seen here as determination to make it to the patient, was the guiding principle. This is reinforced by the pride in receiving not a monetary reward, but simply the praise of the health committee.

> Mind you, the winter … was hard here … One winter, Nurse B. and I did what we could on our own, and then we heard from the roadmen which way it was best to go. There were two roads to Duns … and there was an old lady right in the middle, who needed daily

Figure 17.1 Not all Queen's nurses were reticent about using a motorbike to get about on their rounds. District nurse on her motorbike, *c*1920. Source: Reproduced courtesy of the Royal College of Nursing Archives, P26/1.

> treatment. That was one of Nurse B's patients. And we found out from the roadmen which way was easier. And … we were actually walking on top of the hedge, all of a sudden one would go away down. And it was so bad that the health committee wrote to every nurse thanking us for the way we had carried on during the winter. Oh, we used to get an awful lot of snow.[19]

Although rural areas could encompass many miles and a widely dispersed community of patients, those on the islands bore an added burden of isolation. With no doctor on many of the islands, emergency cases had to be transported to the mainland quickly for hospital treatment. Emergency referrals to hospital had to be authorised by the doctor. However, one Shetland nurse proved an exception to this rule. She recalls attending the confinement of the local schoolteacher's wife, who had suffered bleeding during her pregnancy. Having eventually delivered the baby, the nurse called the doctor who was on a nearby island at the time:

> The doctor sent me word that he would come in the next day. So he came in with a fishing boat … it was 12 miles between Skerries and Whalsay and it took the boat an hour and a half to go and back again

> that was three hours … I had written him a letter telling him that Mrs W. had had her baby, but he didn't know and he said 'What! Well I was wanting to get to her', because she had an APH [ante-partum haemorrhage] during her pregnancy. So he took his bag up to the manse which was also the school house … by then she was dried up … and when he came the next morning I told him I couldn't take the risk another time, I would just send her to Lerwick. He said 'No it'll not happen again' … and he wasn't half way back to Whalsay when I got this frantic call and the Earl [the local transport ship] was coming up that day and she was lying at Baltasound … and I arranged for them to pick her up and I went with her … and I sent a message to the doctor to meet the Earl when it came to Whalsay … and when he came he'd been on the phone to the surgeon … but from that day onwards I got permission that if I needed to send a patient to hospital I could send her.[20]

This quote is particularly revealing, as it is more than just a travelling story. It also exemplifies the competence of the nurse in making professional decisions, even in contradiction to the doctor, and describes a little of the nurse–doctor relationship.

The city district nurse had her own regular travel difficulties, going mostly on public transport or on foot, often over widespread areas up and down tenement stairs. It was not until the late 1960s that urban nurses were given a car allowance as commonplace. In a district of Hamilton where several nurses shared a nurses' home, the local council used to send a taxi to the nurses' home to take the gas and air equipment to midwifery cases, while the nurse on call for confinements had to make her own way either on foot or by public transport. Only between the hours of 11 p.m. and 7 a.m. was the nurse allowed the luxury of taking a taxi to a case.[21]

Whatever the means of transport, and whether in town or a country district, the district nurse of the past was more visible to the public. This was partly due to the fact that, before the nurse's car became commonplace, the nurse walked in her area and became commonly known and recognisable. One of her distinguishing features – the change to which is now much lamented by the retired generation – was the Queen's uniform.

> We had the most beautiful uniform … it was blue dresses and a blue apron of the same material. And a little blue cap and, of course, when you got your own, when you went onto the roll you got … a badge, and you wore it round your neck on a blue and white cord … And we were terribly proud of ourselves when we got that. And you got the Queen's Victoria Jubilee badge for your cap. And you got flashes, QNDIS in silver there and here and epaulettes on your shoulders. And you were measured for your coat. You went and you got your, you were sent, we were sent along to Miss Fellows, and you got measured for this coat. And we got these beautiful coats. And we were told by Miss Gill said that we had to be proud of our uniform. We had to be

Figure 17.2 District nurses outside the Queen Victoria's Jubilee Institute, Edinburgh District Nurses Home, 1930. Source: Reproduced courtesy of the Royal College of Nursing Archives, P26/55.

> proud of ourselves. We had always to be smart, always to be neat. Our shoes had to shine. And, you know, we were very proud, we really thought we were something because we were Queen's nurses. And, of course, the uniform was lovely.[22]

> You know the worst thing they ever did to the Queen's nurses was take away our uniform, our epaulettes ... as far as I was concerned, and any of the other Queen's nurses, they've all said the same to me, that it was, [it] took away what we were really meant to be doing. And it took away what we were doing, a sort of, what would you call it ... your uniform was like something in the community. You were, you were a nurse, your profession, it was your profession. And they respected you.[23]

NON-NURSING CARE

Almost all elderly nurses interviewed, most of whom retired in the 1970s or '80s, spoke of non-nursing care as a feature of the past and not the present. Although this is an aspect which has been used to undermine the professional image of past district nurses, it is one which remains at the heart of district nursing for many of those who practised in the Queen's nurse era. Regardless of service organisation, most nurses visiting people's homes found themselves doing things over and above strictly nursing tasks, thereby extending the definition of nursing care. This, too, was influenced by geographical factors, which in turn influenced the nursing experience. The nurse in the rural or remote district was often the only person available to provide extra help and care, and she was susceptible to professional isolation, while the city nurse had a wider network of colleagues to call upon. The rural nurse was much more likely to find herself helping with domestic chores while stranded in someone's home by bad weather, while the urban nurse had more involvement in the alleviation of social problems such as debt or poor housing conditions. One city nurse recalled organising a repayment scheme for a family to help them out of a spiral of debt; for others, understanding and adaptability were crucial in households where basic material goods were in short supply.

> You learned to be very adaptable. Patients probably wouldn't be able to give you what you were asking for, you might have to wait weeks before they would see a relative that would buy, say basins and face-cloths and towels like that … But you could be adaptable. And we always kept bits and pieces in the nurses home, so as we were able, sheeting and that sort of thing, and people who died, [their relatives] often sent in new stuff that they hadn't used, all new sheets and that. Well, we always made a point of keeping that for someone else who wouldn't have it … you couldn't even get a piece of newspaper in some of their homes, but you, it was nothing to put a newspaper in your bag.[24]

One younger district nurse who was still practising in the late 1980s suggested that the nurse's response to the needs of the patient may not have changed dramatically. It was not uncommon for her to carry out simple domestic tasks for a patient, such as replacing a lightbulb, but, strictly speaking she should not have done this.[25] In the past such things were done at the nurse's discretion (and were very common), but in more recent times nurses have been told explicitly not to do them. This is influenced largely by a concern to maintain a high professional status for nursing, as well as insurance considerations regarding liability if the nurse were to be injured while performing a non-nursing task. But even in the past there were dissenting voices on the matter.

> Well, making maybe cups of tea and stuff like that, you know. And occasionally lighting a fire … I didn't tend to do an awful lot that was beyond nursing, because it's something I disagree with, very strongly, that the nurses, that people expect the nurses should be everything. Why? You know. I mean, other carers could have been in and the doctor could have been in, or anybody, and they wouldn't ask them to light the fire or anything, so why? And then, well, of course, you did it because they put you in an awful position, you couldn't leave somebody … Well, it was very difficult to say no. But, I learned after a while to be a bit more, strong willed about it. Because, you know, it was ridiculous.[26]

Nevertheless, an abundance of testimony confirms the point that district nurses of the past tended to respond in a personal way to patient need, that they continued to do so until the very recent past, and that they may still do so regardless of regulations. With the development of health and social care services since the 1970s, the range of people providing home care has extended throughout the country. Since then more routine tasks, such as bathing, have been delegated to lesser-qualified auxiliaries, leaving the nurse free to exercise her clinical nursing skills. Now, as a complement to home nursing, nurses with specialist training deliver nursing care in the home for specific conditions, or provide palliative care to the terminally ill. Home helps undertake light housework and shopping, 'tuck-in' services check up on the elderly and infirm in the evening, and social workers offer advice and practical help with a range of financial and social difficulties. Hence, tasks which were once discretionary but often perceived as a moral obligation for the district nurse are now the legitimate responsibility of other services. By the mid-1970s professional boundaries were more closely guarded, and all staff were obliged to respect procedures for dealing with patient/client needs. For nurses who had been used to both identifying and fulfilling the needs of their patients in a direct and personal way this new bureaucratic approach was felt to be heavy-handed and laborious.

> Until the change-over in '75 … we could do some of the social work and everything, … but then, when that change took place, you were told definitely that … if you wanted any social work done or anything like that you applied to the appropriate department, you didn't do any of it … I found it frustrating … in fact I overstepped it an odd time … One of the times, there was somebody and there was nothing happening and I contacted the appropriate department and I got into trouble … I was carpeted more or less and told that it wasn't my job to do that sort of thing, that it was social work department … and the head of the social work department had been in contact etc. etc … so please do not do that again.[27]

This story indicates a change in approach to district nursing in the 1970s that challenged the autonomy of the district nurse and offered a more prescriptive form of practice. Coupled with the diminished role of the QNIS, this period of organisational change from the mid-1970s onwards represented a transitional period in district nursing which no longer included the Queen's nurse. Almost all of the interviewees, all Queen's nurses, incorporated non-nursing tasks as an essential quality of district nursing. To them, the whole process of nursing tended to be seen in terms of a relationship with patients which entailed a holistic response to their needs, as opposed to a situation in which clearly delineated professional tasks determine the pattern and manner of care.

The prevailing narrative of these accounts is one of the selfless nurse responding equally to nursing and social need. This pattern represents a key issue in any discussion of the ethical character of district nursing and the ways in which it is perceived to have changed. It may be argued that the professionalisation of nursing (which took hold amidst the reorganisation of the 1970s and set the agenda for the 1980s) has contributed to the creation of an apparent schism between the era of the Queen's nurse and that of today. However, far from distancing themselves from the professional status of nursing, Queen's nurses, as the voice of district nursing between 1940 and 1970, also incorporated professional skill into their nursing stories and had a clear sense of professional autonomy.[28] What is emphasised by them in their stories is not nursing per se but the particular culture of nursing in which they worked: a culture moulded by the exigencies of thrift prevalent since the interwar period, by the confining conditions imposed by Scottish weather and landscape, by the nature of urban and rural Scottish communities, and by a deferential respect for central figures within these communities, such as the nurse. It is this culture which shaped the lives and the sustained identities of nurses who were locked into the communities in which they both lived and worked.

NOTES

1. From interview QNIT 57/2, 10. All recorded interviews and relevant documentation are held in the Royal College of Nursing (RCN) Archive.
2. From interview QNIT 21/2, 9.
3. The report being read in the absence of the nurse appears to have been the practice in the Comrie and Crieff Nursing Association in the 1930s. See RCN Archive, Comrie District Nursing Association Minute Book 1920–1941, QNI/D.2/1.
4. From interview QNIT 24/2, 19.
5. Scotland's Gardens Scheme was begun in 1931 and was part of a national scheme begun in England in 1927. It was a scheme in which interesting gardens, including that of Balmoral, were opened to the public for a fee. The proceeds were given to the QNI[S] to help fund the expansion of district nursing. See Baly, 1987, 136–7.
6. From interview QNIT 13/2, 10–11.
7. From interview QNIT 13/2, 12.
8. From interview QNIT 13/2, 8.
9. From interview QNIT 12/1, 15.
10. From interview QNIT 13/2, 9.
11. From interview QNIT 13/2, 9.
12. From interview QNIT 21/2, 12.
13. From interview QNIT 15/1, 4.
14. Curnow, 2000, 125.
15. From interview QNIT 31/1, 8.
16. From interview QNIT 14/2, Summary, 8.
17. From interview QNIT 11/1, 7.
18. From interview QNIT 11/1, 10.
19. From interview QNIT 11/1, 19.
20. From interview QNIT 50/3, 7.
21. From interview QNIT 40/2, Summary, 2.
22. From interview QNIT 42/2, 16.

23. From interview QNIT 38/2, 29.
24. From interview QNIT 19/2, 26
25. This was discussed in an informal conversation which does not form part of the QNI oral history collection held in the RCN Archive.
26. From interview QNIT 26/1, 21.
27. From interview QNIT 25/2, 20.
28. See Ferguson, 2001, 10–17.

BIBLIOGRAPHY

Baly, M. *A History of the Queen's Nursing Institute: 100 Years 1887–1987*, London, 1987.

Curnow, J. The provision of healthcare in remote communities. In Nottingham, C, ed., *The NHS in Scotland: The Legacy of the Past and the Prospect of the Future*, Aldershot, 2000.

Ferguson, R. Autonomy, tension and trade-off: Attitudes to doctors in the history of district nursing, *International History of Nursing Journal*, 6 (2001), 10–17.

Records of the Queen's Nursing Institute of Scotland, Royal College of Nursing Archive.

18 Workers' Bodies: Occupational Health in Scotland

RONNIE JOHNSTON and ARTHUR MCIVOR

> I was getting a bit slower and a couple of times I noticed that my wife was walking faster than me ... The health aspect has had more impact than the financial aspect actually. The health aspect has stopped us going anywhere and daen things. We used to be running about all over Scotland. Everywhere we went we made friends. We could still do that even though we're skint; but we cannae dae it because of the ill-health.[1]

This is how one Scottish worker interviewed in 1999 described what it was like to live with an occupational disease. Within two years he had died of asbestosis, caused by exposure to asbestos dust while working at an asbestos cement factory in Clydebank in the 1960s. By the time of his death the factory had long disappeared, and so too had most of Clydeside's shipbuilding and engineering industries, which used asbestos extensively in the 1950s and 1960s. The long latency period of his disease meant that his brief encounter with asbestos thirty years previously had a catastrophic impact on his plans for a long and happy retirement. Sadly, as happened to this worker, many Scots are suffering from the legacy of an industrial past based on heavy industries, which, although bringing Scotland worldwide fame as a leading industrial nation, in many ways made for both a dangerous and an unhealthy working environment. Similarly there are currently people working in newer industries in which unforeseen occupational health problems are only beginning to be understood. It is the continuities and changes in work-related ill-health in Scotland that will be explored in this chapter.

In Scotland, as elsewhere, work has always been important as a means of securing a living. Throughout history, the majority of Scots spent most of their waking hours in paid or unpaid physical toil, their bodies being subjected to the rigours which employment often entailed, whether in the fields, in the home, in workshops, down mines or in the factory. Rewards from work dictated standards of living and were a major influence on the health and wellbeing of individuals and families. Work also defined status within the community, influenced political consciousness and brought structure and meaning to life, whether through the hard graft of labouring, the daily grind of domestic service or the pride and identity forged through craft and artisanal work.[2] Employment, however, was always a double-edged sword. The human body could thrive through steady work, which could directly and indirectly give physical and psychological satisfaction. Losing work could lead to a downward spiral into poverty, demoralisation and despair. This is well documented, for example, in the interwar economic recession of Scotland in the 1920s and 1930s.[3] However, many occupations brought their own particular hazards, and as Scotland became increasingly industrialised in the nineteenth century there were few occupations that did not harm workers in one way or another.

THE 'BONE-WEARINESS' OF RURAL AND PRE-INDUSTRIAL TOIL

Prior to the nineteenth century, Scotland was predominantly a rural society in which most people either worked the land or were dependent upon it. One of the most common themes in the literature on pre-industrial work is the fatigue and overstrain that went along with employment in an agrarian society, when ordinary workers were subject to back-breaking labour over very long working hours.[4] It is, however, true that many farm workers relished hard work, such as lifting full sacks of grain, as a demonstration of their physical strength. Communities largely accepted this energy-sapping life without question – this was the way things had always been done – and there was some compensation in terms of the extent of independence and control that such folk had over their work rhythms. Whilst forced clearance played its part, the unrelenting toil to make a living, the drudgery, limited opportunity and lack of leisure time on farms were factors which contributed to the flow of migrants from the country into the towns and abroad during the nineteenth and twentieth centuries. A traditional Fife ploughman's rhyme encapsulated something of the essence of the unrelenting work regime:

> Six days shalt thou labour and do all
> That you are able;
> On the Sabbath-day wash the horses' legs
> And tidy up the stable[5]

In a period when there were few sophisticated mechanical aids, the land and sea had to be worked and produce moved by human graft. The *Statistical Accounts* include many vivid descriptions of such heavy manual labour, such as this description by Thomas Carlyle of the fishwives and women 'carriers' of 1795 who trudged miles from the east coast into Edinburgh carrying fish, salt and other products to sell:

> The women who carry sand to Edinburgh have the hardest labour and earn least. For they carry their burden, which is not less than 200lb weight, every morning to Edinburgh, return at noon, and pass the afternoon and evening in the quarry, digging the stones and beating them into sand. By this labour, which is incessant for six days in the week, they gain only about 5d a day.[6]

Rather than identify the deleterious impact on the health of these workers, Carlyle chose to deplore the way in which 'work of men' made the 'manners' of these women 'masculine'. This extended, he noted, to women playing golf and football. Similar comments on the loss of femininity in fish gutters were made by Charles Weld in 1860, who also reflected on the impact of heavy workloads:

> The Wick gutters – I timed them – gut on average twenty-six herrings per minute. At this rapid rate you no longer wonder at the silence that prevails while the bloody work is going on, nor at the incarnadined condition of the women. How habit deadens feeling![7]

In such agrarian and fishing communities, children were expected to start work young, commonly before the age of ten years. Dr Findlay of Fraserburgh noted in 1789 that from the age of eight in Aberdeenshire 'the boy keeps cattle or sheep, the girl spins linen yarn, and earns 6d per week, some more'.[8]

The nature of work on the land was diverse, and whilst it could be debilitating and insecure (subject to the vagaries of the weather), the other side of the coin was that it was mostly undertaken in the open air, in a relatively healthy, robust working environment. The same could not be said of pre-industrial work in the coal and salt mines (where there was an element of compulsion in the system of mining 'serfdom'), the charcoal ironworks and in many of the small urban craft workshops where work took place in confined enclosed spaces. The health problems of working in such environments were recognised fairly early. For example, Bernardino Ramazzini's *De Morbis Artificum* (Treatise on the Diseases of Tradesmen, 1700) translated into English in 1746, highlighted how such workers could be exposed to a range of airborne dusts and toxic materials, including silica and lead. Earlier still, in 1556, Agricola had identified the deleterious impact of dust in mines.[9] Lay knowledge of the causal agents and the pathology of such diseases was limited at this time, although communities had their own names for a wide range of such ailments, such as 'black spit', 'grinders' rot', 'bad breath', 'potters' lung', 'millers' creeping pneumonia' and 'miners' asthma'. The dangers of working with lead were also recognised by the early 1800s. Plumbers were exposed to lead when fitting and repairing ornamental cisterns and lead pipes – although, as Fraser points out, there was a widespread reluctance amongst plumbers to use lead substitutes – while 'painters' colic', caused by exposure to white lead in paint, was accepted as part and parcel of the job.[10] Bookbinding was also an unhealthy occupation: the working of the leather had to be undertaken in damp cellars to ensure that hides remained malleable; intricate gilding work necessitated working with closed doors and windows to prevent dust; and artisans were frequently burned by the sulphuric acid required to produce blue mottled lining pages.[11] Hence a significant number of workers were exposed to acute and chronic occupation-related diseases prior to the acceleration of industrialisation in the nineteenth century.

INDUSTRIALISATION AND THE DAMAGED BODY

Whilst mechanisation and the application of scientific methods to production could alleviate human physical effort, there is no doubt that rapid and virtually unregulated industrialisation in Scotland (as elsewhere) generated severe pressures upon the human body, exacerbating problems of child exploitation and fatigue, whilst substantially increasing the incidence of traumatic injury and industrial disease. Scotland's industrialisation process was distinctive because of its rapidity, and this was matched by urbanisation in the central belt of the country taking place faster than in any other European region. Forces such as these affected the health of the workers involved, and this was perhaps most evident in the rapidly expanding textile factories and the coal mines. In his classic text on public health in the Victorian period, Anthony Wohl noted how the factory provided 'the perfect nexus for aggravating or accelerating ill-health'.[12] Whilst work regimes varied considerably, the first generation of Scots who entered such establishments in the late eighteenth and early nineteenth centuries generally found their bodies assaulted by high temperatures, excessive humidity levels, dust, and high accident rates associated with fast-moving machinery (drive bands leading to revolving overhead shafting and other hazards). The accelerating pace of work, draconian supervision – sometimes involving corporal punishment – inexperience in working with machines, and excessively long and more tightly regulated working hours all increased the susceptibility of workers to injury and disease, including respiratory disease caused by the inhalation of textile dust, later recognised as byssinosis.

The cotton mill manager and factory reformer Robert Owen commented on the system prevailing in the 1790s in which poorhouse authorities sent orphan children as young as six years old to mills, including at New Lanark under David Dale, to work from six in the morning until seven in the evening. 'Many of them,' he commented, 'became dwarfs in body and mind, and some of them were deformed.'[13] The Scottish evidence given to the Factories Commission of 1833, provides much material on the impact of work upon the human body. A comment by one of the commissioners (Stuart) identified Scotland as one of the worst areas in the UK for child exploitation in factories, noting that 'the greatest number of bad cases occur in the small obscure mills belonging to the smallest proprietors'.[14] Another commissioner (Barry) reported on insanitary conditions and excessive heat and dust in the Monteith and Co. mill at Blantyre in 1833, noting: 'There are no dust fans in any of the rooms, nor have any improvements tending to promote health been effected in this mill since the year 1786.'[15] After weighing the evidence, the 1833 Factory Commission concluded 'that the effects of labour during such hours are, in a great number of cases, permanent deterioration in the physical constitution ... [and] the production of disease wholly irremediable'.[16]

In coal mining, policy makers concentrated on the immorality of the situation in which women worked underground alongside adult men, as well as the bodily damage incurred by vulnerable young children working in this dangerous environment. The Mines Commission of 1842 provided much evidence of the work–health interaction in mining, as well as some evocative line drawings depicting the body at work in the Scottish coalfields. An East of Scotland coal-bearer aged eleven (Janet Cumming) was one of those who provided evidence to the commissioners.

> I gang with the women at 5, and come up at 5 at night; work all night on Fridays, and come away at 12 in the day. I carry the large bits of coal from the wall-face to the pit-bottom, and the small pieces called chows, in a creel; the weight is usually a hundredweight … it is some work to carry … The roof is very low; I have to bend my back and legs, and the water comes frequently up to the calves of my legs; I has no likening for the work; father makes me like it.[17]

An even younger worker (Margaret Leveston, aged six) commented on carrying 56 lbs, noting 'the work is na guid; it is so very sair'.[18] Isobel Hogg, a coal-bearer aged fifty-three with four married daughters working underground, informed the commission that working in the pits eroded the health of all workers – men and women – noting the dusty work atmosphere, the energy-eroding work regime and the prevalence of miscarriages at work. She made an appeal direct to the monarch to intervene:

> Collier-people suffer much more than others – my guid man died nine years since with bad breath; he lingered some years, and was entirely off work eleven years before he died. You must tell the Queen Victoria that we are guid loyal subjects; women people here don't mind work, but they object to horse-work.[19]

The result of the exposure of such lurid conditions in the factories and mines was the Factory Act of 1833 and the Mines Act of 1842. The latter banned the employment of children and women underground, whilst the Factory Act regulated child labour for those aged 9–13 to a maximum 48-hour week, and for those aged 14–18 years to a 69-hour week, introducing inspectors for the first time to enforce the legislation. These were key landmarks in state intervention in occupational health and safety in the UK. Thereafter, the legislation was widened and extended to include adult male workers, other industries and other occupational hazards. There remained, however, a wide gap between the legislation and actual workshop practice – something which persisted well into the twentieth century. Moreover, many other dangerous trades went virtually unregulated. Nonetheless, injury rates in textile factories appear to have peaked around the 1830s and 1840s, when the chief factory inspector, Leonard Horner, was reporting a UK injury rate incidence of around 9 per cent of workers annually. By 1914, the rate of reported injuries was down to around 2 per cent per year.[20] Similarly, as Benson has shown, injury rates declined in coal mining from the mid-nineteenth century onwards.[21]

As industrialisation developed through the nineteenth century it was evident that most work involved some risk to health and wellbeing in one way or another. The landmark publications of Britain's foremost occupational health specialist (a Scot, Thomas Oliver), around the turn of the century, are a testament to the myriad ways in which occupations could impact upon the body.[22] By the early years of the twentieth century the statistics compiled under the Workmen's Compensation Acts clearly identified the occupations that took the greatest toll upon workers' bodies. In terms of occupational mortality from work injuries, the industries at the top of this grisly league table were, in order of most deadly: shipping; dockwork; mining; railways; quarries; engineering and metal smelting. These sectors – transport, mining and metalworking – were dominated by men and employed about a third of all adult male workers in Scotland in the late nineteenth century. Scottish seamen and fishermen were exposed to a dangerous natural environment; though the lack of investment in good port facilities could exacerbate the risk, as in the Eyemouth fishing disaster, when 199 were killed in Scotland's worst single fishing tragedy in 1881.[23] Kenefick's work on the Clydeside dockers has shown how the work regimes imposed upon these workers led to a massive toll of injuries and work-related deaths.[24] Elsewhere, excessive working hours and payments-by-results wage systems increased injury rates. One man called to give evidence to the Departmental Committee on Factory Accidents in 1911 reflected on the high accident rate in the Clyde shipyards, noting that 'excessive haste ... is one of the principal reasons ... they take risks on piecework which they would never dream of taking if they were paid by time'.[25] Similarly, when working hours rose to a twelve-hour day accident rates increased proportionately.[26] The iron and steel works were also notoriously dangerous, as Ballantyne has shown, though the injury risk differed markedly across the metalworking community.[27] Hot splashes and the incessant inhalation of silica dust were amongst the hazards identified in Patrick McGeown's evocative autobiography, *Heat the Furnace Seven Times Over* (1965).

Mining communities remained amongst the most afflicted by work injuries. Underground explosions attracted the greatest attention, though more were killed and disabled by smaller roof falls underground and a range of industrial diseases, including respiratory ailments (pneumoconiosis, bronchitis and emphysema) and the degenerative eye disease nystagmus. A Scottish doctor was amongst the first to identify coal workers' pneumoconiosis – then termed anthracosis – in 1831.[28] Others commented on the gnarled and twisted frames of miners, caused by the contortions necessary for working in narrow seams. Not surprisingly, of all working-class communities it was in the mining villages that impaired and damaged bodies were most in evidence.[29] Preventative policy in the nineteenth century remained geared much more towards restricting the work of women and children, and emphasising safety over occupational health and disease.[30]

However, the attention of medical professionals and the state gradually shifted from a focus on traumatic injury towards more chronic disease associated with work. We have no space to explore this in any detail here.[31]

Suffice to say that a plethora of occupations involved contact with toxic and carcinogenic materials and, most important of all, the inhalation of dust. The Scottish socialist Keir Hardie was involved in a widely publicised exposé of grim work conditions in J J White's chemical works in Rutherglen in 1893. As Walker has shown, despite the use of crude cloth 'bandages' around their mouths and noses, the nasal septum of many of the workers in this chemical works was eroded by inhaling corrosive chemicals.[32] Two-thirds of all Glasgow chemical workers were estimated to have chronic bronchial impairment due to the inhalation of toxic dust and fumes.[33] Chrome holes in the flesh of women working at the United Turkey Red Textile Dyeing Co. in the Vale of Leven were also the topic of press reports before World War I.[34] In a similar way, workers at Pullars' textile-cleaning factory in Perth at the turn of the century suffered frequent burns through working with chemicals – indeed, over the 1903–5 period, over 600 employees required treatment.[35]

A clutch of so-called 'dangerous trades' were regulated in the years immediately prior to World War I, and several specific occupational diseases were identified and prescribed for compensation purposes, including working with lead, phosphorus, mercury, arsenic, anthrax and anklyostomiasis – a mining disease which was rare in Scotland. However, there remained serious flaws in the coverage and effectiveness of the regulatory mechanisms and a wide gap between the discovery of a hazard and anything actually being done about it.[36] Despite the well-publicised exposure of the high death rates of file, cutlery and other metal grinders from the 1870s, silicosis, the first of the prescribed dust diseases, was not added until 1918. Asbestos was not prescribed until 1931 despite the fact that the official annual Factory Inspectors Report had included a statement on the hazards of working with asbestos as early as 1898. Byssinosis and coal workers' pneumoconiosis had to wait until the 1940s to be recognised. As a consequence, coughing and breathlessness, chronic bronchitis and emphysema remained endemic within working-class communities dependent upon such dusty jobs until well into the twentieth century.[37]

WORKERS' HEALTH AND SAFETY IN A CHANGING INDUSTRIAL LANDSCAPE

The twentieth century witnessed a number of interrelated developments which impinged upon the work–health interaction in Scotland. The belated modernisation of the Scottish economy, with the fundamental change from manual to non-manual work, saw a marked decline in the proportion of the labour force that was exposed to the worst hazards and industrial toxins. This was no unilinear trajectory, though, as the basic industries continued to take their toll on the working population for some time to come, much of this having been recorded in autobiographies and oral histories. For example, in his autobiography, Ralph Glasser referred to the stories of death and mutilation he was told relating to Dixon's 'Blazes' Ironworks in the Gorbals.[38] An ex-steel worker from the West of Scotland noted that few could tolerate the work at the furnaces: 'About one steel worker in ten could stand up to them successfully.'[39] One occupational nurse described conditions in the foundry at the North British Locomotive Works in Glasgow in the 1960s thus: 'The air was very black; the men were absolutely black. I was absolutely shocked and I said to somebody "it's like *Dante's Inferno*".'[40] As late as the mid-1970s, a Scottish occupational hygienist, Ian Kellie, commented that conditions in the Scottish steel industry were 'appalling' and that he was 'astonished' at the high incidence of silicosis.[41]

As in so many other areas, Scottish experience increasingly converged with UK patterns in the twentieth century. However, the economic structure of Scotland fossilised in the first half of the twentieth century, and the failure to modernise meant that a somewhat larger proportion of Scottish workers remained confined within the more dangerous, dirty and dusty heavy industries (see Table 18.1).

Table 18.1 Employment in the most/least dangerous occupations, 1951 (percentage total employed)

	Clydeside	*Scotland*	*England*
Most dangerous*	42.2	38.0	31.0
Least dangerous**	34.2	36.6	39.5

* Includes coal mining, metals, mechanical engineering, shipbuilding, textiles, timber, construction, transport.

** Includes distribution, insurance, banking, professionals, administration, other services.

Source: Lee, C H. *British Regional Employment Statistics, 1841–1971*, Cambridge, 1979.

This, combined with a particularly persistent machismo work culture (especially evident in the most industrialised region of west Scotland), where the 'hard man' discourse prevailed, accounts for higher injury, mortality, disability and disease rates north of the border.[42] Data from the Annual UK Factory Inspectors Reports show injury and fatality rates from work accidents running at more than 25 per cent higher in Scotland during 1950–75.[43] Standardised mortality rates from mesothelioma (the main asbestos-related disease) in Scotland in the 1980s were running at 31 per cent above the UK average, with Clydeside almost double the UK average. Deeply entrenched attitudes proved hard to erode, as one Clyde shipbuilding safety officer commented in 1977: 'Safety is always uphill work in a traditional industry like shipbuilding where men are set in their ways … It takes time.'[44]

Moreover, the carnage from industrial accidents and chronic disease continued to fall unequally across different socio-economic groups. Not surprisingly, there were wide differences in disability rates across social classes, with industrial injury and disease falling heavily upon the manual working classes, with relatively low incidences amongst white-collar professional groups, managers and employers. This was clearly the case, for example, with asbestos-related disease.[45] There was also a marked gender imbalance, a reflection of segregated labour markets for women and men and of related patriarchal attitudes, which saw men dominate the more dangerous trades. Official figures suggest 80–90 per cent of all serious work-related disabilities and disease at the middle of the twentieth century afflicted male workers, although this ratio changed as the basic industries declined.[46] There also appears to have been some ethnic and racial segregation operating, with vulnerable immigrant and ethnically marginalised groups clustered in the very worst, most hazardous, jobs. In Scotland in the nineteenth century, dangerous jobs such as dockwork and navvying were undertaken predominantly by the Catholic Irish. Later in the twentieth century, as Bashir Maan recalled, it was the new wave of immigrant Pakistanis who were the only ones to work in the most dangerous processes in Scottish chemical factories.[47]

Oral testimony and personal reminiscence is particularly illuminating in reconstructing the attitudes and perceptions of workers in dangerous occupations like the heavy industries. Such evidence evokes the grim working conditions in mining, ironworks and shipyards, as well as the heavy toll such work exacted upon the body and the limited influence in practice that the state had in the workplace. Oral histories challenge the idea that technology was universally positive in its impact and that statutory controls were everywhere effective. What emerges is the impression that the law was frequently flouted as employers and managers put profit and production before workers' health and wellbeing. Moreover, on occasions, workers themselves colluded in this process – sometimes for personal gain, sometimes because they were not aware of the dangers to their health, and sometimes because of entrenched custom and practice and the fact that the risk seemed acceptable and the outcomes too far ahead in the future to worry about.

For example, insulation engineers, known as laggers, in the Clyde shipyards recalled working amongst a fog of asbestos fibres in the 1950s and 1960s and the blatant flouting of safety precautions by both management and men. Most of the asbestos in the yards was used for insulation purposes, and it was the laggers who were most closely involved with its use and amongst the first to recognise its toxicity. The nature of this work, though, normally required that they worked alongside other tradespeople. One lagger remembered how his work put other tradesmen in danger: 'We used tae insulate the boilers actually on the boat, and the place was covered in asbestos when we were dain that … the dust just a' floated. It floated round and everybody got their share.'[48] Also, a retired ships' plumber, who is now suffering from pleural thickening, could remember how he had got his daily exposure to asbestos: 'I was working in amongst it. Engine rooms, boiler rooms … it used to come down like snow.'[49]

There was a widespread lack of understanding amongst the workforce regarding the extreme health risks inherent in asbestos from the 1930s through to the 1970s. It was a common sight in the shipyards to see young workers playing with asbestos cuttings, and an ex-shipyard labourer recalled how, in the 1960s, 'they were throwing this 'monkey dung' about and that, and hitting folk in the passing just for a game you know. Nobody knew how dangerous it was. These blokes were laggered in it head-tae-foot.'[50] A 74-year-old fitter – whose asbestos-damaged lungs only became apparent in 1997 – described to us the sheer amount of asbestos dust in the 1950s in this way: 'I've seen times when you couldnae see the other side of the boat. That's only what? 40 or 50 feet away. You couldnae see it for the dust.'[51] The wearing of masks was eventually made compulsory when asbestos was present. However, many employers continued to issue their workers with unsuitable masks, many of which were uncomfortable to wear for any length of time – usually because the straps cut into the back of the worker's head or they restricted breathing excessively. Other masks were ineffective, as one lagger recalled: 'They gave us wee Martindale masks. It was like a wee paper thing you put over your mouth. It's no worth a monkey's.'[52] A heating engineer recalled that, just because masks were provided, it did not mean that they were always available on site: 'If you waited for the mask coming you would never get done ... But eh, we accepted it ... It was a general trend in the building trade that you just carried on with the job, you know.'[53]

Under the 1974 Health and Safety at Work Act, insulation firms had a duty to ensure the safety of their workers and those working near them. This act also brought in the principle that responsibility for health and safety was shared equally between worker and employer. However, once again, having the regulations in force and adhering to them were two different things. For example, a heating engineer could remember slipshod procedures in his firm well into the late 1980s:

> If we went down tae strip a boiler we just took it [the asbestos] off with a hammer and chisel, you know. There was nae masks or anything at that time, you know. If you came out for a breather they were asking you what you were dain sitting outside, you know. You were spitting up black for maybe a week, you know, when you came out.[54]

Similarly, coal miners recalled a postwar work regime within the newly nationalised mining industry where production was routinely prioritised over health and safety, where technology alleviated some problems but caused others, and where statutory provisions were widely ignored or subverted. Witness, for example, Bob Smith's sensitive evaluation in his autobiography of the mixed benefits of mechanisation in the pits:

> It seemed that all improvements in coal getting brought new problems. The new machines which cut and automatically loaded coal in a continuous cycle were extremely noisy and produced masses of dust. Ventilation was never good, and the air always hot and humid. In those conditions men sweated heavily, and the dust settled on them thick and black. Only their eyes and teeth stayed white. The air bags which were supposed to control the ventilation and bring fresh air in

> to us were often torn by debris or falling coal, and did not really do their job well. We were supplied with masks but at first they were crude and inefficient. Later models were better, but without doubt they did hinder a man's breathing, and a lot of men found they could not wear the masks when doing heavy work. And in spite of all the mechanisation it was still, as always, very heavy work.[55]

The miners' leader Abe Moffat emphasised in his autobiography that miners were normally more concerned about the possibility of injury than they were about the long-term risks (for example associated with dust inhalation), and he commented on how families could be afflicted by the bodily damage caused in a dangerous and dusty work environment. His brother died aged fifty, 'a physical wreck', with pneumoconiosis.[56]

Oral testimonies flesh out these narratives.[57] Miners' bodies were distorted by working in thin seams. One miner who normally worked in the Lanarkshire coalfield in Scotland remembered working for a while in the Fife pits, where he was struck by the difference in height of the coal seams:

> We went up to Fife. Fucking 8 feet fucking high man. You thought 'What the fucking hell here.' … The Fifers cried us 'the Jimmies' 'cause we came from Glasgow, well down this side you know. We wore knee pads. They didn'ae. They stood and fucking shovelled man because it was that high you know. But we … We couldn'ae shovel standing up because the old back was knackered, you know.[58]

The National Coal Board initiated a range of measures in a proactive campaign from 1947 to tackle what Zweig described as 'enemy number one' in the pits: dust. Some miners tried to protect themselves, chewing tobacco and using silk stockings over their mouths and noses as makeshift masks. At the same time, deaths through pneumoconiosis in Scotland – caused by the inhalation of coal dust – peaked in the 1950s and 1960s. However, miners recalled that preventative methods were very limited in practice, with much subversion of the statutory controls at the coalface as output rather than safety remained the priority. One Ayrshire miner observed that it was not always possible to work safely, ''cause … the management was on top of me for production. Production, production, production.'[59] The water-suppression techniques designed to keep dust down in the pits also didn't always work correctly. Referring to the water jets on coal-cutting machinery, one miner noted: 'These things were a' clogged up … I couldnae see you fae here tae there.'[60] Some also remembered that many miners were opposed to dust suppression, including water infusion into the seam, primarily because it made for a more difficult working environment and reduced their productivity. Hence some miners turned the water off.[61] Others could remember water hoses attached to the machines being too short, which meant many faces had to be cut dry, and some recalled that dust-measuring equipment was tampered with to show favourable readings. Such evidence challenges the idea that NCB and state initiatives were effective, suggesting a significant gulf between what was laid down by law and actual workplace practice.[62]

However, from the late 1950s Scotland's industrial structure was changing. The old heavy industries of the central belt such as coal mining and heavy engineering atrophied and efforts to attract newer industries, such as microelectronics, began in earnest. The painful transition, though, was made even worse by the legacy of older work processes which continued to impact on the working/retired population into the 1980s and beyond, as Table 18.2 makes clear.

Table 18.2 Trends since 1980 in new cases of occupational disease which justify disablement benefit

Decreasing	*Increasing*
Pneumoconiosis	Mesothelioma and asbestosis
Dermatitis	Occupational asthma
Tenosynovitis and beat conditions	Occupational deafness
TB and Hepatitis	Hand/Arm Vibration Syndrome
Leptospirosis	–

Source: Snashall, D. *ABC of Work-Related Disorders*, London, 1999.

One of the most promising of the newer industries to the economic life of Scotland was North Sea oil. Unfortunately, though, the productionist ethos of this industry was not conducive to 'best practice' occupational health and safety considerations. The discovery of North Sea oil meant that for the first time in Scotland's history the main thrust of industrial activity became located along the north-east coast. Montrose, Dundee, Aberdeen and Peterhead all became, to varying degrees, vital service centres for the expanding oil industry. Oil rig construction sites also appeared at Methil in Fife, at Ardersier and Nigg near Inverness, as well as at Kishorn in the north-west Highlands, while over 200 drilling and production platforms were put into operation in the Scottish sector of the North Sea. By the end of the 1980s, there were around 3,000 people employed directly in the North Sea, and an estimated 100,000 more in industries which serviced the industry or depended on the North Sea oil market. However, as Woolfson, Foster and Beck have illustrated, what made this 'new' industry unique was the fact that in many cases the production regime was transplanted almost intact from the USA. This included most of the technology, many of the management structures and, crucially, a dominant 'gung ho' attitude towards health and safety on many of the platforms.[63] It was also the case that it was the Department of Energy, and not the Health and Safety Executive, which was responsible for safety on most of the oil platforms. These factors, it has been argued, combined with the emphasis placed on extracting the oil as quickly as possible, led to a neglect of health and safety offshore in many installations, culminating in the Piper Alpha disaster of 1983.

Electronics companies also began to appear in Scotland from the 1960s as the government began to offer financial start-up packages and custom-built sites in the new towns which were easily accessible by the motorway network. It was in this way that the Scottish electronics industry – or 'Silicon Glen' as it came to be known – became established. Electronics was planned to be the driving force of the Scottish economy, and by the 1990s there were around 50,000 people employed in around 400 companies. However, this industry, which quickly became a big employer of female labour, brought its own health risks, including the exposure to toxic chemicals. In a way similar to some asbestos-related diseases, the health effects of exposure to such chemicals could take a long time to show up. In addition, many of the chemicals used in some of the processes were carcinogenic. Amongst these were chromic acid, trichloroethylene, carbon tetrachloride, and other chemicals such as arsenic and zinc oxide, which directly affected the main organs. On top of this were new risks from exposure to gases and vapours, and the even more threatening danger of exposure to ionising radiation. Finally, performing detailed repetitive work in electronics under strict time pressure resulted in a high incidence of stress-related complaints, which is discussed in greater detail below (see page 361).[64]

One of the most important microelectronics plants in Scotland was the IBM plant at Greenock, originally built in the late 1950s to produce typewriters, keyboards and punch-card readers. The plant was taken over by National Semiconductors in the late 1960s, and in the 1970s special chemicals crucial to the production of silicon chips were introduced to the work process. A proportion of the workforce in the fabrication room was constantly exposed to these chemicals, and before long many began to suffer health problems, particularly severe migraines, nausea and vomiting. However, even more disturbing was the high number of women who began to suffer miscarriages. The TV programme *Frontline Scotland* broadcast an exposé on the problem and spoke to women on both sides of the Atlantic who feared that they may have lost their babies because of their work. One American woman who used to work for IBM miscarried six babies in total and was not told at the time that the glycol ethers which were used in the industry were suspected of being one of the causes of her miscarriages. By the end of the 1970s, suspicions were being raised among the workforce in Greenock that the chemicals were in some way responsible for the apparently high rate of miscarriage and other illnesses in the Inverclyde area. Tests conducted on laboratory animals in the USA in the early 1980s confirmed that contact with ethylene-glycol ethers could lead to serious reproductive problems, and by 1988 the industry and the Health and Safety Executive was forced to respond to increasing concerns and initiate a major investigation into the problem.[65]

It is clear that new industries brought with them unforeseen hazards, and the impact upon health of the high-stress, high-labour-turnover call centres which have become quickly established throughout Scotland (capitalising on the sales effect of the Scottish accent) are now being assessed. Concomitant with Scotland's changing industrial landscape, old illnesses were gradually being replaced by 'new' illnesses, such as repetitive strain injury (RSI), sick building syndrome, and the inchoate complaint of stress. After musculoskeletal disorders, stress-related complaints formed the second most commonly reported group of work-related ill-health conditions in the UK at the end of the twentieth century, with the rate of reported stress at work doubling between 1990 and 1999.[66] In 1998 a major study in the UK called the Stress and Health at Work Study (SHAW) found that one in five of the UK working population believed their jobs were extremely or very stressful, to the point of making them ill.[67] Two processes were at work here. For one thing, the increasing preponderance of white-collar-related occupational diseases reflected the country's changing industrial structure. On the other hand, the slow recognition of these new industrial diseases demonstrated the historical reluctance of employers and successive governments to acknowledge that certain ailments were work-related and therefore potentially liable for compensation. This has been the pattern with, for example, lead poisoning, asbestos-related diseases, coalworkers' pneumoconiosis and, more recently (in fact, not until 1998), with emphysema/bronchitis caused through working in mines. The case of RSI, suffered by many typists and keyboard operators throughout Scotland, illustrates the contested nature of such work-related disorders. Although not known as RSI, the condition was acknowledged by Ramazzini in 1700: 'The incessant driving of the pen over paper causes fatigue of the hand and the whole arm because of the continuous and almost chronic strain of the muscles and tendons, which in the course of time, results in failure of power in the right hand.'[68]

Despite these early observations, throughout the nineteenth century most medics favoured a psychological rather than physiological explanation to what many began calling 'occupational neuroses'. In the early years of the twentieth century a New York doctor, Austin Flint, wrote about writers' cramp, and a further 100 cases were studied and recorded by another American doctor in 1920, including writers, Morse code telegraphers, stenographers, typists, tailors and musicians. Therefore, although RSI is supposed to be a new condition, it really is far from new. This condition was probably tolerated by the Wick gutters mentioned earlier, although it was not until the 1990s that it was finally accepted as an occupational disorder.

However, to maintain a proper sense of perspective, despite the rise of newer industrial ailments, certain factors have made the Scottish workplace safer and healthier over the course of the twentieth century. Certainly, the growth of a mass labour movement provided a much more extensive and effective protective matrix for workers, both at the workplace and the policy-making level. Moreover, whilst a commitment to preventative strategies varied between trades unions and a compensation culture appeared to prevail before World War II, the trades unions did become more health-conscious over time.[69] The state also increasingly stepped in to regulate and control industrial hazards, significantly extending the under-resourced Victorian regulatory framework in the process. The two world wars provided important watersheds, drawing the state further into the workplace and regulation of the labour contract during periods of high demand for labour. Amongst other key developments were: the growth of occupational medicine; the incremental recognition of the pneumoconioses – probably the biggest occupation-related killer in Scotland before the 1970s; the sharp reduction in working hours; the virtual elimination of child labour; the introduction of retirement; the extension of financial compensation for work injuries and disease (notably the 1948 Industrial Injuries Act); the fundamental transformation in occupational health provision with the introduction of the Health and Safety at Work Act in 1974; and legislation in 1977 which allowed workers in trade-unionised workplaces to appoint safety representatives and joint employer–employee safety committees. On the other hand, a key policy weakness in the UK was the failure to integrate the occupational health services within the NHS. This was an important lost opportunity to prioritise workplace health in the UK.[70] Finally, it is also lamentable that despite the long history of ill-health resulting directly from employment, under the devolution settlement responsibility for workplace health and safety remained the responsibility of Westminster, representing another lost opportunity to prioritise workplace health in Scotland.

CONCLUSION

Work clearly affected health in diverse ways, and at the centre of this relationship was a persistent tendency within Scottish industry to place production and profit above personal health and wellbeing. Bodies were damaged in the process. The result was the destruction of individuals, families and even communities reliant upon hazardous work. This is most evident, perhaps, in the trauma of Scottish coal mining or fishing communities when a disaster occurred, or in the dynasties of father–son miners and insulation engineers in shipyards and construction who were killed or severely disabled from inhaling dust at work. For much of the period before World War II, moreover, workers and surviving dependents were left with little support to cope with such sudden or encroaching disability, the loss of the main breadwinner and the physical impairment and mutations in lifestyle this entailed.

What is also evident is that Scottish workers were occasionally complicit in this rather than simply being passive victims of exploitative capitalism. Many male workers accepted and embraced danger, partly as an expression of their masculinity, whilst also opposing and campaigning against hazardous working conditions. Oral evidence from Scottish workers convincingly demonstrates that a wide gap existed between statutory health and safety provision and actual workplace practice where customary practices and a high-risk culture proved persistent and difficult to eradicate. To varying degrees in different industries, the trades unions also colluded in this failure to recognise adequately the importance of health, instead being preoccupied with the fetishism of the wage packet and the protection of jobs. Indeed, the negotiation of 'danger', 'dirt' and 'dust' money by some trades unions provide examples of the institutionalised legitimisation of risky work practices. However, employers and managers frequently did little to change things, and in their position of power and authority they must bear the brunt of culpability for the heavy toll of bodily damage caused by work. From the late eighteenth-century textile millowners to the recently established multinational corporations such as IBM, profit and production were almost always prioritised over health and wellbeing. Whilst nationalisation did lead to a marked improvement in occupational health standards in mining, steel and the railways, negligent and irresponsible practices which damaged workers' health continued in the public sector into the second half of the twentieth century, as the 1998 bronchitis and emphysema in coal mining litigation indicates.[71]

As the form that industry took in Scotland altered over time, workers' health was directly affected by the changing economic circumstances which that alteration represented. The initial rapid pace of industrialisation was an important cause of work-related ill-health in Scotland, and the same could be said of the mirror image of rapid deindustrialisation from the 1960s onwards, in which the decline of the heavy industries meant occupational health and safety usually came a poor second to maintaining productivity in diminishing markets. New industries brought new industrial diseases, illustrating perhaps that many forms of work could be life threatening if not properly regulated. Testimony to this is the fact that many in Scotland are now campaigning for legislation which would allow companies to be charged with a new crime of corporate killing. This new offence would mean there would be no need to locate and charge the 'controlling mind' of a company, but that the company itself would be held responsible, with the possibility of its directors being sent to prison. Such legislation, then, if it is ever enacted, might save Scotland from a future Piper Alpha or from an industrial health disaster on the scale of the asbestos tragedy. Clearly, though, as far as occupational health in Scotland is concerned, the valuable lessons which the past can teach us regarding the potential health risks of work – many of them clearly articulated through oral history testimony – need to be carefully learned for the sake of workers' health and safety in the future.

NOTES

1. Scottish Occupational Health Oral History Project (SOHOHP), Interview A19 (Scottish Oral History Centre, History Department, University of Strathclyde).
2. The most comprehensive survey of work in Scotland is Knox, 1999; see also Whatley, 2000; Gordon, 1991.
3. See Kirkwood, 1935, 151–5; Edwin Muir, cited in Berry and Whyte, 1987, 210–12; Knox, 1999, 190–5.
4. See Thomas, 1999, 308–49. Also Smout, 1969; Fenton, 1976; Devine, 1984.
5. Cited in Smout, 1987, 83.
6. Cited in Pike, 1974, 247.
7. Weld, C. *Two Months in the Highlands, Orcadia and Skye*, 1860, cited in Thomas, 1999, 348.
8. Pike, 1974, 186.

9. Agricola, *De Re Metallica*, 1556, translated by Hoover, H C and Hoover, L, New York, 1950, 214.
10. Fraser, 1988, 28–9.
11. Fraser, 1988, 30.
12. Wohl, 1983, 257–8.
13. Pike, 1974, 38–9.
14. Pike 1974, 64.
15. Pike, 1966, 67.
16. Pike, 1966, 142.
17. Pike, 1966, 171.
18. Pike, 1966, 17.
19. Pike, 1966, 257.
20. McIvor, 2001, 120.
21. Benson, 1980, 43.
22. Oliver, 1902; Oliver, 1908.
23. See Aitchison, 2001.
24. Kenefick, 2000, 140–66.
25. Departmental Committee on Accidents in Places Under the Factory and Workshop Acts, Minutes of Evidence, Cmd 5540, 1911, 464. This remained an important factor in the industry later in the twentieth century. See Johnston and McIvor, 1999, 74–92.
26. Miss Vines, one of the female factory inspectors in Scotland, in Annual Factory Inspectors Report for 1913, 97.
27. Ballantyne 2001; also 2004.
28. Gregory, 1831.
29. Zweig, 1948.
30. See Bartrip, 2002; also Bartrip and Burman, 1983.
31. For more, see McIvor, 1989, 47–67; Wohl, 1983, 257–84; McIvor, 2001, 111–47.
32. See Walker, 2004.
33. Departmental Committee on Compensation for Industrial Diseases, Minutes of Evidence, 1907, Cmd 3496, 138.
34. *Forward*, 9 December 1911.
35. Davies, 1993, 27–43.
36. Harrison, 1996; Bartrip, 2002.
37. The Socialist Medical Association recognised this and campaigned in the 1950s to shake workers out of a fatalistic attitude towards dust at work and the cough. See Socialist Medical Association, 'Dust in the Lungs', pamphlet, n.d., *c*1954; 'Stop that Cough' (*c*1953); 'Challenge of the Rhondda Fach Survey' (*c*1952–3).
38. Glasser, 1987, 5.
39. Cited in Fraser, 1969, 56–7.
40. SOHOHP, Interview B6.
41. SOHOHP, Interview B1.
42. For more on masculinity in Scotland see Johnston and McIvor, 2004.
43. See the annual reports of the chief inspector of factories, 1950–75.
44. *Scott Lithgow House Magazine* (Summer 1977), cited in Bellamy, 2001, 74.
45. See Johnston and McIvor, 2000a, 20–8.
46. In 1950, of 777 fatal work-related accidents in the UK, 769 involved men; whilst men accounted for 86 per cent of the 179,000 non-fatal accidents. See the Annual Factory Inspectors Report for 1950, 47–50, 63.

47. Interview transcript: Neil Rafeek interviewing Bashir Maan, 12 May 2003, Scottish Oral History Centre Archive, University of Strathclyde.
48. SOHOHP, Interview A13.
49. SOHOHP, Interview A2 (see also A14).
50. SOHOHP, Interview A18 (see also A19).
51. SOHOHP, Interview A5.
52. SOHOHP, Interview A5.
53. SOHOHP, Interview A5.
54. SOHOHP, Interview A6.
55. Smith, n.d., 87–8.
56. Moffat, 1965, 232.
57. For more detail see McIvor and Johnston, 2002, 111–33.
58. SOHOHP, Interview C10.
59. SOHOHP, Interview C16.
60. SOHOHP, Interview C10. See also interview C4.
61. SOHOHP, Interview C2.
62. See McIvor and Johnston, 2007.
63. Woolfson et al., 1996, 56–63.
64. Geiser, 1996, 38–49.
65. Health and Safety Executive, 2001, 128. The results were somewhat inconclusive.
66. Health and Safety Executive, 2001, 130.
67. HSE, 1996.
68. Huskisson, 1992, 15.
69. For more detail see the chapter on trades unions in this volume.
70. Johnston and McIvor, 2000b.
71. British Coal Respiratory Disease Litigation, Summary Judgement of Mr Justice Turner, 23 Jan 1998, 5; 8.

BIBLIOGRAPHY

Aitchison, P. *Children of the Sea: The Story of the Eyemouth Disaster*, East Linton, 2001.

Ballantyne, N. Ironmasters and steelmen: Authority and independence in Lanarkshire's iron and steel industries, 1870–1900. Unpublished PhD thesis, Strathclyde University, 2004.

Ballantyne, N. The Lanarkshire Puddlers, 1870–1900, *Scottish Labour History*, vol. 36 (2001), 6–19.

Bartrip, P. *The Home Office and the Dangerous Trades*, Amsterdam, 2004.

Bartrip, P and Burman, S. *Wounded Soldiers of Industry*, Oxford, 1983.

Bellamy, S. *The Shipbuilders*, Edinburgh, 2001.

Benson, J. *British Coalminers in the Nineteenth Century*, London, 1980.

Bromet, E, Amanda Dew, M, Parkinson, D K, Cohen, S and Schwartz, J E. Effects of occupational stress on the physical and psychological health of women in a micro-electronics plant, *Social Science and Medicine*, vol. 34 (1992), 12, 1377–83.

Davies, J McG. Social and labour relations at Pullars of Perth, 1882–1924, *Scottish Economic and Social History*, vol. 13 (1993), 27–43.

Devine T M, ed. *Farm Servants and Labour in Lowland Scotland, 1770–1914*, Edinburgh, 1984.

Fenton, A. *Scottish Country Life*, Edinburgh, 1976.

R. Fraser, ed. *Work*, London, 1969.

Fraser, W H. *Conflict and Class: Scottish Workers 1700–1838*, Edinburgh, 1988.

Geiser, K. Health hazards in the microelectronics industry, *International Journal of Health Services*, vol. 16, no. 1 (1986), 105–20.
Glasser, R. *Growing up in the Gorbals*, London, 1987.
Gordon, E. *Women and the Labour Movement in Scotland, 1850–1914*, Oxford, 1991.
Gregory, J C. Case of anthracosis or black infiltration of the whole lung, resembling melanosis, *Edinburgh Medical and Surgical Journal*, vol. 36 (1831), 389–94.
Harrison, B. *Not Only the Dangerous Trades: Women's Work and Health in Britain, 1880–1914*, London, 1996.
Huskisson, E C. *RSI Repetitive Strain Injury: The Keyboard Diseases*, London, 1992.
HSE. *Health and Safety Statistics 1995–1996*, 1996.
HSE. *Cancer Among Current and Former Workers at National Semiconductors (UK) Ltd*, Greenock, 2001.
Johnston, R and McIvor, A. *Lethal Work: A History of the Asbestos Tragedy in Scotland*, East Linton, 2000a.
Johnston R and McIvor, A. Whatever happened to the occupational health service. In Nottingham, C, ed., *The NHS in Scotland*, Aldershot, 2000b, 79–106.
Johnston R and McIvor, A. Dangerous work, hard men and broken bodies: Masculinity in the Clydeside heavy industries, *Labour History Review*, vol. 69, no. 2 (August 2004), 135–51.
Johnston R. and McIvor, A. Incubating death: Working with asbestos in Clydeside Shipbuilding and Engineering, 1945–1990, *Scottish Labour History*, vol. 34 (1999), 74–92.
Kenefick, W. *Rebellious and Contrary: The Glasgow Dockers, 1853–1932*, East Linton, 2000.
Knox, W. *Industrial Nation: Work, Culture and Society in Scotland, 1800–Present*, Edinburgh, 1999.
McGeown, P. *Heat the Furnace Seven Times Over*, London, 1967.
McIvor, A and Johnston, R. Voices from the pits: Health and safety in Scottish coal mining since 1945, *Scottish Economic and Social History*, vol. 22, pt 2 (2002), 111–33.
McIvor, A. Work and health, 1880–1914, *Scottish Labour History Society Journal*, no. 24 (1989), 47–67.
McIvor, A. *A History of Work in Britain, 1880–1950*, Basingstoke, 2001.
McIvor, A and Johnston, R. *Miners' Lung*, Aldershot, 2007.
Moffat A. *My Life with the Miners*, London, 1965.
Oliver, T. *Dangerous Trades*, London, 1908.
Oliver, T. *Diseases of Occupations*, London, 1902.
Pike, R. *Human Documents of Adam's Smith's Time*, London, 1974.
Pike, R. *Human Documents of the Industrial Revolution in Britain*, London, 1966.
Smith, B. *Seven Steps in the Dark: A Miner's Life*, Ayrshire, n.d.
Smout, T C. *A History of the Scottish People, 1560–1830*, London, 1969.
Smout, T C. *A Century of the Scottish People, 1830–1950*, London, 1987.
Snashall, D. *ABC of Work-Related Disorders*, London, 1999.
Thomas, K. *The Oxford Book of Work*, Oxford, 1999.
Walker, D. Chrome dust at J.J. White's of Rutherglen, 1893–1920. Unpublished paper presented to the *Dust at Work Conference*, Glasgow Caledonian University, 28–9 May 2004.
Whatley, C. *Scottish Society, 1707–1830*, Manchester, 2000.
Wohl, A. *Endangered Lives*, London, 1983.
Woolfson, C, Foster, J and Beck, M. *Paying for the Piper*, London, 1996.
Zweig, F. *Men in the Pits*, London, 1948.

PART FOUR

•

Words and Imagination

19 Theatre Work in Scotland

ADRIENNE SCULLION

> I can always remember the first show I worked on – the play was *Great Expectations*. It was the 1960–61 season – and I sat in the wings. I was fascinated by it. I used to go home and tell people – I'm working in the theatre. The smell was different. When I went home – my wife used to go – you can smell that theatre aff ye – there's a different smell.[1]

INTRODUCTION

The fabled romance of the 'smell of the greasepaint' is something of a rarity in recorded experiences and memories of working in theatre in Scotland. More common are stories of rivalry between companies, tense relationships with authorities and colourful accounts of the business of performing. Such stories encompass legitimate theatre practices and reflect mainstream cultural values. They describe theatre-making with political purpose and that which demonstrates a socially engaged use of drama and performance. They record popular theatre and the multitude of genres, forms and practices associated with it. But, despite a cast list of busy practitioners, theatre in Scotland has, for much of its history, struggled towards sustainability. Faced with topographic and economic challenges, as well as religious and social opposition, the experience of making – and even of viewing – theatre has often been a marginal or even dangerous one and, at times, a radical and countercultural one too. As a consequence of this, Scotland's theatre practitioners emerge as a motley crew of quixotic attitude and distinctive ambition.

Some of Scotland's theatre-makers were bold, some opportunistic, some decidedly heroic, some just plain lucky. Reflecting this, what follows is something of a mixed bill of (allegedly) factual and (clearly) fictional impressions and accounts of working in legitimate, popular and political theatre in Scotland. The particular emphasis is on touring and peripatetic theatre, because this is a form of theatre distribution of distinctive, even of 'national', purchase within Scotland. That said, from the eighteenth century until now, the major cities have had great theatre buildings that sustained companies of longevity and significance. And these theatre companies and theatre-makers often have deep and productive roots with their local communities, both large and small. But, more than any other form of theatre production and distribution, it is the touring form that drives the experience of working in and for Scottish theatre. The touring theatre tradition – the itinerant players and cultural entrepreneurs in the eighteenth and nineteenth centuries, the stock companies, geggy troupes and star performers in the nineteenth century, and the small touring companies of the twentieth century, as well as the nascent National Theatre of Scotland, deliberately conceived with no home theatre base and, therefore, focused on touring within and without Scotland – reveals a distinctive and rich through-line of cultural praxis.

TOURING AND COMMUNITY

In late April 1973 a company of nine actors, musicians and stage management from the newly formed Edinburgh-based 7:84 Theatre Company (Scotland) packed themselves and a theatre set into two Transit vans, left the capital and began a tour of the Highlands and Islands of Scotland that was to last for five weeks. John McGrath's new play *The Cheviot, the Stag and the Black, Black Oil* was performed twenty-nine times during those early summer weeks, mostly in village halls and community centres.

In a theatre system significantly shaped by peripatetic companies *The Cheviot* tour is the most recollected, the most celebrated and the most mythologised, certainly of modern times. Key members of the creative team – including McGrath, Elizabeth MacLennan and John Bett – have written and talked about the experience of creating the piece, and of life on the B-roads of 1970s Scotland touring it.[2]

In form, content, the collaborative process of creation and its polemical message *The Cheviot* was a socialist text, offering a politically committed analysis of social inequality and demanding audience participation and action through both message and form. However, for all its emphasis on the collective, McGrath identified himself as the key author.

> Obviously I, as a writer, had a very clear idea of exactly how I wanted the show to be ... But I also wanted everybody in the company to be intimately involved in the actual process of creating it. I had always fought shy of group-writing before, and still do. This wasn't to be a free-for-all, utopian fantasy ... The company didn't expect to write the play. My contribution was my experience as a writer and director, and I was to be used. But there were two things we could do to break down the insane hierarchies of theatre. Firstly, we could all respect each other's skills and at the same time lay them open for collective discussion and advice. Secondly, we could work as equal human beings, no skill being elevated over another, no personal power or superiority being assumed because of the nature of the individual contribution: no stars, of any kind. And no recourse to the 'I'm an artist' pose to camouflage either power-seeking or avoidance of responsibility to the collective.
>
> So we all sat down, with blank note-pads. I outlined the sixteen main areas or blocks of the play, and how I thought we should approach each one. There was a huge pile of books, cuttings Elizabeth

> [MacLennan] had kept, and other material on the next table. Everyone was given one or two areas to be personally responsible for, check what we said, and answer to in public discussion.
> ...
> After two weeks huddling round a table, occasionally leaping over to the musical instruments to try something, or moving into the open space to see how a scene might move, we had something to begin to work with. Not the purist's improvised theatre. Not 'collective creativity', or group therapy. But a written text that all the company were part of, and deeply involved in, and excited about. Nobody had anything to do that they thought was wrong, everybody knew exactly why every word was there.[3]

In the Scotland of 1973 this was a novel mode of theatre-working that emphasised the collaborative nature of theatre in both its creation and delivery, even extending to the operational aspects of life on the road, which were also a team effort.

> Everybody worked on the get-ins, which became faster and easier the more we did. Then a quick tech, for lighting and sound levels, then a short company talk about changes, new verses, new jokes, etc, for that night or that place. We rehearsed the changes, checked props and costumes, and if we were lucky found the digs or some fish and chips. The show went up at eight. Some liked to be in at seven, to get organised. Others spent the time in the pub, getting to know the audience, turning up alarmingly near time to start.
> ...
> After the end of the show, everybody struck their own costumes and props, I did my roadie bit with the band while the stage was dismantled and the chairs shoved away and the floor swept. In twenty minutes the dance was under way ... We tried to finish at 1.30, but didn't always succeed. During the dances, those not playing in the band packed and wrapped the lights, costumes, props, stag, etc, quietly, and at the end, we wrapped the band gear, and everything was ready for loading the next morning ... By ten [the next morning], everybody was back at the hall to load, and wearily did so ... It's a great way to shatter your constitution.[4]

Speaking in 2002, the actor John Bett agreed that touring in general – and *The Cheviot* tour in particular – was tough but rewarding.

> You need to be young and fit really ... It was very tough and because it had never been done before, the vagaries of booking these halls were such that we tended to make the most incredible kinds of leaps, going from Thurso to Oban to Islay to Fraserburgh – that sort of crazy geography. We were dotting back and forth. And with the dancing, of course, the show went on till one o'clock in the morning. So then you had to pack up the van, travel back to your digs and

> then you were off somewhere else the next day, often quite early. And then on arrival at the new venue, there would be the setting up with the rostra and lights and everything else. And of course a lot of these halls weren't really equipped for this kind of thing. They'd never had theatre ... But the halls themselves, I'd never seen anything like it. I mean the whole village turned up, kids and all. These people in the Highlands were hearing their own history and some of it in their own language.[5]

Although the ideological and political aspects of the processes of making and delivering *The Cheviot* were specific to it and its context of 1970s' Scotland, many aspects of the work of touring theatre remain familiar today. For example, in May 2003 – almost exactly thirty years to the day after the start of *The Cheviot* tour – a company of seven actors and stage management from the Traverse Theatre in Edinburgh similarly left the capital, with themselves and a theatre set packed into vans, and began an eight-week tour. *Outlying Islands*, a play by David Greig, was performed twenty-five times during those summer weeks. Just as happened in 1973, these performances were mostly one-off performances in village halls. Significantly, the working model that was established by 7:84 during its first two Highland tours in 1973 is still discernible in Scottish theatre provision today: a company arrives in a van, puts up the set in a village hall, performs the show, takes down the set, perhaps joins the audience in the pub, stays in bed and breakfast accommodation locally, and then gets up the next day to do the same thing in another village. Reflecting on the experience of the 2003 *Outlying Islands* tour, one young actor summed up the rough and the smooth of presenting theatre in Scotland:

> It's hard. No, it's not hard, that's rubbish. Two hours work a night, and you're staying in beautiful places, being driven around and it's all paid for – it's the easiest job in the world. But it's hard living out of a bag, and the travelling makes me tired.[6]

And that was without the get-ins, music-making and strikes that John Bett and his colleagues had to do as members of *The Cheviot* company.

For both emergent talent and for more established companies, a variation on this touring model is the dominant model of theatre provision in Scotland. Indeed the form is so distinctive of theatre-making and distribution in Scotland that it is a variation on the touring model – no permanent home, but a network of partners, including receiving houses and other venues – that was the model adopted for the new National Theatre in Scotland (NTS) when it was launched in 2006.

Although often the domain of smaller-scale companies and productions, the NTS and other semi-regular visitors such as the Royal Shakespeare Company (RSC) can attempt a larger canvas. The 'eventfulness' of a large-scale company coming to town puts the experience of the 7:84 or Traverse operations into the shade, but the companies' shared experiences have a commonality distinctive of collaborative creation and cultural engagement, a classic manifestation of the performance event creating a new, albeit transient, community of cultural engagement and creativity. For example, one aspect of the RSC's recent tours has been to set up a bespoke temporary theatre in either a sports arena or in a large-scale tent. It was the latter arrangement that provided the home for its production of *The Merchant of Venice* in Forres in 2002; although as Nick Fearne, the arts officer in Moray, described it, the scale of the operation seems very distant from 7:84 and its two Transit vans: 'There was the whole thing of five articulated trucks and building the auditorium and turning [it] into the space.' However, Fearne proposes continuity between the two delivery mechanisms. 'The principles were the same,' he argues, implying that the impact on the community of this event – and the experience for this company – only repeats the mythology of community proposed by McGrath:

> When you think about it 33 people staying a week in local accommodation ... They are ... eating, drinking, hiring cars. Some of them stayed in the pub in Forres [and it] became, like, the 'RSC club'. After the first night, on the Tuesday – usually two men and a dog in there on a Tuesday – it was packed ... In Forres they felt part of the community. When [they were] getting their messages [shopping], check-out ladies are asking 'How are you getting on?' There was a karaoke night with them. It was just great.[7]

Working in touring theatre, then, is most valued when it makes a connection with the local community.

The Cheviot tour rediscovered touring as the quintessence of theatre production and distribution in Scotland. Although touring was a delivery mechanism with a heritage much further back than the 1970s, it can be legitimately argued that the production of *The Cheviot* encouraged fundamental shifts in the attitudes of modern funding bodies and development agencies towards investment in touring theatre, specifically to geographically and economically marginalised areas.

ITINERANT THEATRE

While theatre can now been utilised as part of a regional development plan – and that has certainly been the case in the Highlands and Islands of Scotland – earlier practitioners had more troubled relations with the establishment and its agencies, at even the most local of levels. At the end of the eighteenth century there was a well-established network of fit-up stages in Scotland's smaller towns, and a string of touring companies wandered through the countryside playing versions of Shakespeare as well as various 'Scottish plays', providing entertainments in small temporary booth theatres, at markets and fairs and other impromptu sites. Although reflecting on the impact on audience rather than the professionals themselves, in his fictionalised account of life in small-town Ayrshire, *The Annals of the Parish* (1821),

John Galt's Reverend Balwhidder gives a vivid description of the visit of one such company to the parish of Dalmailing:

> Another thing happened in this year [1795], too remarkable for me to neglect to put on record, as it strangely and strikingly marked the rapid revolutions that were going on. In the month of August, at the time of the fair, a gang of play-actors came, and hired Thomas Thacklan's barn for their enactments. They were the first of that clanjamfrey who had ever been in the parish, and there was a wonderful excitement caused by the rumours concerning them. Their first performance was Douglas Tragedy, and the Gentle Shepherd; and the general opinion was, that the lad who played Norval in the play, and Patie in the farce, was an English lord's son, who had run away from his parents, rather than marry an old cracket lady, with a great portion. But, whatever truth there might be in this notion, certain it is, the whole pack was in a state of perfect beggary; and yet, for all that, they not only in their parts, as I was told, laughed most heartily, but made others do the same; for I was constrained to let my daughter go to see them, with some of her acquaintance, and she gave me such an account of what they did, that I thought I would have liked to have gotten a keek at them myself. At the same time, I must own this was a sinful curiosity, and I stifled it to the best of my ability. Among other plays that they did, was one called Macbeth and the Witches ... But it was no more like the true play of Shakespeare the poet, according to their account, than a duddy betherel, set up to fright the sparrows from the pease, is like a living gentleman.[8]

Such events were increasingly common and were reflected by the interest some local communities had in staging their own theatricals for pleasure and improvement. *The Gentle Shepherd* was performed annually by Pentland villagers well into the nineteenth century, and the tradition of schools and colleges staging drama continued and spread out of the academies into the professional and upper classes in the Lowlands.

A vagabond player rather similar to that described by the Reverend Balwhidder is described in 'The Spouter', by the Paisley weaver–poet Alexander Wilson, who provides a more ribald account of touring players than that given by Galt's douce minister.[9] Wilson's long poem was published in 1876 and in it he gives a distinctive account of the touring entertainments which visited the towns and villages of Scotland. Again the perspective is that of the audience, but it still captures the precarious nature of working in the lower echelons of Scottish theatre.

Wilson tells of the visit made to a small village by a strolling player, one Mr Main, allegedly 'come direct frae Drury Lane,/Where baith their Majesties, the king an' queen,/Had aft wi' his performance pleaséd been' (74–6), and his junior colleague, a 'raggy laddie, ca'd Adolphus Sprat' (164). The pair takes up residence in a local barn and advertise a performance of 'Wondrous novelty' (71). Wilson describes an excited, boisterous and very raucous audience who banter and gossip amongst themselves and are not averse to hectoring Adolphus and passing comment on the subject and the content of Mr Main's stories, debating their veracity and degrees of implausibility, and offering suggestions and potential improvements: 'Stop that damn'd fiddle!' (169); 'Saves! that's an awfu' bluidy tale' (260); 'Think ye that's true?'' (363); 'Encore! encore! ... Come, gie's that sang again!' (512, 515); and 'Wha threw that turnip! curse yer blood!' (554).

However, the next day, while the populace is still keen to talk over the merits of the players' tales, Mr Main and his apprentice have disappeared, and in true vagabond tradition

> had forgot to pay
> The debt they had contracted yesterday.
> An' Willie Watson swore like any Turk
> That it had been a thievish piece o' wark;
> An' if he could the Spouter get, that he
> The inside o' a jail wad let him see.
> Although poor Willie said to us, — 'I trow,
> To sic a rascal 'twad be nothing new'
> ...
> An' then poor Will began an' swore again,
> What he wad do when he got Mr Main;
> When some auld wives said, 'Man, ye should think shame,
> For ye hae nae ane but yersel to blame,
> For they wha mak' an' meddle wi' sic a crew,
> Aye meet with something they hae cause tae rue.'
> An' Willie clawed his head an' said, 'Atweel,
> They wad need a lang spoon wha sup kail wi' the deil.'
>
> (785–93, 801–8)

Although fictional, the kind of entertainment Wilson describes and the kind of work done by Mr Main and Adolphus, along with travelling rope dancers and acrobats, was typical of the experience of theatre that was most familiar to most people in Scotland. It was an experience of an intimate and immediate performance, extravagantly interactive and excessively ribald. It is a tradition that leaves little by way of evidence, but filters through to be part of the nineteenth-century tradition of penny geggies and music hall, where the working life of the theatre-maker could be quite as precarious as that experienced by Wilson's 'spouter'.

CITY THEATRE

Away from the barns and fit-up stages, eighteenth-century theatre was driven by star actors: West Digges and Sarah Ward were amongst the popular actors active in Scotland in mid-century, as was the celebrity actor George Anne Bellamy, whose career took her to engagements across the UK, including spells in Edinburgh and in Glasgow. According to her less than reliable autobiography, Bellamy made her Edinburgh debut in May 1762 and, at the conclusion of her month-long contract there, she accepted an engagement for a summer season in Glasgow. A group of rich Glasgow gentlemen had decided to build and equip a simple theatre at Alston Street in Grahamston which, at this time, was just outside the city boundary, for the specific purpose of engaging the beautiful young star.[10] Bellamy recorded the season in her autobiography.

> Upon my first engagement in Edinburgh, the gentlemen of Glasgow offered to build a theatre by subscription, if our company would promise to perform there in the summer. To this we readily consented, as the inhabitants were not only opulent, but liberal to a degree. The theatre being now ready, we formed very agreeable ideas of the jaunt; and that, not only from the views of profit that it presented, but from the favourable ideas we entertained of the place and people … The next day at noon, we saw the delightful city to which we were going, at a little distance before us. The magnificence of the buildings, and the beauty of the river, which the fineness of the day caused to appear, if possible, to greater advantage, elated my heart … When we arrived at Glasgow one of the performers exclaimed, 'Madam, you are ruined, for you have nothing left but what you have in the chaises.'[11]

The newly constructed theatre that contained her wardrobe and properties ('which had cost many, many hundreds of pounds')[12] was partially destroyed by a fire seemingly set by religious fanatics. Nevertheless, Bellamy was determined to fulfil her contract, satisfy her devoted fans and recover something of her financial loss. She performed, therefore, on a hastily repaired and temporary stage, with costumes and properties lent by Glasgow's fashionable society. Bellamy continues:

> From not being mistress of one gown, I found myself in possession of above forty; and some of these almost new, as well as very rich. Nor did the ladies confine themselves to outward garments only. I received presents of all kinds … together with invitations and parties for the whole time of my residence in their neighbourhood.[13]

As well as suggesting the multiple risks involved in working in the eighteenth-century Scottish theatre, the story captures something of the extremes of opinion generated by actors and acting in Scotland in that period.

Whilst the currency of leading and celebrity actors continued to grow, the force behind theatre-making in Scotland through the eighteenth century and into the nineteenth was a powerful subset of the actors, the actor–managers, some of whom led touring companies, while others led resident companies and programmed touring companies into their theatre buildings. A pillar of the British theatre system, even into the twentieth century the actor–managers were creative entrepreneurs who by force of personality, as much as market demand, forged a regular theatre culture and even laid the foundations for British repertory. One of the most successful in nineteenth-century Scotland was J H Alexander, who ran a company that was more or less resident in Glasgow for many years, building a loyal audience and providing stable employment for actors, including the popular comedian Mr Lloyd and, indeed, Alexander himself.

Alexander was a colourful and popular figure who understood the role of manager as being much more than programming and accounting; he led his company from the stage both as an actor and as a favourite personality. Audiences delighted in his eccentric behaviour, which included him arguing with the audiences, reprimanding his actors whilst on stage, rearranging scenery, counting his audience from the stage in the midst of a performance to ensure that his doorkeepers' returns were accurate, and performing 'Alexander's jig' – an impromptu dance – when the gallery boys called out for it. Colourful though this may be, something of the cut-throat nature of theatre management is captured in a notorious fight over the Dunlop Street Caledonian Theatre, the venue that had been the city's first legitimate house before the Letters Patent transferred to the new Theatre Royal in Queen Street in 1805.[14] Alexander had lost out on the lease of the Caledonian to rival manager Frank Seymour. However, in retaliation, Alexander leased the basement and cellars of the theatre and began a dramatic war of attrition. The story goes that while a sombre tragedy or reflective romance was being performed in one hall, in the other a loud band or dance event would be scheduled. If Alexander's 'Dominion of Fancy', as he named his basement theatre, was producing a play with special pyrotechnic effects, then the company above was quite prepared to pour water down through the floorboards. The antics of the rival managers attracted huge audiences – to the cost of the new Theatre Royal – and, eventually, the magistrates were called upon to calm the situation. It was ruled that the Caledonian and the Dominion should play only on alternate nights. Alexander was able to sustain his business more effectively, however, and when fire destroyed the new Queen Street Theatre in 1829, Alexander bought the patent from Seymour – who had become lessee there – and transferred it back to Dunlop Street, where it remained the city's legitimate theatre until 1869. However, Alexander's attitude towards the work of the theatre gradually and then catastrophically changed from the energetic eccentricity of earlier days to something much more sombre. The quality of his operation fell badly as he grew ever meaner in the fees and wages paid. He sought to fill gaps by continuing to perform himself, sometimes in wholly inappropriate roles, including attempting the youthful Romeo when he was quite old. Finally, though, his interest in the stage waned when, in 1849, a false fire alarm in the theatre led to panic, some seventy people being trampled to death.[15]

GEGGY THEATRES

Much of the theatre system of the stock companies and their actor–manager leaders was abandoned by the mid-nineteenth century, when industrialisation and improvements in transport links via the railways usurped locally based production and legitimate theatre became something that arrived by train one week, only to depart for another receiving house the next. Native talent was marginalised into the booth theatres of the fairs, or ended up in the emergent music halls.

One distinctive aspect of the Scottish theatre scene linked to this loss of an independent indigenous theatre was another touring form, the itinerant geggy theatres, which were fit-up stages of wood and canvas that toured the countryside well into the twentieth century with a repertoire of popular melodramas and reduced or even bowdlerised classics, including versions of the National Drama, the stage adaptations of the novels of Walter Scott that had been a standard part of the Scottish repertory in the early and mid-nineteenth century. An early practitioner, David Prince Miller, remembered acting in a production of *Richard III* in a fit-up theatre at Glasgow Fair and performing it twenty-five times in the space of seven hours.

One of Scotland's favourite comedians, Will Fyffe, came from a theatrical family who owned and operated a geggy theatre. Reminiscing with popular historian Jack House, Fyffe remembered that:

> he started acting in his father's geggie about the age of twelve, though he'd appeared in children's parts before that. By the time he was fifteen he was a fully fledged member of the company, taking the parts of old men, young women and any character his father decided he should play. Charlie Kemble, the great music hall veteran, said he saw Will Fyffe appearing as Polonius in a forty-minute version of Hamlet. He was sixteen at the time.[16]

He also recollected the risks of working in a canvas and wood fit-up, and of a particular incident at Perth:

> Running a geggie was rather like running a small travelling circus. As soon as the company arrived at the selected site, the material would be unloaded and the whole cast, actresses included, would proceed to erect it.
>
> …
>
> At the final performance on a site the whole company would dismantle the geggie and pack it on the truck for its journey to the next place. Will Fyffe was a small and wiry boy and just the perfect person to sclim[17] to the top of the geggie and attend the unbolting and unscrewing of the roof from its walls.
>
> On this last night in Perth there was a terrible storm … Fyffe got a leg up to the geggie roof, holding on like grim death. There came an even bigger blast [of wind and] the portable theatre disintegrated and most of it, including the boy on the roof, was swept into the river.[18]

Fyffe, along with the remains of the geggy, were swept on to an island in the middle of the river:

> 'The only trouble,' he said to me many years later when he told me this story in his Theatre Royal pantomime dressing room, 'was that,

by the way they were all pulling the bits of the geggie out of the river, it was more important than I was!'[19]

Although predicated on an ageing theatrical repertoire – and prone to the excesses of Miller's experience – the geggies were popular, successful and often did work of a high standard. The level of investment required to kit out and maintain a geggy was significant – they could seat up to 400 paying customers warmed by braziers – and no canny manager would risk that investment by hiring anything but the best talent available. And, distinctively, the actors available tended to be highly competent professionals, often the very same Scottish performers who had made a good living in the independent stock companies of earlier days – usually playing popular characters from the National Drama – but whose Scottish accents and acting style did not suit the London-orientated contemporary stage. In a brief account of popular theatre in Scotland, historian Alasdair Cameron recognised the skills and resourcefulness of the geggy actors:

> As most plays had to have a local or at the very least a Scottish setting, actors from the geggies used to visit the larger theatres and adapt for their own purposes any play which they saw and which seemed suitable. They would also specialise in a particular skill. Whereas Geordie Henderson was famed for his 'dying fall', Johnny Parry made a feature of dangling by the neck from the end of a rope in hanging scenes ... the Scottish dramatist Joe Corrie remembered a geggy performance of *East Lynne* in which the angel representing the soul of Little Willie turned round on its upward path to heaven revealing an advertisement for washing powder.[20]

THE 'SCOT' ON THE ROAD

Whilst the geggies maintained in Scotland some kind of grassroots acting business, the downside was that by the end of the nineteenth century, the Scot on stage was almost entirely to be found doing comic turns or singing sentimental songs on the popular, rather than the legitimate, stage. Through time the free-and-easies gave way to formalised music hall, which was itself usurped by the variety theatre – reflecting major financial investments and, often, operated by major entertainment conglomerates out of London and their networks of refurbished and newly built theatres and constantly touring turns and stars. Some major Scottish talents flourished in and benefited from this new, highly capitalised cultural economy, in particular the tartan-clad Scotch comics including Harry Lauder, who balanced national and international tours with pantomime and summer seasons in some of Scotland's first houses, and the older and more neglected figure of W F Frame, whose national and international tours, as well as his autobiographical writing and published songbooks, anticipated Lauder's by many years. Both men were essentially character-based comedians, drawing on the images and sentiments of Scottish popular culture, the iconography of the Highlands, the sentiment of the lost Highlands and the immediacy of the urban experience. Generally dismissed as narrow, pawky and inferiorist today, in their heyday such performers were international stars.

Frame built a career predicated on touring both within and without Scotland. Early tours were to the north of Scotland, including Nairn, which he dubbed 'the Brighton of the North'. Following the Scottish fishing fleets, he travelled to Great Yarmouth and to the Isle of Man: 'I met thousands of them [Scottish fishermen] in Great Yarmouth, where, on my opening night, they crowded Gilbert's Circus and joined in the chorus of 'Hielan Rory' to the surprise and delight of the natives.'[21] He also worked in Ireland, where he found exiled Scots in Belfast and the so-called 'caledonophiles' in Dublin, and he toured to New York in 1898, performing in Carnegie Hall, which 'was crowded long before the concert was timed to begin. It was estimated that over 3,000 people were present, which included the elite of New York-Scotch.'[22] The financial rewards possible by exporting the comic songs and exaggerated costumes of the stage Scotsman were significant and represent a distinct cultural export that still shapes the image and identity of Scotland today.

A NEW HOME – REPERTORY

As well as the call on popular forms – the shift of Scottish-identified performers to the geggies, music halls and variety stages – there was a second, later, kind of reaction to the dominance of London-produced touring theatre. The repertory theatre sought to establish a modern 'stock company', forging sustainable relationships between theatre and municipality. Reacting to the Glasgow Repertory Theatre (1909–14), the popular Glasgow magazine *The Bailie* described the kind of working relationships that repertory valued:

> Much is hoped from the establishment of repertory theatres … While the dramatist is thankful for the wider horizon he enjoys when he writes for a repertory company – one untrammelled by a 'star,' the actor in the repertory company knows that his small part of to-day will be succeeded by an important part to-morrow, and, further, that there will be no long run to identify him with a particular line of business and so stunt his growth in the practice of his art. Consequently all branches of the drama feel they all experience the beneficial influence of the repertory system. Then the gain to the public is equal in its own way to the gain to writers and actors, inasmuch as a company provides them with opportunities for studying the art of acting which the 'touring company' system never allows.[23]

The goal for repertory theatre, for the music hall stars, for geggy theatre, for the local stock companies, for 7:84 in the twentieth century, even if achieved in a wide range of different ways, was to make a distinctive connection with the audience.

Whilst this first phase of repertory did not quite hold in Scotland, the postwar chapter was more secure with James Bridie's Citizens' Theatre (established 1943) at the heart of repertory's multifaceted story. Bridie's repertory proved itself particularly flexible, adapting its modus operandi with each new directorate or even ideological climate. Just as 7:84 reacted to and with the political climate of the 1970s, so too in its way did the Citizens', which, under Giles Havergal's leadership, pursued highly contemporary experiments of text, gender play and even collective decision making.

> Actors never see scripts of new plays, new translations, or adaptations before the first day of rehearsal. This undoubtedly colours the type of acting and the type of actor at the Citizens'. It has contributed to the emergence of a style of actor and presentation very much at odds with the rest of British theatre. This style is criticised or admired, but never denied.
>
> ...
>
> In [Robert David MacDonald's] *Webster* in 1983, the loquacious protagonist had many long speeches in act I at the end of which he was shot in the jaw and said nothing in act II. This 'coup de theatre' was in some way connected with the fact that the actor had to learn the part in a rehearsal period of ten days.[24]

The pragmatics of theatre work can, in such a context of mutual respect, lead to remarkable creativity and aesthetic reward.

CONCLUSION

The risks of working in theatre are, demonstrably, physical and emotional as well as financial, but the rewards may be similarly diverse. From George Anne Bellamy to John McGrath, theatre-makers write about the pleasures of working in the theatre as well as the intellectual and even political satisfactions to be derived. The playwright Robert David MacDonald reflects on the aesthetic as well as personal rewards of creative work across a number of plays, most particularly in *Chinchilla*, his bold 1977 recreation of both the *Ballets Russes* – appropriately enough imagined on tour at Venice Lido – and the Citizens' Theatre Glasgow, with which he was long associated as a key member of the creative team. His Diaghilev figure, the eponymous Chinchilla, reflects on his relationship with his friends and colleagues, and on the transformative potential of their work:

> We are five people who have known each other since before we all started dyeing our hair. We have gone through trials and tortures that make the cellars of the Cheka look like a sanatorium, just as we have known pleasures which would make the ecstasies of St Francis seem a mere faint itch. And we have laughed – alone – at each other – all the time ... And we work. We make revolutions, we make fashions, we make scandals. Many reasonable people are appalled, many despicable people delighted, but none of that matters. It comes from us.

> It is a passion, a disease, a lust ... And because it is all too ecstatic, absurd, miserable, happy, horrible and holy to contain within myself, I will show what I love, and tell what I love, with ardour, style and impeccable bad taste, whether it is Utopia, or the death of kings, or simply those beautiful young men without whom my life is as dry as a nut; so that for a moment we can see them, created in our image, in the glare of arc lamps, as we should; beautiful, clever, wise, just and alive, and for that moment forget that we are ugly, crass, guilty, foolish and dying.[25]

The transformative potential of the work in, as much as of, theatre in Scotland should not, then, be underestimated despite the rigours of this highly distinctive workplace.

However, at its most basic, perhaps, working in the theatres of Scotland is just about having the confidence to perform in front of a paying audience – certainly the performers encountered in this essay have no lack of confidence or ego. But, as one participant proposed, in a community-based performance project at the Citizens', the theatre and, showing off in it, is not for everyone:

> I don't like being on the stage
> oh no
> ye need a brass neck.
> that's whit I used tae say
> tae some ae the actors
> all you've got
> is a brass neck
> stauning up there.[26]

NOTES

1. Citizens' Theatre *Theatre Memories Project 2003: Citizens' Theatre, 1943–2003, 60 Years*, Glasgow, 2003, 14.
2. See, for example, McGrath, 2002; McGrath, 1990; and MacLennan, 1981.
3. McGrath, J. The year of *The Cheviot*. In McGrath, 1981, viii–ix, xi.
4. John Bett, interview: Working with John [McGrath]. In Bradby and Capon, 2005, 201.
5. McGrath, 1981, xxiv.
6. From research – including recorded interviews – pursued by staff of the University of Glasgow's Department of Theatre, Film and Television Studies in respect of rural touring in Scotland. See Hamilton and Scullion, 2004.
7. From *The Same, But Different* research project. See Hamilton and Scullion, 2004.
8. Galt, 1821, 241–4.
9. Wilson, A. 'The Spouter'. In Wilson, 1876, 319–44. All quotations are given with line references to this edition.
10. In relation to modern Glasgow, Alston Street is lost beneath the structure of Central Station, more or less at the junction of Hope Street and Argyle Street.

11. Bellamy, 1785, v. 4, 13.
12. Bellamy, 1785, v. 4, 13.
13. Bellamy, 1785, v. 4, 13.
14. In relation to modern Glasgow, Dunlop Street is a short cul de sac that runs south from Argyle Street, ending at the St Enoch Centre. A useful guide to old Glasgow theatres, and some of the stories attached to them, including those of Alexander and Seymour, is the leaflet *See Glasgow, See Theatre* prepared and published by the University of Glasgow's Department of Theatre, Film and Television Studies in 1990.
15. Alexander retired in 1851 and died just months later. His grave in Glasgow's Necropolis is marked by an elaborate monument designed round the central motif of a proscenium arch.
16. House, 1986, 42.
17. House uses a Scots word meaning 'climb'.
18. House, 1986, 42.
19. House, 1986, 42–3.
20. Cameron, 1992, 9.
21. Frame, *c*1900, 59.
22. Frame, *c*1900, 82–3, 92–3.
23. *The Bailie*, 8 September 1909, 1.
24. Giles Havergal, Choosing plays: The conditions of artistic choice at the Citizens' Theatre, Glasgow, 1969–85. J F Arnott Memorial Lecture, Tenth World Congress of the International Federation of Theatre Research, September 1985.
25. Chinchilla in *Chinchilla*, 1977, reprinted in MacDonald, 1991, 140–1.
26. Fred McGowan and Billy Findlay, edited by Davey Anderson. *A wee job in the theatre*. Unpublished typescript, 2004, 1.

BIBLIOGRAPHY

Bellamy, G A. *An Apology for the Life of George Anne Bellamy, Late of Covent Garden, Written by Herself in 6 Volumes*, London, 1785.

Bradby, D and Capon, S, eds. *Freedom's Pioneer: John McGrath's Work in Theatre, Film and Television*, Exeter, 2005.

Bruce, F, Foley, A and Gillespie, G, eds. *Those Variety Days: Memories of Scottish Variety*, Edinburgh, 1997.

Cameron, A. *See Glasgow, See Theatre*, Glasgow, 1990.

Cameron, A. Popular theatre and entertainment in nineteenth century Glasgow: Background and context. In Marshalsay, ed, 1992, 5–12.

Cameron, A. and Scullion, A, eds. *Scottish Popular Theatre and Entertainment: Historical and Critical Approaches of Theatre and Film in Scotland*, Glasgow, 1995.

Campbell, D. *Playing for Scotland: A History of the Scottish Stage, 1715–1965*, Edinburgh, 1996.

Coveney, M. *The Citizens': 21 Years of the Glasgow Citizens' Theatre*, London, 1990.

Devlin, V. *Kings, Queens and People's Palaces: An Oral History of the Scottish Variety Theatre*, Edinburgh, 1991.

Dibdin, J C. *The Annals of the Edinburgh Stage*, Edinburgh, 1888.

Findlay, B. ed. *A History of Scottish Theatre*, Edinburgh, 1998.

Frame, W F. *W F Frame Tells His Own Story*, Glasgow, *c*1900.

Galt, J. *The Annals of the Parish; or, The Chronicle of Dalmailing*, Edinburgh, 1821.

Hamilton, C and Scullion, A. *The Same, But Different. Rural Arts Touring in Scotland: The Case of Theatre*, Stroud, 2004.

House, J. *Music Hall Memories: Recollections of the Scottish Music Hall and Pantomime*, Glasgow, 1986.

Irving, G. *The Good Auld Days: The Story of Scotland's Entertainers from Music Hall to Television*, London, 1977.

Littlejohn, J H. *The Scottish Music Hall, 1880–1990*, Wigtown, 1990.

MacDonald, R D. *Three Plays: Chinchilla, Webster and Summit Conference*, London, 1991.

MacLennan, E. *The Moon Belongs to Everyone: Making Popular Theatre with 7:84*, London, 1981.

McGrath, J. *The Cheviot, the Stag, and the Black, Black Oil*, London, 1981.

McGrath, J. *The Bone Won't Break: On Theatre and Hope in Hard Times*, London, 1990.

McGrath, J, edited by N Holdsworth. *Naked Thoughts That Roam About: Reflections on Theatre*, London, 2002.

Mackie, A. D. *The Scotch Comedians: From the Music Hall to Television*, Edinburgh, 1973.

Maloney, P. *Scotland and the Music Hall, 1850–1914*, Manchester, 2003.

Marshalsay, K, ed. *The Waggle o' the Kilt: Popular Theatre and Entertainment in Scotland*, Glasgow, 1992.

Stevenson, R and Wallace, G, eds. *Scottish Theatre since the Seventies*, Edinburgh, 1996.

Wilson, A. *The Poems and Literary Prose of Alexander Wilson*, Paisley, 1876.

20 The High Road or the Low Road? Tourism and Employment in Scotland

DENNIS P NICKSON and TOM G BAUM

INTRODUCTION

The significance of tourism employment to the Scottish economy is such that it is often held to be the most important industry in Scotland.[1] However, despite generating £4.5 billion in gross revenue in 2002,[2] being the fourth largest employer in the economy and a sector that pays the wages of more employees than the oil, gas and whisky industries combined,[3] there remain concerns about the quality of many of these jobs. For example, Douglas Coupland, the notable cultural commentator, has for many captured the zeitgeist when he talked pejoratively of the 'McJob', which he has described as: 'A low-pay, low-prestige, low-dignity, low-benefit, no-future job in the service sector. Frequently considered a satisfying career choice by people who have never held one.'[4] The collection of essays by MacDonald and Sirianni recognises the challenges of living and working in a service society which, according to them, is characterised by two kinds of service jobs: large numbers of low-skilled, low-pay jobs, and a smaller number of high-skilled, high-income jobs, with few jobs being in the middle of these extremes.[5] Such a situation leads many to ask what kinds of job are being produced in the post-industrial era and who is filling such jobs. This question is pertinent within the Scottish tourism industry. Many are concerned as to whether these jobs are 'real' or 'good', especially when compared to work opportunities offered by what many would consider to be the emblematic areas of work, such as the manufacturing and extractive industries.[6] Related to this issue is the gendered nature of much tourism employment, with many perceiving it as being largely 'women's work'.[7] From a more prescriptive point of view, for those involved in the promotion of Scottish tourism there may be concerns over the extent to which those entrusted with portraying Scotland to visitors are able to offer something that is meaningful or authentically Scottish. For example, do those who work in tourism feel responsible for communicating to visitors what Scotland and the Scots are and have been?

Recognising these issues, this chapter will seek to consider a range of topics surrounding the nature of tourism employment in Scotland including:

its scale; its relationship to Scotland's cultural integrity; the notion of 'Scottishness'; the quality of work that is offered within the sector; and the contribution or otherwise of tourism work to social inclusion in Scotland.

THE QUANTITY AND DIVERSITY OF SCOTTISH TOURISM EMPLOYMENT

As has been noted, the importance of tourism to Scotland is widely acknowledged as the economy of the country, along with most other developed economies, has shifted towards increasing domination by services. There are no definitive figures on how many people are employed by the Scottish tourism industry. For example, a recent report from 'People 1st' (the Sector Skills Council for the hospitality, leisure, travel and tourism sector) suggested that tourism accounted for 187,500 jobs in Scotland.[8] Others such as 'Futureskills Scotland'[9] and the Scottish parliament[10] put the figure closer to 200,000.

Losekoot and Wood,[11] writing about tourism employment in Scotland, see real difficulties in securing reliable and wholly explanatory statistical data on tourism employment. They argue that the figures for tourism employment are based on flawed interpretation of extant tourism employment statistics. Consequently, Losekoot and Wood question the extent to which jobs can be disaggregated as being created by touristic activities, particularly if a wholly inclusive view of what connotes tourism jobs is accepted. In that sense, they concede that there is broad agreement as to what constitutes the core tourism industries, which include: hotel trade; restaurants, cafes and similar eating places; public houses and bars; nightclubs and other licensed clubs;[12] other forms of tourist accommodation; tourist offices and similar services; and travel and related sectors, such as travel agencies and airport services. More controversially, according to Losekoot and Wood, other areas that are added are: libraries; theatres; museums; sport and related leisure provision; and, the final and most challenging element, a proportion of retail employment. Criticism of the inclusion of this sector lies less in the fact that a proportion of retail employment is reliant on touristic activity, than the assessment of the *number* of jobs that are created in retail by the tourism industry. Losekoot and Wood are concerned to raise substantive and searching questions in their suggestion that:

> In countries like Scotland where tourism has been frequently viewed as a panacea for employment decline in manufacturing, all-inclusive definitions of tourism and tourism potential raise unrealistic hopes that the quantity and quality of jobs in tourism industries can, in the short-term, compensate across the economic board for erosion of the nation's manufacturing base.[13]

Arguments about the true extent of tourism employment in Scotland, such as that developed by Losekoot and Wood, arise in part because of the relatively crude outcomes of macro-economic analysis of employment impact. Essentially, this analysis leads to the conclusion that tourism's workforce ultimately consists of everyone in the country whose employment and livelihood is dependent on the spending of the tourist pound, euro, dollar or yen, in terms of direct and indirect expenditure. This analytical process, as Baum points out, is a blunt instrument and fails to address qualitative aspects of work in the sector.[14] One of the notable features of a tourism destination such as Scotland is the range of perceptions of the work held by those employed in the sector. Workers in Losekoot and Wood's core subsectors (accommodation, transport, and so on) generally will see themselves as working in tourism, with little perceptual ambiguity. By contrast and with the exception of specialist outlets, those in retail (supermarkets, garage forecourts, banks, etc) are unlikely to see themselves in such terms and are, therefore, less likely to offer the type of tourism-oriented service sought through training initiatives such as Welcome Host.[15] Given this failure on the part of many who serve tourists to self-identify as being employed in the tourism industry, is it legitimate or realistic to have ambassadorial expectations of those workers? If people do not see themselves as tourism workers, should they, as a consequence, be excluded from enumeration of the sector's employment impact?

Clearly the shifting nature of Scotland's economy, and the increasing reliance upon areas such as tourism for job creation, raises many fundamental issues for policy makers and academics alike, a point considered later in the chapter. At this juncture, however, the important point remains that tourism is a key source of employment in Scotland.

Within the broad classification of tourism, there is massive diversity in the types of job generated. This diversity exists in relation to the technical and skills demands, educational requirements, terms and conditions, and the type of person likely to be attracted to this type of employment. One consistent theme, though, regardless of the employee, is the increasing recognition of the role played by those employed in the Scottish tourism industry in 'making or breaking the tourist experience'.[16] For many tourists coming to Scotland, the quality of their experience is likely to be reliant to a large extent on the interactions they will have with the variety of front-line staff in the tourism industry. These 'moments of truth'[17] are therefore crucial for organisational effectiveness, success, competitiveness and profitability. Indeed, within an industry that is characterised by diversity of purpose, size, ownership and demands of the enterprise, the only real points of commonality are the delivery of service to customers and the need to manage people in such a way that they offer a quality service. The corollary of this would be the belief that such front-line staff would therefore be sufficiently well paid, trained and motivated to offer high quality service. The reality, however, is that often such staff have the lowest status in the organisation, are the least trained, and are the poorest paid employees, a point which is returned to below.

WHITHER 'SCOTTISHNESS'?

A further issue to be considered is the increasing importance of migrant employees, who are becoming crucial to the provision of Scottish tourism. The marketing of Scotland as a tourist destination and Scottishness in both product and service terms has long been typified by the traditional symbolism of kilts, bagpipes, heather, mist, monsters, shortbread and whisky, with service delivered by smiling Scots culturally and physically attired to conform to stereotypes consistent with these symbols. This stereotype of Scotland and the Scots is challenging for many seeking to project the country with images of modernity, dynamism and economic opportunity, but it is one that persists in the minds of many who are responsible for promoting tourism. This is illustrated in a letter to the *St Andrews Citizen*, in which an American correspondent noted:

> I have just finished contacting 22 American golfing friends, golf writers and other acquaintances who have played the Old Course. Every single man and woman that I contacted has stated that they want an experienced Scottish caddie when they play the St Andrews Links. Several have been assigned American caddies, Germans and even a Chinese caddie. All of them felt the same as me in that the Old Course experience was diminished, rather than increased, as a result.[18]

This expectation of stereotypical Scottishness is challenged by the increasing multiculturalism of Scottish society, especially in its main urban centres. This challenge is added to by the attraction of the country to a wide range of people from outwith Scotland who aspire to live and work here and for whom tourism is a natural environment within which to seek work. As shall be seen, this is not a debate that is confined to Scotland.

Frewin reports on a recent survey from recruitment agency Chess Partnership. They surveyed 230 foreign hospitality students or managers, 65 per cent of whom wished to work in the UK. London was the most popular destination, followed by Scotland, which recorded 17 per cent of the vote. Specifically, overseas workers were attracted to Scotland by its reputation for good work–life balance (43 per cent), its coastal areas (40 per cent), the quality of life (27 per cent) and its countryside (24 per cent). Edinburgh was the preferred location, cited by 64 per cent of respondents, whilst Glasgow was suggested by 37 per cent, and 42 per cent said they would work anywhere in the country.[19] Generally, employers seem positive about foreign workers, particularly as they are willing to fill roles that local workers seem reluctant to fill. Indeed, a real feature of Scottish tourism employment in recent years is the increasing recognition of the key role played by migrant workers. To date, there has been no systematic attempt to research the employment experience of these workers, though a number of anecdotal accounts do offer some insight.

Langlands, for example, notes the relatively large number of

Australasian workers working in Scottish tourism, and in recognising this point suggests: 'Strewth! That will be the word on many holidaymakers' lips around Scotland this summer when, expecting to encounter kilted Scots and bonnie lasses staffing the front lines of the country's tourist industry, they instead find surprising numbers of bronzed Australian Shielas and Bruces.'[20] Although there are no definitive figures for migrant workers in Scottish tourism, Langlands quotes a representative of the Scottish Tourism Forum, who suggests that between 7,000 and 10,000 Australians, New Zealanders and South Africans are working in the Scottish tourism industry. The reason for such large numbers is, in part, explicable by the fact that indigenous Scots do not see tourism as a worthwhile career. More recently, Meiklem notes how a number of Eastern Europeans have taken advantage of the ability to work legally within Britain since 1 May 2004.[21] Likewise, Ross reports on the role played by international workers in the development of the Aviemore resort in the Highlands, where a multicultural team combine employment whilst undergoing training.[22]

Generally, migrant workers seem to be viewed in a positive manner by employers. For example, Peter Lederer, the chairman of VisitScotland, the body responsible for promoting Scottish tourism, suggests: 'It doesn't matter if you're from Australia, New Zealand or South Africa, a smile and a warm handshake is what's most important ... As long as they can offer our visitors a quality product, their nationality is not important.'[23] Individual employers seem equally effusive about migrant workers. Jean Urquhart, owner of the Cellidh Place, an arts hotel in Ullapool, is quoted as saying: 'The whole tourism industry would collapse without them.' Similarly, a manager from Pizza Hut in Inverness suggests that his five Polish workers are 'never sick, never late, they just work away and we value them very, very highly', even to the extent of suggesting that they are better than Scottish workers.[24] Indeed, Langlands notes how the Crieff Hydro sent representatives to Australia to interview potential employees, meeting 90 applicants from a total list of 300 over a period of nine days.[25] There is also evidence that migrant workers in tourism in Scotland are highly aspirant in terms of education and careers when compared to their local counterparts. Applications to undertake part-time postgraduate courses in tourism and hospitality at Strathclyde University from Polish and other new accession state nationals who are working in the tourism sector in Glasgow increased significantly in 2005.

Whilst many employers seem to welcome migrant labour and many of the employees interviewed speak positively about their experience of working in Scottish tourism, there are also difficulties facing migrant workers. Frewin notes some of the obstacles in recruiting immigrants, including cultural differences and a lack of affordable housing.[26] She also notes how some Scottish employers had detected signs of prejudice from both staff and customers, especially towards Eastern Europeans, who were sometimes almost regarded as second-class citizens. Indeed, an academic from the Highlands suggests that rural racism can be a concern. She is quoted as saying: 'We are only half-tolerant of migrants. We put up with them because we need them but we aren't completely welcoming.' The same academic does recognise the key role played by the migrant workers, whilst offering the caveat that they should not be taken for granted in low-skilled jobs in particular: 'Migrant workers are not the panacea to Scotland's economic ills. They are part of the wider solution.'[27] Similarly, the SNP MSP for Central Fife, Tricia Marwick, is quoted as suggesting that Scotland is best sold and represented by Scots, as 'in terms of selling Scotland, nobody does it better than the Scots'.[28]

This is an interesting point which raises the question of authenticity and whether the Scottish tourism product is, indeed, best presented and represented by Scots. This argument is, of course, not unique to Scotland. In the Irish Republic, McManus, for example, looks at the contradictions that have emerged between modern, urban Ireland (the Celtic tiger) and the pastoral images of the west that still dominate in formal tourism marketing and in the images of film and television.[29] This is a contrast that is also pertinent in the Scottish context. McManus points to the impact of the commodification of heritage and culture for purposes of tourism in both rural and urban Ireland as running counter to the desires and aspirations of an increasingly multi-ethnic country. In particular, McManus questions the appropriateness of tourism market imagery, relating as it frequently does to people as a core dimension, in the light of growing multi-ethnicity in the tourism workforce. She argues: 'None of the images presented (in the tourism marketing literature) seem to recognise the increasing levels of multiculturalism and diversity within Ireland, probably reflecting our own identity confusion.'[30]

This is a theme that touches highly emotive and sensitive political and cultural strings and is one that has received little academic consideration in Scotland. It is, however, worthy of reflection and further investigation.

THE QUALITY OF TOURISM EMPLOYMENT

As already noted, there may be considerable diversity within sub-sectoral aspects of tourism employment, and of occupations within these various sub-sectors. Therefore, this discussion can be seen as a 'snapshot' of some key issues in the way that organisations manage their human resources. Notwithstanding this caveat, we can contextualise our discussion by recognising a number of complex and inter-related themes that Baum has termed 'universal themes' in international tourism. These themes are of concern to both human resource professionals within the industry and academics researching and writing within this area, and are likely to be apparent to a greater or lesser extent according to the destination or enterprise context.[31]

- Demography and the shrinking employment pool/labour shortages, particularly in Western Europe and North America.
- The tourism industry's image as an employer.

- Cultural and traditional perceptions of the industry.
- Rewards and benefits/compensation.
- Recruitment, retention and staff turnover.
- Education and training, both within colleges and industry.
- Skills shortages, especially at higher technical and management levels.
- Linking human resource concerns with service and product quality, and especially a limited recognition of the importance of human resource development in the provision of high-quality products and services.
- Poor management and planning information about human resource matters in the tourism industry.
- The tendency to develop human resource policies, initiatives and remedial programmes that are reactive to what is currently happening rather than proactive as to what is likely to occur.

A number of these issues can be seen to underpin the questions posed by Choy in his seminal work on the quality of tourism employment.[32] Having first noted that 'the development of a tourism industry creates new employment opportunities', he nonetheless goes on to recognise that 'critics of the industry contend that tourism provides primarily low-paying, low-skilled jobs which are demeaning'.[33] Accordingly, Choy seeks to investigate four commonly held beliefs about tourism employment, these being:

- Tourism generates primarily low-skilled jobs.
- Tourism generates low-paying jobs.
- Tourism jobs do not offer high levels of job satisfaction.
- Tourism offers limited opportunities for advancement for local residents.

We can now consider several of these aspects within the context of Scotland, focusing specifically on the issues of skills and remuneration.

RECONCEPTUALISING SKILLS

The aspiration of the Scottish government (alongside the governments of virtually all developed economies) is to create an environment in which a significant proportion of the country's citizens are engaged in what can be called high value-added, high-skills employment, often described as the knowledge economy and articulated through policy documents such as 'A Smart Successful Scotland'.[34] This vision of an economy driven by creativity and enterprise is typically seen in terms of work that is technology-driven, based on high research and development input or which bases its wealth-creating capability on creative dimensions unavailable elsewhere. Tourism, as a sector, does not automatically match what we expect of high-skills work when we attempt to analyse its skills requirements. Moreover, beyond the often rather superficial and wish-fulfilling descriptions of the knowledge economy and the high-skills jobs that it putatively creates, there is a secondary issue of the conventional views of skills, filtered through accounts of 'traditional' work, in areas such as the manufacturing and extractive industries.[35] This has often led to characterisation of much work in tourism as being low-skilled to, at best, medium-skilled. However, there is a need for consideration of the changing nature of skills in the contemporary economy, and especially the emergence of the importance of 'soft skills' in areas such as tourism, a point that can be considered in reviewing some recent research on the nature of skills demanded by tourism employers in Glasgow.

Glasgow was once an industrial city. Now, over 82 per cent of the city's jobs are in services.[36] Aiming for the city-break tourist market, the city promotes its retail, cultural and hospitality attractions. Between 1994 and 2000, the number of major hotels in the city increased from forty-two to eighty-nine, with twenty-seven more planned. Glasgow has approximately 1,000 bars and restaurants and is second only to London as Britain's culinary capital.[37] Similarly, Experian acknowledges Glasgow as the second largest retail centre in the UK outside London.[38] The city also now has a well-developed niche of designer retailers, boutique hotels and style bars, cafes and restaurants, such that 'leisure shopping and a new café culture have been fused with the city's famously friendly atmosphere to create an almost tangible buzz'.[39] Not surprisingly, the city was recently described by US magazine *Travel and Leisure* as 'the UK's hippest and most happening city'.[40] Three million tourists visit the city each year, generating £670 million annually in the local economy.[41] In recognition of this new economic success, the city rebranded itself as 'Scotland with Style' in 2004.[42] With the shift to a service economy, the type of skills demanded by employers has also shifted. Employers in hospitality and tourism in Glasgow increasingly desire employees with the 'right' attitude and appearance.[43] The right attitude encompasses aspects such as social and interpersonal skills, which are largely concerned with ensuring employees are responsive to, courteous towards and understanding with customers; or, in simple terms, that they can demonstrate emotional labour. In her work on airline stewardesses, Hochschild sees emotional labour as the selling of a state of emotion or mind, and recognises how:

> This labour requires one to induce or suppress feelings in order to sustain the outward countenance that produces the state of mind in others … This kind of labour calls for a co-ordination of mind and feeling, and it sometimes draws on a sense of self that we honour as deep and integral to our personality … I use the term 'emotional

labour' to mean the management of feeling to create a publicly observable facial and bodily display.

In simple terms, service employees are expected to use emotional labour to display the 'right' attitude and emotions in interacting with customers.[44] However, it is not only the right attitude that employers seek. Nickson et al. (2001) have developed the term 'aesthetic labour'[45] – the ability to either 'look good' or 'sound right'[46] – which points to the increasing importance of the way in which employees are expected physically to embody the company image in tourism and hospitality.

In an analysis of 5,000 job advertisements across a number of different occupations and sectors in the UK, Jackson et al. found that the skills stated as necessary by employers were 'social skills' and 'personal characteristics'; only 26 per cent of organisations mentioned the need for educational requirements. Within personal services, this figure was less than 10 per cent.[47] Rather like the review of job adverts reported in Warhurst et al.,[48] Jackson et al. found numerous instances of advertisements for front-line service jobs which sought attributes that referred less to what individuals could *do* than what they were *like,* such as being 'well-turned out' or 'well-spoken', or having 'good appearance', 'good manners', 'character' or 'presence'.

Nickson et al. report evidence from a survey of nearly 150 employers in the Glasgow retail and hospitality industry.[49] On the question of what employers were looking for in customer-facing staff during the selection process, Nickson et al. found that 65 per cent suggested that the right personality was critical, with the remainder of respondents suggesting this aspect was important. Similarly, 33 per cent of the employers surveyed felt that the right appearance was critical, with 57 per cent seeing it as important and only 2 per cent of respondents feeling it was not important. These figures can be compared to qualifications, with only one respondent seeing qualifications as critical, 19 per cent of employers feeling it was important and 40 per cent suggesting it was not important at all for selecting their customer-facing staff. In terms of the skills deemed necessary to do the required work, employers placed a far greater emphasis on 'soft' skills for customer-facing staff. Ninety-nine per cent of respondents felt that social or interpersonal skills were of at least significant importance, and 98 per cent felt likewise about self-presentation, or aesthetic, skills. Conversely, 48 per cent of employers felt that technical skills were important in their customer-facing staff, and 16 per cent stated they were not important at all. The skills that matter to employers in customer-facing staff in tourism and hospitality are generally 'soft', including aesthetic skills, rather than 'hard' technical skills, which will often develop through training when people join the organisation

This list is slightly puzzling. None of the 'skills' being demanded by employers would have been regarded as skills in the past, when being 'skilled' was associated with successful completion of an apprenticeship involving formalised training that mobilised physical dexterity and technical 'know-how'. However, these days, whatever employers say is a skill seems to be regarded as a skill. As a consequence, Grugulis et al. note that what constitutes a skill has evolved over time, with 'the growing tendency to label what in earlier times would have been seen by most as personal characteristics, attitudes, character traits, or predispositions as skills'.[50] In the search for 'skilled workers' within this context, employers are really seeking to employ the 'right kind' of person, someone able to manage their emotions and, seemingly, their corporeality.

Nickson et al. recognise how many of the particular 'skills' in personal presentation, self-confidence, grooming, deportment and accent that Glaswegian service-sector employers are seeking are liable to be linked to the parental social class and educational background of the job applicants.[51] As such, the style labour market tends to draw on young workers from Glasgow's middle-class suburbs, which also partly accounts for the attractiveness to employers of students. By virtue of their cultural capital, these workers are perceived to be more appealing to consumers and, through their work, create a point of reference for customers. For example, Nickson et al. recognise how one boutique hotel based in Glasgow sought customer-facing workers who were typically in their twenties, graduates and well-travelled.[52] The company deliberately placed job advertisements in the *Sunday Times* rather than the local evening newspaper in order to recruit the sons and daughters of the middle class. Aesthetic labour thus tends to be sourced from the middle rather than the working class.

This search for a certain type of outgoing middle-class sociability has obvious implications for how the Scottish tourism product may be represented within certain niche areas, such as the style niche in Glasgow. Broadening the argument, though, there may also be some resonance with our earlier identification of the increasing use of migrant labour. Thus, as product and producer are conflated in interactive services, so the social background or nationality of employees becomes an issue and becomes important in creating employability and the capacity to do work, so that having or contriving to have 'middle-classness', or a sense of being distinct by dint of being Australian, South African and so on, becomes key in both getting and doing these jobs. Beyond issues of representation of Scottish tourism, there is also the vexed issue of social inclusion and whether certain groups in the labour market, most obviously young working-class men and women, are being excluded from what some may consider as the more desirable jobs in tourism.

CONDEMNED TO LOW PAY?

Generally, when we are talking about remuneration in the tourism industry we can start with the fairly negative observation that relative to other industries the majority of jobs and occupations within the tourism sector are poorly remunerated. Baum and Riley outline a number of structural features of the tourism industry which are likely to have downward pressure on wage levels, the most important of these being:

- Small unit structure of the industry: the industry in most countries is highly fragmented and heterogeneous, being an amalgam of small to large businesses. However, the majority of businesses are small and medium-sized enterprises (SMEs).[53]
- Fluctuations in levels of business activity: there is constant fluctuation in consumer demand across long and short time periods. For example, whilst the foot and mouth outbreak in 2001 did not have a significant impact on Scottish tourism overall, in certain geographical areas the impact was much more pronounced. Dumfries and Galloway, for example, saw a 24 per cent decrease in visitors to their attractions, and this had a major impact on the community in social and economic terms.[54]
- Cost pressures induced by competition.
- A reliance on vulnerable and so-called 'marginal'[55] workers: for example, drawing on sections of the labour market that have little bargaining power, such as young people, students, married women returning to work, ethnic minorities and migrant labour.

As a result, low pay is a very real issue for many tourism employers, especially those in the hospitality sub-sector. For example, the *New Earnings Survey 2002* reported that average gross earnings for full-time adult employees in the hotel and restaurant sector were £289.10 a week in 2001, the lowest of the main seventeen industrial categories. Similarly, the *New Earnings Survey 2003* found that hotel and restaurant employees were the lowest paid in the country. Average gross annual pay for full time restaurant and hotel employees was just £16,533, compared to a UK average of £25,170. More recently the *Annual Survey of Hours and Earnings*, which has replaced the *New Earnings Survey*, suggests that waiters/waitresses, kitchen porters and kitchen hands are amongst the lowest paid of all UK employees. The same is also true for managerial jobs, with managers in tourism jobs often earning less than their managerial counterparts in other industries.

The relatively low pay levels outlined above are very much thrown into sharp relief with the recognition of the important work of Lindsay and McQuaid on the perceptions of the unemployed in Edinburgh and Glasgow to service work.[56] As they note, the main thrust of government policy under New Labour is to emphasise the importance of paid work as the best form of welfare. In this sense, social inclusion is increasingly equated with labour market inclusion. Resultantly, government has attempted to intervene to create at least a minimum infrastructure of decency and fairness in the workplace. This policy can be seen in the introduction of initiatives such as the national minimum wage (NMW), improvements in the tax system to encourage those with children to access employment, new legislation on leave entitlement and working time, and enhanced rights for part-time workers. With the nature of work and labour market characteristics that we have outlined above, all of these initiatives would ostensibly seem set to improve the lot of many of those working in tourism. At one level these various initiatives have had a very positive impact on the employment experience of many service workers, and especially those in tourism and hospitality-related occupations. However, it is the *relative* nature of this improvement that it is important to note. In that sense, despite the introduction of the NMW, tourism jobs generally remain relatively low-paid compared to other sectors. Taken in conjunction with other broader aspects of terms and conditions, it is here that many would compare tourism jobs unfavourably with more 'traditional' jobs in manufacturing or extractive industries.

Of course it is important to recognise the differences in the various sub-groups interviewed by Lindsay and McQuaid. For example, older males who had previously worked in manual, craft-related or machine operations work were very unlikely to consider working in retail and hospitality. This finding should come as no great surprise and is in concordance with some of the debate outlined above by Losekoot and Wood. Beyond this sub-group of older males, younger men were more open to the idea of working in the service sector, and that tends to be especially true for students.[57] Part of the reason for some not considering work in the tourism sector is therefore cultural – for example, it can be perceived as 'women's work' – but equally there is very much the issue of simple economics for many. Tourism is seen as not providing a large enough income to make it a viable sector to work in. Thus, in sum, as Lindsay and McQuaid note:

> It would appear that assumptions about the supposed 'gendered' nature of some forms of service work and the skills required to do these jobs, combined with concerns over pay rates may have led many job seekers (and especially men) to reject entry level retail and hospitality positions.[58]

CONCLUSIONS

This chapter has attempted to consider a number of important issues in examining work and employment in the tourism industry in Scotland. Whilst it has considered a number of areas, space constraints mean that such a review is inevitably rather episodic. For example, we have not considered the more strategic aspects of developing Scottish tourism and the implications for training and development.[59] Equally, we have rather concentrated on front-line work in tourism, with much less focus on back-of-house employees, such as those who work in kitchens. This points to the difficulties of writing a chapter of this nature on tourism, as it is such a disparate industry with the potential for significant variations between sub-sectors and occupations within the various sub-sectors.

Nevertheless, a key issue identified is the important role that tourism plays in wealth creation and employment provision in the Scottish economy, sustaining many small business and rural communities as well as providing significant opportunities in the urban conurbations such as Edinburgh and Glasgow. In this sense, the weak labour market characteristics of the sector facilitate access in that the requirement for formal qualifications and pre-entry skills may be somewhat limited. Unsurprisingly, then, policy makers view service industries, such as tourism, which is largely characterised by labour intensity and what have traditionally been thought of as low- to medium-skilled work, as important in future job creation. Nevertheless, in September 2004 there were over 5,000 vacancies in core tourism occupations, with the largest number of vacancies being for kitchen and catering assistants, waiters and waitresses, chefs, cooks and bar staff.[60] An important theme in this chapter is the contention that tourism does not create 'real' or 'good' jobs. Thus, although tourism provides easy access to work, this may be offset by concerns about making a living and developing sustainable career paths. Many would argue that there is an increasing perception that entry-level service jobs do not need to provide for essentials such as housing and food, as they are likely to be a means of securing additional income for groups such as students, second earners (in this case, women) and migrant workers (who may be perceived as likely to be transient). For some, this would exemplify the manner in which service employment in Scotland is less valued culturally and economically. In part, the shifting nature of skills and rising importance of soft skills may go some way towards affecting a cultural change in perceptions of service work, particularly in niche labour markets, such as the style labour market that was considered in this chapter. Equally, though, it is important to recognise that too often these skills remain unaccredited and unrewarded, as evidenced by the relatively low levels of pay in much tourism work. In sum, if tourism is to become accepted as an important industry in Scotland, then jobs need to offer realistic wages, decent conditions and opportunities for personal development and advancement; a task which is rather easier said than done.

NOTES

1. Burnside, 2002.
2. Dewar, 2004.
3. Burnside, 2002.
4. Coupland, 1993, 5.
5. MacDonald and Sirianni, 1996.
6. See for example Lindsay, 2005; Lindsay and McQuaid, 2004.
7. Purcell, 1997.
8. People 1st, 2005.
9. Futureskills Scotland, 2003.
10. Cited in Dewar, 2004.
11. Losekoot and Wood, 2001.
12. In reality, these sub-sectors – usually characterised generically as the hospitality industry – account for the majority of tourism-related jobs in Scotland,

as with the UK generally, at around 60 per cent (see People 1st, 2005a; 2005b). A consequence of this is that the bulk of academic work on employment issues in tourism focuses primarily on the hospitality sub-sector.

13. Losekoot and Wood, 2001, 96.
14. Baum, 1993.
15. The Welcome Host scheme is based on a Canadian hospitality programme called Super Host. This scheme was introduced in British Columbia in 1986 to support the growth of tourism around the World Expo in Vancouver. Sweeney, 1995, 8, describes Welcome Host as 'an ongoing, comprehensive, community-based programme designed to upgrade the standards of service and hospitality provided within the tourism industry ... By involving the whole community, the scheme provides access to more formal training for the smaller operator who may also come into contact with the visitor.' The basis of Welcome Host is 'People helping People' and its objectives are 'Professionalism and Pride'.
16. Baum, 1996.
17. Carlzon, 1987.
18. St Andrews Caddies, 2004.
19. Frewin, 2004.
20. Langlands, 2002, 9.
21. Meiklem, 2004, 9.
22. Ross, 2005.
23. Cited in Langlands, 2002, 9.
24. Both cited in Meiklem, 2004, 9; see also Ross, 2005.
25. Langlands, 2002.
26. Frewin, 2004.
27. Both quotes from Meiklem, 2004, 9.
28. Cited in Langlands, 2002, 9.
29. McManus, 2005.
30. McManus, 2005, 247.
31. Baum, 1993, 9–10.
32. Choy, 1995.
33. Choy, 1995, 129.
34. Scottish Executive, 2001
35. See Baum, 2002 for a review of this issue.
36. Scottish Enterprise Glasgow, 2005.
37. Glasgow Development Agency, 2002; 2003.
38. Experian, an information solutions company, produce The Retail Ranking Index, which assesses the relative vitality of retail centres in the UK. The index is the result of physical surveys of more than 1,100 separate shopping locations and over a third of a million UK retail outlets. Experian's Retail Ranking lists the top 250 locations for shopping in the UK; see also the Glasgow Development Agency, 2002.
39. Glasgow Tourism Development Group, 2002, 5.
40. Cited in Glasgow Tourism Development Group, 2002, 7.
41. Glasgow Tourism Development Group, 2002.
42. See, for example, Stewart, 2004.
43. Nickson et al., 2005.
44. Hochschild, 1983, 7.
45. Nickson et al., 2001.

46. Warhurst and Nickson, 2001.
47. Jackson et al., 2005.
48. Warhurst et al., 2000.
49. Nickson et al., 2005.
50. Grugulis et al., 2004, 6.
51. Nickson et al., 2003.
52. Nickson et al., 2001.
53. Baum, 1995; Riley, 1993.
54. Yeoman et al., 2005.
55. Wood, 1997.
56. Lindsay, 2005; Lindsay and McQuaid, 2004.
57. See also Canny, 2002.
58. Lindsay and McQuaid, 2004, 313.
59. For further discussion of developments in the strategic direction of Scottish tourism generally, and issues surrounding work and employment issues specifically, see Futureskills Scotland, 2003; Scottish Executive, 2000; Scottish Executive, 2002; Scottish Tourism Co-ordinating Group, 1994; Watson and Drummond, 2002.
60. People 1st, 2005a.

BIBLIOGRAPHY

Baum, T. Human resources in tourism: An introduction. In Baum, T, ed., *Human Resources in International Tourism*, Oxford, 1993, 3–21.

Baum T. *Managing Human Resources in the European Tourism and Hospitality Industry: A Strategic Approach*, London, 1995.

Baum, T. Making or breaking the tourist experience: The role of human resource management. In Ryan, C, ed., *The Tourist Experience: A New Introduction*, London, 1996, 92–111.

Baum, T. Skills and training for the hospitality sector: A review of issues, *Journal of Vocational Education and Training*, 54/3 (2002), 343–63.

Burnside, R. *Tourism Scoping Paper*, Edinburgh, 2002, available online at: http://www.scottish.parliament.uk/business/committees/historic/x-enterprise/papers–02/elp02–17.pdf.

Canny, A. Flexible labour? The growth of student employment in the UK, *Journal of Education and Work*, 15/3 (2002), 277–301.

Carlzon, J. *Moments of Truth*, Cambridge, Mass, 1987.

Choy, D. The quality of tourism employment, *Tourism Management*, 16/2 (1995), 129–37.

Coupland, D. *Generation X: Tales for an Accelerated Culture*, London, 1993.

Dewar, J. *Area Tourist Boards*, Scottish Parliament Information Centre (SPICe): Edinburgh, 2004, available online at: http://www.scottish.parliament.uk/business/research/briefings–04/sb04–36.pdf.

Frewin, A. Scotland lures workers – but could do better, *Caterer and Hotelkeeper*, 28 October 2004, 8.

Futureskills Scotland. *Tourism Sector: Scottish Sector Profile*, Glasgow, 2003.

Glasgow Development Agency. *Glasgow Economic Monitor*, Glasgow, 2002.

Glasgow Development Agency. *Glasgow Economic Monitor*, Glasgow, 2003.

Glasgow Tourism Development Group. *Glasgow Tourism Action Plan 2002/07*, Glasgow, 2002.

Grugulis, I, Warhurst, C and Keep, E. What's happening to 'skill'? In Warhurst, C, Grugulis, I and Keep, E, eds, *The Skills That Matter*, London, 2004, 1–18.
Hochschild, A. *The Managed Heart*, Berkeley, 1983.
Jackson, M, Goldthorpe, J and Mills, C. Education, employers and class mobility, *Research in Social Stratification and Mobility*, 23 (2005), 1–30.
Langlands, E. Staff wanted: Only Aussies need apply, *Sunday Herald*, 4 August 2002.
Lindsay, C. 'McJobs', 'good jobs' and skills: Job seekers' attitudes to low-skilled work, *Human Resource Management Journal*, 15/2 (2005), 50–65.
Lindsay, C and McQuaid, R W. Avoiding the 'McJobs': Unemployed job seekers and attitudes to service work, *Work, Employment and Society*, 18/2 (2004), 297–319.
Losekoot, E and Wood, R. Prospects for tourism employment in Scotland, *Scottish Affairs*, 34 (Winter 2001), 91–106.
MacDonald, C and Sirianni, C, eds. *Working in the Service Society*, Philadelphia, 1996.
McManus, R. Identity crisis? Heritage construction, tourism and place marketing in Ireland. In McCarthy, M, ed., *Ireland's Heritages: Critical Perspectives on Memory and Identity*, Aldershot, 2005, 235–50.
Meiklem, P. Highland hospitality … courtesy of Eastern Europeans, *Sunday Herald*, 14 November 2004.
New Earnings Survey 2002, London, 2002.
New Earnings Survey 2003, London, 2003.
Nickson, D, Warhurst, C, Witz, A and Cullen, A M. The importance of being aesthetic: Work, employment and service organization. In Sturdy, A, Grugulis, I and Wilmott, H, eds, *Customer Service: Empowerment and Entrapment*, Basingstoke, 2001, 170–90.
Nickson, D, Warhurst, C, Cullen, A M and Watt, A. Bringing in the excluded? Aesthetic labour, skills and training in the new economy, *Journal of Education and Work*, 16/2 (2003), 185–203.
Nickson, D, Warhurst, C and Dutton, E. The importance of attitude and appearance in the service encounter in retail and hospitality, *Managing Service Quality*, 15/2 (2005), 195–208.
People 1st, *The Hospitality, Leisure, Travel and Tourism Sector in Scotland*, London, 2005a.
People 1st, *Hospitality, Leisure, Travel and Tourism: A Skills and Labour Market Profile*, London, 2005b.
Purcell, K. Women's employment in UK tourism: Gender roles and labour markets. In Sinclair, M T, ed, *Gender, Work and Tourism*, London, 1997, 35–59.
Riley, M. Labour markets and vocational education. In Baum, T, ed., *Human Resources in International Tourism*, Oxford, 1993, 47–59.
Ross, D. The youngsters changing the face of Aviemore, *The Herald*, 7 March 2005, 12.
Scottish Enterprise Glasgow, *Glasgow Economic Monitor*, Glasgow, 2005.
Scottish Executive. *A new strategy for Scottish tourism*, Edinburgh, 2000. Policy document available online at: http://www.scotland.gov.uk/library2/doc11/sfst.pdf.
Scottish Executive. *A smart successful Scotland: Ambitions for the enterprise network*, Edinburgh, 2001. Policy document available online at: http://www.scotland.gov.uk/library3/enterprise/sss.pdf.
Scottish Executive. *Tourism framework for action 2002:2005*, Edinburgh, 2002. Policy document available online at: http://www.scotland.gov.uk/library3/tourism/tfar.pdf.
Scottish Tourism Co-ordinating Group *Scottish Tourism Strategic Plan*, Edinburgh, 1994.

St Andrews Caddies' Letter to the Editor, *St Andrews Citizen*, 20 August 2004.
Stewart, S. Black to the future as Glasgow does it with style, *The Herald*, 10 March 2004, 3.
Sweeney, A. Welcome Host Wales to Welcome Host UK. Paper presented to the Fourth Annual Council for Hospitality Management Education (CHME) Research Conference, Norwich, April, 1995.
Warhurst, C, Nickson, D, Witz, A and Cullen, A. Aesthetic labour in interactive service work: Some case study evidence from the 'new' Glasgow, *Service Industries Journal*, 20/3 (2000), 1–18.
Warhurst, C and Nickson, D. *Looking Good, Sounding Right*, London, 2001.
Watson, S and Drummond, D. A strategic perspective to human resource development in Scottish tourism, *International Journal of Contemporary Hospitality Management*, 14/5 (2002), 253–4.
Wood, R C. *Working in Hotels and Catering*, London, 1997.
Yeoman, I, Lennon, J and Black, L. Foot-and-mouth disease: A scenario of reoccurrence for Scotland's tourism industry, *Journal of Vacation Marketing*, 11/2 (2005), 179–90.

21 Printing in Scotland

WILLIAM G D WATSON

INTRODUCTION

If you ask a member of the general public the question What is a printer?, the majority would reply that it is an output device for a PC which can produce black-and-white or multi-colour prints. To a certain extent today that answer is correct, but to someone like this author, who has spent a considerable part of his life working in the printing industry, a printer is a highly skilled employee working in the mass communications industry using one of the many production techniques[1] available to produce multiple quantities of visual materials.

The world of print confronts us in our daily lives from the moment we open our eyes in the morning until we shut them at night. It includes the images on the dial of our alarm clock or mobile phone, the cereal packet and fruit juice cartons we open at breakfast, the newspaper or magazine we read on our way to work, and even the bus, train or parking ticket we collect is printed. The advertising materials which influence us to buy goods, and the paper money or plastic card we use to purchase them, are all examples of printed media.

Most of us use printed material every day, yet very few of us know anything about it. To most of us, the printing industry – or the 'black art', as it is often referred to – and its production techniques are a mystery.

HISTORICAL BACKGROUND

The invention of both printing and papermaking are credited to the Chinese some 2,000 years ago. Initially they used a stencil-printing process similar to today's screen process to transfer images onto silk. Around 1050 AD, another Chinese alchemist is recorded as having invented a method of printing using baked clay tablets of typematter. It is believed that this system, along with a method of printing from relief-image wood engravings devised simultaneously in Korea, were the earliest attempts at transmitting the printed word[2] onto paper using a process similar to latter-day letterpress printing.

Up until the fifteenth century there is little evidence of printing having taken place in Europe, with most documents being transcribed onto vellum by monks in monasteries. Haarlem in Holland became known as a centre for European printing in the early fifteenth century. However, most credit for the introduction of printing in Europe is given to the German goldsmith Johannes Gutenberg, who in 1448 invented a method of engraving and casting individual metal letters which were composed into words. These words were then formed into lines, and the lines were finally constructed into pages from which multiple copies could be reproduced. This typesetting method, in conjunction with the use of a converted wine press which was used to transfer the prints onto paper, became known as letterpress printing[3] and was in one shape or form the mainstay of print production well into the twentieth century. If illustrations were to be included to enhance the printed page, they were engraved into wooden blocks, which were incorporated with the typematter into the printing forme.

Letterpress printing was first introduced into England from Holland in 1476 by William Caxton, and from there it spread northwards to Scotland. The history of printing across Europe at this time was heavily influenced by both church and state, and such was the case in Scotland. This can be seen in the fact that the first licence[4] to establish a printing press was granted by King James IV in 1507 on the instigation of one Bishop Elphinstone of Aberdeen, whose main objective was to have a Scottish Service Book available for use in Scottish churches rather than that of the 'Salisbury Use' which was used at that time. This licence was granted to Walter Chepman and Andro Mylar, who set up their press in the Cowgate, Edinburgh, to produce 'the books of laws and other books that might be required' by the church and state at that time. The *Aberdeen Breviary* which Elphinstone commissioned was produced in two volumes during the period 1509–10 by Chepman. Using this method of licensing, the number of printing presses in operation was controlled by the crown until the end of the seventeenth century.

Printing presses were established in the towns of Dundee in 1547, St Andrews in 1552, Stirling in 1571, Aberdeen in 1622 and Glasgow in 1638. It should be noted that most of these print shops were very small, normally operated by one or two men at most, with some of them being established by the same person who had travelled from one town to another to establish a new press. This movement to other towns was not only confined to employers, but employees over the centuries have also moved from company to company and town to town seeking employment. This in later years became known as the 'tramping' system.[5]

After the invention of photography[6] in the mid-nineteenth century and the production implications it had on the print industry, a dramatic technological advance took place. The use of photographic techniques for the reproduction of both black-and-white and coloured illustrations allowed more visual images to be produced. In turn, this also enabled offset lithography,[7] which had been invented by Alois Senefelder in 1798 as a method for reproducing quality illustrations and art reproductions, to become the dominant production process of the late twentieth century. The other highly specialised print production processes of gravure, screen process and flexography evolved and developed on the basis of the photographic techniques which had now become available.

INFLUENTIAL SCOTS

The history of Scottish printing is littered with many names and personalities who have had influence not only at home but also abroad. The following is a representative sample of some of the more notable, the list being descriptive rather than prescriptive.

In the late seventeenth and early eighteenth century, Andrew Anderson and family of Edinburgh secured a patent from Charles II, lasting forty-one years, as printer to the College of Edinburgh. This monopoly almost crippled printing in Scotland by giving the family privilege to print and reprint all and sundry books in any language, learned or vulgar, including acts of parliament, proclamations and edicts from the king, and bibles including the psalms for the Churches of England, Ireland and Scotland. During this period the quality of typography and printed matter produced by them deteriorated to what many described as its lowest ebb. After Anderson's death, his widow and heirs were notorious for controversy and legal battles within the industry.

An adversary of the Andersons was one James Watson of Edinburgh, who secured the gift of king's printer after the expiry of the Anderson monopoly. He is credited with the production of several early newspapers, including the *Edinburgh Gazette* in 1699. However, his most notable production was his *History of Printing*, dated 1713, which is claimed to be the first history of printing published in Britain. Watson was concerned with the quality of the work being produced, and his *History* is described as a creditable example of good workmanship of that period.

A new era in printing came to Glasgow in 1738 when the Foulis brothers, Robert and Andrew, were given permission by the college (university) to establish the trade of booksellers in that city. In 1741, Robert Foulis took up the trade of printing, and he was appointed as printer to Glasgow University in 1743. His name is associated with the production of classical works, which were noted for their quality and clarity of reproduction. Foulis waged war against the London bookselling trade on the question of copyright law for a period extending over thirty years.

In 1765 the Edinburgh company Neill & Co. was founded after Patrick Neill's partner, James Balfour, had retired from the business. This company was one of Edinburgh's most prominent printers for the next 200 years, until its demise in 1973. An apprentice of Patrick Neill was one William Smellie, who himself eventually became printer to Edinburgh University. His greatest claim to fame, however, is that of having been editor of the first edition of the *Encyclopedia Britannica*, which when published for the first time in 1771 had been printed by Carruthers and Bell, Edinburgh, who had used copper-plate engravings for the reproduction of the illustrations contained within it.

Up until the middle of the eighteenth century, Scottish printers had to purchase their typematter from London foundries, as there were no typefounders in Scotland. In 1740 Alexander Wilson opened a type foundry in

St Andrews, which he later transferred to the Camlachie area in Glasgow. It was type manufactured in this foundry that Robert Foulis purchased and used for his many publications. During the twentieth century the well-known Glasgow book printer–publisher William Collins and Sons Ltd used the trade name Fontana for their paperback book series. The style of lettering used in the design of the Fontana trademark was based on a typeface originally produced by Alexander Wilson in his foundry in Glasgow in the 1850s.

Around the same period as Wilson was establishing his foundry, another Scot, William Ged, an Edinburgh goldsmith, invented the process known as stereotyping. This process had a huge influence on the industry's ability to produce the cheap editions of books necessary to meet the demands of a growing population. Stereotyping was a method of creating duplicate printing plates from typeset pages, thereby allowing the original type to be reused on other work. The method necessitated the production of a papier mâché mould of the page to be printed, into which molten metal would be poured to create the duplicate printing plate. Plates could be produced either flat for use in letterpress flatbed printing, or curved for use on letterpress rotary presses. These curved stereos became the mainstay of newspaper plate production up until the late 1970s.

The advantage of this system meant that the printer no longer had to keep standing type, which had long been a source of dispute between the printer and his customer. Standing type was the type used by the printer to create a specific image for a customer, and as long as that customer believed the image would be reprinted the printer was expected to keep the image standing, rather than have to reset it again if a reprint was required. For this service the customer paid an additional charge. At the same time, this also meant that the printer would be required to purchase additional typematter from the type-founder for use on other work whilst the original type was standing. Unfortunately, for this reason Ged's invention was not well received by the industry, especially type-founders, and it was not until after his death that its benefits were truly realised.

As already stated, photographic techniques played a significant role in the development of printing in the nineteenth century. Prior efforts to produce relief-image plates of illustrations by chemical etching had been largely unsuccessful until the development of the half-tone principle. In this system, the continuous tone image[8] is photographed through a grid of finely drawn cross-lines. The resulting image is broken down into a pattern of dots of varying size based on the amount of light reflected back through the grid from the original during photographic exposure. The grid, known as a screen, has a pattern of crossing lines in a measured area such as inch/centimetre, the quality of the image to be produced and the substrate it is to be printed on determining the grid used. The half-tone principle is used today by all printing processes for the reproduction of continuous tone images.

From this development, the Scottish scientist and inventor Mungo Pontin is credited with the invention of photo-engraving. It was his experiments into the light sensitivity of certain chromium compounds that enabled others, such as Fox-Talbot, to develop chromium-treated colloids, which could be used as acid resists when etching metal-relief images for letterpress printing and intaglio images for gravure printing.

The new process of lithography was developing as a commercial printing process in Glasgow in the early part of the nineteenth century, but the name of the first printer to use the process is somewhat obscure. In a presentation to the Old Glasgow Club in 1902, entitled *The Early History of Lithography in Glasgow* Thomas Murdoch (of the firm J and J Murdoch) names not only his own company but that of Hugh Wilson, Argyle Street, Glasgow, as well as other companies such as MacLure and MacDonald and Gilmour and Dean as being early exponents of that process.[9] In 1851, MacLure and MacDonald, who had been established in the Trongate, Glasgow in 1835 as engravers and lithographers, installed a Sigl lithographic press from Germany. This press was the first in the UK to be driven by steam power, and it produced 600 copies per hour.

Since the beginning of the nineteenth century and the industrial revolution that ensued, printing became both an established industry and a major employer throughout Scotland, with most towns of any size having their own print works. This was usually coupled with the publication of a weekly local newspaper such as the *Govan Press*, the *Kirkintilloch Herald*, the *Buteman*, and so on. Many of the companies established at that time survive or, through amalgamations, survived until the late twentieth century; for example, both the Trinity Mirror Group (Hamilton) and the Johnson Press (Falkirk) each continue to produce a number of well-known local titles under a licence agreement.

The production of daily newspapers as we know them stems from the repeal of stamp and paper duties in the mid-nineteenth century. The first newspapers printed in Scotland were reprints of the London *Mercurius Politicus* in Leith in 1653. The first legitimate Scottish newspaper, the *Mercurius Caledonis*, was printed in Edinburgh in 1660. Glasgow did not publish its first newspaper, the *Glasgow Courant*, until 1715, but by 1793 it had five newpapers, one of which was the *Glasgow Advertiser and Evening Intelligence* founded in 1783. This newspaper became the *Glasgow Herald* in 1802, then the *Herald* in 1992, and it is deemed to be the oldest continuously produced newspaper in the English-speaking world, pre-dating the London *Times* by two years.

At the turn of the twentieth century, production of weekly newspapers was flourishing. From amongst their number came one of the few printing presses invented in Scotland. Tom Cossar, the son of the founder of the *Govan Press*, invented the Cossar web-fed flatbed rotary. This hybrid machine was unique in that it adopted the web-fed rotary paper-feed principle found on today's newspaper and magazine presses, in combination with a letterpress flatbed printing unit which was commonly used at that time.[10] Tom took his invention to the printing machine manufacturer Dawson, Payne and Elliot in Otley, Yorkshire, who over a sixty-year period manufactured and sold 500 of these machines worldwide. Although there are no records of any of these machines still in use in the UK, this author personally came across incidences of their use in both the Indian subcontinent and East Africa during the late 1990s.

THE INDUSTRY

Printing is a service industry in that most of its products are produced to support other manufacturers of goods or services. To this end, print production can be identified with the four main industrial areas it supports:

Publishing: This includes the production of daily and weekly newspapers, magazines in all formats to suit all age groups and interests, books (both fact and fiction), journals, diaries, calendars and greeting cards of all types.

Packaging: This incorporates wrapping papers, paper and plastic bags, labels of all sorts, cartons both flexible and rigid, food sleeves, glass, metal, plastic and cloth containers, etc.

Commercial Print: Although some of the products in this grouping are of a highly specialist nature, this is the most general of the four service areas and is where most of today's companies compete for business. It includes products such as publicity fliers and leaflets, brochures and instruction material, stationery of all types, posters, billboard and window advertising, and maps and road signs. This area of the trade offers its customers a complete range of services from the concept of the initial design to the completed product. Originally these companies were sited within the confines of the town centre, but with the arrival of the industrial estates and the high city-centre business taxes, many moved to the periphery of the cities. This left a void in the marketplace, which was filled in the late twentieth century by enterprises such as the copy shop – a walk-in, one-stop franchise organisation using the latest technology either in house or in collective groupings, offering a complete on-demand service a 24–48-hour turnaround for work that they cannot handle in house.

Industrial: This is a highly specialist area of print production, with most work completed on the manufacturing site of the product supplier – for example, the production of printed circuit boards and fascia panels for the electrical and automotive industries. When metal such as tin and/or aluminium foil is used for food packaging, printing unit(s) are often incorporated at the end of the metal manufacturing production line. Similarly, this occurs when printing directly onto glass or plastic bottles, and when wood or tiles for decorative purposes are being produced for the building industry.

PRINT PRODUCTION PROCESSES

Just like the various industries it supports, the production processes used to produce print also differ. The following are the traditional printing methods[11] which were used almost exclusively up until the late twentieth century:

- Letterpress printing
- Offset lithographic printing
- Gravure printing
- Screen process printing
- Flexography

However, with the advent of the electronic age and computer technology, newer printing methods[12] have been devised to suit special requirements, including inkjet printing and digital laser printing. Table 21.1 sets out the relationship between the production process used and the products of the industry it supports.

Table 21.1 Print Media Reproduction Methods

Printing processes				
Letterpress Offset Lithography Flexography	Gravure	Screen Process	Inkjet	Digital Laser
Product groupings by process(es)				
Newspapers Books Magazines Publicity matter Stationery Packaging Metal Decoration	Colour supplements Packaging Art reproductions Magazines	Posters Signs on glass, metal, cloth, plastic, etc. Industrial printing Speciality products	Advertising and display matter Signs on cloth, glass, metal, etc. Industrial printing	Limited editions of books/ magazines/ publicity materials Personalised printed products, including packaging

At the present time, offset lithography is the dominant production process, holding approximately 50–60 per cent of the market share. Inkjet printing and digital laser printing have over the last decade grown from nothing to somewhere between 15 and 20 per cent of the market. They have also created new markets that were not previously financially viable using the traditional printing processes. The specialist processes of gravure, flexography and screen process share the remainder. Letterpress, the former dominant process until the mid-twentieth century, has fallen away almost completely, although it is still used on multi-process equipment for some highly specialist products. These figures are reflective of the worldwide market; figures for Scotland follow a similar pattern.[13]

In the industrial printing market, screen process was formerly the most widely used process. Recently, the computer-driven versatility of inkjet printing has enabled it to take a more dominant role, due to the highly specialist nature of some of the products.

EMPLOYMENT IN THE INDUSTRY

Just as there are various production processes, people are employed to complete different jobs within each process. Each job is directly related to the production operation for which they are responsible. In that manner production can be subdivided into the following areas: pre-press production; machine printing; and post-press production and paper converting.

Pre-Press Production

This area is concerned with all the operations up to and including the production of the printing plates ready for the press. Before the introduction of computer technology and desktop publishing (DTP) software, this was a labour-intensive, male-orientated operation consisting of a series of separate highly skilled tasks which culminated in the production of the printing plates. Some female labour had been employed in the industry during the late nineteenth and early twentieth centuries;[14] today these tasks are completed by a small number of very highly multiskilled individuals of either gender utilising computer technology to its fullest extent.

Traditionally the person who sets the type was called a compositor, originally setting the type by hand, or later in the technological evolution utilising one of the 'hot' metal typesetting or latterly photo-typesetting techniques.[15] Compositors worked in the area of the print works known as the case-room, as this is where the type was stored, each different typeface and size of letter being stored in its own individual case. Letter size is based on the point system,[16] with text lettering ranging in size from 6 points to 14 points. Display lettering normally has a point size of greater than 14 points.

The type-case was subdivided into compartments which held the individual letters, numbers, spaces and all other characters, which are referred to as sorts. The top half of the case was known as the upper and the bottom half the lower, and from that derived the different description of typematter into upper case (capitals) and lower case (small letters).

After the typesetting was completed, printed proofs were taken and passed to the proofreader, who checked it for accuracy against the original copy. If there were any errors, it was sent back to the compositor for correction.

Whilst the typesetting was being completed, any illustrations that were to be included were being prepared separately. This consisted of photographing the original illustration, processing it to produce a film negative, retouching it for quality if necessary, printing down the negative onto specially prepared metal, developing the image on the metal, then etching it to produce the printing image. The people employed for these tasks were referred to as camera operators, photo-retouchers and photo-engravers. When the corrected type and the illustrations were ready, the next production operation involved another compositor, known as a stone-hand, bringing all the images together according to the designer's rough, first into individual pages, which were then imposed into formes ready for plate-making or printing.[17]

In the middle-to-late twentieth century, when offset lithography had become the dominant printing process, the pre-press tasks described above were performed using a paper paste-up technique (rather than metal type) onto specially prepared grids, to create the image prior to photographing it to film for plate-making.

Today's pre-press workshops generally consist of either data-transferring words and spaces from a wordprocessing package prepared on another computer, or scanning the hard copy into the company's DTP system via an optical character reader, then checking the transfer for accuracy. The text-matter is then converted to the typeface and size to be used for the finished product. Illustrations are scanned in, brought to the working size and retouched for any quality issues. Pages are made up using an on-screen grid system to bring the text and illustrations together according to the graphic designer or sub-editor's rough. Imposition of the pages into their printing formes and outputting the image either to film for plate-making on a photo-setter, or direct to the photosensitive plate on a plate-setter, is the final operation.

On reflection, it can be seen that most of the tasks required today are similar to those which were necessary by the traditional method. However, the application and versatility of computer technology and DTP software has made this operation far less labour intensive, with one or two people now handling the whole pre-press stage. This technological change means that the number of people employed in this area of the industry has reduced dramatically over the last twenty years.

Machine Printing

Machine printing is concerned with transferring the printed image onto the substrate, which can be almost any material from paper and board through glass, metal and plastic to the finest gossamer cloth, and uses any one of the previously mentioned production processes. This area has traditionally been the domain of male employees; possibly this was related to the physical strength required to lift the heavy printing formes that were used by the letterpress printing process.

Before the introduction of paper-feed mechanisation, many women were employed both as machine paper feeders and auxiliary labour in the pressroom. Since the mid-twentieth century, when offset lithography became the dominant production process, women have been employed in many companies as small offset press operators, i.e., operating presses with a printing sheet size up to and including the paper size A3.

Presses can be sheet-fed, where individual cut sheets of the substrate pass through the printing unit, or they can be web-fed, like those used in the newspaper industry, where a continuous ribbon of the material is passed through the press during and after printing, until it reaches the on-line finishing unit. They are designed to print single colour or multi-colour in one pass through the press, and can print on either one side or both sides of the substrate at the same time. Printing on both sides at one time is referred to as 'perfecting'.

It is imperative that today's press operators have a fine eye for detail and are not colour blind, as they not only have to maintain the quality of the printed colour during the press run but may also be required to mix and match the coloured inks to a pre-set standard beforehand. They must be computer literate, as nearly all press manufacturers have adapted computer technology to assist the pressman to achieve his production targets. Modern presses cost vast capital sums: a typical sheet-fed, single-colour small offset lithographic press that prints one colour on one side in one pass through the press produces 5,000–10,000 copies per hour and would cost in the range of £40,000 to purchase new, whereas the installation of a web-fed newspaper/magazine production unit, printing full colour throughout and producing 50,000-plus complete newspapers or folded sections per hour could cost in the region of £10–30 million, depending on the degree of sophistication of the press and the add-ons required.[18] It can be seen therefore that any mistake even for a brief period of time can have serious financial implications.

Post-Press Production

This area of production, although now highly mechanised, still tends to be labour-intensive, with mostly female labour completing both skilled and semi-skilled work under the supervision of a number of highly skilled male or female operatives employed to set up the production equipment. It is split into two distinct units, depending on what post-press production operations are required to complete the finished product.

Unit A, Print Finishing and Binding, is concerned with products for the publishing and commercial print industries. It includes the print-finishing operations such as paper cutting, folding, leaflet and booklet makeup, wire-stitching, trimming, hole punching and drilling, numbering and perforating, padding, etc; and the binding operations such as folding, gathering and collating, book makeup, thread sewing, case making, book-binding and trimming for both paperback and hardback books.

Unit B, Paper Converting, is concerned with the completion of products for the packaging industry and includes operations such as label cutting, shaping and punching, wrapper slitting, paper and plastic bag making, carton and sleeve manufacture including cutting and creasing, forming and glueing, embossing, foil blocking, film laminating, window punching and plastic insertion.

Normally only one of the units A or B would be part of the print works. Only in exceptionally large companies covering the requirements of all printed products would both groupings be found in house. Similarly, due to the aspirations of today's graphic designers and marketing departments, some of the technology from one group has been adopted to suit products from the other, for example, embossing, foil blocking, laminating and window punching from the paper-converting group is widely used for paperback book covers and greetings cards from the publishing industry. Print finishing and bookbinding units have been established as separate entities offering specialist trade services to the printing industry in general.

PRINTING EDUCATION

The need for printing education was originally based on the employment and 'on the job' training of apprentices within the workshop. The increased mechanisation of the industry in the late nineteenth century and the need for technical training to support it led to formalisation of training. To this extent, further education to supplement training has been available in one format or another since the early 1900s.

First attempts at establishing further education are recorded in Glasgow in 1903 and Edinburgh in 1912, but it was not until the establishment of the Apprenticeship Training Committee in 1928 – a joint venture consisting of representatives of both employer and employee organisations – that the first printing schools were established in the 1930s. Four printing schools were established at Heriot-Watt College – later transferred to Napier College, Edinburgh – Stow College of Printing, Glasgow, the Dundee Institute of Art and Technology – later renamed Duncan of Jordanstone College – and Aberdeen Technical College.

Over the period of the middle to the late twentieth century, responsibility for printing education was reduced to two establishments: Napier College in Edinburgh, which later became Napier University, covered the east of the country, offering print education up to honours degree level; in Glasgow a dedicated printing school, Stow College of Printing, was established in the Cowcaddens area prior to World War II. In the early 1960s this institution was renamed the Glasgow College of Printing when it moved to new premises in the city centre. An amalgamation in the early 1970s with the college of building, with which it shared a site, saw the formation of the Glasgow College of Building and Printing. More recently, via another series of amalgamations the School of Printing is now an integral part of Glasgow Metropolitan College.

The apprenticeship training committee also brought forward a method to assist employers in the selection of young people coming into the industry, thereby ensuring that the apprentice entered the industry with the necessary educational and psychological qualifications. This scheme, after various revisions and amendments, is operated on behalf of the industry by the Scottish Print Employers Federation (SPEF) and continues until the present day.

The government education revisions of the 1950s included the formalisation of further education provision, and the Industrial Training Act of 1964 took the responsibility for industrial training away from the individual employer, transferring it instead to an industry-wide training board. These changes culminated in the establishment of the Printing and Publishing Industrial Training Board (PPITB) in 1968. The aim of the Industrial Training Act was to ensure that enough workers with the required skills were available in the right place at the right time, and to provide the opportunity for individuals to develop their skills and use their abilities to the full.

The PPITB changed the format of printing education by introducing day release/block release classes to all apprentices across the whole country, thereby affording them the opportunity to attain a nationally validated certificate. Achievement of certification was via formalised craft assessments and theoretical examinations taken at the end of year two for Intermediate level and year four for Advanced Craft level. Until the formation of the Scottish Vocational Education Council (SCOTVEC) in the 1980s, this certification was validated and awarded by the City and Guilds of London Institute.

Today, certification of a similar nature is awarded by the Scottish Qualifications Authority (SQA) to apprentices in the Scottish printing industry. It requires successful completion of a series of compulsory and optional modules at Scottish Vocational Qualification (SVQ) level 2 and SVQ level 3. SVQs are designed to assist the learner to reach the National Occupational Standard for their industry, with each award being designed in consultation with the relevant Sector Skills Council. Assessment is based on the competence of the candidate in matching the skills necessary to do his or her job.

Print-related educational qualifications at Higher National Certificate (HNC) and Higher National Diploma (HND) level validated by the SQA can be taken either by part-time or full-time study at the Department of Printing Studies, Glasgow Metropolitan College. The honours degree in graphic communications provided at Napier University is no longer available to prospective students after the university opted out of the provision of printing education in June 2005.

The technological changes of the late twentieth century, with the advent of computerisation and desktop publishing, has seen a dramatic reduction in the number of opportunities for young people entering the industry, especially apprentices in the pre-press area. People are being employed in pre-press in the industry after having taken a full-time course in a related subject area at a college or university. Most apprenticeships, although not exclusively, are offered in the machine printing and post-press production areas.

These changes in the technology within the industry has seen the demise of the printing schools in Aberdeen (1950s) Dundee (1980s) and, more recently, Edinburgh (2005). Only Glasgow Metropolitan college now offers dedicated programmes of study in printing education. Other FE/HE institutions throughout Scotland offer courses in the related subject of graphic design.

INDUSTRY ORGANISATIONS

Since its early beginnings in Scotland at the start of the sixteenth century, printing has been a fluid industry in that owners moved from one town to another to establish businesses. Similarly, as the industry grew, employees also travelled from firm to firm and town to town in search of work. In the days prior to trade societies this led to the 'tramping' system. Under the tramping system an unemployed printer with proof of his credentials could travel from town to town in search of work. He was usually afforded food and lodgings paid for by contributions from the local workforce, and if no work was found he would return to his home town. This system gave rise to the friendly societies, which in later years led to the establishment of both the masters' federation and the employee trades unions. Even into the late twentieth century, an employee of the printing trade union had to supply proof of identity and clearance of their credentials if they wished to move from one employer to another in a different area of the country.

The Scottish Print Employers Federation (SPEF)

This organisation, originally known as the Scottish Alliance of Masters in the Printing and Allied Trades, was founded at a meeting in Edinburgh on 10 November 1910 at the instigation of several local associations, of which the Edinburgh Printing and Kindred Trades Employers Association was perhaps the oldest. Its original remit was to represent the interests of printing employers throughout Scotland in its dealings with trade unions. In 1952 it became the Scottish Alliance of Master Printers, then in 1960 the Society of Master Printers of Scotland and finally, in 1991, it adopted its present title, the Scottish Print Employers Federation. Along with the Society of Master Printers of Ireland, it is affiliated to the British Printing Industry Federation (BPIF).

Membership is open to all companies involved in printing and related activities on payment of an annual fee based on the size and payroll of their business. Membership affords an advisory service on a range of issues, including employment relations; education and training; technical and legal advice; health and safety; and pensions and insurance schemes.

Through its affiliation with the BPIF, it represents the industry's interests with the governments in Edinburgh, London and Brussels, as well as other external bodies. It also has affiliation to other related employers' organisations, such as the Newspaper Society, the Newspaper Publishers Association and the Screen Printers Association.

Young Managing Printers (YMP)
This group provides training and professional development for young managers in the printing industry. It comes under the auspices of SPEF, who provide it with operational guidance. The group was established as the Young Master Printers of Scotland at a conference in St Andrews in 1928. Its remit at that time was to offer a forum for the education and training of the sons (family) of master printers entering the industry. In the early 1960s this group, in conjunction with senior educationalists like Bill Morris at the Glasgow College of Printing, devised and operated a full-time programme of study which was the forerunner of today's higher education HNDs and degrees.

Over the interim period, the group has developed and gone through many changes, not least being the change in name from Young Master Printers to Young Managing Printers. This is in part due to the changing management structure of the industry, with fewer families retaining outright ownership. Membership is by application and payment of an annual fee. The current remit is to organise activities complementary to workplace experiences, thereby developing the skills necessary to be a successful manager; help and advise on career development; and organise social and sporting activities which will encourage interaction amongst the group. Unfortunately, due to the changing nature of the industry, this group is presently dormant.

Amicus: Graphic, Media and Print (GMP)
The organisation of workers within a printing company into a 'chapel' is a custom as old as the industry itself. The origin of the use of the word chapel in this context is uncertain: some explain it as the relationship to the original monks who transcribed the ancient texts, others to the fact that William Caxton, the English printer, set up his press in Westminster Abbey. The original chapel embraced all workmen in the company, including the master and his employees, but later only the employees were included. The father of the chapel in the days of the guilds, when the chapel acted as a disciplinary body, was the oldest freeman in the company. Today, the father or mother of the chapel is the elected official appointed after a vote of the registered members within the group has taken place. He or she is their representative to district and national associations, at the same time working closely with the employers' representatives to resolve local disputes. Mothers of the chapel are normally found in sections of the workforce such as print finishing or paper converting, where the majority of the employees are female.

Trades unions were made illegal by the Combination Act of 1799. In spite of this act, the organisation of labour in the printing industry continued taking the form of trade societies. One such society was the Edinburgh Letterpress Printers Society, which was incorporated in 1758. Early attempts at establishing a national workers' organisation for the Scottish print industry date from 1836, but it was not until 1853 that the Scottish Typographical Association was founded. At its formation, the association consisted of five branches – Glasgow, Edinburgh, Dumfries, Kilmarnock and Paisley – and over the next two decades branches were established across the whole country.

Having established and developed a Scottish print union, the organisation now looked outwards to other print unions across the UK in an attempt to establish a UK-wide body. This took some considerable time and effort, and in fact it was not until 1902 that the National Printing and Kindred Trades Federation, of which the STA was a founder member, was formed. This body was not a trade union in itself but a federation of trade union representatives who, with similar problems, were looking for support and guidance.

During the next century many of the skilled and specialist unions merged to form the National Graphical Association (NGA) and the Society of Graphical and Allied Trades (SOGAT). In 1991 the Graphical, Paper and Media Union (GPMU) was formed by a merger of the NGA and SOGAT, to represent all grades of workers in the print, paper and media industries across the UK. Finally, in 2004 the GMPU merged with AMICUS, helping to create the biggest private-sector trade union in the UK.

Over the years of their existence, the printing trades unions have made a notable contribution to the labour movement across the whole of the UK, including assisting in the formation of the TUC and the Labour party, as well as being in the forefront of workers' rights and conditions of employment.

THE FUTURE

Printing as an industry was needed due to the population explosion which accompanied the industrial revolution and the people's desire to become more educated. It has survived for 500 years because it is a dynamic, vibrant, flexible industry which is capable of changing as and when necessary. It is an essential part of today's society, with more printed matter being produced than ever – so much for the doom-and-gloom projections of the 1980s and the 'Goodbye Gutenberg' slogan predicting the demise of the printing industry. Today it is one of a number of media focusing on communication. It competes with and is complemented by other media such as the internet, television, radio, CDs and DVDs, the concept of multimedia products being common in both the publishing and packaging industries. Printed materials are easy to use and do not need to be booted up, like a computer program. They can take the occasional hard knock, can be taken and used almost anywhere, and when finished with, they can readily be recycled into other products.

Today the industry is more compact and focused: gone are the days of the very large labour-intensive companies. Print shops have become digital factories, and integrated digital production has had an impact across all areas of the print industry. The profession of printer, whilst retaining the highly skilled elements, is more that of a data manager. Cross-media and electronic information processing has seen the adoption of new, faster pre-press production techniques where the customer can become part of, and have greater control of, the production process.

The use of electronic-driven, computer-controlled presses, where multiple print processes are undertaken on the one production line, is now common, especially for the packaging industry. The introduction of digital printing and inkjet printing from their conception in the 1990s to mainstream processes in their own right, and as supplementary processes to the more traditional production methods, has added a further dimension to the range of products being produced. These two processes have opened up new markets which just a few short years ago were inconceivable either via the production techniques available at that time or from the cost of production. At the same time, they have enabled traditional printing companies who have embraced this technology the opportunity to offer their customers extras to their print media that were not feasible until recently.

As the printing industry in Scotland enters its sixth century, it has much to be proud of when historically reviewing its place in the evolution of this country. It now finds itself at a very important and exciting time, with competition in the communications market perhaps greater now than at any time previously. However, if it adopts and embraces the challenges of 'new' digital technology in the same way as it has absorbed other changes over the centuries, there is no apparent reason why it should not continue to flourish as one of the country's major industries.

NOTES

1. Barnard, 2000, 192–235.
2. *Encyclopedia Britannica,* 1995, vol. 26, 71–4.
3. Durrant, 1989, 2–6.
4. Scottish Printing Archival Trust, 1990, 1–6.
5. Gillespie, 1953, 77–83.
6. *Encyclopedia Britannica,* 1995, vol. 25, 761–4.
7. Durrant, 1989, 8.
8. A continuous tone image is an original photographic print in black and white, or full colour, or the artist's original pencil/charcoal drawing, or painting where the image graduates continuously from white through the mid-tones to black.
9. Murdoch, 1902.
10. Barnard, 2000.
11. Barnard, 2000, 192–225.
12. Barnard, 2000, 226–35.
13. UK Industry Review 2004/2005, *Printweek.*
14. Gillespie, 1953, 101–8 and 203–7.
15. Durrant, 1989, 71–9.
16. Barnard, 2000, 31–5.
17. Imposition is the technique used by the printer to lay out the pages in a sequence ready for printing which will enable them to run consecutively after printing and folding has taken place.
18. *Prepress News; PrintWeek; Printing World; British Printer; Seybold,* 2005.

BIBLIOGRAPHY

Barnard, M, ed. *The Print Production Manual*, 8th edn, London, 2000.
Boyter, I. *Robert Smail's Printing Works*, Edinburgh, 1990.
British Printer, 2005.
Durrant, W R. *Printing: A Guide to Systems and Their Uses*, Oxford, 1989.
Encyclopedia Britannica, 15th edn, vols 25 and 26, Chicago, 1995.
Gennard, J. *A History of the National Graphical Association*, London, 1990.
Gillespie, S C. *A Hundred Years of Progress*, Glasgow, 1953.
Murdoch, T. *The Early History of Lithography in Glasgow – Being the substance of a paper read before the members of the Old Glasgow Club*, Glasgow, 1902.
Prepress News, 2005.
PrintWeek, 2005.
Printing World, 2005.
Reynolds, S. *Britannica's Typesetters: Women Compositors in Edwardian Edinburgh*, Edinburgh, 1989.
Spiers, H M. *Introduction to Prepress* and *Introduction to Printing and Finishing*, PIRA/BPIF, London, 1998.
Scottish Printing Archival Trust. *A History of the Edinburgh Printing Industry*, vol. 1 of the trust's A Reputation for Excellence series, Edinburgh, 1990.
Scottish Printing Archival Trust. *A History of the Glasgow Printing Industry*, vol. 2 of the trust's A Reputation for Excellence series, Edinburgh, 1994.
Scottish Printing Archival Trust. *A History of the Dundee and Perth Printing Industries*, vol. 3 of the trust's A Reputation for Excellence series, Edinburgh, 1996.
Scottish Printing Archival Trust. *A History of the Aberdeen and Northern Counties Printing Industries*, vol. 4 of the trust's A Reputation for Excellence series, Edinburgh, 2000.
Seybold, 2005.
Smail, J C. *Printing in Scotland 1507–1947*, Dundee, 1953.

Websites
Amicus: http://www.amicustheunion.org
Pira International Limited: http://www.pira.co.uk
Scottish Print Employers Federation: http://www.spef.org.uk
technotrans AG: http://www.globalprint.com
The Scottish Printing Archival Trust: http://www.scottishprintarchive.org

22 The Lives of Scottish Book Traders, 1500–1800

ALASTAIR J MANN

From the arrival of the printed book in Scotland in the late fifteenth century and the advent of the press in 1508 to the great flowering of print culture in the Scottish Enlightenment, the progress in Scottish intellectual culture depended on a diverse band of book merchants and book makers. This group varied in wealth, capacity for inventiveness, political and religious beliefs, and links with the establishment, but nevertheless had much in common. Not least of these common bonds was the requirement to ply their trade in the same 'national crisis' of Scotland's early modern period. From Flodden in 1513 to Culloden in 1745, warfare, religious revolution, civil war and economic collapse battered Scottish society. These political and religious upheavals presented a rigorous challenge for the Scottish printer and book trader before the outstanding successes of the Scottish Enlightenment. After all, print merely precipitated and reflected the qualities of national history. But even before the Enlightenment, we should marvel at the success and resilience of Scottish print culture and its mediators.[1]

Printing itself arrived in Scotland in 1507–8, a few decades after England but before Russia and some of the Scandinavian countries. Walter Chepman (*c*1473–*c*1528) and Andrew Myllar (fl.1503–8) began printing in Edinburgh in 1508 under a licence provided by James IV. Nothing survives of their press after 1510, and though there are a few significant productions from a scattering of other presses, notably by Thomas Davidson (fl. 1532–42) (see Figure 22.1), printing was not firmly established in Edinburgh until after 1560. As print slowly expanded in the capital, before some faster growth in the early seventeenth century, presses were established in Aberdeen (1622) and Glasgow (1638).[2]

Slow economic growth and slow local patronage explain this tardiness outside Edinburgh, there being no government or trade restrictions on press proliferation. In its first century the small Aberdeen press proved remarkably innovative in music publishing and almanac printing, but the eighteenth century also saw the coming of age for the Glasgow press, which looked to supply the demand of the Americas and Ireland. Glasgow also supplied Europe, and in particular the press of Robert and Andrew Foulis (fl. 1746–76) became famous throughout the continent for quality editions of Latin and Greek classics.[3] Therefore, the 'golden age' of the Enlightenment

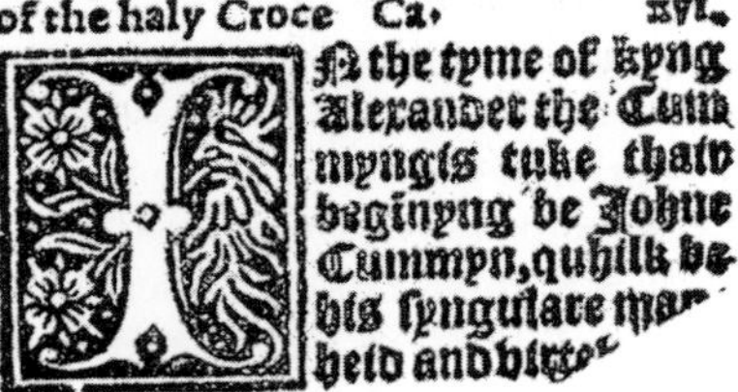

¶Hovv kyng Dauid past to the huntis on the Croce day in heruest. Hovv he vvas doung fra his hors be ane vvyld hart. And hovv he foundit the abbay of Halyrudhous be myracle of the haly Croce Ca. xvi.

In the tyme of kyng Alexander the Cummyngis tuke thair beginnyng be Johne Cummyn, quhilk be his syngulare man-heid and virtu

Figure 22.1 A woodcut of the Crucifixion from Thomas Davidson's fine printing of Hector Boece's *The History of the Croniklis of Scotland* (c1536–41), translated from Latin to Scots by John Bellenden. This high-quality early domestic printing shows that Scotland's relatively late start and uncertain establishment of the press did not necessarily mean poor workmanship.

was not merely an Edinburgh event, with the press throughout Scotland playing its part. Nonetheless, although printing spread out to other burghs during the eighteenth century, the domestic press was never able to meet domestic demand. For this reason booksellers were more important mediators than printers to the development of Scottish print culture before 1800.

Who were these early modern book traders of Scotland? They regarded themselves variously as book merchants, stationers, printers, booksellers and bookbinders. Essentially, however, there were two types of book trader: the printer who may also have been a bookseller; and the bookseller who may also have been a bookbinder. The only complete specialists were some printers. In Scotland the term 'stationer' was always synonymous with bookseller until the mid-eighteenth century, when it began to take on its modern meaning. In fact it was one of the quirks of the Scottish trade that from the 1670s to 1690s 'stationer' became a fashionable label. This fashion reflects the self-conscious and polite intellectualism that developed in the Restoration period and supports Houston and Allen's views on the early foundations of the Scottish Enlightenment.[4]

The detailed accumulation of references from testaments, inventories, wills, council registers and burgess and apprentice rolls, supplemented with bibliographical and biographical data, has enabled estimates to be made of the numbers of book traders active in Scotland between 1500 and *c*1750.[5] The number of traders in the sixteenth century was small but growing, rising from half a dozen throughout the century to over twenty in the 1580s and 1590s, with the greatest expansion being in the number of booksellers rather than the more expensive business of printing. The first period of take-off in numbers was the 1630s and 1640s, when the printing of scripture and religious controversy fueled trade expansion. Further dramatic increases in numbers took place after the Cromwellian period, but growth was by no means constant. Following the Restoration boom from the late 1660s to 1680s, the number of Scottish printers more than doubled from twenty to forty-five, while over the same period the number of booksellers rose from about forty to seventy. The number of printers was slight before the 1650s, but bookselling, whether by printers or specialist book vendors, was obviously well established. The 1650s represented an interesting watershed. While this was a period of contraction in the book trade of Edinburgh, we see from the 1650s the expansion of bookselling into the corners of the kingdom. Inasmuch as there was a Cromwellian recession, it was based in the capital as far as the book trade was concerned. New booksellers were to be found in many burghs, reaching a high point in the 1670s. They were operating in such places as St Andrews and Perth, and for the first time appear in Dundee, Ayr, Dumfries, Lanark and Kilmarnock, and even the northerly burghs of Banff and Forres. The spread of domestic bookselling, along with slowly increasing literacy, was the great catalyst for the demand side of the book trade of Scotland when the domestic press could not meet the needs of the Scottish reader. This all sounds promising from a commercial point of view, yet the activity figures show that the recession of the 1690s hit the book trade throughout Scotland in both printing and bookselling, even though a recovery set in after the Union of 1707.

These figures confirm that the scale of the Scottish book trade was greater than pessimists would have us believe. In fact in total over 200 printers and press partnerships and 500 booksellers and bookbinders were active from 1500 to 1730.[6] The Scottish book trade of 1707 was about four times the size it was in 1603, while surviving editions recorded by Aldis in his pre-1700 catalogue show press output to have increased by perhaps eight times, a factor created by the poor survival rate of earlier productions but also resulting from increased productivity.[7] Nonetheless, before the late eighteenth century the geographical spread of booksellers and book makers differed: the former increasingly spread throughout the country; the latter mostly focused on the printing burghs of Edinburgh, Aberdeen and Glasgow.

Were these book traders book specialists? In fact, many were merchants of other goods and products. Specialist or 'permanent' booksellers did not emerge until the late sixteenth century, where in Edinburgh people like Robert Gourlaw (fl. 1580s) and Andro Hart (fl. 1587–1622), later a major printer, were very active.[8] However, many were related to other trades. The Edinburgh bookseller Edward Cathkin (fl. 1585–1601) was formerly a skinner, a link to the related trade of bookbinding, and so also was another Edinburgh bookseller, Richard Lawson (fl. 1603–22). Lawson clearly saw the commercial prospects of bookselling, as did Cathkin's brother James (fl. 1601–22), who took over from Edward after the latter's death in 1601. Specialisation in specific genres appeared in Edinburgh from the 1630s trade expansion, as seen with Andrew Wilson (fl. 1634–54) and John Vallange (fl. 1678–1712), school and law book specialists respectively.

Booksellers from the smaller burghs and towns were more likely to diversify. In Perth the bookseller Andrew Watt (fl. 1678–85) was also a barber, and the other Perth booksellers, James and Patrick Black (fl. 1680–90s), were glaziers. The Kelso bookbinder Robert Cathcart (fl. 1694) sold medicines as well as books. Even in the eighteenth century, trade diversity was a feature. The Glasgow bookseller John Greig (fl. 1730–41) was also a saddler. The testament of the Aberdeen printer and bookseller James Nicol (fl. 1710–49) reveals a large quantity of haberdashery, cloth and household goods indicative of a general store. Nicol's widow auctioned his book stock in 1749–50, and auctioneering also became associated with some book traders, including the Edinburgh booksellers John Tennant (fl. 1690–1718), David Freebairn (fl. 1689–1714) and James Davidson (fl. 1719–40), and the Aberdeen bookseller David Angus (fl. 1739–48). Book-stock auctions, mostly in the capital, became common from the 1690s.[9]

Another eighteenth-century diversification was in papermaking or paper sales, but only for the most wealthy of book traders. Papermaking, a natural cousin of book making, took time to develop as a viable industry in Scotland. Indeed, Scottish papermaking was primitive until the 1690s, even though Scotland's first paper mill, located at Dalry in Edinburgh, was set up in 1590 by Mungo and Gideon Russell.[10] Foreign labour and expertise became essential. After the Restoration, Scotland was indebted to the German Peter Breusch (Bruce), who became a printer as well as a papermaker, and to Frenchmen like Nicolas Dupin who, in 1694–5, brought financial and practical expertise to the establishment of the Society of the White-Writing and Printing Paper Manufactory.[11] The granting of a charter to this joint-stock company was a deliberate attempt by the government to put domestic paper production on an economically viable and qualitative footing. Nonetheless, most Scottish press papers still had to be imported. In spite of these unpromising circumstances some printers entered into papermaking in the early eighteenth century, including the wealthy royal printer Agnes Campbell (fl. 1676–1716), who in 1709 acquired land from Sir John Clerk of Penicuik and there established the Valleyfield Mill on Esk Water. Five years after her death in 1716 the mill, now run by her daughter, was in financial trouble, a clear sign of the risks in such enterprises.[12] A more common diversity was paper wholesaling, where some larger printers held stock to supply smaller and provincial presses. The Edinburgh printers John Moncur (fl. 1707–29) and Gavin Hamilton (fl. 1730–64) both acted as paper wholesalers, though they also became partners in various paper mill schemes.

These book traders had to be educated to a degree, literacy being a requirement, even though the level of learning ranged from such people as Henry Charteris (fl. 1568–99), the most important printer and bookseller of the sixteenth century and publisher of George Buchanan, down to the simple necessities for the humble press journeyman.[13] Some book traders were very proactive over education, and the Foulis brothers in Glasgow went as far as establishing an academy of fine arts within Glasgow University in 1753, though it had to close in 1776.[14] Robert Foulis began his working life as an apprentice barber, although individuals usually entered the book trade either through family ties or through apprenticeships. The apprenticeship system was maintained by the burghs at a level beneath their control over burgess and guild membership. Both booksellers and printers practised the system, mostly over seven or five years. Although the general relaxation in the use of apprenticeships was a feature of the freeing-up of trade in the

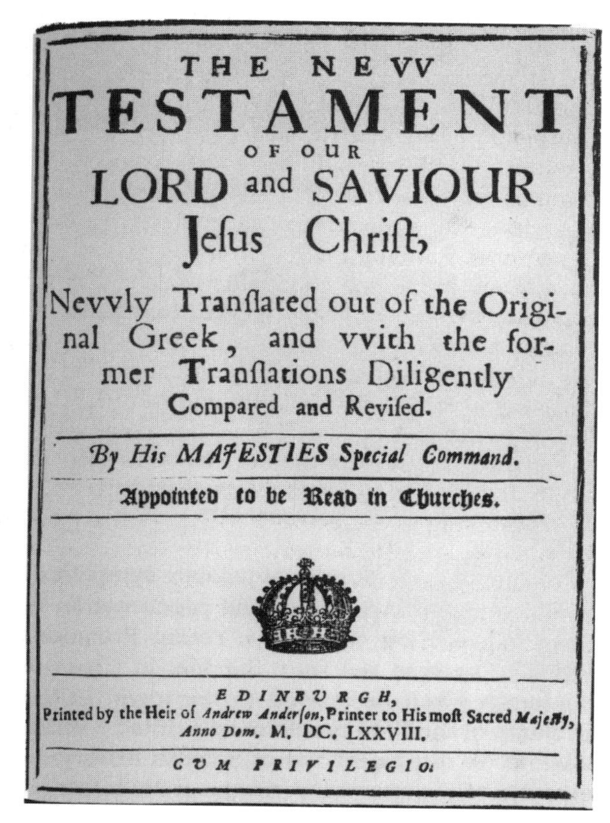
THE NEVV
TESTAMENT
OF OUR
LORD and SAVIOUR
Jesus Christ,
Nevvly Translated out of the Original Greek, and vvith the former Translations Diligently Compared and Revised.

By His MAJESTIES Special Command.

Appointed to be Read in Churches.

EDINBURGH,
Printed by the Heir of Andrew Anderson, Printer to His most Sacred Majesty, Anno Dom. M. DC. LXXVIII.

CUM PRIVILEGIO.

Figure 22.2 The title page of the *New Testament*, printed by Agnes Campbell in 1678. Campbell dominated the Scottish printing industry between 1676 and 1711 and was one of the wealthiest female merchants of the early modern period.

post-Restoration period, the apprenticeship system was surprisingly resilient. When, in 1725, the Stirling merchant guild set out new and strict rules for apprenticeships, they were reflecting the approach of most burghs.[15] Some of the book-trade apprentices were the sons of baxters, shoemakers, tailors, maltmen and gardeners, but many came from educated families. The apprentice rolls do not always cite fathers' occupations, and some fathers are merely described as 'indwellers' or appear to be lairds, but the Edinburgh rolls from 1583 to 1800 show at least twenty sons of ministers apprenticed to the book trade, including successful individuals such as the Edinburgh printer John Wreittoun (apprenticed 1609) and Edinburgh bookseller Alexander Ogstoun (apprenticed 1651). Strangely, the last minister's son in the roll is George Meikle, apprentice in 1748 to Gideon Crawford, bookseller of Edinburgh, and son of Alexander, minister of Langholm. Sons of clergy seem by 1750 to have regarded a life in trade as below their station. However, the book trade continued to attract the educated. The second largest identifiable occupation group are the sons of writers, and these sons of notaries and clerks continued to be attracted by a career in the book trade throughout the eighteenth century, a century in which the sons of schoolmasters and army officers also joined the world of books.[16]

Although the burgh magistrates controlled trade matters, sometimes the privy council and court of session had to step in to regulate trade apprenticeships. Agnes Campbell used the courts and council to restrict the activities of her Edinburgh apprentices. In 1680 Patrick Ramsay and John Reid, senior, were prevented by Campbell, via the privy council, from setting up their own press, as they had not completed their full apprenticeships. James Watson the younger (fl. 1695–1722), used the court of session in 1714 to prevent the premature departure of two apprentices. Although in 1633 the young bookbinder David Robeson was supported by the privy council and freed from forced servitude with the bookbinder Monases Vautrollier, it must have seemed to the apprentices of early modern Scotland that very little 'freeing-up' was taking place.[17]

Entering the trade as part of the family business was very common, and various family dynasties existed from the 1590s onwards. For example, in Edinburgh Henry Charteris was succeeded in 1599 by his son Robert (fl. 1599–1610); in Glasgow the burgh printer Robert Sanders (fl. 1661–94) was succeeded in 1694 by his own son Robert the younger (fl. 1695–1730); and in Aberdeen the printer–bookseller John Forbes (fl. 1650–75) was succeeded in 1675 by his son John the younger (fl. 1662–1704), though in this case the son printed for a period while the father sold books and edited and reissued his famous musical collection *Cantus, Songs and Fancies*, first published in 1662. Sometimes sons were sent elsewhere to learn the trade, as in 1667 when William Kerr, the Aberdeen journeyman printer, sent his son Andrew to be apprenticed to Joseph Storie, printer in Edinburgh. We have no knowledge of a connection between William Kerr and Storie, but contact by marriage, time served in apprenticeship or business partnership would be typical linkages across the trade as a whole.[18]

The Edinburgh Presbyterian book-trade network that existed from the 1580s to the 1640s provides us with the best illustration of that combination of business and personal relationships. This was a group of men and women committed to trade, yet also to strongly felt religious beliefs. This network can be traced from the clerical subscription crisis of 1584–5, when there was controversy over the imposition of an oath of obedience to the so called 'Black Acts' (1584) sustaining crown supremacy over the Church, to the covenanting revolution of 1638–39.[19] The key participants in this line were Andro Hart (fl. 1587–1621), the most wealthy and most significant bookseller and printer/publisher before the Restoration, and his third wife, Janet Kene. However, the first book traders to appear in the nonconformity movement were the brothers Edward and James Cathkin from the 1580s. These 'Melvillians', Presbyterian followers of the divine Andrew Melville, were banished in the summer of 1584 for refusing to subscribe to James VI's episcopalian policies. Also both, along with Hart, were arrested in the Edinburgh Presbyterian riots of December 1596. It was after these riots that King James cowed the Edinburgh town council into giving up Presbyterian clergy and its own overt opposition following his threat to move the capital elsewhere. But King James's victory over Edinburgh Presbyterianism, and his re-emphasis on the importance of bishops, did not end the activities of the nonconformist clergy and their associated printers. The historian David Calderwood (1575–1651) became a champion of Presbyterianism and employed presses at home and abroad, especially out of Leiden and Amsterdam, from whose presses over a dozen of his Presbyterian tracts were produced from 1619 to 1624. In Scotland in June 1619 there was a detailed investigation into the printing and distribution of Calderwood's anonymously published *Perth Assembly*, a tract which railed against the Five Articles of Perth, the new and quasi-Anglican ritual forced through the general assembly that met in Perth in 1618. The Presbyterian book traders of Edinburgh were the prime suspects, and the houses and booths of Hart and Richard Lawson were searched and ransacked, and both were arrested. The bookseller James Cathkin, happening to be in London, was interrogated by the king himself. In the end, for lack of evidence, little action was taken against these merchants, but undoubtedly the Presbyterian network had distributed Calderwood's works and much else of a Presbyterian hue.[20]

These book traders were connected by religion and ink, yet the linkages provided by their wives was of special significance. After James Cathkin died in 1631, his wife Janet Mayne, sister-in-law to Richard Lawson, continued bookselling until her death in 1639. Janet Kene (fl. 1621–41), Hart's widow, with the help of her sons, actively maintained her husband's press until it passed to James Bryson in 1639. On her death, her bookshop fell into the hands of John Threipland who had been apprenticed to and worked for James Cathkin. In addition, Janet's sister Margaret Kene married the printer John Wreittoun, who had been operating a press from at least 1624. This extensive and expanding range of book-trade and nonconformist connections provided the print lubrication for dissent leading up to the revolution

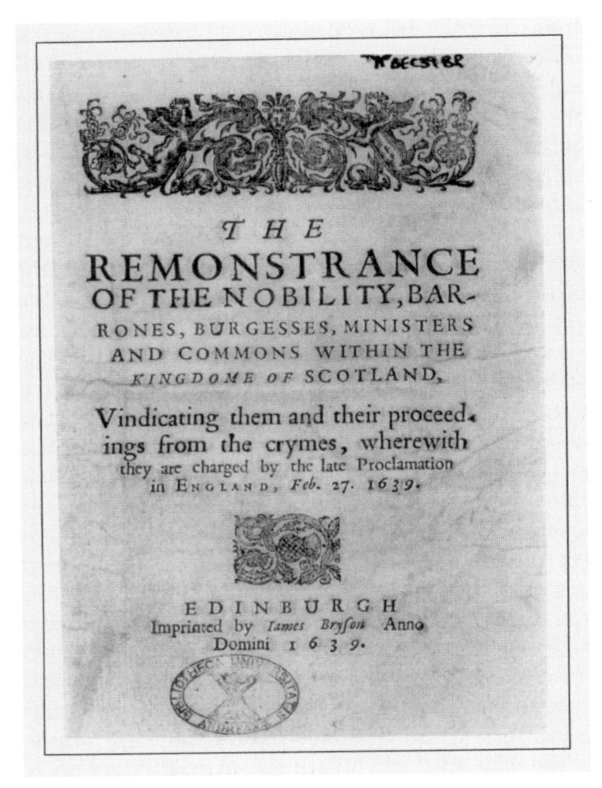

THE

REMONSTRANCE

OF THE NOBILITY, BAR-
RONES, BURGESSES, MINISTERS
AND COMMONS WITHIN THE
KINGDOME OF SCOTLAND,

Vindicating them and their proceedings from the crymes, wherewith
they are charged by the late Proclamation
in ENGLAND, *Feb.* 27. 1639.

EDINBURGH
Imprinted by *Iames Bryson* Anno
Domini 1639.

Figure 22.3 The title page from James Bryson's *The Remonstrance of the Nobility, Barrones, Burgesses, Ministers and the Commons of Scotland* (1639), one of the printings commissioned by the covenanters explaining their rebellion against the king to their brethren in the parliament of England. Bryson was part of the Presbyterian print network so important to the spread of covenanting propaganda. Source: courtesy of St Andrews University Library.

of 1638–9. Furthermore, when the covenanting government required to justify its policies and actions to the parliament of England and the wider world, it turned to the presses of Bryson (see Figure 22.3) and Wreittoun, along with, from 1638 to 1640, a new flood of tracts imported from the presses of Amsterdam and Leiden.

Therefore, this distinct book-trading community was held together by the five families: Hart, Cathkin, Bryson, Kene and Mayne. Two sets of sisters, Mayne and Kene, straddled the bookselling and printing branches of Edinburgh book commerce. In political terms, they also bridged the gap between the old Melvillian religious nonconformity and the robust declarations of the National Covenant. This coincidence of political and social connections went beyond mere links arising from normal trade intermarriage yet nonetheless underscores the character of book-trade networks in the early modern period.[21]

Marriages were not necessarily especially political and, no different from other trades in the social mixture of peers and fellow craftsmen, marriages were frequently within trade. For example, Elizabeth Brown, daughter of the Edinburgh printer, bookseller, bailie and council treasurer Thomas Brown (*c*1658–1702), married the Edinburgh law bookseller John

Vallange. In addition, Thomas Brown's first wife was Marian Calderwood, a relative of the stationer John Calderwood (fl. 1676–82), and Brown himself entered into a series of deeds and contracts with his son-in-law and Agnes Campbell in the 1680s and 1690s. Also, Campbell's eldest daughter Issobel married the Edinburgh bookseller William Cunningham in 1676. In an act common to the book trade, the following year Issobel facilitated her husband's elevation to a burgess of Edinburgh 'by right of his spouse Issobel, daughter to umquhyle [late] Andrew Anderson'.[22] These family networks were complex and interwoven, and the dowry of a 'burgess ticket', allowing an individual to trade as a burgess, was an added attraction to prospective husbands.

The contribution of wives, daughters and especially widows to the Scottish early modern book trade has only recently received much-needed attention.[23] There are two major figures. Janet Kene, as we have already noted, continued Andro Hart's press and was a formidable operator in her own right. It was she and not the king's printer who was chosen by the magistrates of Edinburgh to produce the special edition of poems *Escodia Musarum Edinensium in Carole* presented to Charles I to commemorate his coronation visit in 1633, and her appeal to the lords of exchequer in 1632, against the licence as Scotland's royal printer being granted to the Englishman Robert Young and not a Scot, was respectfully listened to if not successful.[24] Agnes Campbell would become not only royal printer but early modern Scotland's most wealthy female merchant in any trade, not having inherited significant wealth. Her husband Andrew Anderson's legacy was his patent as royal printer and not a strong financial position – he was in debt to the tune of nearly £7,500 at his death in 1676. Campbell went on to develop a large trading network supplying book stock, paper and capital to the printers and booksellers of Glasgow, Aberdeen, Belfast, Londonderry and Newcastle, and trading extensively with the London trade. At her death she had accumulated a fortune of £78,000 Scots (wealthy in English as well as Scottish terms) and operated a print shop with many presses and apprentices. She became notoriously litigious in defence of her privileges and patent.[25] She also fought a bitter war of words and printed slanders against her trade enemy James Watson the younger, author of the first print history published in the British Isles, *The History of the Art of Printing* (1713) (Figure 22.4). Eventually this dispute reached the court of session in the case *Watson, the younger* v. *Freebairn, Basket and Campbell* (1713–18), concerning the validity of co-partnerships over the gift of king's printer.[26] However, as women in the book trade, the high-profile Kene and Campbell are more representative than we might think.

It would be wrong to think that women were merely a cheap, 'informal' labour force in Scotland's book or print shops. From 1600 to 1750, perhaps up to thirty Scottish women were professional book traders. Women printers traded in the names of fathers, husbands and sons, although some booksellers did so under their own names.[27] Before 1600, very few are known to have actively joined the family businesses, but post-1600 the list is dominated by widows, with some exceptions, such as the daughters of the bookseller James Harrower (fl. 1638–51), and Janet Hunter (sometimes Mrs Brown) (fl. 1722–35), a co-printer with a number of partnerships of Glasgow printers in the 1730s who, along with the booksellers Martha Stevenson (fl. 1690–1732), Anne Edmonstoun (fl. 1733–44) and Jean Smith (fl. 1722–31), traded before widowhood. Those widows or relatives who, after their husbands' deaths, kept printing and bookselling businesses turning over for a short period before sale by auction, before sons coming of age or until a suitable second marriage was agreed, make up an expanding list of book trade 'professionals'. Second marriages were frequently sought to carry on the family business and to transfer assets into competent hands for the greater benefit of the family. For example, Beatrix Campbell, widow of the bookseller and printer Archibald Hislop (fl. 1670–8) and sister to Agnes Campbell, maintained her husband's bookselling business for at least

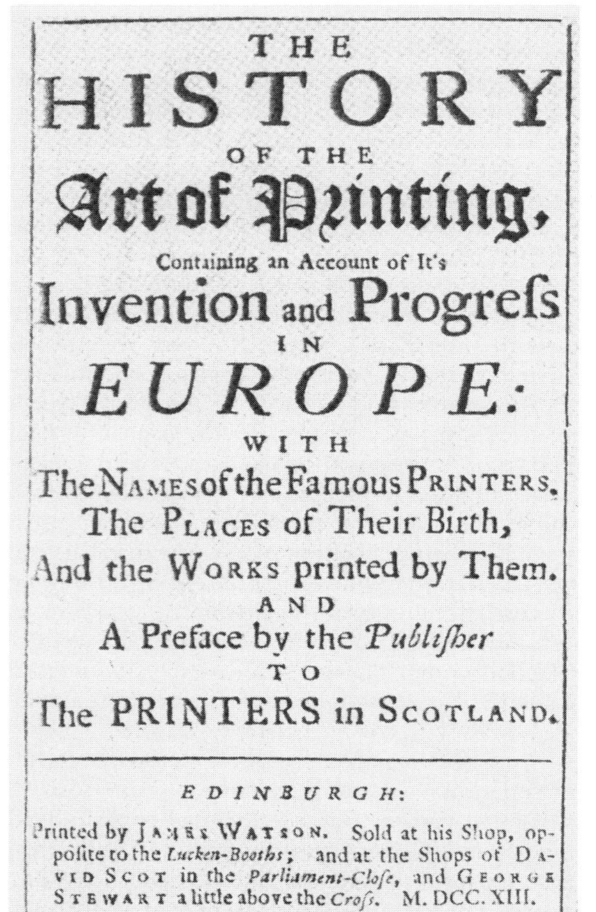

THE
HISTORY
OF THE
Art of Printing,
Containing an Account of It's
Invention and Progreſs
IN
EUROPE:
WITH
The Names of the Famous Printers,
The Places of Their Birth,
And the Works printed by Them.
AND
A Preface by the *Publiſher*
TO
The PRINTERS in Scotland.

EDINBURGH:
Printed by James Watson. Sold at his Shop, oppoſite to the *Lucken-Booths*; and at the Shops of David Scot in the *Parliament-Cloſe*, and George Stewart a little above the *Croſs*. M. DCC. XIII.

Figure 22.4 The title page of James Watson's *The History of the Art of Printing* (1713), the first print history published in the British Isles.

twelve months following his death and until her marriage in 1679 to the 'wryter'-turned-stationer Robert Currie. Issobel Harring (Herron), widow of the printer Robert Bryson (fl. 1637–45), acted in a similar manner to Beatrix Campbell before her second and judicious marriage to the printer Gideon Lithgow. Issobel printed as the 'Heirs of Robert Bryson' in 1646. These women together represent perhaps ten per cent of Scotland's identifiable book traders in the early modern period. Interestingly, the number of these participants, full-time and part-time, major or minor, suggests that women book traders were more common in Scotland than in England. This activity is highlighted by the independent Margaret Reid (fl. 1712–20), daughter of the Edinburgh printer John Reid, senior (fl. 1680–1712). After the death of her father in August 1712, Margaret took over the use of some of her father's type, acquired the printing office of the deceased Andrew Symson and then set up on her own. This appears to be the only example of a Scottish woman setting up a printing press without the 'partnership' of a male, dead or alive, and accounts no doubt for the anonymity of her printings. The last we know of her printing activity is a dispute with the famous poet and bookseller Allan Ramsay (1686–1758). Her printing of one of his poems without permission led him to mock her in his 'Elegy on Lucky Read'. However, in spite of such highlights, female involvement declined in the second half of the eighteenth century as the onset of joint-stock companies and business partnerships, along with new social attitudes about correct female behaviour, saw women leave the stage. Before then the daughters of book men came with added value: useful book-trade experience.

Levels of profit from the business of books and the value of estates passed on to widows fluctuated enormously. An analysis of nearly 100 surviving notarial testaments and inventories of printers and booksellers from 1577 to 1766, coming from the 'traditional' book burghs of Edinburgh, Aberdeen and Glasgow as well as the smaller towns of Perth, Lanark and Dumfries, confirms an interesting and varied picture. There were a few 'super rich', such as Andro Hart (d. 1621) and Agnes Campbell (d. 1716), with estates of £20,000 and £78,000 Scots respectively. However, Scotland's wealthy book merchants were not wealthy individuals in English terms. Hart and other Scottish printers such as Robert Bryson (d. 1642) and James Watson the younger (d. 1722) would be placed in the lower reaches of the middling wealthy, perhaps £1,000 to £5,000 sterling (using £12 Scots to £1 sterling). Only the royal printer Agnes Campbell would fall into the £5,000 to £10,000 sterling band for substantial London merchants. Earlier merchants, and in particular the great vernacular publisher Henry Charteris (d. 1599, estate worth £7,000), were also of substantial means before the last formal devaluation of the pound Scots in 1601. Charteris's estate located him amongst his contemporary merchant élite, while his status as a bailie of Edinburgh placed him above some of greater wealth. Having both become printers after earlier careers as booksellers and commissioners of print, the careers of Charteris and Hart show that printing was the more profitable career path. Two of Scotland's most successful specialist booksellers, the Edinburgh pair

Andrew Wilson (d. 1654, estate worth £15,000), and John Vallange (d. 1712, estate worth £12,500), were some way behind major individuals who focused more on printing. The phenomenon of wealthy copyholding booksellers did not develop in the prominent way it did in England.[28] There it was encouraged by the patenting system managed by the Stationers' Company of London, the merchant guild that tightly controlled the English press from its Elizabethan foundation.[29]

The book trade was not merely the preserve of the most wealthy. Excluding the large estate of Agnes Campbell, the surviving testaments reveal an average estate value of £2,500 Scots. Including Campbell, less than one in ten had estates of over £10,000 and a third had estates of very small value, but a large middling group existed, consisting of about 13 per cent, with between £2,500 and £5,000 stock and personal wealth. A great number of book traders within and without of Edinburgh, over a third of the total, had estates under £500 in value. The wealth of book traders, craftsmen or merchants, did not in fact differ greatly from other trades, such as the wealth of apothecaries and surgeons, as revealed in MacMillan's study. However, a higher incidence of bankruptcy is found in the book trade, at a level almost double that of craftsmen as a whole. At death over one in ten book traders were bankrupt, and there is a clear impression that a considerable sector of the book trade was engaged in very marginal activity, with small binders and occasional vendors making very little money indeed. Also, debt fell especially heavily on the poorest third (with estates valued up to £500), and over half of these poorest book traders were in debt, while close to half of the same group were owed money by customers and smaller suppliers. The middling wealthy (£2,500 to £5,000) were more likely to be creditors themselves, while the more wealthy book merchants, with estates over £5,000, were as inclined to be in debt as the middling group but were much less likely to be victims of customer bad debt. These aspects of indebtedness confirm the grim financial position of the poorest sector, often in debt but having difficulty retrieving money owed to them, due to the consignment trading methods of the larger merchants and their ability to retrieve debts, and also the necessity of middling booksellers to allow credit to customers. That so much of the book trade survived on shaky credit explains the trade's fragility during the economic recessions of the 1650s and 1690s and before the more sure nationwide growth from the 1730s.[30]

Bankruptcy could have grim repercussions for some desperate book vendors. The town councils did what they could to help the needy in such cases, but the lack of guildry for booksellers or printers, even in Edinburgh, reduced the prospects for group assistance. Edinburgh's magistrates intervened on various occasions: in 1597 alms were given to the poor Edinburgh bookseller James Brown, and in 1649 a sum of 200 merks (£135) was provided for the destitute printer William Marshall, a fall indeed for a former apprentice to Andro Hart. The tides of political fortune had a direct impact on the welfare of printers, as the Edinburgh Presbyterian printer James Glen would discover in the 1680s. In November 1687 Glen was arrested for printing

The Root of Romish Rites, an anti-Catholic pamphlet offensive to James VII. Although Glen gave a spirited defence, asking provocatively if he could now sell the bible, as it was patently 'anti-papist', he was ruined by the case. However, after the revolution of 1688–9, Edinburgh town council, in view of Glen's great poverty, awarded him a small pension.[31] Yet desperation could take on levels beyond the reach of any authority, and in the winter of 1674 the bankrupt Edinburgh bookseller John Mason committed suicide by drowning himself. Days later Mason's Scottish, English and Dutch creditors coldly lodged their rights with the privy seal in Edinburgh.[32]

The authorities provided financial assistance but also policed book traders in a range of other areas. The bailie courts, especially that of Edinburgh, resolved small-level debt disputes between printers and booksellers. Occasionally, the magistrates also imposed sanctions for unacceptable

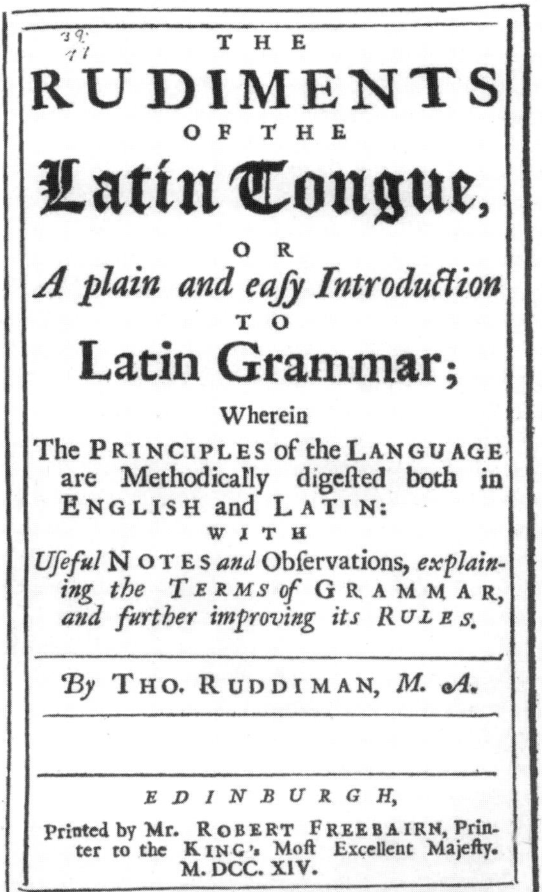

THE
RUDIMENTS
OF THE
Latin Tongue,
OR
A plain and easy Introduction
TO
Latin Grammar;
Wherein
The PRINCIPLES of the LANGUAGE are Methodically digested both in ENGLISH and LATIN:
WITH
Useful NOTES *and* Observations, *explaining the* TERMS *of* GRAMMAR, *and further improving its* RULES.

By THO. RUDDIMAN, *M. A.*

EDINBURGH,
Printed by Mr. ROBERT FREEBAIRN, Printer to the KING's Most Excellent Majesty.
M. DCC. XIV.

Figure 22.5 The title page of Ruddiman's *The Rudiments of the Latin Tongue* (1714) printed by Robert Freebairn only months before he joined the Jacobite rebellion and so lost his position as king's printer. Ruddiman and Freebairn were political as well as publishing bedfellows, and so the violent confrontation between Ruddiman and Freebairn's brother remains something of a mystery.

personal behaviour. In 1639 Aberdeen's first printer Edward Raban (fl. 1622–50) and his wife were imprisoned for a drunken brawl with their neighbours, and in 1663 the Edinburgh bookseller Robert Lindsay (fl. 1655–63) was arrested for keeping a 'baudie house'. Book traders appeared in their share of petty offence cases. In 1721 Edinburgh bailie court heard the extraordinary case of an assault on the famous grammarian and printer Thomas Ruddiman (fl. 1712–57) by James Freebairn, brother of Robert Freebairn (fl. 1701–47), the Jacobite printer and friend and publisher of Ruddiman. The cause of the dispute is not clear, although it was unlikely to be politics, given the Jacobite sympathies of both families.[33]

From the government's point of view, censorship mattered more than disputes over debt, and this had been the case from the banning of the works of Martin Luther in the 1530s to the arrest for sedition of the republican Edinburgh bookseller, Alexander Leslie, in 1798.[34] Before 1707 a system of licensing existed, but it was loosely policed to the point where a licence was only sought where a product was controversial, though legal, or worthy of copyright protection. No licences were sought for seditious material, of course. We might expect censorship to increase as press output grew, though it was not so. After a brief flurry in the 1620s a definite increase in book proscription occurred from the 1660s, followed by particular peaks in the 1680s and the first decade of the eighteenth century. The years 1680 to 1690, a period of anti-covenanting measures and of James VII's unconvincing authoritarianism, and 1700 to 1705, years of anxiety over the succession after the deaths of William and James VII, were the busiest years for banning books. As the supply of books expanded, both imported and home-produced, so the early notions of licensing the entire press were replaced by a focus on printings dealing with specific, offensive topics. This produced at the end of the seventeenth century a more targeted government effort. Indeed, censorship under William and Mary and Queen Anne was the most comprehensive of the early modern period, even though by now punishments were less severe. For book merchants, sentences which impeded the ability to trade hit hard at the commerce that supported themselves and their families. The closure of the booth of the bookseller John Calderwood in 1680 and the shutting of the press of John Reid in 1691 were dramatic examples.[35] All stock was confiscated. However, the fact that not a single printer was executed for illegal printing from the arrival of the press in 1508 shows that the authorities were not excessive in their zeal. It is true, nonetheless, that a few authors paid the ultimate price.

The privy council and court of session also became involved in other aspects of trade regulation, although sometimes trade and family disputes came together. Two of the main book cases that came before the court of session between the 1670s and the early eighteenth century exhibited that mixture of commercial and family crisis. The case *The Heirs of Hislop* v. *Robert Currie and Agnes Campbell* (1678–90) arose after the Hislop press was, according to James Watson, sold off to John Cairns on Hislop's death. Thereafter the Hislop children and what had been a large family bookselling business went on to suffer hereditary injustice and protracted litigation. Hislop's widow Beatrix Campbell married Currie, a writer, but she died soon afterwards and the case was brought against Currie for failing to manage the bookselling business in the interests of his two young step-children, and against Agnes Campbell for failing to deliver a bond her sister Beatrix had passed to her for the benefit of these children. The whole affair dragged on until 1690 at great legal cost, and there is a strong suggestion of collusion between step-father and aunt.[36] The case *Robert Sanders, the younger* v. *Bessie Corbett*, his mother (1694–1705) was also a family squabble in which the two Robert Sanders, senior and younger, appear to have fallen out over the implementation of a marriage contract the father had agreed to not long before his death in 1694. Subsequently the son sued his own mother and sisters for disposing irresponsibly of some stock and print materials, and not handing over others. In addition, they had made alterations to the family home in such a way as to reduce its value and his inheritance. The bitterness of the dispute is highlighted by the fact that not once did Robert refer to Corbett as his mother, and the dispute only ended with her death in 1705.[37]

If Scotland's book traders were occasionally disunited within their families, they were also inclined to commercial wrangling. These disputes were carried out through both legal and illegal means. As early as 1509 the privy council upheld a complaint from the king's printer Walter Chepman to prevent other merchants from illegally importing the Salisbury breviary. A century later, in 1618, the council prosecuted the Presbyterians Andro Hart, Richard Lawson and James Cathkin for breaching the right to print a catechism licensed to the Episcopalian bookseller Gilbert Dick.[38] Until 1708 the Scottish privy council was the main copyright-granting agency in Scotland and so it took seriously its duties to protect literary property. However, the legal and practical complexities over patents and licences increased markedly after the Restoration. Much of this arose from the wide and unprecedented monopoly powers granted to the king's printer Andrew Anderson in 1671 and taken up by his widow Agnes Campbell in 1676. Extra-legal behaviour became a feature from 1671. In October of that year the Glasgow printing house of Robert Sanders the elder was raided and looted, and his workmen driven off by Anderson and his partners. Anderson claimed that Sanders was printing without authority, but the privy council ordered the immediate release of Sanders and his men, and that the case should be heard before the council. In the end, in spite of a petition by Sanders and other outraged printers attacking the Anderson monopoly, Anderson and his partners had their rights confirmed, even though the rights of other printers and booksellers to import bibles was asserted. In fact, over the next decade the council gradually eroded Anderson's rights in favour of greater freedom of trade, although in 1681 Anderson's widow was still the most litigious of book merchants. In January of that year the privy council ordered the magistrates of Edinburgh to release from prison John Reid, senior, following his incarceration on Campbell's initiative. Again his 'crime' was apprenticeship absenteeism, and although he was released the judgment of the lords of council was that he should return to toil at the widow's printing house. However, when two years later Campbell had Reid's premises searched without legal authority, accusing him of stealing her type, the council gave her a severe reprimand.[39]

Some printers could call on their burghs for commercial protection. An example of this is found in the copyright history of the Aberdeen almanac, first published by Edward Raban in 1623 and the most successful Scottish almanac of the seventeenth century.[40] In October 1667 the magistrates of Aberdeen responded to a petition from the burgh printer–bookseller John Forbes the elder, protesting at the activities of the chapman Alexander Gray, who had brought into Aberdeen 1,000 copies of an 'alien' almanac, breaching the market for Forbes's own edition. The council upheld Forbes's complaint.[41] However, it was John Forbes the younger who in the 1660s and 1670s developed the reputation of the 'Aberdeen almanac' as the most prestigious edition in Scotland. The jealousy from Edinburgh was considerable. Andrew Anderson, having acquired his wide-ranging royal licence in 1671, went on deliberately to attack the printing rights of Forbes in early 1672 with a view to strangling Forbes's most valuable asset, the 'Aberdeen almanac'. The threat of legal action by Anderson and his Edinburgh cartel had to be taken seriously by Forbes, who knew that Sanders' press in Glasgow had been ransacked only a few weeks before. Anderson argued the same case, that Forbes had printed without permission, but he failed to take account of the strong views of Aberdeen town council, who were outraged at this attack on their independence and right to license within their own environs. The magistrates were prepared to start a book-trade war with Edinburgh, yet before Aberdeen took their case to the privy council, Anderson realised he had pressed matters too far, and in February 1672 he conceded Forbes's right to print under licence of the town, universities and bishop of Aberdeen.[42] Nevertheless, such was the success of the 'Aberdeen almanac' that various pirated editions were subsequently produced in Glasgow and Edinburgh.[43] Whatever can be said about the early modern book trade of Scotland, it was not averse to illegal activity in search of profit, even at the wealthy end of the market.

Religious prejudice was a constant backdrop to the book trade before the 1720s and helped foster social cleavage. During the Marian Civil War (1567–73) the printer Robert Lekpreuik took the side of King James and the Reformation, while the Episcopalian Thomas Bassandyne printed for Mary, Queen of Scots and her party, though when peace was declared and their peripatetic presses settled in Edinburgh, the two men came to terms. During the reign of Charles I, the Edinburgh trade was split between Presbyterians led by Hart and Episcopalians led by Gilbert Dick. Yet the book trade became a more serious focus for religious rivalry and illegal behaviour after the accession of James VII. Incidents of printers' premises being attacked resulted from a combination of hatred of foreign workers and a fear of 'papists'; the former relatively uncharacteristic in Scottish print history, the latter all too common. In 1684 the magistrates of Edinburgh closed the press of Dutchman Jan Colmar and his partners. Colmar claimed that the bookseller Charles Lumsden and others had obtained a warrant from the magistrates of the burgh 'without any ordor or law or proces against the petitioner'. The privy council reversed the decision of the burgh magistrates in October 1685, and ordered all stock and materials to be returned to the Dutchmen, who unfortunately soon went bankrupt.[44] By February 1686 the Dutchmen's press had been bought by James Watson, senior, a Catholic printer and father of the aforementioned James Watson the younger. The Dutchmen were now employed by Watson, but a few days later they were assaulted at Watson's rented premises by a crowd of fifty or so rioters. The government believed that anti-Catholic elements in the capital were responsible. Only two weeks before, the home of Peter Bruce, the German and Catholic engineer and future 'household printer' to the king, had been attacked by a crowd, led by some soldiers of the burgh. In spite of a privy council investigation, Bruce's house was again under siege a few weeks later, and the printer John Reid the younger was accused of involvement.[45] The hopes of James VII and his government of maintaining good order in the book trade were hampered by the religious politicisation of all aspects of public and commercial life. This continued into the eighteenth century, as the Edinburgh trade split into Hanoverians and Jacobites, though the number of committed book-trade Jacobites was few by the time of the 1745 rebellion.[46]

The 'brotherhood' of the book trade was clearly not always cohesive, and internal burgh commercial disputes often set one group of tradesmen against another. In a group action by the Edinburgh booksellers, a petition was put before the town council in December 1683 complaining of the bookselling of 'cramers' (stall salesmen) throughout the city, most of whom were 'not in the leist frie aither as burgess or gild breither', charged low prices that 'undersold the said stationers' and, while paying only a little for their stalls, were not subject to the burgh taxes that free burgesses had to pay.[47] The decision of the bailies was that cramers should open proper shops, and that straightaway those 'unfree' should become paid-up freemen of the burgh. However, in September 1710 the council was forced to concede that the erection of the 'paper cryers' or chapmen into a society had failed, with many printers complaining of the cryers' scandalous manipulation of prices. As a result, the council then agreed to dissolve this society and to allow anyone to sell in the streets printed papers, pamphlets, ballads and story books.[48]

But if the paper cryers were allowed to form a society, why not printers and booksellers? Why did Edinburgh not establish an equivalent of the Stationers' Society in London? Partnerships existed from the sixteenth century, and Andrew Anderson gathered a larger group of partners between 1671 and 1675, though this was never a society as such. Indeed, even though by the 1680s over sixty book traders operated in Edinburgh, there are several reasons why they were not incorporated. First, like the hammermen – the guild that caused the burgh most difficulties – the book traders were a disparate group. They consisted of wealthy stationers, moderately comfortable printers and booksellers, small and large bookbinders, journeyman printers, and street traders and chapmen. The hammermen, a mixture of metal workers of all kinds from blacksmiths to goldsmiths, were a similarly divided group and therefore difficult to control. Second, by the Restoration, clear indications were emerging of specialisation between printers and booksellers and, as a result, what was in the interests of one was not always to the benefit of the other. For example, import controls would benefit printers, not booksellers. Last, for those book traders who were ambitious to become members of the council or magistrates, there were opportunities without the need for a specific society, and for these wealthy traders membership of the merchant guild was almost automatic, such as Henry Charteris, council member of Edinburgh, and William Dickie, bookbinder, a council member of Glasgow before the Union of 1707.[49]

In spite of this, book makers made some efforts to establish a society. In 1681 the printers Patrick Ramsay, John Reid and Hector Aysoun incorporated themselves into the hammermen without permission of the Edinburgh town council, and as punishment all three were instructed to give up their burgess tickets. A more co-ordinated effort was made in 1722, when fifteen printers petitioned the council to form a society. The proposal was shelved, although by 1759 the journeyman printers had formed their own benevolent society, and in 1758 an Edinburgh letterpress printers' society was incorporated.[50] But always the most basic control of commercial activity exercised by the burghs were the keys to craft burgess or merchant guild membership. The 'ticket' was a badge of merit as well as of trade, and if necessary a means of coercion. More importantly for the history of the book in Scotland, the lack of an equivalent of the Stationers' Company, or an Edinburgh society, provided Scotland with a loose and decentralised system of book-trade regulation. Essentially, the smaller scale of the Scottish press, and the tradition of burghs having equal status and rights to develop commerce independently, prevented the formation of such a centralising society in Edinburgh.

Scottish printers and booksellers before 1800 were indeed a diverse group, but as book demand became intoxicated with the Enlightenment age, they were able to meet Scotland's expanding need for book supply. The brothers and sisters of the book were not a homogeneous grouping with necessarily shared commercial, private, religious and political interests. They were, however, all in trade and all in families, and each of them struggled in some unpromising economic circumstances until the trade benefits of union with England emerged from the 1730s. In fact, the Anglo-Scottish book trade began badly in the eighteenth century, marred as it was by a clash between the booksellers of Scotland and England over copyright. This concerned 'illegal' reprinting by Scotland and culminated in the infamous *Donaldson* v. *Becket* case, and final judgment of the house of lords in 1774, which mostly favoured the Scottish interpretation of limited copyright in the interests of freedom of trade and wider access to learning.[51] This case is a metaphor for the life of Scotland's book traders in the early modern period: argumentative to the last, but equally conscious of the contribution their trade could make to the welfare of the Scottish people.

NOTES

1. For an overall account of the early modern book trade in Scotland see Mann, 2000a; and for a summary, Mann, 2001.
2. Mann, 2000a, 7–33.
3. Murray, 1913; Napier, 1991, 40–4.
4. Allan, 2000; Allan, 1993, 9; Houston, 1993; Houston, 1994.
5. For details of sources and method see Mann, 2000a, 214–24.
6. Mann, 2000a, 219; Mann, 2001, 198.
7. Aldis, 1970.
8. For Hart see Mann, 2000a; Cowan, 1896; Mann, 2008.
9. Mann, 2000a, 261–69; National Library of Scotland (NLS), SBTI, 2005.
10. National Archives of Scotland (NAS), PS/1, 61, 84v.
11. The society and its antecedents continued papermaking until the 1830s. Thomson, 1974, 18.
12. NAS, Clerk Muniments, GD/18, 889, 1317, 1320, 1323; Thomson, 1974, 120.
13. For Charteris see Mann, 2000a; Dickson and Edmond, 1975, 348–76; MacDonald, 1998, 93–5.
14. Murdoch and Sher, 1988, 134–5.
15. Cook and Morris, 1916, 90.
16. Grant, 1906; Watson, 1929a; Watson, 1929b; Wood, 1963. For the print culture of clergy see Mann, 2005.
17. Mann, 1998, 142–3; Fountainhall, 1759–61, I, 104; Register of the Privy Council (RPC), iii, 3–4, 7 and 31–2; NAS, Court of Session (CS)/29, box 443 (Mackenzie); RPC, ii, 5, 174–5, 182, 580.
18. Dickson and Edmond, 1975, 490–508; Couper, 1914; Edmond, 1886; Watson, 1929a.
19. MacDonald, 1994.
20. Mann, 1999, 142–3; Mann, 2000a, 69, 86, 88–9, 171; Calderwood, 1842–9, iv, 78–9, v, 510–2 and 520–1, vii, 348–9 and 382–3; *Calendar of State Papers*, vii, no. 171; *Bannatyne Miscellany*, i, 199–215; *Extracts from the Records of the Burgh of Edinburgh* (*EBR*), 7, 109.
21. Mann, 1999, 142–3.
22. NAS, Commissary Court Records, CC8/8/82, CC8/8/88 and CC8/8/85; Paton, 1905, 93; NAS, Registers of Deeds, 3/77, 297; Fairley, 1925, 19.
23. Mann, 1999.
24. Aldis, 1970, no. 802; Maidment, 1845; Cowan, 1896, 6; Wood, 1936, 109; *Records of the Parliaments of Scotland* [*RPS*], 1633, 6, 57.
25. For an updated summary of Campbell see Mann, 1998; Fairley, 1925.
26. Mann, 1998, 140–1; NAS, CS/29, box 436.1 (Mackenzie); *Journals of the House of Lords*, 21, London, 1891, 609–10; Couper, 1910.
27. Mann, 1999, appendix, 146–7.
28. For the statistics, testaments and methodology see Mann, 2000a, 193–200, appendix 3, 262–69.
29. Myers and Harris, 1997.
30. Mann, 2000a, 196–7; MacMillan, 1992, 107, 115, 289.

31. *EBR*, 5, 178; *EBR*, 8, 206; *EBR*, 12, 22.
32. NAS, PS/3, 2, 493.
33. Aberdeen Council Archives (ACA), Burgh Court Book, 52 (2), 178; *EBR*, 9, 317; Edinburgh City Archives, Bailie Court Processes, box 1, bundle 2 (July 1721). For Raban and Freebairn see Mann, 2002, 266–7 and 278.
34. NAS, Justiciary Court Papers JC/26, 293. *Glasgow Courier*, 29 May 1798. For censorship see Mann, 2000a, 163–91.
35. Mann, 2000a, 173 and 175.
36. NAS, CS/157, 166/2 and CS/96, 306; Grant, 1925.
37. NAS, CS/138, 5219 and CS/158, 445; Couper, 1914, 46–9.
38. For copyright see Mann, 2000a, 95–124; and in summary Mann, 2000b, 11–25.
39. RPC, iii, 3–4, 7 and 31–2; RPC, iii, 8, 250–1; Fountainhall, 1848, ii, 464–5; Mann, 2000b, 22–23.
40. For almanacs see McDonald, 1966.
41. ACA, Aberdeen Council Records, 55, 58–61; Edmond, 1886, iv, xlv; Taylor, 1942, iv, 321–2.
42. RPC, iii, 3, 424; ACA, Aberdeen Council Records, 55, 362–3; Edmond, 1886, iv, xlvi; RPC, iii, 3, 596–9.
43. Mann, 2004a, 46–52.
44. RPC, iii, 11, 196; Mann, 2000a, 132.
45. Watson, 1713, preface, 10–24; RPC, iii, 12, 19–25, 23, 30, 143, 159 and 210; Mann, 2000a, 132–3.
46. Mann, 2002, 276–80.
47. *EBR*, 11, 96; Mann, 2000a, 17.
48. *EBR*, 13, 199–200; Mann, 2000a, 17.
49. Mann, 2000a, 17–18.
50. *EBR*, 11, 18; Edinburgh City Archives, Moses Bundles (MB), 152, 5952 (5 June 1722); Houston, 1994, 99; Gillespie, 1953, 18.
51. Mann, 2000b, 11–13; Feather, 1994, 64–96.

BIBLIOGRAPHY

Aldis, H G, ed. *A List of Books Printed in Scotland before 1700: Including those Printed furth of the Realm for Scottish Booksellers*, Edinburgh, 1970.

Allan, D. *Philosophy and Politics in Later Stuart Scotland: Neo-Stoicism, Culture and Ideology in an Age of Crisis, 1540–1690*, East Linton, 2000.

Allan, D. *Virtue, Learning and the Scottish Enlightenment*, Edinburgh, 1993.

Bain, J, et al., eds. *Calendar of State Papers, Scotland Series*, 13 vols, Edinburgh, 1898–1969.

Brown, P H, et al., eds. *Register of the Privy Council, Third Series, 1661–1691*, Edinburgh, 1908–1970.

Brown, K M, et al., eds. *The Records of the Parliaments of Scotland to 1707*, St Andrews, 2008, http://www.rps.ac.uk/.

Calderwood, D. *A History of the Kirk of Scotland by Mr David Calderwood*, ed. T Thomson, 8 vols, Edinburgh, 1842–9.

Cook, W B and Morris, D B, eds. *Extracts from the Records of the Merchant Guild of Stirling, 1592–1846*, Stirling, 1916.

Couper, W J. James Watson, king's printer, *SHR*, 7 (1910), 244–62.

Couper, W J. Robert Sanders the elder, printer of Glasgow, 1661–94, *Records of the Glasgow Bibliographical Society*, 3 (1914), 26–88.

Cowan, W. Andro Hart and his press, *Edinburgh Bibliographical Society Papers*, i, 12, (1896), 1–14.
Dickson, R and Edmond, J P. *Annals of Scottish Printing: From the Introduction of the Art in 1507 to the Beginning of the Seventeenth Century*, Cambridge, 1890; reprinted in facsimile Amsterdam, 1975.
Durkan, J and Ross, A, eds. *Early Scottish Libraries*, Glasgow, 1961.
Edmond, J P. *The Aberdeen Printers*, 4 vols, Aberdeen, 1886.
Fairley, J A. *Agnes Campbell, Lady Roseburn*, Aberdeen, 1925.
Feather, J. *Publishing, Piracy and Politics: An Historical Study of Copyright in Britain*, London, 1994.
Feather, J. *A History of British Book Publishing*, London, 1988.
Gillespie, S. *A Hundred Years of Progress: The Record of the Scottish Typographical Association, 1853–1952*, Glasgow, 1953.
Grant, F J, ed. *The Register of Apprentices of Edinburgh, 1586–1666*, Edinburgh, 1906.
Grant, J. Archibald Hislop, stationer, Edinburgh, 1668–1678, *Edinburgh Bibliographical Society Papers*, 12 (1925).
Harris, B. *Politics and the Rise of the Press: Britain and France, 1620–1800*, London, 1996.
Houston, R A. Literacy, education and the culture of print in Enlightenment Edinburgh, *History*, 254 (1993), 373–92.
Houston, R A. *Social Change in the Age of Enlightenment: Edinburgh 1660–1760*, Oxford, 1994.
Laing, D, ed. *The Bannatyne Miscellany*, 3 vols, Edinburgh, 1827–55.
Lauder, Sir John, of Fountainhall. *Historical Notices of Scottish Affairs*, 2 vols, Edinburgh, 1848.
MacDonald, A A. Early modern Scottish literature and the parameters of culture. In Mapstone, S and Wood, J, eds., *The Rose and the Thistle: Essays on the Culture of Late Medieval and Renaissance Scotland*, East Linton, 1998.
MacDonald, A R. The subscription crisis and church–state relations, 1584–1586, *RSCHS*, 25 (1994), 222–55.
McDonald, W R. Scottish seventeenth-century almanacs, *The Bibliotheck*, 4, 1:8 (1966), 257–322.
MacMillan, J K. A study of the Edinburgh burgess community and its economic activities. Unpublished PhD thesis, University of Edinburgh, 1984.
Maidment, J, ed. Information anent His Majestie's printers in Scotland (by Robert and James Bryson), *The Spottiswoode Miscellany*, 2 vols, Edinburgh, 1844–5, 1, 299–300.
Mann, A J. Book commerce, litigation and the art of monopoly: The case of Agnes Campbell, royal printer, 1676–1712, *Economic and Social History*, 18 (1998), 132–56.
Mann, A J. Embroidery to enterprise: The role of women in the book trade of early modern Scotland. In Ewan, E and Meikle, M, eds., *Women in Scotland: 1100–1750*, East Linton, 1999; reprinted 2002, 136–51.
Mann, A J. *The Scottish Book Trade, 1500 to 1720: Print Commerce and Print Control in Early Modern Scotland*, East Linton, 2000a.
Mann, A J. Scottish copyright law before the statute of 1710, *Juridical Review*, 1 (2000b), 11–25.
Mann, A J. The anatomy of the printed book in early modern Scotland, *SHR*, 80 (2001), 181–210.
Mann, A J. The press and military conflict in early modern Scotland. In Murdoch, S and Mackillop, A, eds., *Scotland's Military Experience, c1600–c1815*, Leiden, 2002, 265–86.

Mann, A J. Some property is theft: Copyright law and illegal activity in early modern Scotland. In Myers, R, Harris, M and Mandelbrote, G, eds., *Against the Law: Crime, Sharp Practice and the Control of Print*, London, 2004a, 31–60.

Mann, A J. A spirit of literature: Melville, Baillie, Wodrow and a cast of thousands – the clergy in Scotland's long renaissance. In Davis, C and Law, J E, eds., *The Renaissance and the Celtic Countries*, Oxford, 2005, 90–108, and also *Renaissance Studies*, 18 (2004b), 90–108.

Mann, A J. Hart, Andro (b. in or before 1566, d. 1621), *Oxford Dictionary of National Biography*, Oxford, online edition, http://www.oxforddnb.com/view/article/12470, last accessed January 2008.

Marwick, J D, Wood, M and Armet, H, eds. *Extracts from the Records of the Burgh of Edinburgh, 1403–1718*, 13 vols, Edinburgh, 1869–85.

Masson, D and Brown, P H, eds. *Register of the Privy Council, Second Series, 1625–1660* Edinburgh, 1899–1908.

Murdoch, A and Sher, R B. Literary and learned culture. In Devine, T M and Mitchison, R, eds., *People and Society in Scotland, Volume 1: 1760–1830*, Edinburgh, 1988, 127–42.

Murray, D. *Robert and Andrew Foulis and the Glasgow Press*, Glasgow, 1913.

Myers, R and Harris, M, eds. *The Stationers' Company and the Book Trade, 1550–1990*, Winchester, 1997.

Napier University. *Imprints in Time: A History of Scottish Publishers Past and Present*, Edinburgh, 1991.

National Library of Scotland, Scottish Book Trade Index (SBTI), online database.

Paton, H, ed. *The Register of Marriages for the Parish of Edinburgh, 1595–1700*, Edinburgh, 1905.

Taylor, L B, ed. *Aberdeen Council Letters, 1552–1681*, 6 vols, Oxford, 1942–61.

Thomson, A G. *The Paper Industry in Scotland, 1590–1801*, Edinburgh, 1974.

Watson, C R B, ed. *Register of Edinburgh Apprentices, 1666–1700*, Edinburgh, 1929a.

Watson, C R B, ed. *Register of Edinburgh Apprentices, 1701–1755*, Edinburgh, 1929b.

Watson, J. *A History of the Art of Printing*, Edinburgh, 1713; facsimile edition, London, 1965.

Wood, M, ed. *Register of Edinburgh Apprentices, 1756–1800*, Edinburgh, 1963.

PART FIVE

•

Co-operation

23 Trade Unions in Scottish Society

ARTHUR MCIVOR

> Well if you're talking about the unions I always think the union is a mythical body. This branch done a lot. We happened to belong to a trade union. We fought ... I mean we walked the streets for 26 weeks to get conditions. We were the ones that forced them to give us tables and chairs to sit down to have a meal with, made them give us a changing room to hang our clothes up.[1]

This is how one retired asbestos insulation lagger who worked in the Clyde shipyards from 1947 to the 1980s articulated his identification with his trade union – emphasising the role of individual agency, the empowerment which solidarity provided and the importance of the local, branch level. It is also something of a 'heroic', rank-and-filist narrative, putting primacy on the requirement to 'fight' for workers' rights in a conflictual relationship against a reluctant management: 'them' and 'us'. Other personal testimonies suggest that industrial relations were more consensual. Some adopted a more negative attitude towards the unions, focusing on their misogynist neglect of women workers, the distancing of the leaders from their constituency, and their narrow, economistic aims. Unions are thus popularly perceived both as villains and saviours. Interpretations range widely from those at one extreme, who see the trade unions as responsible for most of Scotland's economic and social woes, to those at the other extreme, who exalt such institutions and the role they have played in protecting the interests of a relatively powerless and atomised proletariat.[2]

Whatever one's position in these debates, what is clear is that individuals joining together to protect themselves and advance their economic interests has long been a feature of everyday life in Scotland, and in the twentieth century trade unions came to play a pivotal role in Scottish society. This chapter briefly traces the incubation, growth and development of such associational activity at the workplace from the seventeenth century to the present. Attention is paid to the impact such collective organisations had upon Scottish people's lives, and how and why working people came to form, support and identify with trade unions.

WORKERS' ORGANISATION AND PROTEST IN PRE-INDUSTRIAL SOCIETY

In their classic *History of Trade Unionism*, first published in 1894, Sidney and Beatrice Webb defined trade unions as organisations of wage earners, though their emphasis on the formal nature of such bodies led them to underestimate the extent and the diversity of collective organisation and action prior to the Industrial Revolution. Merchants and craftsmen (the burgess classes) in the larger urban settlements began to organise themselves in guilds and incorporations in the medieval period. For example, out of a total population of around 7–8,000 in 1604, members of the Glasgow merchant guilds numbered some 213, and the 14 Glasgow craft guilds in 1604 mustered some 361 members.[3] Clearly, these organisations represented only the tip of the occupational hierarchy in urban areas at this time. In this period the journeymen's associations were élite bodies which regulated the trade in their districts, enforcing a series of restrictive practices, such as control over entry to the craft and regulation of prices, with artisans frequently negotiating prices directly with merchants.

Recent research has shown that trade unions – that is solely of wage *earners* – existed in many of the main towns in Scotland by the first half of the eighteenth century across a swathe of trades, including handloom weaving, coal mining, woolcombing and tailoring. Whatley has shown how Scottish coal miners formed combinations even earlier, persuasively challenging the orthodox view of servile, deferential colliers pre-industrialisation. Amongst other things, salters and colliers in the early eighteenth century were working shorter hours than that laid down by statute (where a six-day work week was specified), and the taking of extra holidays after fairs and on Mondays was not uncommon.[4] Such collectively endorsed – and hence legitimised – actions within particular occupational communities probably occurred more frequently in the absence of formal trade unions at this time.

The most stable of the early trade unions in Scotland involved artisanal workers, such as the stonemasons, tailors, shoemakers, metal workers, building workers, cabinet makers and printers. Most were local, small and predominantly recruited fully trained skilled men. Some developed from the friendly societies – such as the Journeymen Woolcombers' Society of Aberdeen.[5] Their main job was to provide insurance, on payment of a small weekly fee, to cover members for loss of work, sickness, accidents and, in many cases, for burial expenses – hence the euphemism 'coffin clubs'.[6] Some also attempted to restrict entry to the trade. Common mechanisms to achieve this were the imposition of quotas on the number of apprenticeships and the refusal to allow 'outsiders', or 'interlopers' to work in the trade. Much of the ritual and ceremony – including oath-taking and pledges of secrecy – of the older craft guilds was replicated in the early trade unions. As the gap between employers and craftsmen widened, disputes and confrontation occurred over payment, prices and work conditions. Violent protest by the emerging journeymen's societies in the eighteenth century was not uncommon, and included incendiarism, vitriol throwing, ear-cropping and cutlass fights, sometimes directed towards those workers who breached collective rules and customs, and sometimes aimed at local figures of authority, such as the landowner, employer or manager. Industrial sabotage also occurred, as in the case of the striking Aberdeen seamen in 1792 who boarded ships, removed the rigging and forcibly prevented cargoes being loaded and unloaded.

This does, however, need to be kept in perspective. Whilst collective organisation and strike activity did pre-date industrialisation, relatively few folk within Scottish society would have had any involvement in a trade society of any kind prior to the nineteenth century. This especially applied to the vast majority who worked on the land, and to female workers. It also applied, however, to the majority of 'unfree' and relatively impoverished journeymen and labourers in the towns.[7] The penalties at law for such aberrant behaviour acted as a disincentive, whilst élite paternalism operated as a social cement. Moreover, within this semi-feudal society relatively few folk identified with fellow workers. Rather, vertical ties bound society together. One way in which this reciprocity in social relations was expressed was in the prevailing system whereby local justices of the peace regulated wages.[8] Through the JPs, government came to play an important role in determining wages and working conditions in pre-industrial Scotland, and this regulation persisted longer in Scotland than in England.

INDUSTRIALISATION, TRADE UNIONS AND INDUSTRIAL RELATIONS

The Industrial Revolution may not have created trade unions, but the related processes of mechanisation, the factory system, the 'free market' and virtually unregulated urbanisation acted as catalysts, producing an environment conducive to the rapid expansion of unionisation. As Hamish Fraser has demonstrated, the other key processes were the collapse of state paternalism and the erosion of company paternalism.[9] Neither process was complete, but the spread of the ideas of Adam Smith and the laissez-faire free-market concept critically undermined the vestiges of medieval reciprocity and patronage. With the protective matrix of paternalism disintegrating, and workers exposed directly to the volatility of the boom-and-slump market economy, many looked towards combining with fellow workers and strength through unity. This was associated with the faltering and uneven emergence of working-class consciousness in this period, fuelled further by the rise of the radical press and indicated, at the extreme, by the development of the Owenite and Chartist movements. Two of the sectors where significant trade union development took place at this time were cotton textiles and coal mining. In mining communities, as Alan Campbell's seminal work on Lanarkshire has demonstrated, underground workers organised in local and coalfield unions to advance wages and improve working conditions in a notoriously hazardous industry.[10] One of the tactics of hewers was to exert peer pressure to restrict output – sometimes called 'the darg' – and hence maintain pit-head prices for coal (on which wage levels were usually based). Before the mid-nineteenth century, however, the miners' unions maintained a fairly precarious existence: powerful and influential when labour markets were tight, but usually undermined as trade conditions changed and the coalmasters moved on to the offensive, using victimisation, blacklisting, legal intimidation and the lock-out to root out miners' organisations.

A similar situation existed in the cotton textile industry. Around 7,000 handloom weavers (that is, traditional craft workers) in the East End of Glasgow stopped work in protest at wage cuts in 1787 and maintained a strike for almost ten weeks, in the face of intense poverty. The dispute involved incendiarism, violence and sabotage, much of which was directed against the replacement non-labour or blackleg workforce that the employers recruited. In response, the Glasgow magistrates took a firm line. In one incident on 3 Sept 1787, troops opened fire on stone-throwing strikers near the Drygate bridge, killing three outright, fatally wounding three more and injuring many others. This 'massacre' prompted riots throughout the city. Public order was eventually restored, and folk drifted back to work on the employers' terms, to humiliating and deep wage cuts of 20–25 per cent. Arrests and victimisation of the leaders and activists followed.

Nonetheless, levels of trade unionism and protest remained high, especially amongst the rapidly growing numbers of handloom weavers, and violence erupted periodically thereafter in industrial relations. Indeed, the context in Scotland in the late eighteenth and early nineteenth centuries probably encouraged union growth to a greater degree than in England. The Combination Laws that declared collective organisation illegal in England were looser in Scotland, whilst the persistence of wage regulation by the courts in Scotland long after this dissolved in England also encouraged formal organisation amongst wage earners. The right of the courts to regulate wages was a critical issue in the three-month strike of some 40,000 handloom weavers in the West of Scotland in 1812. This dispute effectively marked the dissolution of state regulation of wages, and in its aftermath trade unionism was declared illegal in Scotland, a situation which lasted until 1824.

The initiative in the early nineteenth century drifted from the traditional craft workers to the new factory workers, especially the cotton spinners. In 1820, male textile factory workers burnt down the Broomward mill in Glasgow after the employers brought in female labour. In 1827, a non-union textile worker was shot dead whilst sleeping, whilst five years later several women workers brought in during a strike were blinded by vitriol thrown into their faces by trade unionists. Concurrently, the Glasgow cotton spinners' developed a more modern form of trade union organisation in the 1820s and 1830s, with elected members, an executive committee (and several sub-committees), membership fees and cards. It paid out unemployment and funeral benefit, assisted emigration to relieve the labour market and restricted the supply of labour, refusing to train those who were not close male kin of a spinner (thus excluding women). These policies were pursued against a backdrop of the rapid and virtually unregulated development of the factory system and the mechanisation of the cotton-spinning process. By the 1830s, the cotton spinners were amongst the best organised groups of workers in Scotland, having a virtually closed shop.[11]

At the same time, the extent of an emerging class consciousness is indicated in the formation of the Glasgow United Committee of Trades Delegates in 1830, which published its own newspapers – including *Herald to the Trades Advocate* – pushed for parliamentary reform and helped local trade unions. This was the precursor to the Glasgow Trades Council. Owenite-inspired moves towards general unionism and co-operation also represented attempts to find alternatives to emerging capitalism in this period. Such developments prompted the élites to take action. Aided by a trade depression, the Glasgow cotton millowners organised to defeat the union in the mid-1830s, to slash wages and to remove all restrictions on their right to manage. In the violent strike of 1837 which ensued, a replacement non-union worker, John Smith, was shot in the back and killed. The following trial saw five members of the spinners' strike committee convicted and sentenced to seven years in a prison colony in Botany Bay, Australia – although they were never actually sent, and were pardoned in 1840. Trade unionism became discredited by its association with violence, and 1837 represented a watershed in industrial relations in Scotland. Thereafter, aided by a sharp economic recession from 1838 to 1842, membership collapsed, and trade unionism in manufacturing in Scotland was set back severely for the rest of the nineteenth century.

In the second half of the nineteenth century, the most stable group of unionised workers in Scotland were the craft artisans in urban areas. These workers developed a strong associational culture tied closely to specific trades such as printing, engineering, boilermaking, carpentry, shipbuilding, bookbinding and tailoring. Around mid-nineteenth century there emerged the so-called 'new model' unions of such workers, so termed because many were national federations of local branches across England, Scotland or the UK. Examples would be the Amalgamated Society of Engineers (UK), formed in 1851, the Operative Stonemasons of Scotland (1852) and the Associated Carpenters and Joiners of Scotland (1861). Amongst the new generation of trade union leaders was Alexander Campbell, a time-served joiner, who was also associated with the Glasgow Trades Council formed in 1858. Historians disagree on the 'core' characteristics of such craft unions in the second half of the nineteenth century. The 'orthodox' view posits that such organisations were essentially conservative, cautious, responsible and non-militant, concerned to establish the respectability of the labour movement after the debacle of the Glasgow spinners' strike and other militant action, such as the entanglement of some trade unions with radical Chartism in the 1830s and 1840s. Hence such organisations prioritised negotiation and collective bargaining, and laid considerable emphasis on achieving recognition by employers and gaining status. Key weapons in their drive to maintain members' living standards were the accumulation, through a relatively hefty membership fee, of large defensive funds, and the provision of a wide range of benefits, frequently including unemployment pay, sickness and accident benefits, superannuation and, in some cases, a grant to facilitate emigration. The kinds of workers who were members were usually those who had undergone the customary five- to seven-year trade apprenticeship, and frequently such unions enforced strict rules on entry to the trade. At the extreme, some commentators have represented such organisations as a kind of betrayal of the working class: an incorporated labour élite or 'aristocracy of labour' which absorbed middle-class values and came to accept capitalism, working only to improve the status of the artisans within the system. Robert Gray's depiction of the craft élite in Edinburgh provides an example.[12] Such 'reformism' was also reflected in the election of trade union leaders into local government and as Liberal MPs into parliament, epitomised, perhaps, in the election of the Scottish coal miners' leader Alexander McDonald as an MP in 1874.

Recently, however, this view of the skilled craft unions of the Victorian period has been modified by a recognition that confrontation co-existed with consensual strategies. Even within the mid-Victorian boom era, such artisanal associations were willing to strike – albeit frequently as a second line of defence – and to press hard for improvement in legislation and state involvement in the workplace, which benefited the wider constituency of workers.[13] The shorter-hours movements, the ameliorative labour rights legislation of the early 1870s and the Factory Acts would be tangible examples of significant and lasting achievements clawed from a parliament in which workers still had little representation. Whilst they were hardly class-conscious warriors in the vanguard of the struggle against exploitative employers, neither were the artisanal unions the deferential lap dogs of Victorian capitalism which some accounts suggest. Moreover, as Richard Price has shown, when threatened, the craft unions responded vigorously, especially where customary craft controls over the labour process were being challenged.[14] The main indictment against them, perhaps, is that by adopting an exclusive mode of organisation, where labourers and semi-skilled workers were shunned, they were partially responsible for creating a sectional structure (as opposed to an industrial form of organisation – the model, for example, in Germany) which ultimately weakened the bargaining power of labour vis-à-vis capital. As Knox has shown, the workplace remained the primary focus for collective organisation, and in this period most trade unionists in Scotland remained members of exclusively Scottish unions.[15]

For most Scottish workers, involvement with a trade union remained rare and frequently temporary in the nineteenth century. Quite literally, in most industrial communities a 'master and servant' relationship persisted. In some cases, as in the Coats cotton mills in Paisley, workers' associational impulses were contained by a judicious strategy of company welfarism.[16] Only at the very peak of the trade cycle, when demand for labour exceeded the supply, did some of the poorer workers find the capacity to organise. The early 1870s and the mid-late 1880s were such periods. Kenefick has charted the sporadic attempts of the Glasgow dockers to organise, identifying the important role of the Catholic church in local welfare activities in dockland communities.[17] Similarly, some agricultural labourers became organised, with the Scottish Farm Servants', Carters' and General Labourers' Union formed in 1885 – though permanent organisation of such workers came only with the creation of the Scottish Farm Servants Union in 1912. The emergence of local trades councils was also important – and probably more so in Scotland than in England – in promoting and co-ordinating collective organisation and action. Edinburgh and Glasgow trades councils were formed in the 1850s, and by the end of the nineteenth century there were fifteen trades councils dotted across the main urban settlements in Scotland.

In the nineteenth century, trade unionism was largely though not exclusively a man's world. Whilst emphasising the male domination of the trade union movement in this period, Gordon has identified the increasing organisation of female workers, especially in the textile factories in the late nineteenth and early twentieth centuries. She persuasively demonstrates how some groups of female workers, including the female jute workers in Dundee, initiated strike action and succeeded, in some cases in the face of opposition and blacklegging by male workers. Women, Gordon shows, could be particularly vociferous in demonstrations during strikes and in picketing, where those breaking strikes would be subjected to a barrage of humiliating abuse, and, occasionally, to violence.[18]

Nevertheless, in the 1890s only a small proportion of male workers were members of trade unions, and a far smaller proportion of female employees. Whilst established in mining, shipbuilding, metalworking and waterfront communities – albeit still vulnerable in times of high unemployment – the penetration of trade unionism was still extremely uneven across the economy as a whole. Scottish union membership was probably even less than the UK average of around one in ten workers at this time. The Webbs survey in 1892 indicated that union membership in Scotland lagged about 20 per cent behind that of England. Most members, moreover, were geographically concentrated in the industrial conurbation around Clydeside. Trade union penetration was also significant in Edinburgh, Dundee and Aberdeen, but much less evident elsewhere in the country.

Associational workplace activity had developed significantly throughout the nineteenth century in Scotland. Nonetheless, trade unions remained weakened by a multitude of factors, including legal constraints, a biased media (which frequently portrayed union leaders as tyrants), poor communications, low levels of literacy, and the coercive anti-trade unionism of some employers and the paternalism of others.[19] Saturated urban labour markets for much of the period, fuelled by significant in-flows from the Highlands, Ireland (especially after the famine) and the surrounding rural hinterland, also undermined workers' collective organisation and could neutralise strike activity. Working-class communities, moreover, remained fractured along lines of occupation, gender and religion. These divisions continued to militate against effective combination in trade unions throughout the Victorian period, and indeed beyond.

TRADE UNIONS AND SCOTTISH SOCIETY IN THE TWENTIETH CENTURY

The first quarter of the twentieth century witnessed a massive surge in trade union membership and strike activity in working-class communities in Scotland. New unions were formed, existing unions expanded and amalgamated, and membership extended from around 10 per cent of the employed population up to almost 50 per cent by 1920. Workers also became more militant and increasingly politicised, and the trade union movement reflected such changes. The Scottish Trades Union Congress (STUC) was formed in 1897 partly as a reaction against the British TUC, which was unhappy with the direct affiliation of the more radical local trades councils. The STUC provided Scottish trade unionism with its distinctive voice in the twentieth century, developing a not entirely unjustified reputation for being more consistently to the left than the TUC.[20] In this period (1900–20) the trade unions in Scotland shifted their traditional political allegiance from the Liberal Party to the fledgling Labour Party (formed in Scotland in 1888 and in the UK in 1900). The funds provided from the trade union movement were critical in the expansion of the political wing of the labour movement. Identities were complex, however, and class consciousness continued to be fractured by divisions based on gender, religion, nationality and locality. Trade unionism never proved to be totally synonymous with a single political party. Indeed, a significant strand of working-class Toryism persisted throughout the twentieth century, whilst some Scottish trade unions supported Marxist political organisations such as the Social Democratic Federation and, later, the Communist Party, which was formed in 1920. Moreover, the Labour Party remained relatively weak in Scotland prior to World War I.

Nonetheless, working-class consciousness sharpened in this period, and one reflection of this was the emergence of a more militant, aggressive and confrontational trade union movement. This has frequently been associated with World War I and the early 1920s, but the process was well under way before this, especially in the Clydeside industrial conurbation. Work intensification, deskilling, a rising awareness of inequality, the socialist press (such as the Independent Labour Party paper *Forward* from 1906), improved working-class education and rising price inflation all helped to cement workers' disgruntlement and draw them into protective workplace organisations in increasing numbers in the decade or so before 1914. Dissatisfaction erupted in the much vaunted labour unrest of 1910–14, when a wave of strikes paralysed Scottish industry. This included strikes of Scottish railwaymen, seamen, dockers and the Singer sewing machine workers in 1911; Scottish coal miners, Dundee jute workers and Falkirk ironmoulders in 1912; and dockers, carters, ironworkers and quarryworkers in 1913.[21] A domino effect was evident in these years, with successful strikes encouraging a further expansion of trade union membership and a sense of heightened collective power, stimulating workers to strike for further improvements in wages and work conditions. Trade unionism was changing, with an increasing number of unskilled male workers and female workers brought into membership.[22] Once linked specifically with the years 1889–90, this 'new unionism' is now more correctly viewed as a more drawn-out process over several decades from the 1880s to the 1920s.

The metamorphosis of workplace organisation and politics was most evident in the main cities – Glasgow, Dundee, Edinburgh and Aberdeen – and in the mining areas. Communities in Lumphinians in Fife and the Vale of Leven in Dumbarton developed reputations as 'Little Moscows'.[23] However, the notion of 'Red Clydeside' has been most commonly employed to characterise this mutation of working-class consciousness. Folklore has it that Glasgow was close to a Bolshevik-style workers' revolution in January 1919, when a strike for the forty-hour week culminated in a riot in George Square and the British government responded by sending troops and tanks to restore order. 'Red Clydeside' has become one of the most contentious issues within Scottish labour history.[24] One view has it that 'Red Clydeside' was a myth. It is important not to exaggerate the extent of the dissatisfaction, and few now subscribe in an unqualified way to the idea that Glasgow was a second Petrograd, where a genuine revolutionary situation arose. Nonetheless, a kind of 'Red Clydeside' existed, partly manifested in the changes occurring in trade unionism – the greater propensity to strike, the infusion of socialist ideals into the unions and the warrening of many workplaces with shop stewards who were to become pivotal figures in the trade union movement in the twentieth century. What the establishment also found particularly threatening was the congruence of working-class community and workplace action – as in 1915 in Glasgow when shipyards threatened to strike in support of the rent strikers.[25] World War I acted to incubate trade unionism in Scottish communities and encourage many more workers to associate together to protect their interests in a troubled and uncertain period. Shop stewards became more important, with the pressures of World War I prompting a new trend towards unofficial shop-floor action in response to management techniques and mechanisation. New union structures emerged from the rank and file to challenge both the employers and the official union bureaucracy, notably the Clyde Workers' Committee – a kind of workers' soviet. Rapid price inflation contributed, as did a more conciliatory attitude on the part of the state and employers, who increasingly came to recognise trade unions for bargaining purposes. One historian has seen the decade 1910–20 as the pivotal period when the previously autocratic anti-trade union behaviour of Clydeside capitalists dissolved as they made the transition towards more consensual industrial relations.[26] This liberalisation of attitude was not uniform across Scottish capitalism, but it did play a part in removing the fear of victimisation felt by many workers (and especially the lesser skilled) if they joined a union. The reform of labour law, with the Trades Disputes Act of 1906, also provided a more favourable environment for the trade union movement to flourish.

The 1920s and 1930s have traditionally been viewed as barren years for the trade union movement, when mass unemployment in the economic recession empowered the bosses and castrated the capability of workers to organise, defend themselves and succeed in strike action. Union densities fell sharply as membership dropped by perhaps 50 per cent between 1920 and 1932–3, at the nadir of the interwar depression (see Table 23.1). Union and political activists found themselves picked off by employers keen to press home their advantage and reassert their prerogative to manage their concerns as they thought fit. The shop stewards were amongst the first targets, as one observer noted in 1933: 'In Clydeside engineering shops where there used to be, say, from thirty to fifty shop stewards there are now only a handful, perhaps none at all.'[27] The unions also faced more sophisticated, centralised mechanisms to blacklist and victimise, not least the various branches of the ultra-right wing group the Economic League.

What is perhaps most remarkable, however, is the resilience of collective organisations and the solidarity of working-class communities in the face of these pressures. For example, there was strong support in Scotland for the General Strike of 1926 amidst widespread deprivation and despite an extended period of recent very high unemployment.[28] Whilst weakened, the core concept of trade unionism – strength through unity – survived intact in many working-class communities, with Hutt noting in the early 1930s that 'the spirit of the workers remains unbroken'.[29] This was perhaps particularly evident in single-industry communities, such as the colliery villages, where strike-breaking was notoriously difficult because of the close-knit solidarity of the miners.[30] This was reflected in entrenched attitudes towards 'blacklegs', who were ostracised within such communities throughout the twentieth century. A poem, 'Ode to a Scab' by John Heeps, a Communist miner in Kilsyth, expressed the opprobrium felt towards workers who went against the union in 1926–7:

Oh human skunk, outcast disrespected,
As one with pestilence infected,
Labour's ideals thou hast rejected,
To lick the boots
Of coal lords, landlord's men dejected
Ungrateful brutes

Worms of men, slime of the gutter,
Your actions are of those who murder
Labour's army, who toiled under
The iron heel
Of those who lie, rob and plunder
The common weal[31]

Union membership and confidence were again on the rise from the mid-1930s as the economy geared up for war and unemployment levels dropped. The popular discourse of society during World War II emphasised social harmony and co-operation as everyone pulled together to make sacrifices to prosecute the war effort. In reality, workplace conflict persisted and union power expanded, in part promoted by state policies, much influenced by Ernest Bevin (Minister of Labour) and Tom Johnston (Secretary of State for Scotland), which extended the recognition of unions and further legitimised their existence.[32] Mass Observation, the wartime social anthropological survey, was amongst the groups who registered the changing balance of power in the workplace and the revival of the shop stewards' movement in the 1940s.

The period from the outbreak of World War II through to the late 1970s represented the historic peak of trade union power in Scotland. Unions became accepted as responsible institutions within society. Despite some dissenting voices, they were widely perceived as doing a vital job, protecting the interests of workers and guarding against a return to the 'hungry 1930s'. A closed shop (where union membership was compulsory) operated in most of the heavy industries, whilst the nationalised industries and the public sector – including local and national government – were particularly well unionised. Moreover, unionisation spread rapidly, especially in the 1960s and 1970s, in the expanding service and professional 'white collar' sector, which helped to compensate for declining membership in the traditionally well-organised heavy industries and manufacturing, which continued to experience rapid contraction in the numbers employed as deindustrialisation gathered pace. By the mid-late 1970s, STUC membership reached one million, and again union density nudged past 50 per cent of the employed workforce. Unions became firmly implanted within communities of workers that had previously been poorly organised, including the unskilled (especially in the general unions) and, as Table 23.1 indicates, female workers.

Table 23.1 Scottish Trades Union Congress membership

	Total	*Female*	*Percentage female*
1892	147,000	not known	–
1923–4	324,256	78,740	24.2
1931	244,289	48,125	19.7
1939	382,866	57,047	14.9
1945	613,807	136,879	22.3
1951	745,686	140,189	18.8
1971–2	913,316	277,648	30.4
1979	1,002,841	353,000	35.2
1990	855,000	not known	–
2000	635,000	not known	–

Sources: Figures for 1892 taken from Webb, S and Webb, B. *The History of Trade Unionism*, London, 1950, 743. Remaining data obtained from the STUC *Annual Reports* and data supplied by the STUC research officer.

With growing membership came a heightened commitment to the use of the strike weapon and experimentation with other, more novel, forms of industrial action, such as the factory occupation and work-in. Upper Clyde Shipbuilders and Caterpillar are examples.[33] Studies of the geography of strikes in the UK have identified west-central Scotland as one of the most strike-prone regions.[34] Gall and Jackson's recent research has both qualified and confirmed this.[35] Church and Outram, in a wide-ranging analysis of strike activity in UK mining, identified a more militant 'Celtic fringe' (Scotland and Wales).[36] John Foster has also shown that Scottish workers had a proportionately greater participation rate in the one-day 'political' strikes that punctuated the period from 1969 to 1984.[37] In large part, as Gall and Jackson have persuasively argued, this appears to be a product of differences in industrial/occupational structure between Scotland and England. Other factors which contributed to relatively high levels of strike activity were the persistently authoritarian, inflexible attitudes of many Scottish managers, the insecurity of working in volatile, contracting product markets in the traditional industries, and somewhat poorer working conditions – involving higher injury rates and poorer standards of occupational health – in Scotland compared to south of the border (see chapter 18).

The shipbuilding communities on Clydeside exemplify the growing power and influence of trade unions within Scottish society in this period. This was a tough, macho world in many respects, with a multiplicity of unions in the yards fighting to protect the men's interests, frequently against a hard-nosed, recalcitrant management. Contradictory pressures impinged upon the workforce. On the one hand, yard closures as deindustrialisation proceeded curtailed workers' bargaining power. On the other hand, the unions operated in a generally favourable political, legal and cultural climate. Working-class solidarities were certainly never complete, but relatively few dissented from the union line, and most benefited from union vigilance over work conditions. One shipyard Communist shop steward in the 1960s, Matt McGinn, penned a poem, 'Can O' Tea', to express the solidarity of the community when one union activist was sacked, ostensibly for brewing up.

> But the men said I'd been victimised
> For the Union I had organised
> So when I laid down my can o' tea
> A thousand men marched out with me[38]

The work-in organised by the Upper Clyde Shipbuilders (UCS) in 1971 perhaps best exemplified the determination of Scottish shipyard workers to fight to preserve their jobs. UCS was a consortium of five firms established in 1968 into which the British government invested heavily until UCS declared itself bankrupt in 1971. Collective solidarities were nurtured in the shipyards and cemented in the highly charged atmosphere of the yard mass meeting. Traditional working-class consciousness was much on display here, intersecting with the hegemonic masculinity of the Clydeside 'hard man'.[39] As one of the UCS leaders, Jimmy Reid, said: 'We don't only build ships on the Clyde, we build men.'[40] Not that such attitudes were solely confined to the shipyards. An Ayrshire coal miner expressed similar sentiments, reflecting that: 'Men and management in general were always at loggerheads in the coal mining industry. Men and management ... If you were a weak man you would have did what the boss said.'[41]

That is not to imply, however, that the unions were omnipotent – as some commentators have suggested – in the 1960s and 1970s. Power varied considerably across different industries and within different occupations, whilst to some extent the union movement undermined its own influence through internal divisions and internecine squabbles. This reflected workers' complex, intersecting and evolving identities. Scottish unions were divided by politics and by occupation, with inter-union demarcation disputes particularly prevalent in the shipyards. The traditional male domination of the movement also eroded only very slowly. Consequently, it took time for working women to identify with such institutions, especially when they had so little representation within positions of power in the labour movement. As late as 1980, there was only one woman on the General Council of the STUC, and only 7 per cent of the delegates to the STUC General Congress were women.[42] For much of the twentieth century, the unions absorbed and reflected rather than challenged the prevailing machismo 'hard man' work culture of the heavy industries.

Moreover, with some notable exceptions (including the miners' unions), the movement prioritised organisational survival, protecting wages, jobs and working hours, whilst neglecting some of the longer-term chronic health problems associated with industrial work, especially occupation-related respiratory diseases and cancers, such as asbestos-related mesothelioma.[43] Whilst the employers and state were the key players in defining health and safety standards at work, the evidence also suggests that the trade unions could have done more to prioritise occupational health over production imperatives. Mass Observation noted in 1942: 'In view of its evident importance to production, the extent to which industries and unions concern themselves with the health of their workers is noticeably slight.'[44] Unions continued to prioritise compensation for victims rather than addressing the cause and prevention of health problems at work. For some unions this reflected a fear of members being dismissed on medical grounds. Elsewhere, rank-and-file activists criticised trade unions' absorption of productionist values at any cost.[45] Judging from an examination of the annual reports of the Scottish Trades Union Congress, occupational health and safety was not a primary campaigning issue of the Scottish trade union movement from the 1930s to the 1970s. This situation changed in the last quarter of the twentieth century as the unions modernised, expanded further into the service sector and embraced issues pertinent to workers of both genders, as well as a more health-conscious agenda in which the protection of workers' bodies in production increasingly came to take precedence.[46]

Concurrently, however, unions were becoming increasingly viewed as being responsible for the country's economic problems. The 'new right' ideologies of the 1970s blamed unions for strikes and unrealistic wage levels, paving the way for the Tory administrations of Margaret Thatcher and the passage of a cluster of legislation which undermined union rights to strike and to picket. An anti-trade union discourse emerged, and the sharp rise in unemployment levels from the mid-1970s bolstered the attack on the unions. With governments in the 1980s and into the 1990s prioritising the control of inflation at the price of rising levels of unemployment, the trade unions once again entered a phase of retrenchment, decline and demoralisation.[47] It took some time, however, for the new market and political realities to sink in.

A watershed in industrial relations was the defeat of the miners in the bitter year-long strike of 1984. Scottish miners played an important role in this dispute. Polmaise pit in Stirlingshire was one of the five pits that faced a threatened lock-out and closure which prompted the UK-wide miners' strike, involving some 14,000 Scots miners (and a further 100,000 in England and Wales). Scotland and Yorkshire were the two areas that initially campaigned for wider action, leading to the NUM calling an all-out strike across the industry to defend jobs. The Scots miners exhibited a characteristic display of solidarity, resilience and courage against massive and insurmountable odds, with the Thatcher government marshalling the full force of a myriad of anti-union laws and state power to crush the strike. The dispute was characterised by much violence, especially in clashes between pickets and police at Ravenscraig, Hunterston, and at Bilston Glen and Monktonhall collieries. An important supportive role was played in the strike by miners' wives in pit communities up and down the country. The miners held out for a year in one of the most remarkable episodes of trade union camaraderie in Scottish history. They returned, defeated and demoralised, with many activists (over 200) subsequently sacked by the NCB in Scotland for their part in the conflict, including forty-six from Monktonhall, the greatest number of victimised men from any single colliery in the UK.[48] The industry contracted sharply thereafter, with the last Scottish underground coal mine at Longannet in Fife closing in 2002.

For more than a decade from the end of the miners' strike, strike levels reached all-time lows and trade union membership sharply slumped. The growth of the foreign ownership of manufacturing in Scotland – including US-owned companies who were opposed to the principle of trade unionism – further undermined membership in Scotland, as did the virtual elimination of shipbuilding, steelworking, heavy engineering and mining as significant employers of Scottish labour. STUC membership fell from 1.1 million in 1980 to 855,000 in 1990 and to 635,000 in 2000 (representing around 35 per cent of Scottish employees).[49] By the end of the century most trade unionists were employed in the public sector, whilst only just over one in five Scottish workers in privately owned companies were members of a union.[50] With these transformations, the nature of trade unionism was also changing, As Knox has observed: 'The newly dominant service sector and white-collar unions, with their socially diverse working constituencies, cannot hope to forge such intense solidarities among their members.'[51] What is also evident is that there has been a convergence effect over the second half of the twentieth century, with differences in strike activity between Scotland and elsewhere in the UK dwindling. This has fuelled an ongoing debate over the distinctiveness of the Scottish pattern of trade unionism and collective action, with the most recent contribution arguing strongly for Scotland being 'part of wider phenomena and processes found elsewhere and throughout Britain'.[52] Certainly the number of purely Scottish unions declined sharply from World War II, with only six remaining by the mid-1980s – the largest, significantly, being the Educational Institute of Scotland.[53]

However, despite severe setbacks in the 1980s and 1990s, the trade union movement remains entrenched within Scottish society in the first decade of the twenty-first century and may yet expand over the immediate future under the devolved Scottish government. Whether the emasculated union movement ever comes to exert as much influence over Scottish people's lives as it did in its heyday from World War II to the 1970s remains to be seen. What is clear is that historically these institutions have performed a vital role in protecting the interests of the individual in the workplace through the last century and beyond. Without such collective organisations, individual workers would have been, and would continue to be, subjected without redress to the vicious vagaries of the free, unregulated market and to an exploitative, profit-orientated system, in which labour costs the workers more than their time and exertion, also seriously undermining their health. Workers, through their trade unions, played no small part in transforming the deference-based 'master and servant' relationship of the Victorian period, creating in its stead communities of workplace citizens with an extensive array of inalienable rights and a reflexive commitment to collective organisation and action to defend themselves. This was no mean achievement, and one we would do well not to forget in our hedonistic, consumer-orientated, materialist age.

NOTES

1. Scottish Occupational Health Oral History Project, Interview A22, Scottish Oral History Centre, University of Strathclyde, SOHCA/016. See Johnston and McIvor, 2000, 158–72.
2. For the most comprehensive scholarly overview see Knox, 1999.
3. Smout, 1969, 160–1.
4. Whatley, 2000, 84–6. See also Thompson, 1963.
5. MacDougall, 1985, 24.
6. Knox, 1999, 52.
7. Smout, 1969, 164–5.
8. For an example in Lanarkshire see Ward and Fraser, 1980, 4–5.
9. Fraser, 1988.
10. Campbell, 1979.
11. Knox, 1999, 53–5.
12. Gray, 1976; Gray 1981; Hobsbawm, 1964; Hobsbawm, 1984.

13. Knox, 1999, 118–9.
14. Price, 1980.
15. Knox, 1990, 150.
16. MacDonald, 2000; Knox, 1995.
17. Kenefick, 2000; see also Kennefick, 2007.
18. Gordon, 1991.
19. Melling, 1982.
20. Aitken, 1997; Tuckett, 1986.
21. See Kenefick and McIvor, 1996.
22. Gordon, 1991.
23. MacIntyre, 1980. On Edinburgh see Holford, 1988.
24. See McLean, 1983; Foster, 1990; Melling, 1982; Kenefick and McIvor, 1996.
25. Melling, 1982.
26. Johnston, 2000.
27. Hutt, 1933, 119.
28. MacDougall, 1979.
29. Hutt, 1933, 119.
30. See Campbell, 2000.
31. Carter and Carter, 1974, 13.
32. For World War II see several articles in the special edition of the *Scottish Labour History Society Journal*, 30, 1995.
33. See Foster and Woolfson, 1986; Woolfson and Foster, 1988.
34. Charlesworth et al., 1996.
35. Gall and Jackson, 1998, 97–112.
36. Church and Outram, 1989.
37. Foster, 1998, 230–31.
38. Cited in Bellamy, 2001, 164.
39. See Johnston and McIvor, 2004.
40. Cited in Bellamy, 2001, 199. For an earlier period see McKinlay, 1991.
41. Scottish Occupational Health Oral History Project, Interview C1.
42. McIvor, 1992. See also Breitenbach, 1982.
43. Johnston and McIvor, 2000, 158–72.
44. Mass Observation, 1942, 203.
45. See, for example, Tom Murray's scathing critique of the unwillingness of the TGWU to protect dockers exposed to dangerous chemicals at Leith docks in 1941–2, in MacDougall, 2000, 287.
46. For occupational health and safety see Johnston and McIvor's contribution in this volume, Chapter 18.
47. See Martin et al., 1996.
48. See Duncan, 2005, 259–66.
49. I am grateful to the STUC research officer for this information.
50. Foster, 1998, 233; Gall, 2003, 63.
51. Knox, 1999, 295.
52. Gall, 2003, 71.
53. MacDougall, 1985, 256.

BIBLIOGRAPHY

Aitken, K. *The Bairns o'Adam: The Story of the STUC*, Edinburgh 1997.
Bellamy, M. *The Shipbuilders*, Edinburgh, 2001

Breitenbach, E. *Women Workers in Scotland*, Edinburgh, 1982.
Campbell, A. *The Lanarkshire Miners: A Social History of their Trade Unions, 1775–1884*, Edinburgh, 1979.
Campbell, A. *The Scottish Miners, 1874–1939*, Aldershot, 2000.
Carter, P and Carter, C. The miners of Kilsyth in the 1926 General Strike and lock-out, *Our History Pamphlet*, 58 (Spring 1974).
Charlesworth, A, Gilbert, D, Randall, A, Southall, H and Wrigley, C. *An Atlas of Industrial protest in Britain, 1750–1990*, Basingstoke, 1996.
Church, R and Outram, Q. *The Militancy of British Miners*, Leeds, 1989.
Duncan, R. *The Mineworkers*, Edinburgh, 2005.
Duncan, R and McIvor, A, eds. *Militant Workers: Labour and Class Conflict on the Clyde, 1900–1950*, Edinburgh, 1992.
Foster, J and Woolfson, C. *The Politics of the UCS Work-In*, London, 1986.
Foster, J. Strike action and working class politics on Clydeside, 1914–1919, *International Review of Social History*, 35 (1990), 33–70.
Foster, J. Class. In Cooke, A, Donnachie, I, MacSween, A and Whatley, CA, eds, *Modern Scottish History 1707 to the Present, Volume 2*, East Linton, 1998, 210–34.
Fraser, W H. *Conflict and Class: Scottish Workers 1700–1838*, Edinburgh, 1988.
Gall, G. Trade unionism and industrial relations in Scotland since UCS, *Scottish Labour History*, 38 (2003), 51–74.
Gall, G and Jackson, M. Strike activity in Scotland, *Scottish Labour History*, 33 (1998), 97–112.
Gordon, E. *Women and the Labour Movement in Scotland, 1850–1914*, Oxford, 1991.
Gray, R. *The Labour Aristocracy in Victorian Edinburgh*, London, 1976.
Gray, R. *The Aristocracy of Labour in Nineteenth Century Britain*, London, 1981.
Hobsbawm, E. *Labouring Men*, London 1964.
Hobsbawm, E. *Worlds of Labour*, London, 1984.
Holford, J. *Reshaping Labour: Organisation, Work and Politics. Edinburgh in the Great War and After*, London, 1988.
Hutt, A. *The Condition of the Working Class in Britain*, London, 1933.
Johnston, R. *Clydeside Capital: A Social History of Employers*, East Linton, 2000.
Johnston, R and McIvor, A. *Lethal Work: A History of the Asbestos Tragedy in Scotland*, East Linton, 2000.
Johnston, R and McIvor, A. Dangerous work, hard men and broken bodies: Masculinity in the Clydeside heavy industries, 1930–1970s, *Labour History Review*, 69/2 (August 2004), 135–51.
Kenefick, W. *'Rebellious and Contrary': The Glasgow Dockers 1853–1932*, East Linton, 2000.
Kenefick, W and McIvor, A, eds. *The Roots of Red Clydeside, 1910–1914?*, Edinburgh, 1996.
Kenefick, W. Scottish dock trade unionism, 1850 to present. In Mulhern, M A, Beech, J and Thomson, E, eds, *The Working Life of the Scots* (Scottish Life and Society: A Compendium of Scottish Ethnology, vol. 7), Edinburgh, 2008, chapter 31.
Knox, W W. The political and workplace culture of the Scottish working class, 1832–1914. In Fraser, W H and Morris, R J, eds, *People and Society in Scotland, Volume II, 1830–1914*, Edinburgh, 1990, 138–66.
Knox, W W. *Hanging By a Thread*, Preston, 1995.
Knox, W W. *Industrial Nation: Work, Culture and Society in Scotland, 1800–Present*, Edinburgh, 1999.
MacDonald, C. *The Radical Thread*, 2000.

MacDougall, I, ed. *Essays in Scottish Labour History*, Edinburgh, 1979.
MacDougall, I, ed. *Labour in Scotland*, Edinburgh 1985.
MacDougall, I. *Voices from Work and Home*, Edinburgh, 2000.
MacIntyre, S. *Little Moscows*, London, 1980
Martin, R, Sunley, P and Wills, J. *Union Retreat and the Regions: The Shrinking Landscape of Organised Labour*, London, 1996
Mass Observation, *People in Production*, London, 1942.
McIvor, A. Women and work in twentieth century Scotland. In Dickson, A and Treble, J H, eds, *People and Society in Scotland, Volume III, 1914–1990*, Edinburgh, 1992, 138–73.
McIvor, A. *A History of Work in Britain, 1880–1950*, Basingstoke, 2001.
McKinlay, A. *Making Ships, Making Men*, Clydebank, 1991.
McKinlay, A. Management and workplace trade unionism: Clydeside Engineering, 1945–57. In Melling, J and McKinlay, A, eds, *Management, Labour and Industrial Politics in Europe*, Cheltenham, 1996, 174–86.
McLean, I. *The Legend of Red Clydeside*, Edinburgh, 1983.
Melling, J. Scottish industrialists and the changing character of class relations in the Clyde region, *c*1880–1918. In Dickson, A, ed., *Capital and Class in Scotland*, Edinburgh, 1982, 61–142.
Melling, J. *Rent Strikes*, Edinburgh, 1983.
Price, R. *Masters, Unions and Men*, Cambridge, 1980.
Scottish Labour History (formerly *Scottish Labour History Society Journal*), 1965– .
Smout, T C. *A History of the Scottish People, 1560–1830*, London, 1969.
Smout, T C. *A Century of the Scottish People, 1830–1950*, London 1987.
Thompson, E P. *The Making of the English Working Class*, London, 1963.
Tuckett, A. *The Scottish Trade Union Congress*, Edinburgh, 1986.
Ward, J T and Fraser, W H, eds. *Workers and Employers*, London, 1980.
Webb, S and Webb, B. *A History of British Trade Unionism*, London, 1894.
Whatley, C. *Scottish Society, 1707–1830*, Manchester, 2000.
Woolfson, C and Foster, J. *Track Record: The Story of the Caterpillar Occupation*, London, 1988.
Woolfson, C, Foster, J and Beck, M. *Paying for the Piper: Capital and Labour in Britain's Offshore Oil Industry*, London, 1996.

24 A Different Commonwealth: The Co-operative Movement in Scotland up to 1924

CATRIONA M M MacDONALD

In 1910, the Kilmarnock Equitable Co-operative Society reflected on its position on the eve of the twentieth century:

> The Society was a large, though never a bloated, capitalist. It was, in one direction, a banker, in another an educational authority, in another a considerable employer of labour. It had its brain, its body, its hands, and its feet. Its brain was busy planning, its hands were making, amassing, and distributing its products, its feet were carrying them to their destinations, its body was being nourished. It was a grocer, a baker, a boot and shoe maker, a flesher, a draper, a milliner, a china merchant, a fishmonger, a sausage maker and a ham curer.[1]

The co-operative movement, incorporating hundreds of consumer-owned co-operative societies across Scotland, disrupts the comfortable dichotomies that typically frame our understanding of the workplace and the locus of power in labour relations. Co-operators were both producers and consumers, employers and employees. They were at the interface of significant economic and ideological debates that dominated both the labour movement and radical politics at a key stage of their development at the beginning of the twentieth century, but they were awkward bedfellows for the emergent socialist interest. While central to the lives of thousands of Scottish households from the late nineteenth century, until 1918 (at least) the movement remained on the periphery of Labour politics. Its membership and collective resources eclipsed anything either the trade unions or the Labour party could muster, but they occupied an uncomfortable 'space' in the commercial landscape that could not be easily accommodated in a Marxist world view.

On first appraisal, co-operators may have appeared the ideal body to act as the vanguard for creating society anew in a socialist mould, but their collective operations were typically undertaken in the pursuit of independence – independence from the capitalist trader, the 'middle man', the truck shop, the food adulterator, and the exploitative manufacturer. Unity was not necessarily a goal in itself but a means to liberating each individual so that they, in turn, could seek self-improvement in an environment conducive to respectability.

AIMS AND IDENTITY

Critics of the movement were scornful of co-operative idealism, and drew attention to the movement's vague yet all-embracing agenda. 'R.L.', writing in Chelsea in 1872, noted:

> These societies are to make the poor rich, the rich richer, to reduce the profits of the middle-man, to abolish the retail trader, to find employment for the precious talents of professional and other gentlemen in their leisure hours, to sweeten trade, and purify society, to teach self-reliance and self-supply, to teach the virtue of goods had, and paying cash promptly and before-hand, – and to work a revolution in the morals, manners, and trade of this great commercial country.[2]

Co-operators themselves hardly helped to clarify their position. An appeal from Kilmarnock Co-operators in 1866 highlighted tensions between collective and selfish interests:

> We believe it is possible by means of co-operation, for working men to obtain more wages with less work. Every man who eats may now add the profits of trade to the wages of labour. We look to co-operation – that is, united working for the general good – to achieve labour's emancipation.[3]

Of the founders of this Ayrshire Co-op, it was noted: 'They had no faith in State Socialism; they founded their hopes upon mutual self-help; they did not believe in an equality of condition; what they desired was an equality of opportunity, that they might by their own efforts rise to their rightful position in the social economy.'[4] Some years later, at the annual soirée of the Vale of Leven Co-operative Society in 1915, James Borrowman of the Scottish Co-operative Wholesale Society (SCWS) encouraged his audience to support the movement, not 'based on any sentiment of charity, but on enlightened self-interest'. He went on to explain the success of the movement in terms of 'intelligent selfishness'.[5]

When working men became employers, and prioritised their interests as consumers over their position as producers, the common order of things seemed to be overturned. It is hardly surprising that there was some confusion as to what all this meant. The co-ops could not and would not be typical shops; the co-operative wholesalers could not and would not be conventional manufacturers and distributors. In the day-to-day running of co-operative societies in Scotland the tensions and contradictions inherent in 'co-op' aims and identity are made transparent.

In some ways, there was little to distinguish co-operative business practices from those of independent retailers. Co-operators did not let their commitment to the emancipation of labour blind them to poor service. A few examples must suffice. The Galashiels Waverley Store sacked their foreman baker for bad timekeeping in 1874; the drapery salesman of the neighbouring co-operative Provision Store was disciplined in 1888 for spending too much time reading newspapers in the shoemakers' workshop; a year later the local co-op butchers were reprimanded for 'reckless driving', and in the same year, one co-operative member complained bitterly about the staff of the Eastern branch of the Galashiels Provision Store 'being almost worthless, continually making mistakes and very slow discharging their duties'.[6]

As co-operative societies multiplied, there was also often as keen a rivalry between co-operative 'brothers' as there was among the independent retailers themselves. In East Lothian, the Musselburgh and Fisherrow Co-operative Society rose to the challenge of new competitors, the Musselburgh Industrial Co-operative Society, in 1875.[7] And in the west, some thirty-five years later, the Wishaw Co-operative Society refused to draw up lines of demarcation to distinguish its areas of interest from those of the Cambusnethan and Overtown society branches, 'as it interfered with individual liberty'.[8]

James Borrowman (SCWS) seemed to echo similar sentiments in his address to congress in 1873 when he said: 'The distributive store is only co-operative to its members, and not to its employees, standing to them simply in the relation of capitalists and employers.'[9] But things were not that straightforward. In terms of their origins, influences, internal politics and relationships with other retailers, the co-ops were simply *different*.

ORIGINS AND INFLUENCES

Many co-operative societies across Scotland were greatly influenced by their origins and the outlook of their founders. In Coalsnaughton (est. 1872), and Sauchie (Newtonshaw Co-operative Society, est. 1865), miners dominated the founding 'fathers' of the local societies.[10] In Cupar – as befitting a market town – diverse trades, including a cabinet maker, a blacksmith, a journalist and factory workers, contributed to the founding of the local society (est. 1889).[11] The first twenty-nine members of the St George Co-operative Society (est. 1870) in Glasgow were all connected with the Grove Park weaving factory, and in Galashiels the early membership of the Provision Store reflected the dominance of the wool trade in the area, being overwhelmingly spinners, dyers and weavers.[12]

The motives of such men otherwise unconnected with retailing were not those of the typical independent grocer: profit-making was only to be applauded insofar as it contributed to the wellbeing and progress of the society and the prosperity of its customers. Their work for the co-op often reflected their interests elsewhere. According to Kilmarnock's co-operative historian, the local pioneers were 'the very pick and cream of their class … men … of deep reading and of wide research'.[13] An analysis of the Vale of Leven founders shows this group in a similar light: a good many were elders or enthusiastic followers of local dissenting Presbyterian churches, active supporters of friendly societies, Volunteers, and even parish councillors.[14] These were retailers who were to be open to ideals higher than the profit motive and also more diverse.

In a number of areas the co-operative movement was willing to sacrifice profit and higher dividends for the wider interests of the communities served. Examples abound of co-operative societies attempting to relieve distress when economic depression or industrial strikes threatened the livelihoods of their members. The Wishaw Co-operative Society distributed grants from its benevolent fund to needy members during the economic depression of 1908–9.[15] Stirling Co-operative Society reduced the cost of bread to striking miners and donated £50 for the relief of distress in the area in 1912, discounted all goods purchased for the local soup kitchen during the General Strike in 1926 and made weekly £5 grants to those experiencing acute distress.[16]

As employers, the societies also frequently offered advantageous terms and conditions to their employees. Co-operative employees in Sauchie were awarded a weekly half-holiday in 1868, and employees in the Vale of Leven gained this right in 1872.[17] From 1897, Wishaw co-operators insisted that all contractors doing work for the society must pay trade union wages,[18] and in 1914 the St George Co-operative Society insisted that all its employees should belong to an appropriate trade union.[19] Beatrice Potter noted admiringly that the SCWS started the business of shirt manufacture at an immediate loss 'to avoid dealing with firms who employ labour under bad conditions'.[20] Wider interests were clearly at work.

Tensions between good business sense and co-operative loyalties are clearly illustrated in the movement's ambiguous attitude to credit. In general, co-operators were against offering members credit in their shops. However, dealing as they were with predominantly working-class customers whose occupations were frequently jeopardised by cyclical depressions and temporary unemployment, few societies could resist a credit culture for long. This could and did cause problems. In Sauchie, within eighteen months of launching the new co-op, the debts of purchasers equalled one fifth of total sales.[21] A year later (1869), a cash-only system was adopted. This was certainly in keeping with co-operative aspirations of encouraging responsible consumption. But when, within a short space of time, membership growth gradually declined and many members withdrew from the society, a more moderate policy was adopted, and a credit system was restored in 1870.[22] In Cupar, members' debts acted as a serious brake on the development of the infant society. By 1895, six years after the society's foundation, 'sales, if not going down were not increasing, debts would have to be wiped off in many cases, a sharp drop in dividend was the result, and members began to drop out'. To alleviate matters, the management committee agreed not to draw their salaries until the dividend was back to two shillings.[23]

Interest groups within the membership also made a significant impact on the direction of co-operative enterprise in Scotland. The Scottish Co-operative Women's Guild (SCWG) most usefully illustrates this tendency. Jeannie Murdoch, 'a neat well-dressed woman', is the chief protagonist in Catherine Morison's 1927 co-operative sketch, *Dod! That Beat's A' or, Jining the 'Co'*. She explains to her friends what 'the Guild' is all about:

They are wanting better conditions for women and children; they want better wages for the men; to do away with strikes and lock-outs and unemployment. They want a higher standard of life for everybody, and they want justice for all, and, above all, ... they want to abolish war with all its horrors and to establish peace on earth and goodwill to all.[24]

By the early 1950s, 'the Guild' was the largest self-governing women's organisation in Scotland, with over 33,000 members in nearly 500 branches.[25] From its inception in 1892 it had often acted as the conscience of the movement and as the voice of the typical co-op shopper – a female voice.

At its 'coming of age', the guild's president, Mrs Annie Buchan, claimed that it should be to the wider co-operative movement 'what the wife or mother is to the home ... [a] strong, refining influence, leading (societies) to Co-operative idealism, as the mother leads her husband and children in the right direction'.[26] The 'direction' the SCWG took supported minimum wage legislation, attacked credit trading, developed alliances with women's unions (such as the Women's Protective and Provident League), encouraged the development of co-operative rest homes, and, from 1893, supported the female parliamentary franchise. Theirs was a strong voice, and one that predominantly male co-op leaders up and down the country ignored at their peril after 1918.

Far from being a uniform body, then, the co-operative movement encompassed both competing and complementary interests, and divisions within local communities frequently expressed themselves in the representative offices of the co-op boards. Unlike standard 'private' business interests, co-operative endeavours were often organic expressions of local interests. While this was generally a positive attribute, it also meant that local divisions

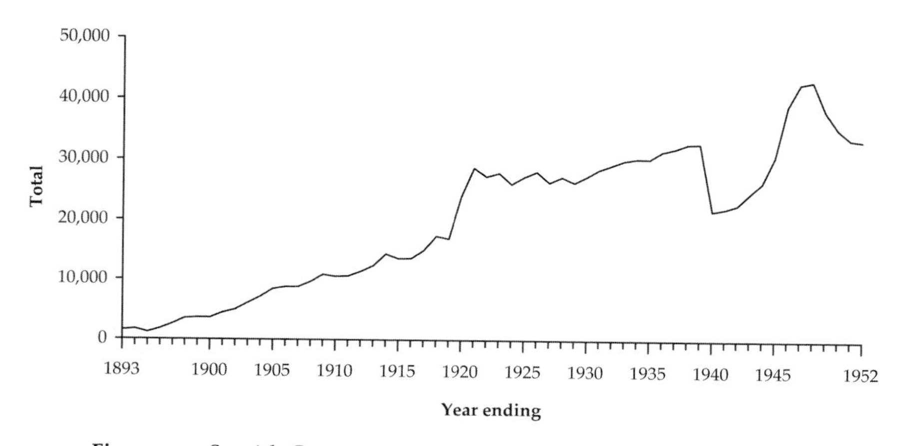

Figure 24.1 Scottish Co-operative Women's Guild (membership, 1893–1952). Source: Callen, K M. *History of the SCWG*, Glasgow, 1952, 31–2. Note: The 1894 figure contains one Irish branch; the 1901 figure contains three Irish branches; in 1906 seven Irish branches seceded.

were often manifest in co-operative structures too. In Sauchie, long-term tensions in the Newtonshaw co-operative board led to the secession of a group of members who, in turn, formed their own co-operative society (the Industrial Society, est. 1880).[27] Similarly, in Kilmarnock, several hundred co-op members left to form their own society (the Kilmarnock Industrial Co-operative Society) in 1881, following a disagreement over the management of the Kilmarnock Equitable Co-operative Society.[28] Still more sinister local tensions found expression in co-operative business elsewhere. In Wishaw in 1908, many Catholic members of the Wishaw Co-operative Society broke away to form the Wishaw Independent Co-operative Society, which was refused membership by the SCWS.[29] Sectarianism was clearly the root cause of this secession.

THE BOYCOTT

The co-op was not a 'conventional' retailer, nor was it treated as such by other independent traders. Suspicion, derision and outright opposition were the most common responses of the 'independents' when a local co-op shop opened nearby. In Perth, a fearful grocer erected a sign bearing the words 'Rochdale Store' above his own shop to 'see off' the new competition when the Perth Co-operative Society opened in 1871.[30] At their height, such feelings expressed themselves in boycott action against co-operative retailers and wholesalers.

In Perth, a society of traders and private dealers was established as early as 1885 to counteract the expansion of the Perth co-operative movement: part of its constitution demanded that no member employ anyone connected with the Co-operative Society.[31] While the trade of the co-op suffered in the short term, by 1888 matters appear to have stabilised, with no long-lasting implications for either the membership or the board. In the west-central belt of Scotland, however, the 1880s merely gave societies there a foretaste of things to come.

A Scottish Traders' Defence Association, with branches throughout the West of Scotland, was established in 1888 to combat co-operative propaganda.[32] Suppliers were encouraged to sack employees with co-op connections.[33] A lull in activity followed after 1889, only to be followed in the years 1895–7 by the longest period of severe anti-co-operative activity that Scotland had witnessed. Attacks on co-operation appeared in the press, and many private bread manufacturers found it impossible to sell their wares in Glasgow until their employees were given the choice of either leaving the co-op or being dismissed. Glasgow's William Reid recounted how, in the years 1896 and 1897:

> spies waited outside of the doors of Co-operative shops and tracked the customers home until their names were ascertained. The next step was that of finding out the employers of the breadwinners and any other members of the families who were at work, when a letter was

sent to the firm calling attention to the fact that the employees or their fathers or mothers, were members of a Co-operative society, and requesting their dismissal.[34]

Similarly, in Wishaw in 1897, the children of co-operative members were discharged from their employment in a local factory, and other co-operators claimed they were being victimised by their landlord on account of their co-op membership.[35]

The co-operative societies fought back by forming their own 'vigilance committees', though these proved of little use in combating the boycott's ultimate expression: a dispute over meat sales in Glasgow's municipal markets. In June 1896, market stallholders and Glasgow's master butchers refused to take bids from or trade with co-op agents, disrupting meat supplies to co-operative societies in the West of Scotland and beyond. Trading at Yorkhill quay in Glasgow was also affected, thus preventing the co-operators from initially sourcing foreign meat alternatives. Indeed, despite a by-law forcing traders to accept offers from all bona fide customers in municipal markets, the salesmen began trading in private, thus effectively blocking co-operative trade.

In the end, the boycott was broken when it became clear that by trading directly with Canadian cattle ranches and alternative local suppliers, the co-ops were getting round it. Indeed, many co-operators had been enthused by this fight, reaffirming their faith in the need for a co-operative commonweal. Nevertheless, it had proved – if, indeed, proof was required – that the co-operators were seen from 'the outside' as a troublesome alternative to traditional retailers. Self-consciously distinctive in terms of their aims, origins, influences and practices, the co-ops failed to find accommodation in conventional trading circles. This would prove their greatest strength, and perhaps their most telling weakness in the long term.

THE SCOTCH

As with many things 'Scotch', the strength of the co-operative movement north of the border in the early twentieth century was generously explained in terms of its native sensibilities. James Lucas, an early historian of the movement, explained in 1920 that:

> Co-operation in Scotland, by its very success, has taken from the air that it breathes something redolent of the mountains, moors and mosses of Caledonia … In its vigour today, in its persistence, in its self-assertion, in a certain roughness of order and method, we see proof of its national character. Its faults no less than its virtues proclaim it no alien. It is no exotic requiring hot-house protection and a gardener's special care.[36]

In its early development of co-operative values, in its numerical strength and successes in manufacture as much as trade, in the national profile of its leadership and in its rhetoric of co-operative idealism, the Scottish movement thought of itself – and was often styled by outsiders – as a powerhouse of co-operation in the UK.

Although some commentators questioned his influence, Robert Owen was adopted by many as Scotland's first co-operative prophet. Through his influence and that of his many disciples in the cause – most notably, Alexander Campbell – the co-operative ideal was introduced to early nineteenth-century Scots.[37] By the late nineteenth century, however, all that was left to remind another generation of these early experiments were memories of failure and a few struggling remnants of the societies founded by these and earlier pioneers.[38] Scotland could boast the oldest established co-operative society in the country – a 'title' alternately shared by the Govan Old Victualling Society (est. 1777) and the Fenwick Weavers' Society (est. 1769) – but few of these early societies weathered the economic challenges of the mid-nineteenth century: by 1910, the oldest co-operative society still trading was the Bridgeton Old Victualling Society (est. 1800).[39] It was not until the 1860s that the movement really took off. Nevertheless, for the Scots, and for the movement in general, evidence of an earlier tradition in co-operative trading was significant: it seemed to establish their movement as a long-standing feature of the industrial landscape, and one that could not be easily dismissed as a whim or a passing fad.

Indeed, the strength of Scottish co-operation by the end of the nineteenth century and beyond would seem to indicate that the movement had learned the lessons of its past (Table 42.1). On most Scottish high streets, and in the iconic solid structures of the Scottish Co-operative Wholesale Society at Shieldhall and St Cuthbert's in Edinburgh, the Scottish co-operative movement at the turn of the twentieth century had become a recognisable feature of a new consumer age. Indeed, in 1968, a SCWS pictorial souvenir recorded:

> Scotland's figure for annual purchases per member is the highest in the United Kingdom – £115 per annum. One Scottish housewife in three takes her milk from a Co-operative dairy; every third spoonful of tea put into Scots' family teapots comes from a Co-operative grocery.[40]

The leaders of this movement also developed a popular public profile. James Deans was an effective and respected propagandist for the movement in Scotland, first joining a co-operative Society in 1872. He was a major influence on the Co-operative Union and the Scottish Section, and took a lead role in the co-operative interest during the boycott of the 1890s.[41] Yet towering above him was (Sir) William Maxwell, president of the SCWS between 1880 and 1908, and president of the International Co-operative Alliance between 1907 and 1921.[42] Even during the interwar years the co-op had no difficulty in recruiting talented leaders: Neil Beaton, a native of Assynt, rose through the ranks of both the SCWS and the Shop Assistants' Union to become president of the SCWS (1932–46), and later a director of the North of Scotland Hydro Electric Board.[43]

Table 24.1 Co-operative strength in Scotland: 1886, 1909

District	*1886*		*1909*	
	No. of Societies	*No. of Members*	*No. of Societies*	*No. of Members*
Ayrshire	29	9,137	36	27,916
Border Counties	12	6,497	13	10,796
Central	43	8,525	43	36,920
East of Scotland	34	22,891	24	73,891
Falkirk	20	8,720	22	20,623
Fife and Kinross	17	6,239	35	31,614
Glasgow and Suburbs	31	11,432	37	92,041
Perth and Forfar	39	24,086	46*	66,530
Renfrewshire	27	11,084	23	30,011
Stirling, Fife and Clackmannan	38	13,829	14**	14,511
Unallotted	20	14,376	–	
Wholesale	1	231	1	276
Totals	311	137,047	294	405,129

Source: Maxwell, W. *The History of Co-operation in Scotland: Its Inception and its Leaders*, Glasgow, 1910, 376–98.
* Includes Perth, Forfar and Aberdeen.
** Includes Stirling, West of Fife, and Clackmannan.

It was largely due to the careers, platform performances and writings of such men that Scotland acquired a reputation as a country alive to the most radical interpretations of co-operation. In his history of the movement in Scotland, which was dedicated to James Deans, Maxwell took a determined stance:

> This great united effort … is training our people to be more sympathetic with each other, to have no connection with shams and shoddy nor the sharp practices of trade and commerce, to be self-respecting and self-reliant. And it is giving the monetary assistance to strengthen them in this great moral and material reform.[44]

For Maxwell – a staunch Liberal – the most obvious way of achieving such 'reform' was through political engagement.

From an early date, Scottish co-operators were considered the most eager in the country to 'embark on the stormy sea of politics', and they appeared more strongly in favour of independent political action than many of their English brethren.[45] Writing in 1920, James Lucas reflected

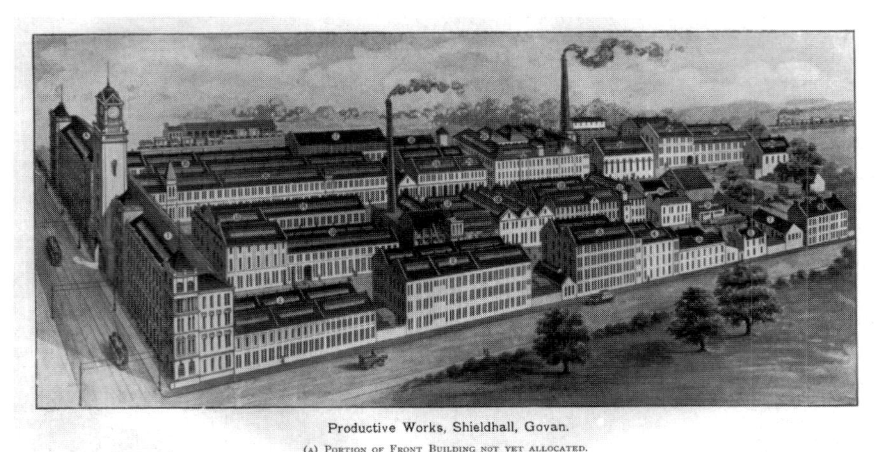

Figure 24.2 Productive works, Shieldhall, Govan. Source: Maxwell, 1910, 270.

that Scottish co-operators continued a national trait in that '(they) followed the polemics of politicians with the same fervour that their ancestors had followed the polemics of theologians'.[46] But despite the efforts of Maxwell and others at successive co-operative congresses at the end of the nineteenth century, the movement proved resistant to their clarion calls.

All things considered, one can clearly identify many features of co-operation in the north to ground Scotland's reputation as a nursery of co-operative activism. But is this the whole story?

POLEMICS AND PRAGMATISM

In 1922, Deans was sceptical of Scotland's commitment to co-operative idealism:

> What we now have is the danger of a large membership that does not understand one thing about co-operation, and thinks of itself only as entitled to receive material benefits and not in any way bound to make sacrifices for the future success of a grand idea. Our membership has grown beyond its own knowledge of principles, and large numbers who buy in the co-operative stores know no more and care no more for the creation of a co-operative system of trade and employment than if the idea had never been proposed.[47]

Throughout the history of co-operation in Scotland, the lived experience, as one might expect, fell far short of the rhetoric of platform speeches, and suggested that in many respects, Scotland's reputation for co-operative idealism was perhaps rooted in little more than inventive sentiment.

Maxwell himself was forced to admit that even the co-operative pioneers of the 1860s lacked an ideological attachment to co-operative principles – 'In vain will one look for anything higher than local interest and immediate reward from purchases.'[48] Indeed, in Musselburgh and Fisherrow, the financial spark that lit the co-operative flame was little more than a chance win on the Derby by a group of net-mill workers.

The minute books of co-operative societies point to the fact that day-to-day business matters occupied the thoughts of members most of the time: the route of delivery vans, the quality of butter and the shape of loaves of bread dominated meetings in Wishaw, for example. Attendance at such meetings was also not always guaranteed, and many societies (including Wishaw, Cupar and Galashiels) experimented with fines for unexplained absences. Lack of interest often plagued the societies' social events: the St George Society in Glasgow ran up a deficit after a series of 'socials' in 1888, and in Perth an evening lecture was cancelled in 1880 when it became clear that the speaker could not hope to compete with the counter-attractions of a 'waxwork and circus in the town'.[49]

Often it did not take much to lure members away from their co-operative loyalties and back into the arms of private enterprise. In Kilmarnock,

Figure 24.3 Sir William Maxwell. Source: Maxwell, 1910, frontispiece.

Figure 24.4 Painting depicting chairmen (past and present) of the Scottish Sectional Board of the Co-operative Union. Source: Maxwell, 1910, 362.

the co-operative committee bemoaned the actions of 'weak and simple [members] whose idea of excellence is expressed in the word "cheapness"'.[50] Other committees fought hard to realise their social obligations to distressed members as frugal co-operators resisted attempts to use co-operative profits for this purpose and cited the limitations imposed by the rule book in their defence.

Elsewhere, the much-vaunted co-operative interest in education was hardly in evidence. While Maxwell lauded the efforts of the St Cuthbert's educational department in Edinburgh which, by 1908, supported a strong musical culture in the society and conducted classes in adult education, book-keeping, industrial history and political economy, other societies were reluctant educators.[51] In Perth, a second attempt was made in as many years to end the regular grant to the educational fund at the thirty-fourth quarterly meeting of the society, and draughts and dominoes seemed to be the major attraction for members using the society's reading room.[52] The Wishaw Co-operative Society actually abolished its educational committee for the duration of World War I, and reduced its grant to the educational fund.[53]

In practice, co-operative actions often seemed to speak louder than words, and those actions were often far from radical. Despite co-operative pride in the growth of the SCWG, Mrs Annie Buchan – a prominent guildswoman – had to be nominated thirteen times before she was accepted on to the otherwise male-dominated board of management of the St George Co-operative Society.[54] Meanwhile, the Coatbridge Co-operative Library in 1921 boasted more entries under 'Samuel Smiles' than 'Socialism', and still retained old 'classics' by Cobden and Bright.[55]

When it came to politics, co-operative societies were frequently more cautious than the movement's leadership. In early 1905, a St George Socialist Group was established within the co-operative society of the same name. Within months, its existence was condemned at the quarterly meeting, whereupon the group seemed to sink without trace.[56] In Wishaw even renting the co-op hall to the Independent Labour party was a matter of debate that rumbled throughout 1918.[57]

Pragmatism rather than polemic motivated the majority of Scottish co-operators at the end of the nineteenth-century and the beginning of the twentieth century. For most members, the dividend mattered more than the fostering of a wider co-operative culture. For the leadership, however, World War I brought challenges that confirmed them in an increasingly political stance.

Figure 24.5 The Shettleston Co-operative Junior Choir, Shettleston, Glasgow, *c*1945. Source: SLA.

Co-operative politics were an exercise in maintaining a somewhat unstable equilibrium, which shifted with the circumstances of the wider political community and were notoriously prone to the disruptive influences of individual localities and personalities. True, a number of co-operative successes in local government in Scotland were recorded in the 1890s, but the parliamentary dimension was controversial.[58]

A Joint Parliamentary Committee from 1892 monitored legislation going before both houses that was pertinent to co-operative interests, and William Maxwell managed to harness the resolve of the national co-operative congress meeting in Perth in 1897 in favour of a greater co-operative role in representative office. But support in subsequent years was not forthcoming. Even in Scotland, the border counties, the eastern district and Renfrewshire all declared themselves opposed to parliamentary representation in 1899, and in other districts members were clearly divided on the issue. Congress would not seriously debate representation again until the Paisley congress of 1905, when the key speaker referred to co-operative indecision as neither 'dignified, consistent nor worthy of a movement like ours.'[59] On the eve of war, the co-operative movement remained in the political shadows of a growing Labour interest, cast at the end of Liberalism's Indian summer. But that would soon change.

In August 1914, Scottish co-operators followed most other Labour interests in quickly moving from scepticism regarding war aims to patriotic support. Whilst on 7 August the *Scottish Co-operator* warned that 'the very civilisation of Europe has been flung into the melting pot of a world war, and may emerge not purified but destroyed',[60] a week later it hesitatingly commented that 'Britain had no choice but to take her part in the struggle if she wished to retain her place of honour among the nations of the world'.[61]

The war was to have a profound impact on the politics of Scottish co-operators and was to be the main factor in convincing many members and leaders alike of the necessity of political action. Due to its loss of employees and resources as a consequence of military service, the role of the state in mobilising the home front, the taxation of co-operative dividends and the preferential treatment accorded private traders in the allocation of scarce commodities, the co-operative movement found that only a political response could address what were – for the most part – very practical problems.

The roll of honour of the SCWS suggests the toll that that war took on co-operators and their employees across Scotland. Over 2,000 SCWS employees served in the armed forces during the war, and of these 315 lost their lives.[62] Across Scotland, other local co-operative societies also recorded serious losses. Regardless, the co-operative motto, 'each for all and all for each', was regularly employed to reaffirm co-operative support for war aims, which found a ready echo in their own movement. But war 'cost' in other ways: £140,000 was spent by the SCWS making up the difference between their employees' military pay and their usual co-op salaries, and co-operative vehicles and horses were commandeered for the war effort: fifteen horses from Coatbridge had found their way into military service as early as August 1914.[63]

In the early months of war, co-operators were clearly more than prepared to 'do their bit'. But war weariness would dull initial enthusiasm. Co-operators came to complain about the presence and influence of traders on local military tribunals adjudicating claims for exemption from service.[64] Too often, it seemed, co-operative employees were being denied exemption, while the experienced staff of independent traders were safeguarded by their bosses. For some this was symptomatic of a state that was ignoring co-operative views and taking their support for granted.

Discontent became most manifest in 1915, when it was declared that the co-operative movement was to face excess profits duty on the dividends it awarded members. Co-operative trade had grown significantly over the course of the war, but this was largely due to an enlarged membership, across which any profits had to be divided. It seemed plain to co-operators that the government either did not understand or chose to ignore the nature of co-operative enterprise. The rules were eventually changed in 1917 so that co-op members were not unduly penalised, but between 1915 and 1917 the duty had seriously eaten into co-operative profits: for example, the amount paid by the SCWS in duty wiped out all the profits from its Shieldhall works, a sum of over £167,000.[65] For individual co-operators, most felt the impact of the tax in reduced dividends at a time when wages were failing to keep up with the cost of living.[66]

The year 1917 was to be a turning point. As scarcities in sugar and cereals across Britain were being recorded, co-operators found themselves excluded from local food committees established by the Ministry of Food to co-ordinate the distribution of scarce supplies. Instead, the committees were dominated by local independent traders who, it seems, safeguarded the interests of private enterprise to the detriment of the co-op. Abuses were evident across Scotland: for example, Kinloch and Butt record how the SCWS failed to receive any of the additional supplies of sugar destined for the West of Scotland in 1917, despite being the largest retailer in this area.[67] In September 1917, the *Scottish Co-operator* ran a story headlined 'Where did the Sugar Go?'.[68] Even at their most even-handed, the co-op was disadvantaged by the food committees' system of distribution, which based allocations on pre-war levels of consumption: co-op membership had increased, but no-one seemed to be listening.

In Paisley, earlier Labour unity over taxation was again reaffirmed when the Trades and Labour Council refused a request by the local authorities to co-operate in a food economy committee and instead supported the Equitable Co-operative Society's food protest demonstration in December 1917. A month later, Labour and co-operative interests in Paisley formed a food vigilance committee. Across Scotland, similar moves towards greater understanding were bringing co-operators and the Labour movement closer together. P Malcolm wrote in the *Scottish Co-operator*: 'The old humdrum repetitions of the benefits of high dividends are a thing of the past.'[69]

The Annual Congress of the Co-operative Movement meeting in Swansea in 1917 affirmed that British co-operators would henceforth be represented by their own political party, known after 1919 as the Co-operative Party. And at an emergency meeting in Westminster later that year, a programme for the infant party was agreed. Scottish co-operators had anticipated the establishment of the national party earlier in the year when they had set up a central co-operative parliamentary representation committee, and Scottish co-operators (including William Maxwell) were to the fore at the Westminster meeting. Indeed Deans, secretary of the Scottish Section of the Co-operative Union, moved the acceptance of the eleven aims which together formed the first programme of the Co-operative Party.

FIRSTS AND FAILURES

As the co-op in all its various manifestations was arguably the most successful workers' movement in Scotland, hopes were high in those early days that the new party would break the mould of Scottish politics and infuse the bitter political environment of the coalition years with the idealism of the movement's founding principles. There was every reason to believe it would be successful. Co-operative membership in Scotland by 1916 numbered over half a million, and it kept rising. In addition, in an era which saw women enter polling booths for the first time, the influence of the Co-operative Women's Guild, it was anticipated, was sure to be an advantage.

Yet, its initial forays into parliamentary politics were disappointing. In the 1918 general election, the Co-operative party contested only three Scottish seats – Paisley, Clackmannan and Kilmarnock – and was unsuccessful in each. Its fight against Asquith in Paisley in 1920 was admirable: the co-op candidate, J M Biggar, improved on his 1918 performance and secured nearly 40 per cent of the vote in a three-cornered contest. But in the 1922, 1923 and 1924 general elections, the results were disappointing: Tom Henderson was successful in Tradeston in 1922, and held his seat in 1923 when he was joined by A. Young in Partick; but in 1924, while Henderson remained in office, Young's brief spell in the Commons came to an end.

Explaining why the Co-operative party was not more of an immediate success is a complex matter, though four areas call for particular attention: the nature of the co-operative movement as both an employees' and employers' organisation; the diversity of political views that remained in the movement even after 1917; the regionalism and local autonomy of a movement which, albeit its greatest strength, could also be a political liability for a party effectively controlled from London; and the relationship which the infant Co-operative party had with the Labour party.

Putting to one side local disputes over candidatures and affiliation, the last of these factors encourages us to return again to fundamental issues surrounding the aims and identity of the co-op. The Labour and Co-operative parties, while sharing many immediate goals in 1917, were at odds over the role of the state. Indeed, it is open to question whether many in the Co-operative party in 1917 had a vision of co-operative politics outwith the context of war. Of the eleven points in the co-op programme of 1917, the first three sought the protection and extension of co-operative enterprise; the fourth, fifth, seventh and eighth offered a series of pledges about the scientific development of agriculture, the provision of light railways and quality housing, the taxation of land values, equal opportunities in education and an increase in death duties; the sixth called for a national credit bank to aid municipalities and (of course) co-operative societies; the ninth demanded parliamentary control of foreign policy; the tenth wanted to see the sensitive demobilisation of military conscripts; and the eleventh vaguely pointed to the 'breaking down of the caste and class systems, and the democratising of state services – civil, commercial and diplomatic'.[70]

At a time when the Labour party stridently embraced a centralising and nationalising state, the co-operators were strangely silent on their vision of government and the control of industry. Even among the co-operative leadership a clear party line was not evident.

As far back as 1898, Maxwell had declared: 'Co-operation seeks the help of no one; it asks neither the help of State or municipality, but demands the help of its membership.'[71] Yet, at the Paisley congress in 1905 it was claimed that 'co-operative action and enlightened State action are but developments of the same ideal, namely, the regulation of our complex social relationships upon the basis of justice and equity'.[72] As the Co-operative party came increasingly under the wing of the Labour party after 1927, and as joint candidatures became ever more the norm, these ambiguities at the heart of Co-operative party policy making were still to be resolved.

CONCLUSION

> The trade unionist, generally speaking, had no use for what he regarded as the milk-and-water gospel of the Co-operator; the Co-operator thought the trade unionist merely a loud-mouthed and foolish disturber of industrial peace. They agreed, however, in thinking that the Socialist was an empty-headed dreamer of vain dreams; and the latter retaliated by cursing them both as impracticable materialists who were losing a kingdom whilst they fought for a few scraps on a dust-heap![73]

The Scottish co-operative movement has achieved much more than 'a few scraps on a dust-heap' during its lifetime, though in the long run it has never fully realised its early leaders' dreams of creating Scotland anew.

Largely due to amalgamations in the first instance, the number of societies affiliated to the SCWS began to decline from the late 1930s. Indeed, in 1943 only the objections of Scottish nationalists blocked proposals for the amalgamation of the SCWS and the English Co-operative Wholesale Society (CWS). The CWS voted for amalgamation. By the 1950s, the Scottish co-ops were clearly feeling the pressure of competition in an increasingly affluent society as consumers prioritised lower prices over the 'divi', and were prepared and – with car ownership increasing – more able to travel to secure the 'best' price. Merger of the SCWS with the CWS was finally brokered successfully in 1973, and thereafter changes in management practices sought to 'modernise' the co-op.

At the turn of the twenty-first century, the co-op in Scotland would have been unrecognisable to its late Victorian sponsors. In 1981, Edinburgh's famous St Cuthbert's Society joined with the Dalziel Co-operative Society of Motherwell to form the Scottish Midland Co-operative Society (SCOTMID). Continued growth and mergers with successive smaller interests meant that even in the competitive trading environment of the new century it boasted 160 outlets, including health and beauty shops and funeral parlours. In 2001, the Co-operative Union – the main sponsor of co-operative idealism and political engagement since 1869 – combined with the Industrial Ownership Movement, merging for the first time the foremost national organs of consumer and employee co-operation. At much the same time, the CWS changed its name to the Co-operative Group Ltd. At that point the Co-operative party had twenty-nine MPs in the House of Commons, co-sponsored with the Labour party. In 2004 the party boasted six MSPs in the new Scottish parliament.

The history of co-operation in Scotland throws into relief 'big' questions concerning the role of workers and consumers in the development of both the country and its politics. We are left to wonder why Labour did not prove broad enough to speak for the consumer as loudly as it did the producer, and why the biggest and arguably most successful grassroots movement in Scotland failed to flourish in the late twentieth century. In the end we retain the beguiling 'what might have been' of a workers' movement that – on paper, at least – could have reshaped the labour history of Scotland, an economic institution that had the potential to democratise the marketplace, and a community that for the most part strove to live its founders' motto, 'each for all and all for each' – a different commonweal.

NOTES

1. Robertson, 1910, 87.
2. 'R.L. of Wandsworth Common', *c*1872, 12.
3. Robertson, 1910, 22.
4. Robertson, 1910, 120.
5. Stirling, 1915, 43.
6. National Library of Scotland (NLS), Acc 4722/15, 18 Jan. 1874; Acc 4772/4, 8 March 1888, 25 March 1888, 22 April 1889.
7. Musselburgh and Fisherrow Co-operative Society Limited, *c*1962.
8. Battison, 1939, 41.
9. As quoted in Kinloch and Butt, 1981, 64.

10. Cook, 1990, 6; Flanagan, 1915, 18.
11. Laing, 1939, 11–13.
12. Reid, 1923, 10; NLS, Acc 4772/1, 25 Sept. 1846.
13. Robertson, 1910, 119.
14. Stirling, 1915.
15. Battison, 1939, 39.
16. Murray, 1930, 49 and 61.
17. Flanagan, 1915, 18; Stirling, 1915.
18. Battison, 1939, 23.
19. Reid, 1923, 100.
20. Potter, 1987, 105.
21. Flanagan, 1915, 28.
22. Flanagan, 1915, 33.
23. Laing, 1939, 21.
24. 'Catriona', 1927, 10.
25. Callen, 1952, 17, 32.
26. Buchan, 1913, xii.
27. Flanagan, 1915, 43.
28. Robertson, 1910, 66.
29. Battison, 1939, 38–9.
30. Willocks, 1892, 132.
31. Willocks, 1892, 227–35.
32. Lucas, 1920, 65.
33. Kinloch and Butt, 1981, 245–67.
34. Reid, 1923, 64.
35. Battison, 1939, 23–4, 26.
36. Lucas, 1920, 10.
37. See Fraser, 1996.
38. Here the Orbiston community, 1826–8, springs to mind. The co-operative community of Orbiston employed 100 men at its height, and housed them and their families in basic accommodation on an estate of roughly 290 acres in Lanarkshire. Holyoake notes: 'An ill-assorted random collection of most unsuitable persons flocked to the spot, which speedily acquired the emphatic name of "Babylon" from the surrounding population, a title, we should imagine, most applicable alike to its inhabitants and its proceedings.' Holyoake, 1875, 272–6.
39. Maxwell, 1910, x. (Note: Govan Victualling Society closed its doors in 1909.)
40. Scottish Co-operative Wholesale Society Limited, 1968, Introduction.
41. Deans, 1922.
42. Obituary, *Glasgow Herald*, 11 February 1929.
43. Wardlaw, 1946.
44. Maxwell, 1910, 373.
45. Lucas, 1920, 72. See also Cole, 1944, 312.
46. Lucas, 1920, 72.
47. Deans, 1922, 55.
48. Maxwell, 1910, 234.
49. Reid, 1923, 40; Willocks, 1892, 205.
50. Robertson, 1910, 29.
51. NLS, Acc11650, 1908.
52. Willocks, 1892, 187 and 216.

53. Battison, 1939, 47 and 65.
54. Callen, 1952, 24. Note: Even in 1949 the board of directors of the St Cuthbert's Co-operative Society (Edinburgh) had only four female members out of a total membership of twenty-one. St Cuthbert's Co-operative Association, 1949, 11.
55. Coatbridge Co-operative Society Limited, 1921.
56. Reid, 1923, 89–90.
57. Battison, 1939, 67–8.
58. See Macdonald, 2000; Smyth, 2000.
59. Tweddell, 1905, 7.
60. *Scottish Co-operator*, 7 August 1914.
61. *Scottish Co-operator*, 14 August 1914.
62. Scottish Co-operative Wholesale Society Limited, 1920, iii.
63. Scottish Co-operative Wholesale Society Limited, 1920, iii; *Scottish Co-operator*, 21 August 1914.
64. See Laing, 1939, 34.
65. Kinloch and Butt, 1981, 275.
66. Kinloch and Butt, 1981, 272 (Table).
67. Kinloch and Butt, 1981, 276.
68. *Scottish Co-operator*, 28 September 1917.
69. *Scottish Co-operator*, 5 October 1917.
70. *Scottish Co-operator*, 19 October 1917.
71. Maxwell, 1909, 332.
72. Tweddell, 1905, 9.
73. Murray, 1930, 47.

BIBLIOGRAPHY

Battison, G. *Wishaw Co-operative Society Ltd: A Record of its Struggles, Progress and Success from its Inception in 1889*, Glasgow, 1939.

Buchan, A. *History of the Scottish Co-operative Women's Guild*, Glasgow, 1913.

Callen, K M. *History of the Scottish Co-operative Women's Guild: Diamond Jubilee 1892–1952*, Glasgow, 1952.

'Catriona' [pseud., Catherine Morison]. *Dod! That Beat's A' or, Jining the 'Co': A Humorous Scottish Sketch for Five Females*, Glasgow, 1927.

Coatbridge Co-operative Society Limited, *Catalogue of Books in the Library of Coatbridge Co-operative Society, Limited*, Coatbridge, 1921.

Cole, G D H. *A Century of Co-operation*, London, 1944.

Cook, W. *Co-operation in Coalsnaughton, Golden Jubilee: A Historical Sketch of the Coalsnaughton Society Limited, 1872–1922*, Alloa, 1990.

Deans, J. *Co-operative Memories: Reminiscences of a Co-operative Propagandist*, Manchester, 1922.

Flanagan, J A. *Co-operation in Sauchie: A Retrospect, 1865–1915*, New Sauchie, 1915.

Fraser, W H. *Alexander Campbell and the Search for Socialism*, Manchester, 1996.

Holyoake, G J. *History of Co-operation*, London, 1875.

Kinloch, J and Butt, J. *History of the Scottish Co-operative Wholesale Society*, Glasgow, 1981.

Laing, H C. *History of Cupar and District Co-operative Society Ltd*, Cupar, 1939.

Lucas, J. *Co-operation in* Scotland, Manchester, 1920.

Macdonald, C M M. *The Radical Thread: Political Change in Scotland, Paisley 1885–1924*, East Linton, 2000.

Maxwell, W. *Address at the Opening of the Chancelot Flour Mill, Edinburgh, August 1894*, Glasgow, 1909.
Maxwell, W. *The History of Co-operation in Scotland: Its Inception and its Leaders*, Glasgow, 1910.
Murray, R. *Stirling Co-operative Society: A Historical Sketch of its Fifty Years of Progress*, Glasgow, 1930.
Musselburgh and Fisherrow Co-operative Society Ltd. *One Hundred Years of Progress: Musselburgh and Fisherrow Co-operative Society Ltd (1862–1962)*, c1962.
National Library of Scotland, Acc 4722/15, Galashiels Waverley Co. Ltd, Minutes, 18 Jan. 1874.
National Library of Scotland, Acc 4772/1, Galashiels Provision Store Co., Minutes, 25 September 1846.
National Library of Scotland, Acc 4772/4, Galashiels Provision Store Co., Minutes, 8 March 1888, 25 March 1888, 22 April 1889.
National Library of Scotland, Acc 11650, William Maxwell MS, Notebook, 'Co-operation in Great Britain', 1908.
Potter, B. *The Co-operative Movement in Great Britain*, Aldershot, 1987; first published 1893.
'R.L. of Wandsworth Common', *Diddledom: Or Tonics for the Co-operative Society Fever*, Chelsea, c1872.
Reid, W. *Fifty Years of the St George Co-operative Society Ltd, 1870–1920*, Glasgow, 1923.
Robertson, W. *Kilmarnock Equitable Co-operative Society Ltd: A Fifty Year's Record*, Kilmarnock, 1910.
St Cuthbert's Co-operative Association. *1859 to 1949: Festival Cavalcade Exhibition, Waverley Market, Edinburgh 20th August to 3rd September, 1949*, Edinburgh, 1949.
Scottish Co-operator, 1914 and 1917.
Scottish Co-operative Wholesale Society Limited. *Roll of Honour, 1914–1919: Employees of S.C.W.S. Who Served with His Majesty's Forces in the Great War*, Glasgow, 1920.
Scottish Co-operative Wholesale Society Limited, *Co-operation in Scotland: Pictorial Souvenir of Scotland and Some Co-operative Enterprises*, Glasgow, 1968.
Smyth, J J. *Labour in Glasgow, 1896–1936*, East Linton, 2000.
Stirling, T. *History of Vale of Leven Co-operative Society*, Alexandria, 1915.
Tweddell, T. *Direct Representation in Parliament*, Manchester, 1905.
Wardlaw, J. *Neil S. Beaton: An Appreciation*, Glasgow, 1946.
Willocks, J. *The City of Perth and its Co-operative Society*, Glasgow, 1892.

Websites

The National Co-operative Archive is based at the Co-operative College in Manchester, available online at: http://archive.co-op.ac.uk/.
Heriot-Watt University's Martindale Library (Scottish Borders Campus) also holds a significant range of relevant titles, donated to the university by the Co-operative College, available online at: http://www.hw.ac.uk/sbc/library/collect.html.
The archives of the Scottish Co-operative Wholesale Society are to be found in the Mitchell Library, Glasgow, available online at: http://www.glasgow.gov.uk/en/Residents/Leisure_Culture/Libraries/Collections/ArchivesandSpecialCollections/wholesalesocieties.htm.

25 The Organisation of the Employer Classes in Scotland

RONNIE JOHNSTON

As far as the history of social relations in Scotland goes there has been a tendency for historians to concentrate on the activities of the working class. To some extent, this focus reflects Karl Marx's dismissal of the middle class as being transient, destined to be eclipsed by the impending workers' revolution. There has been some research into the early formation and composition of the middle classes and the activities of the landed classes, which goes some way towards addressing this imbalance.[1] However, much more work is required, and it is still the case that although the historiography bulges with research on the rise of the Scottish labour movement, the changing nature of work, the extent of working-class consciousness and so on, we still do not know enough about the collective activities of Scotland's employers. As this chapter will try to illustrate, such organisational activity played just as crucial a role in Scotland's economic development as the rise of the labour movement and the growth of trade unionism. The aim here is to illustrate some of the most important aspects of collaboration amongst Scotland's employer classes, focusing on three main areas: the regulation of trade; collective organisation for labour relations' purposes; and collective activities which sustained social bonding – although there was frequent overlap between these categories.

MERCHANT GUILDS AND CRAFT INCORPORATIONS

Ancient guilds of merchants and craftsmen were vitally important for the early economic development of Europe's towns and cities, with the London guilds being the oldest. The rise of long-distance exchange between trading cities evolved under very complex and uncertain conditions, and the rise in the number of merchant guilds was, to some extent, a response to the need for leadership, networking and information channels that were crucial for sustaining merchants' confidence during the early phases of capitalism.[2] The spread of these organisations resulted in networks of shared norms and values throughout Europe, which were to prove vital in the development of trading links and the growth of internal and overseas commerce.

It is not clear whether Scotland's merchant guilds evolved independently or if the basic model was copied from those that had already come into being in England. But we know that the emergence of these organisations from the twelfth century onwards was directly linked to the growing influence of a merchant class in Scotland's burghs. Although towns and proto-urban settlements undoubtedly existed before the reign of David I (1124–53), it was in that period that the granting by the crown of foundation charters to 'burghs' – an idea which paralleled similar patterns throughout western Europe – became a key trigger in the stimulation of Scotland's economy. The burgh, it has been said, was 'a community organised for trade';[3] royal burghs were granted extensive rights of monopoly over their surrounding rural hinterlands as well as the unique privilege of engaging in overseas trade in staple goods. As such, burgesses, who by the sixteenth century made up at most only 30 per cent of all adult male inhabitants, had extensive economic powers as employers both within the town and in the countryside around it.[4] By 1700 there were over 300 burghs of barony throughout Scotland, and about 80 more prestigious royal burghs.[5] The royal burghs had comprehensive trading privileges, which included the right to trade overseas, although this privilege was eventually removed by the Scottish parliament in 1672. The importance of the royal burghs was also reflected in their having representation in the Scottish parliament, and in the fact that from the 1550s they had their own umbrella organisation in the Convention of Royal Burghs, which set the relative contributions made by each to national taxation and generally regulated most aspects of local, regional and overseas trade.

By 1400, at least thirteen towns had an organisation known as a guild merchant, and by the end of the century the figure was nineteen. These organisations stemmed from a series of twelfth- and thirteenth-century enactments, including the Laws of the Four Burghs and the Statutes of the Guild, the latter emerging from Berwick, then Scotland's largest trading port; their purpose was social and religious as well as mercantile. The immediate stimulus was from England: the burgh laws owed much to the 'customs' of Newcastle-upon-Tyne and other English towns, including Winchester and Northampton, but they, in turn, had extensively borrowed from the customs of Breteuil in Normandy. Initially, as was the case throughout Europe, there was no dividing line between merchants and craftsmen, and burgh craftsmen were allowed to join the guild merchant on an equal footing with the merchants as long as they owned land of a certain value: in other words, as long as they were burgesses.[6] However, from the thirteenth century onwards there was a gradual separation of craftsmen from merchants throughout Europe as craftsmen began challenging the rights of the merchants over the running of town affairs.

In Scotland, this process seems to have occurred much later. In one view, it was not until the fifteenth century that a clear separation between craftsmen and merchants began, triggered by a shift in economic fortunes of the two groups stemming from a serious decline in overseas exports and the lingering, knock-on effects of falling population which had begun with the Black Death. In this climate, merchant-dominated town councils increasingly intervened to hold down prices and wages.[7] From this period onwards, though, effective control of the larger burghs increasingly fell into the hands of merchants, normally the richest of the burgesses, and the result was a growing conflict between craftsmen and merchants, bringing sharp tensions, a struggle for burgh office and riots, involving either craft master or 'craftis childer', their unruly apprentices. The later fifteenth century saw a good deal of parliamentary legislation directed against burgh unrest, but it is more likely that this was a reaction to problems in specific burghs, such as Aberdeen in the 1480s and Perth in the 1550s, rather than general legislation against a widespread problem.

In another view, the 'problem' of tension between merchants and crafts has been greatly exaggerated. In the bulk of cases – some thirty-one out of fifty-six – and almost invariably the case in smaller burghs, the foundation of a merchant guild, giving monopoly powers to an inner élite of merchants, belonged to the period after 1560.[8] It was in the larger towns with more complex economies that separate craft organisations first emerged in the last quarter of the fifteenth century. Edinburgh's first craft to incorporate was the skinners, in 1474, and by 1530 the town had fourteen incorporated craft guilds. Glasgow had the same number by 1605, but in other towns where the economy was less diversified the number of incorporations was smaller: Dundee and Perth had nine, and there were seven in both Stirling and Aberdeen.[9]

Official recognition came in the form of seals of cause issued by a burgh council to a craft. This gave the new incorporation a deacon, a deacon's council which oversaw standards, internal discipline and the right to collect dues from all members from masters down to apprentices, and a craft altar, saint and chaplain in the burgh church. The idea of fraternity underpinned much of the new arrangements: the craft worked and worshipped together and fiercely defended its own privileges.

Even after the Reformation took away chaplains and altars, the sense of fraternity persisted: Edinburgh's large hammerman craft referred to itself as the 'house' throughout the seventeenth century. The seal of cause meant that the deacons were now able to look after the interests of their respective crafts, both in the burgh courts and in their own internal courts. This was done in several ways: by imposing the rules and regulations of the craft; controlling who could work in the craft and regulating the number of apprentices; and ensuring strict entry qualifications by insisting that apprentices successfully complete a piece of craft work known as an essay. Relatively little attention has been given to the sharp increase in economic and social power which incorporation gave to an élite of craft masters. More stress has usually been laid on the continuing tensions between crafts and merchants: Fraser points out that the crafts in Scotland never managed to dislodge the merchants from their privileged positions in the towns as completely as craft guilds in some other parts of Europe – although there were regional variations.[10]

The original intention of the craft guilds was, like their merchant equiv-

Figure 25.1 Seal of the Incorporation of Hammermen of Dundee. Source: Dundee City Archives.

alents, to ensure the privileged position of members within the burghs, and to protect their trade from those who were deemed unqualified. Yet the craft élite could and did also belong to the merchant guilds: in Edinburgh skinners, tailors, baxters and the more prestigious parts of the hammermen trade – such as cutlers, pewterers and saddlers – did so in significant numbers. These were craftsman employers who did not soil their own hands with craft work, or they might be wholesalers dealing in grain, or retailers catering for a more sophisticated market. Camouflaged behind the confusing terminology of 'merchants' and 'crafts', a craft aristocracy was emerging which enjoyed much the same economic rights and powers as their 'merchant' counterparts. In the process, a good deal of the old structure of burgh legislation was ignored in practice. The burgh laws insisted that a craft master should be restricted to a maximum of three 'servants', a term that embraced journeymen and day labourers as well as apprentices. On the face of it, this restriction was complied with: the 399 craft masters listed in an Edinburgh muster roll of 1558 had a total of 333 servants, and the 285 merchants had 411 working for them. But there were huge variations. Merchants – from humble booth owners to wholesalers and overseas exporters – ranged from having seven servants to none. The same was true of some of the élite crafts. The muster detailed ten bakehouses in the capital, and six of them had between six and nine workers. Their owners were already running large-scale workshops. It is likely that the seventeenth-century development of manufactories – especially in cloth manufacture – was an extension or rationalisation of a much older system of putting out, in which the town was the finishing centre of a long-established, largely rural-based, industrial process.[11]

As well as their economic functions, the incorporations also played important ceremonial roles. Upon incorporation, each craft was assigned a specific place and role in civic ceremony, which might range from full-scale religious processions like that of Corpus Christi, first traced to Aberdeen in 1451, set-piece royal entries or a simpler parade of the craft on the feast day of its own saint. Civic ceremony continued after the Reformation, if in a muted form. For example, when James VI (1567–1625) visited Perth in 1617 he was entertained by a sword dance put on by the town's glovers and an 'Egyptian dance' by the bakers. Interestingly, such ritual was still in evidence in Perth over 200 years later when Queen Victoria made her official visit and was greeted by a glover walking in the procession of citizens 'with the cap on the head and bells jingling at every step'.[12] Yet the size and significance of crafts and craftsmen varied from one town to another. Both Glasgow and Aberdeen had relatively small numbers of craftsmen – Glasgow 361 in 1605 and Aberdeen a mere 127 in 1637 – whereas Edinburgh had 602 in 1634, despite a flight to the suburbs of poorer crafts to escape taxation.[13] Yet in the case of both Edinburgh and Glasgow, the importance of their crafts was reflected in impressive architecture: in the Tailor's Hall, built in Edinburgh's Cowgate in the 1621 and raised to four storeys in 1757, or in the building in Glasgow of a purpose-built Trades Hall in Glassford Street, designed by Robert Adam and now opened to the public.

Increasingly, the merchant guilds in Scotland concerned themselves with maintaining their members' privileged position regarding the organisation of trade and commerce. However, from the late seventeenth century, the tight restrictions which had been imposed by the guilds over the right to trade in the burghs began to break down as commercial activity throughout Scotland increased. In 1681, for example, non-burgesses in Stirling were allowed to engage in business if they made an annual contribution to the town's funds, and a similar trend was apparent throughout Scotland. The number of prosecutions of non-burgesses caught trading declined, and by the 1720s in Glasgow and Edinburgh, burgess-ship was looked upon more as a social badge than a necessary qualification to trade; the importance of such a social badge continued well into the industrial period.[14]

Devine points out that further evidence of the lessening of merchant guild control over Scottish trade and commerce is evident in the slackening of the bonds of the merchants' apprenticeship system. Initially, those aspiring to become overseas merchants had to serve a seven-year apprenticeship before being granted burgess status. This term was reduced to three to five years by the early eighteenth century in the main burghs. Concomitant with this, though, was the fact that the number of apprentices also fell, indicating that mercantile trading was being opened up in the face of changes in the economic system, whereby guild protectionism was becoming less necessary.[15] As the Scottish economy began to grow rapidly – and especially so from the 1750s – the older forms of merchant restrictions, which earlier had proved to be so necessary, became out of step. From this period, although membership of the merchant guilds continued to be a prerequisite to membership of the urban élite, it was not a prerequisite to success in trade.

Much more important, and especially for those involved in overseas trade, was the network of personal and family connections which was expanding as the middle classes in Scotland expanded and developed.

The important institutional and social separation of merchants from craftsmen, or of craftsmen employers from humbler craft masters, in and after the fifteenth century was replicated to some extent from the eighteenth century onwards in the separation of masters from their workers. Fraser tells us that 'there was no definite date when journeymen came to see themselves as a group distinct from their employers'.[16] However, from the beginning of the eighteenth century the relationship of journeymen to masters in the incorporations became increasingly difficult. Older paternalistic relationships between masters and men – which were once accepted without question – were changing. In particular, the older practice in which apprentices were formally indentured to a master and lived in the master's home was giving way to one in which the relationship was bound through the payment of wages alone. It was changes such as these that began to untie the bonds which had held the masters and men in the incorporations.

These changes came about in a very uneven fashion – some Glasgow bakers, for example, were still being paid in kind up to the early 1750s – and well into the nineteenth century the distinction between master and journeyman was blurred in many crafts.[17] However, as was the case with the merchants, the eighteenth century brought a significant transformation to the lives of many Scottish urban workers. The growth of the economy and the impact of intensifying urbanisation meant that some trades expanded while others atrophied. At the same time, controlling entry into the crafts became increasingly difficult. This brought growing uncertainty about status. Some incorporations expected journeymen to be subservient to the masters and work as instructed, whilst increasing numbers of journeymen looked upon themselves as independent craftsmen in their own right and thereby able to dictate their own patterns of work.[18] The demise of the old certainties extended to skilled status, too. As Fraser points out: 'To be a tradesmen in 1700 meant something clear and understandable to those around. At the end of the century, it could mean anything from the most skilled of mechanics to the most unskilled of weavers.'[19] Therefore, in such circumstances the close identification of master and skilled man, which had been the bedrock on which the trade incorporations rested, began to fracture.

The end result of these upheavals was that increasingly the trade incorporations became far more orientated towards protecting the interests of masters from the activities of their journeymen than on representing the interests of both groups. This was especially the case when journeymen began to organise to protect their own interests. Initially this type of association took the form of charitable organisations – previously an important role of the trade incorporations.[20] However, the masters' fears that workers' charitable organisations could become industrial organisations turned out to be well founded, and as Fraser illustrates, most trades' mutual aid societies were indeed the precursors of trade unions designed for labour relations

Figure 25.2 Flag of the Hammermen of Leith. Source: The People's Story Museum, City of Edinburgh Museums and Galleries.

purposes.[21] A process was under way, therefore, in which older paternalistic methods of labour control were giving way to a far more conflict-ridden system characterised by separate associations for masters and workers. Trade incorporations, which had originally been designed, in part, to protect craftsmen from the hegemony of the merchant élites, now became orientated towards protecting the employers' right to manage.

Unlike the situation in England, where the power of the trade incorporations dwindled as the state matured, trade incorporations remained a significant force in Scotland, due primarily to the fact that they were bolstered by the force of the magistracy in their efforts to control the workers – this being another incentive to trade union growth. One of the consequences of the Parliamentary Reform Act of 1832 was that the running of town and burgh affairs became more broadly based – Glasgow's Merchants' House, for example, no longer supplied all the town's councillors. More significantly, the passing of the Burgh Trading Act of 1846 curtailed the influence of Scotland's incorporated trades and merchant guilds even more by removing almost all of their privileges.

However, in several towns and cities, these organisations continued to play an important role as agencies of middle-class identity well into the industrial age, and membership of the old guilds remained important for social reasons. This was certainly the case in Glasgow, where the incorporations continued to attract members throughout the nineteenth and twentieth centuries. Moreover, for some time the hammermen, wrights, masons, gardeners, etc, retained some links with their respective workplace crafts. However, increasingly, they began to attract a mix of individuals from across the industrial spectrum.

EMPLOYERS' ASSOCIATIONS

The first evidence of a British employer combination orientated exclusively towards industrial relations was in 1745, when Manchester cotton manufacturers banded together to bring down wages to enable the town to pay the reparations demanded by Bonnie Prince Charlie's army. However, ad hoc organisations of this kind were probably fairly common throughout the UK, and became even more common as the economy developed and competition increased. Certainly, thirty years after the Manchester example, Adam Smith indicated that he was well aware of the trend when he wrote in the *Wealth of Nations* that employers were everywhere in a sort of 'tacit combination', to raise prices and keep wages low.[22] Moving on another twenty years or so, we know that there were formal combinations of Scottish employers across a band of trades, including printing, coal mining and cotton manufacture. The ending of serfdom in the Scottish coal mines in 1799 also meant that coal and iron masters had increasingly to deal with a far less compliant workforce, so here too employer organisation began to develop.[23]

Collective action, underpinned by the need for employers to deal with a potential labour threat, followed on from the changing relationships between masters and workers. After the repeal of the Combination Acts in 1825, more employers' associations were formed throughout the UK. However, as was the case with the early trade unions, there was a tendency for such combinations to dissipate when the conflict with labour was over. Moreover, like the trade unions – which did not in any case pose a significant threat at this time – the early development of employers' associations was retarded by poor communications and by the fact that most masters still had significant freedom over how they ran their businesses. There was a combination of Glasgow's cotton weaving employers in 1766, and we know that the masters were refusing to employ weavers who did not have certificates provided by their former employers – a form of labour control that was to remain a constant of employer organisations in general to the 1920s and beyond.[24] However, such organisation was sporadic and there were no long-lasting associations of employers in the cotton industry at this time. Indeed, even by the 1830s, although the main cotton spinners in Glasgow had their own secretary and were meeting regularly in the city's Exchange Sales Rooms, there were no written rules to guide the association.[25] Rules or no rules, though, it was still the case that seven years later the masters were sufficiently united to orchestrate a successful lock-out of their employees, which set back the development of trade unionism in the Scottish cotton industry for the next fifty years.[26] However, although the cotton masters' ruthless lock-out stands out as an extreme example of employer organisation power being used in Scotland at this time, the key motivating factors which were to fuel the growth of employer collective action were still not in place.

It was from the 1850s that effective employer combination developed in Scotland. By this time Britain was at its peak as the 'workshop of the world', and from this time the industrial regions of Scotland were to play a leading role in producing goods for domestic and imperial markets. Heavy industrialisation and significant urbanisation occurred in Scotland a generation later than England. However, when the process got under way in Scotland it was more intense, with Glasgow, for example, urbanising faster than any other European city. Within this rapidly changing economic and social climate the fissures between masters and workers, which had opened up in the seventeenth century, rapidly widened, and although a few employers still utilised paternalism as a labour control strategy, for the most part confrontational industrial relations became part and parcel of the new social order.

The growth of trade unionism in Scotland has been well documented (see Chapter 23 of this volume), and there is a degree of agreement that from the defeat of the Glasgow cotton spinners in 1837 up to the 1880s, trade union strength fluctuated considerably, with industrial relations being far from volatile.[27] This was a pattern which continued, and even by 1892 the estimated 147,000 trade unionists in Scotland represented only 3.7 per cent of the population, compared to 4.9 per cent in England. Trade union strength was most potent in the heavy industries, but even here too many unions representing too few members weakened the labour challenge.[28]

Concomitant with fluctuations in the level of trade union activity, after 1850 Scottish employers also had to contend with the growth of socialism and increased government intervention in labour relations and working conditions; this was an interference which intensified from the 1880s onwards. By 1897 the board of trade had limited powers of conciliation in labour disputes, and the same year the Royal Commission on Labour reported on its survey of employers and workers across the UK regarding the nature of industrial relations. On this evidence it was quite clear that those Scottish employers who gave evidence to the commission were more hostile to the principle of collective bargaining than those who were interviewed from elsewhere.[29] One possible reason for this was that the development of the Scottish industrial economy was at an earlier stage than that of the more mature economy in England, where collective bargaining was a more accepted part of industrial relations. However, another factor which may account for this hostility is that in line with employers throughout the UK, Scottish capitalists were having to trade in increasingly difficult markets as intensifying domestic and – from the 1870s onwards – foreign competition stripped away their profits. Things were so bad across the UK that a Royal Commission on the Depression of Trade and Industry was initiated in the mid-1860s, and it is interesting that the Glasgow chamber of commerce's statement to the royal commission identified the labour problem and foreign competition as two sides of the same coin: 'As regards competition in manufacturing with foreign countries, the question of the provision of the Factory Acts of this country require careful consideration and revision.'[30]

Pressures such as these acted to bring employers together, especially those in industries which faced the twin forces of tightening markets and increased trade union growth. A process was set in motion in which employers in industries such as mining, building, shipbuilding and engineering increasingly began to bond together in organisations. In Clyde shipbuilding, for example, the Clyde Shipbuilders' Association (CSA) came into being in 1866 to represent shipbuilding companies and engineering firms in the west of Scotland. By 1892, the Clydeside engineers had formed their own association, too, the North Western Engineering Trades Employers' Association (NWETEA). These two organisations became the most important heavy engineering groupings in Scotland, and by the early 1900s they were further strengthened by being linked with a broad band of similar organisations representing employers in the likes of ironfounding, sheet metalworking, and tin plate works. Membership of these organisations meant that inquiry notes or blacklists could be circulated amongst firms very efficiently, that lock-outs could be orchestrated quickly (on several occasions these were supported by the shipbuilding employers in the north-east of England) and, increasingly from 1900 onwards, that collective bargaining agreements could be enforced firmly in the employers' favour.

It was a similar story in the east of Scotland. Dundee, with its economy based on shipbuilding and jute production, had its own Shipbuilders' Association, an Association of Engineers and Ironfounders, a Bobbin and Shuttle Manufacturers' Association, a Spinners' and Manufacturers' Association, and a Master Calenderers' Association – representing those employers involved with bleaching and textile dyeing. In Edinburgh there was an East of Scotland Engineers and Ironfounders' Association, while Aberdeen had its own Iron Trade Association. Coalmasters were well organised, too, with five employers' associations, formed purely for labour relations reasons, representing the Lanarkshire, Ayrshire, Lothian and Fife coalfields. The mineral-oil producers of the central belt were also organised by the 1880s.

No matter how small the company, being a member of one of these organisations meant increased clout in the relationship with organised labour. Employers were now able to quickly share information on potential labour relations issues and respond accordingly; the fact that the major engineering employers' organisations on the Clyde used the same secretary made this especially straightforward.[31]

By the turn of the twentieth-century there were over 150 local and national employers' associations throughout Scotland.[32] Moreover, it was not just in the heavy industries that employers found collaboration a good thing. Organisations came into being across a wide range of trades and industries, including painting and decorating, plumbing, boot and shoe manufacture, carting, fishing – a good example being the Aberdeen Steam Trawler Association – tailoring, printing – both Edinburgh and Glasgow had a master printers' association, which eventually formed the nucleus of a Scottish Printers' Alliance in 1913 – and baking and confectionery. There was also the establishment of the National Farmers' Union of Scotland in 1913, although strong individualism amongst the farmers meant this organisation was susceptible to internal conflicts for some time to come.[33] There is no space in this chapter to survey these in any detail. However, an outline of the course of events in the Scottish building industry during this period provides a good illustration of how employer organisation is known to have worked.

Before 1920 the building industry across Scotland was made up of a patchwork of small firms. Increasingly, employers' associations became orientated towards collective bargaining, and trade regulation became important in bonding these disparate firms together and providing some cohesion to the industry. Most of these associations performed several roles at the same time. They controlled competition by fixing wage scales and job prices; they acted as pool-purchasing services to cheapen the costs of materials; and, at the same time, they strengthened the employers' hands in labour relations. By the 1900s there were building trade organisations operating in every town and city in Scotland, and some were beginning to band together into federations – most notably the Scottish Building Employers' Federation (SBEF), which came into being in 1895. Edinburgh and Aberdeen had master builders' federations by this time, as did Peterhead, whilst the companies who helped construct the granite city were organised in the Aberdeen Granite Association. In Edinburgh there was the Edinburgh and Leith Master Builders' Association; there were associations of master plasterers in Aberdeen, Edinburgh, Dundee and Glasgow – with an unsuccessful attempt made to form a Scottish-wide association in the early 1900s; and there were local associations of master slaters in Aberdeen, Edinburgh and Glasgow.

The Glasgow Master Masons' Association was a typical locally based building trade grouping. One of its main roles in the late nineteenth century was to set wage scales and co-ordinate lockouts. However, it also performed other crucial services, such as operating a price list to ensure that master masons charged equal rates for jobs, organising annual excursions for members, and bringing members together in regular social evenings. Many employers' associations across Scotland provided social activities such as these. For example, the annual dinner of the Glasgow and West of Scotland Master Plumbers' Association was one of the biggest social events of the season; the Glasgow Paper and Stationery Trades Association had an annual golf tournament; members of the Scottish Paper Makers' Association held an annual dinner; and the Greenock Master Wrights' Association held annual soirées and dinner dances, as did the West of Scotland Furniture Employers' Association.

It should be borne in mind that not all Scottish employers became involved in organisations such as these. For example, in the late nineteenth-century leather industry, although tanners frequently met for social purposes, they remained in cut-throat competition for most of this period, and it is interesting to note that over the same period their English counterparts operated a very efficient price co-ordination policy but hardly ever met together socially. Moreover, it is important to remember that price lists and wage scales were only recommended rates, and that it is impossible to determine the extent to which they were adhered in practice by the members. Finally, even when firms became members of organisations, in many cases they still retained the right to determine how they conducted their business and what the quality of their service should be. However, despite these caveats, many employers preferred a degree of co-operation rather than outright, fierce competition.

Scottish capitalists also frequently banded together to control product prices, and this continued throughout the twentieth century. For example, during the nineteenth century, the West of Scotland coalmasters collaborated at times by withholding their stocks to force up the price of coal. In 1902 this ad hoc trading network was institutionalised when twenty-four coal companies formed themselves into the United Colliers to better regulate competition between them. A similar process took place amongst the coal merchants, as in 1909 twenty-one of them formed the Coal Merchants' Association of Scotland, which could boast 400 member firms by 1920. Many Scottish companies also became involved in Scottish and UK-wide price associations, such as the Scottish Colour, Paint, and Varnish Association, the Scottish Lead Sellers' Association and the Soap Makers' Association. Indeed, by 1920 the government was so concerned at the number of cartels springing up that it initiated an inquiry, the Standing Committee on Trusts.

CHAMBERS OF COMMERCE

Although chambers of commerce can be linked to the evolution of merchant guilds, the history of the movement really dates from the mid-seventeenth century, when a chamber of commerce was formed in Marseilles. The first chamber in the UK came into being in Jersey in 1768, while the first on the British mainland was the chamber which opened in Glasgow in 1783. This had come into being as a reaction by the town's principal merchants to the loss of the American colonies, and reflected their deep concerns over the sudden disappearance of lucrative markets. Initially, the membership of the Glasgow chamber of commerce comprised 153 merchants and 38 manufacturers, from the Glasgow, Paisley, Greenock and Port Glasgow areas. However, a chamber was formed in Greenock in 1813 to cater for merchants and industrialists from the lower reaches of the Clyde.

From its inception the Glasgow chamber of commerce acted as a watchdog over the interests of West of Scotland trade and industry. It very quickly established a Scotch Commercial Agency in London to promote Scottish goods and to monitor parliamentary legislation that may prove detrimental to commerce in the West of Scotland. However, as was the case with the merchant guilds, the chamber also increased the social capital of the members and became – along with the city's Merchants' House – an important link in the chain of developing middle-class identity. Only two years after the appearance of the Glasgow chamber a similar one was set up in Edinburgh. Like its Glasgow equivalent, the Edinburgh chamber was quick to ensure that the interests of the town's trade and commerce were paramount; one of its first actions was to improve the lighthouse on the Isle of May in the Firth of Forth. Five years later the chamber suggested that to improve mail deliveries letters should be date-stamped, and it can also claim responsibility for recommending the firing of a gun from the castle at one o'clock as an audible signal of the time kept at Calton Hill Observatory.

By the 1920s, along with the Glasgow and Edinburgh chambers there were chambers of commerce in Dundee (1862), Greenock (1813), Leith (1897), the South of Scotland (1864), and Kirkcaldy (1926).

These organisations performed important trade, regulatory and political roles. Moreover, although they did not become directly involved in labour relations issues, their voices in Westminster meant they were able to exert an influence on work-related legislation. For example, in 1874 the Greenock chamber petitioned against the Factory Amendments Bill, framed to reduce the hours worked in factories from sixty to fifty-six hours a week. Nineteen years later, the chamber was making its voice heard again regarding the 1892 amendment to the Factory and Workshops' Act, and the following

Figure 25.3 Dundee chamber of commerce. Source: Dundee City Council, Central Library, Photographic Collection.

Figure 25.4 Directors of Glasgow chamber of commerce pictured with officers of the 17th Batallion, Highland Light Infantry at Troon in 1915. Source: Mitchell Library.

year it was petitioning against the Employers' Liability Bill. Memorials and petitions such as these were passed by the chamber to local MPs, and particular care was taken to ensure that the member was amenable to the business interests represented by the chamber. In such a way, chambers of commerce further strengthened a commercial and business network that was now stretching across the country. This was strengthened even more from 1860 onwards, when the Association of British Chambers of Commerce was formed, all of the Scottish chambers being affiliated by the 1920s.

The capitalist net was reinforced even more by employer participation in other status-reinforcing organisations. The Rotary Club is a good example. Glasgow and Edinburgh's Rotary Clubs were both established in 1912, with the founder members of the Glasgow club representing 173 different businesses and professions from across the second city of the empire. The Edinburgh club, although not industrially focused, was responsible for setting up the British Association of Rotary Clubs in 1913. Clearly, then, by the early years of the twentieth century, the organisation of Scotland's employer classes was well developed. A few more organisations from the Glasgow area could be mentioned in passing: there was the Glasgow City Business Club, which had 600 members from a wide range of industries by 1923; the Govan Weavers' Society, which still managed to attract 178 members by 1920; and the Glasgow Athenaeum, which was formed in 1847 'to provide an agreeable place of resort in the intervals of Business' and had 1,600 members by 1918. In the years before state welfare, there was also a broad spectrum of philanthropic activity with which the business classes became involved. And of course there were other interlocking networks. Many employers were involved with freemasonry; and there was shared religious affiliation, too, with many employers subscribing to evangelicalism in the 1830s and 1840s – when Glasgow became known as 'Gospel City' – while many became members of the Free Church of Scotland, and especially so after interdenominational rivalry between it and the united Presbyterian Church ended with the merging of the organisations in 1900.[34]

TOWARDS THE PRESENT: EMPLOYER COLLECTIVISM IN THE TWENTIETH CENTURY

From the 1900s, and more significantly during the period of industrial unrest during 1910–14, trade union membership levels increased markedly, and once again the increase was most obvious in the heavy industries. To meet this threat, employer organisation strength increased, with the number of employer associations in the UK peaking in the 1920s. However, after this period there was, for several reasons, a contraction.

For example, in many industries the small-scale structure of family-run concerns changed as larger companies emerged. To some degree, the need for collective trade organisation, which had fuelled the growth of employers' associations, declined as companies became more confident. In the Scottish building industry the number of small-scale master masons, slaters, wrights and so on declined as larger multi-trade building firms arrived on the scene. Increasingly, it was these larger companies which steered the Scottish Building Employers' Federation from strength to strength. At the outset, the aim of the SBEF was to 'secure united action in promoting and advancing general contracts;[and] cooperation in dealing with demands of operatives'.[35] With such a mandate the SBEF increasingly took over the role of the large number of smaller organisations that constituted much of the building trade in Scotland. Moreover, in 1963 the federation entered into an affiliation with what was then the UK-wide Federation of Building Trades Employers – now the Building Employers' Federation. Clear evidence of the strength of this UK-wide building-trade employer combination was apparent in 1977 when the Labour government's manifesto suggested that the construction industry be nationalised. This provoked a concerted response from building trades employers' associations throughout the UK, and their banding together in the Campaign Against Nationalisation of the Building Industry (CABIN) successfully saw off the challenge. Moreover, further indication of the continued strength of the employer movement in Scotland was the formation of the Scottish Construction Industry Group in 1980 to act as the main conduit of communication between the industry and the Scottish Office.[36]

The trend towards the national centralisation of employer activity was clearly apparent across a wide range of industries from the 1920s onwards, and it became increasingly clear that an employers' movement in the UK was growing. Many Scottish companies became represented by the Federation of British Industry (FBI), formed in 1916, and the National Confederation of Employers' Organisations (NCEO), which was formed two years later. Nationalisation of the coal industry in 1948 brought an end to the employer associations in that industry which had played such a vital role in the era of private ownership. Moreover, by the time of denationalisation in the 1990s, the Scottish coal industry – under the name of Scottish Coal – had been rendered moribund. By that time, the Scottish shipbuilding industry had virtually disappeared, taking with it the once powerful Clyde Shipbuiders' Association and the similar organisation in Dundee. Heavy engineering in Scotland followed a similar pattern, with deindustrialisation from the 1950s altering the industrial landscape, sometimes beyond recognition.

One of the most significant developments of employer organisation was the formation of the Confederation of British Industry (CBI), which came into being in 1965 following a merger of the FBI with two other associations: the British Employers' Federation, and the National Association of British Manufactures. By this time the labour-relations role of employers' associations had evolved from the days of outright confrontation. Although levels of labour militancy fluctuated throughout the twentieth century, collective bargaining was now tightly woven into the fabric of Scottish and UK society, and the days of employers being forced into banding together in order to blacklist or to organise lock-outs were now over. Increasingly, from the 1950s onwards, a more significant role for employer associations lay in providing advisory and consulting services to member firms on a range of concerns, such as employment legislation and trade issues.[37]

Notwithstanding the altered industrial landscape, employer organisation remained, and still remains, a significant force in Scotland. Although employer involvement in philanthropy declined with the advance of the welfare state, middle-class bonding continued to be an important by-product of membership of employers' organisations, chambers of commerce, Round Tables, Rotary Clubs and the like. Moreover, in many towns and cities throughout Scotland the old merchant guilds and incorporated trades continue to attract members. The old industrially forged class boundaries may have become blurred in the twentieth century following deindustrialisation, urban regeneration and the expansion of house ownership. However, the evidence suggests that although never constituting a homogeneous employer class, a significant number of Scotland's employers became involved in trade-related organisations specifically designed to set those who employed labour apart from those who did not.

NOTES

1. Devine, 1995, 54–73 and 182–96; Nenadic, 1991a; Nenadic, 1991b; Morgan and Daunton, 1983; Morgan and Trainoir, 1989.
2. Adelman, 2001.
3. Dickinson, 1945–6, 224; Maitland, 1897, 193.
4. Palliser, 2000, 719–30; Smout, 1969, 27.
5. Pryde, 1965; Palliser, 726–8.
6. Lynch, 1988, 7–9, 24, 233–4, 245–60; Palliser, 720–22; Ewan, 1990, 58–63.
7. Grant, 1984, 85–7; Fraser, 1988, 17.
8. Lynch, 1988, 9, 258n, 271.

9. Lynch, 1988, 261–86, 274.
10. Fraser, 1988, 17.
11. Lynch, 1988, 278–9.
12. Bain, 1887, 29.
13. Lynch, 1988, 274; Smout, 1969, 161.
14. Devine, 1995, 18.
15. Devine, 1995, 20–1.
16. Fraser, 1988, 39.
17. Fraser, 1988, 23.
18. Fraser, 1988, 22.
19. Fraser, 1988, 38.
20. Fraser, 1988, 41.
21. Fraser, 1988, 39–56.
22. In Clayre, 1982, 193.
23. Significantly, many coal and iron masters were united against the passage of the 1799 bill, and were supported by the Glasgow chamber of commerce, the Merchant's House, and the deacons of Glasgow's fourteen incorporated trades. Campbell, 1979, 16.
24. Fraser, 1988, 60.
25. Knox, 1995, 140.
26. Fraser, 1976.
27. Knox, 1999, 114–21.
28. Knox, 1999, 159 and 156–62.
29. Johnston, 2000, 169–99.
30. Johnston, 2000, 177.
31. Johnston, 2000, 169–99.
32. *Directory of Industrial Associations in the UK*, 1902.
33. Anthony, 1997, 111–18.
34. Brown, 1997, 195.
35. Campbell, 1995, 22.
36. For a good account of the rise of the SBEF, see Campbell, 1996, 22–31.
37. Bain, 1983, Chapter, 5.

BIBLIOGRAPHY

Adelman, J and Aron, S, eds. *Trading Cultures: The Worlds of Western Merchants*, Tunhout, 2001.

Anthony, R. *Herds and Hinds*, East Linton, 1997.

Bain, E. *Merchant and Craft Guilds: A History of the Aberdeen Incorporated Trades*, Aberdeen, 1877.

Bain, G S. *Industrial Relations in Britain*, Oxford, 1983.

Brown, J and Rose, M B, eds. *Entrepreneurship, Networks and Modern Business*, Manchester, 1993.

Campbell, A B. *The Lanarkshire Miners*, Edinburgh, 1979.

Campbell, B. The history of the SBEF, *Building Matters* (Spring, 1996), 22–31.

Clayre, A, ed. *Nature and Industrialisation*, London, 1982.

Devine, T M. *Exploring the Scottish Past*, East Linton, 1995.

Devine, T M and Finlay, R, eds. *Scottish Elites*, Edinburgh, 1994.

Dickinson, W C. Burgh life from burgh records, *Aberdeen University Review*, xxi (1945–6), 214–26.

Ewan, E. *Townlife in Fourteenth-Century Scotland*, Edinburgh, 1990.
Fraser, W H. *Conflict and Class: Scottish Workers 1700–1838*, Edinburgh, 1988.
Fraser, W H. The Glasgow cotton spinners, 1837. In Lythe, E and Butt J, *Scottish Themes*, Edinburgh, 1976, 80–97.
Fraser, W H. *A History of British Trade Unionism*, London, 1999.
Gordon, E and Trainor, R. Employers and policymaking: Scotland and Northern Ireland, 1880–1939. In Connoly, S J, Houston, R A and Morris, R J, eds, *Conflict, Identity and Economic Development: Ireland and Scotland, 1660–1939*, Preston, 1995.
Grant, W P and Marsh D. *The CBI*, London, 1977.
Grant, A. *Independence and Nationhood: Scotland, 1306–1469*, London, 1984.
Ilersic, A R. *Parliament of Commerce*, London, 1960.
Johnston, R. *Clydeside Capital: A Social History of Employers*, East Linton, 2000.
Knox, W W. *Hanging By a Thread*, Preston, 1995.
Knox, W W. *Industrial Nation*, Edinburgh, 1999.
Lumsden, H and Aiken, H. *History of the Hammermen of Glasgow*, Paisley, 1912.
Lynch, M, Spearman, M and Stell, G, eds. *The Scottish Medieval Town*, Edinburgh, 1988.
Lythe, E and Butt, J, eds. *Scottish Themes*, Edinburgh, 1976.
McIvor, A. *A History of Work in Britain*, Basingstoke, 2001.
Maitland, F W. *Domesday Book and Beyond*, Cambridge, 1897.
Morgan, M and Trainor, N. The dominant classes. In Fraser, W H and Morris, R J, eds, *People and Society in Scotland Volume 2, 1830–1914*, London, 1990, 103–37.
Morgan, N J and Daunton, M J. Landlords in Glasgow: A study of 1900, *Business History*, 25 (1983), 264–86.
Nenadic, S. Businessmen, the middle classes and the dominance of manufacturing in nineteenth century Britain, *Economic History Review*, 44 (1991a), 66–85.
Nenadic, S. The rise of the urban middle classes. In Devine, T M and Mitchison, R, eds, *People and Society in Scotland, Volume 1*, Edinburgh, 1991b, 109–26.
Palliser, D M, ed. *The Cambridge Urban History of Britain: Volume I, 600–1450*, Cambridge, 2000.
Pryde, G S. *The Burghs of Scotland: A Critical List*, Oxford and Glasgow, 1965.
Scott, J and Hughes, M. *The Anatomy of Scottish Capital*, London, 1980.
Sessions, M. *The Federation of Master Printers: How it Began*, London, 1950.
Smith, A M. *The Nine Trades of Dundee*, Dundee, 1995.
Smith, A M. *The Three United Trades of Dundee*, Dundee, 1987.
Smout, C. *A History of the Scottish People*, London, 1969.
Whatley C. *Scottish Society 1707–1830*, Manchester, 2000.

26 The Social and Associational Life of the Scottish Workplace, 1800–2000

IRENE MAVER

This chapter considers the changing nature of work-related associational culture from the context of Scottish industrialisation as it rose and then contracted between 1800 and 2000. Such culture was evident in Scotland in a variety of ways. It could be expressed through voluntary organisations for self-protection, such as friendly societies. Operating outside the trade union and cooperative sphere, they often had local roots and a highly individual profile. Some were occupation-based or could be connected with a particular employer; others, like the Free Gardeners, Oddfellows and Rechabites, were more generic in character and formed part of an extensive nationwide network. All were a popular way of providing for financial support in times of hardship, especially illness and bereavement. A particular attraction was the sense of convivial companionship generated by the societies; sociability and security were neatly intertwined. They were also lauded for exemplifying the nineteenth-century virtues of thrift and independence, albeit in a collective context. For Samuel Smiles, the Scottish medic-turned-journalist and Victorian guru of 'self-help', their culture was a vital step forward in the direction of 'the social emancipation of the working classes'.[1]

This was very different from paternalism and employer amenity welfarism, which nevertheless provided scope for a range of workplace associational activities, ranging from literary societies to instrumental bands and sports clubs. In some respects the sturdy independent spirit of the early friendly societies yielded to the growth of welfarism from the late nineteenth century, and there could often be a moral and improving edge to employer-sanctioned pursuits. Whether this constituted social control is a moot point, given that trade unions, among other collective organisations, were campaigning for better working conditions. And workplace social life was often worker-led, reflecting the growth of recreational opportunities through the general expansion of leisure time and technological improvement, especially from the late nineteenth century. A further crucial factor was that divisions between workplace and community were often hazy. For instance, mining, shipbuilding and textile towns were dominated and defined by their industries, which could create difficulties in disentangling the social and working experience.

What follows is a selective analysis of Scottish workplace 'clubbability' between 1800 and 2000, starting with friendly and other provident societies associated with employment, especially in the nineteenth century. The chapter then looks at the growth of workplace welfarism and the eclectic range of associational activities that emerged and sometimes even survived the process of deindustrialisation at the end of the twentieth century.

GUILDS AND INCORPORATIONS

In 1925 Richardson Campbell – a Scot who had formerly been high secretary of the Independent Order of Rechabites – wrote a history of what he called 'provident and industrial institutions'. Although the book was intended for consumption beyond Scotland, Campbell placed heavy emphasis on the tradition of guilds and incorporations north of the border. Their early organisation bore 'remarkable evidence of the outstanding national characteristic thrift of the Scottish people', and though exclusive in their constitutions they set high standards for the mutual assistance of their members.[2] According to Campbell, friendly societies served a similar purpose during the nineteenth and early twentieth centuries, although from the outset they were democratically structured, 'instituted by workmen who knew the requirements and wants of workmen'.[3] Belief in the twin virtues of 'Self Help' and 'Self Government' underpinned Campbell's crusading attitude, at a time when the dominance of the voluntary ethos was being challenged by encroaching state welfarism through national insurance.[4] He used history to trace a line of continuity in forms of associational activity that were deeply ingrained, especially in urban Scotland. As will be seen, this catch-all approach misrepresented the reality. On the other hand, at the time he was writing, guilds, incorporations, fraternities and friendly societies remained a highly visible component of local culture.

Merchant guilds and craft incorporations had (and have) a long pedigree and were originally the product of Scotland's burghs. Under the feudal system, which emerged from the twelfth century, crown, landowner or ecclesiastical jurisdiction defined the nature of a burgh through its charter. Commercial life was monopolised by the merchant and trade burgesses, who were granted exclusive privileges by the burgh's feudal superior. Merchant guilds had more prestige, especially those that dealt with overseas trade, while craft incorporations were more specialised. The first of Edinburgh's fourteen incorporated trades, including the skinners, weavers and masons, were each given official burgh sanction through their 'seal of cause' in the late fifteenth century.[5] Essential crafts were similarly recognised across Scotland, before and after the Reformation. These institutions were politically important, as until the Scottish municipal reform legislation of 1833 representatives of the burgess institutions could often form the local government of a town. Moreover, Scotland's exclusive trading privileges were not abolished until 1846. Before that time, membership of the relevant guild or incorporation was, in theory, mandatory.[6] All this meant that burgess institutions were able to develop deep roots in their communities – a status that was maintained long after their formal disconnection from civic politics during the 1830s.

As the nineteenth century progressed, guilds and incorporations concentrated on the benevolent and charitable role that Richardson Campbell praised so highly. This involved the protection of 'decayed brethren', financial support for the dependents of deceased members and a range of philanthropic activities for the common good of the town. Education was a particular priority. Edinburgh's wealthy Merchant Company, chartered by the crown in 1681, managed some of the city's most academically successful schools, including George Watson's College.[7] There was also, inevitably, a convivial dimension. Convenery dinners, trades' balls and, over time, the patriotic allure of Burns' suppers, were regular features of the social calendar. Legitimacy and tradition were visibly reinforced by using heraldic symbols and elaborate ceremonial. For instance, the Blue Blanket, or Craftsmen's Banner, was a gift to Edinburgh's incorporated trades by King James III in 1482. Up to the mid-nineteenth century this relic would be prominently displayed at the annual trades' dinner, where it was committed to the charge of the newly elected deacon convener.[8] Burgess institutions were also an integral part of the common ridings that proliferated in the Borders burghs and beyond. This was hardly surprising, as these festivals flamboyantly celebrated burgh autonomy. In Selkirk's procession, the incorporations are still headed by the hammermen, and their banners are important emblems in linking the town's past with the present.[9]

Although guilds and incorporations had their status conferred by burghs, there were anomalies in this pattern. The Society of Free Fishermen of Newhaven had been given a charter by King James VI in 1573, although the town was never an independent burgh and was ultimately absorbed into Edinburgh in 1920. One Newhaven historian has argued that in the nineteenth century the society was effectively the town's local government, providing funds for environmental improvement, harbour maintenance and welfare.[10] Socially, the society was also a force for community cohesion. Its annual elections were a gala event, where the focal point was a torchlight procession formed for the ceremony of 'lifting the box'. This was literally a strongbox, conveyed to the residence of the newly elected boxmaster, the society's equivalent of deacon. The box symbolised the society's assets and, given its ubiquity in local affairs, the common good of the town. Similar associations flourished outside the bounds of the burgh system, reflecting the need for expanding communities to provide for self-protection. In Lanarkshire, the Govan Weavers' Society was founded in 1756 to counter the effects of economic insecurity in an erratic trade.[11] Socially, there were strong parallels with the Newhaven experience. A procession celebrated the election of the society's deacon, and was held annually up to 1881 during the Govan Fair. In place of the fishermen's box was a decorated pole topped by a sheep's head, with elaborately curling horns.[12] The sheep was an emblem of the weaving trade and appeared on the society's coat of arms.

Trade monopoly and the right to enforce it were among the factors that originally distinguished incorporations from fraternities. Yet even before free-trade campaigners helped to secure the abolition of trading privileges in 1846, there were already societies that eschewed incorporation for the practical reason that their work was not always conducted within burghs. The Fraternity of the Gardeners of East Lothian was one of the earliest recorded examples, dating from the 1670s.[13] Starting as individual craftsmen's or 'operative' organisations, the Free Gardeners combined to become a fully fledged friendly society. The scale of growth in Scotland was due to their inclusive membership, to the extent that non-gardeners came to dominate. By the 1840s, lodges were well established across the central belt, and multiplied rapidly during the second half of the century.[14] In 1883, along with other friendly societies, Aberdeen's Thistle and Rose Lodge formed an integral part of the official procession celebrating the grand municipal opening of Duthie park.[15] For decades Airdrie's annual Gardeners' parade, organised and headed by the fraternity, was a local holiday.[16]

The Free Gardeners offered a range of attractions. As friendly societies, lodges provided sickness, death and other benefits. Perhaps in anticipation of mortality, there was a lively social life. The *Scotsman* newspaper reported regularly on the activities of Edinburgh's St Cuthbert's Lodge after its formation in 1824. Favourite pastimes included balls, dinners, excursions and flower-festooned street processions. In 1826 one journalist described a gathering of the brethren in Hepburn's Tavern, High Street, where 'the evening was spent with that mirth which has always distinguished the members of this ancient fraternity'.[17] The fraternity demonstrated an additional dimension that has been explored by Robert Cooper, a historian of Scottish freemasonry. Ritual, based closely on masonic precedents, became more elaborate as lodge culture developed.[18] Free Gardeners were required to go through initiation ceremonies based on the three degrees of apprentice, journeyman and master. They wore masonic-style regalia, including aprons, sashes and jewelled insignia. Significantly, style was also borrowed from the guilds and incorporations that were such a visible feature of Scotland's burgh system. The society was anxious to stress its pedigree, and so used readily identifiable models

Simon Cordery, a historian of Britain's friendly societies, has shown that this highly expressive approach to mutuality was by no means unusual in the nineteenth century. A focus on secrecy, ceremony and conviviality gave a unique colouring to associations that operated, more mundanely, as insurance societies.[19] Before the arrival of state welfarism, they blended traditional and contemporary ideas of self-protection through unity. While members may have been attracted by the notion of fellowship bonded by ritual, their associational activities also exemplified the spirit of Burns' oft-quoted lines about 'the glorious privilege of being independent'.[20] This was especially so from the 1790s, when Westminster successively legislated for the regulation of friendly societies, which in turn necessitated greater centralised control and bureaucratic structures. During the course of the nineteenth century, the societies came to operate at two levels: one that demonstrated their financial efficiency and accountability, and another that catered for more sensuous membership expectations. Although the level of arcane intensity varied, Free Gardeners, Foresters, Oddfellows, Rechabites and Shepherds were among the friendly societies that successfully adopted the branding.

This exotic distinctiveness found expression in community ceremonial, which survives in the twenty-first century through the Oddfellows' 'Old Year' procession in Newburgh, Fife. Like much in friendly-society culture, the name Oddfellows is intended to intrigue; it reputedly derives from the Latin *omnium gatherum*, to signify universality.[21] With associational roots in the mid-eighteenth century, Oddfellows' societies grew and diversified from the 1800s as orders emerged across the United Kingdom. By 1900 at least five separate orders were based in Edinburgh, three of them with headquarters at Oddfellows' Hall in Forrest Road.[22] Like other friendly societies, Scottish membership contracted sharply as the twentieth century progressed. Uniquely, the Newburgh Caledonian Oddfellows survived, thanks to the resilience of their annual Hogmanay procession. Whatever the roots of this ceremony – and Newburgh Oddfellows are reluctant to pin down their history – it retains a rumbustious image, with strong echoes of pre-Reformation festivals of misrule.[23] The Oddfellows wear fancy dress and follow their newest member, who rides backwards, bareback, on a piebald pony. One veteran summed up the modern role of the Newburgh Lodge: 'Today we're the only Oddfellows left in Scotland. We're affiliated to nobody, we just look after ourselves and the Burgh, and that's no bad thing – and we keep the traditions going.'[24]

OCCUPATIONAL SOCIETIES AND EXCLUSIVITY

The largest friendly societies – the affiliated orders that were under the control of a central association, or Grand Lodge – catered for workers across a range of industries. They may have evolved from a specific 'operative' trade like the Free Gardeners, but they did not pursue closed entry. Nevertheless, there were societies identified with particular occupations, such as the Free Colliers, a fraternity whose heartland was in east Stirlingshire. In his 1860s' survey of Scotland's coal districts, journalist David Bremner commented on the need for benefit societies, as 'the great majority [of miners] have yet to acquire provident habits'.[25] He looked favourably on the recent emergence of Free Colliers' lodges, which helped to fill the vacuum. Formed in 1863, the movement was structured along masonic lines, and their Grand Lodge was named after William Wallace, the medieval Scottish 'Liberator'.[26] Their mid-Victorian heyday coincided with a revival of interest in the Wallace phenomenon, centred on the soaring new monument at Abbey Craig, Stirling. The miners regarded Wallace's spirit of patriotism, independence and egalitarianism as an appropriate role model. Scottish-wide support for the Free

Colliers declined as mining trade unionism was gradually consolidated. However, they retained an enduring base around the village of Redding, near Falkirk. In the twenty-first century the surviving Grand Lodge members are no longer working miners, but they determinedly maintain the tradition of their annual procession to Wallacestone in celebration of 'Wallace's Day'.[27]

Free Colliers initially operated in a culture of secrecy, in order to maintain their integrity against employer intimidation in a notoriously difficult trade. Yet while they espoused trade union objectives, elements of their association reflected a pre-industrial profile. Such links were more obviously apparent in the rural-based Brotherhood of the Horsemen, or Horseman's Word, described by writer Timothy Neat as 'a form of primitive Trades Union ... a half-professional, half-tribal group'.[28] Generally identified with the east of Scotland, especially Angus and the Mearns, the roots of this phenomenon have been traced to the late eighteenth century, when horses began to replace oxen as draught animals.[29] There were echoes of masonic ritual in the initiation ceremonies for apprentice horsemen, albeit in a subverted and idiosyncratic form. The remote nocturnal setting, preferably at the time of a full moon, heightened the drama of the occasion, although it also facilitated post-initiation celebrations through whisky drinking and songs. Structurally the society was not centrally organised, and it served as a focus for promoting occupational skills rather than encouraging friendly-society models of worker thrift. After initiation, apprentices were privileged with essential information about their craft, especially the handling and care of horses. This learning process was often aided by rhymed incantations. Memory was maintained through the society's oral traditions, while the skills were protected by members' strict allegiance to the Horseman's Word.[30]

Of course, the Horseman's Word has evident parallels with the Mason Word. Although the latter was absorbed into the mysteries of modern freemasonry, in Scotland its origins lay in 'operative' masonry, as a device for identifying initiated craftsmen.[31] However, from the seventeenth century freemasonry became detached from the practical craft associations retained by groups like the horsemen, and the organisation did not develop into a friendly society, like the Free Gardeners. Instead, freemasonry evolved into what one historian calls 'an elite convivial-club', an image that the organisation encouraged from the 1800s in an effort to play down early accusations of political sedition and religious unorthodoxy.[32] In the nineteenth century, masonic emphasis on the harmoniously integrated society was intended to promote class cohesion and the notion that all men were brothers, whatever their social background. Much inspiration derived from the example of Robert Burns, initiated to the Tarbolton St James' Lodge in 1781. David Willox, a well-known Liberal town councillor in Glasgow during the 1890s and 1900s, paid tribute to his hero by producing copious emulative verse about the masonic 'sons of light'.[33] Willox's origins in the city's working-class East End, from a long line of handloom weavers, typified the strong artisan following for 'the Craft'. Support was consolidated into the twentieth century. According to one estimate, by the 1920s around a quarter of Britain's masonic lodges were Scottish-based, while 10 per cent of adult males in Scotland were members.[34]

Despite the organisation's self-proclaimed tradition of egalitarianism, membership could often arouse hostile emotions. In the Clydeside shipyards freemasonry came to be associated with Protestant-only recruitment policies in the skilled trades. Speaking of the 1920s, one miner from Newtongrange, Midlothian, expressed his conviction that the local lodge constituted a surveillance network on behalf of the coal company.[35] The Roman Catholic Church, disturbed by the radical, secular edge to continental freemasonry, prohibited its adherents from joining. Quasi-masonic organisations like the Free Colliers were consequently shunned by Catholics, a factor that undermined worker unity in the coalfields.[36] Instead, Catholic benevolent and friendly societies catered for the growing number of Irish migrants to Scotland's industrial districts, especially in the west.[37] Irish associational culture could also have a distinctly political profile, outside church control. The National Brotherhood of St Patrick, founded in 1861, soon had branches operating in Dundee and Glasgow. Beginning as a literary society for working-class migrants, the organisation was increasingly identified with militant nationalism.[38] Protestant Irish incomers imported the Orange Order, with Ayrshire handloom weaving communities forming the initial growth base from the 1800s. For all the order's assertive political stance in defence of Protestantism, at the grassroots level it acted as a friendly society, offering sickness and funeral benefits. One early example was the Glasgow Orange Union Funeral Society, established by local lodges in 1834 to provide for weavers who had settled in the city's East End.[39]

As well as sectarian divisions, there were gender divisions. The thousands of societies, clubs, fraternities and lodges that emerged during the nineteenth century pitched their appeal predominantly at skilled working men. Organisations like the Freemasons were noted bastions of brotherhood, while fraternities like the Horseman's Word positively celebrated their masculinity. Friendly societies were generally more inclusive, with the Free Gardeners and Oddfellows encouraging women-only and even juvenile lodges. Temperance friendly societies, such as the Rechabites, were particularly welcoming. They argued that women and children had a civilising impact on the drouthy habits identified with associational culture, especially male socialising in taverns and public houses.[40] Barriers began to erode from the 1890s as competition intensified among friendly societies to insure women. The growing public profile of women, especially in politics and education, also helped to change attitudes. Yet it remained difficult for women to overcome the deep-rooted traditions that defined the identity of certain institutions. Only in the late twentieth century did the surviving guilds and incorporations begin to alter their masculine character. After her election in 1960, Glasgow's first female lord provost, Jean Roberts, made history when she attended previously all-male functions like the Trades House dinner.[41] Much later, in 2006, the city's Trades House finally buried its exclusive image by announcing that two pioneering women, Annemieke

Cunninghame of the barbers and Maria McKellar of the weavers, had been elected deacons of their respective trades.[42]

WORKPLACE WELFARISM

During the nineteenth century, friendly and other societies that were formed for mutual assistance ostentatiously displayed their independence through the cultivation of sometimes bizarre ceremonial and ritual. The notion of welfare paternalism had ambiguous connotations, and for active Oddfellow Samuel Smiles the virtues of independence were enshrined in the notion that 'help from without is often enfeebling in its effect, but help from within invariably invigorates'.[43] Yet workplace paternalism had early roots in industrial Scotland, and is most immediately identified with Robert Owen's New Lanark textile manufacturing community, where work and leisure time were strictly regulated. Owen's management and then ownership of the mills between 1799 and 1824 exemplified his firm belief that education and environmental planning shaped character and ultimately contributed to workplace harmony and productivity. A more critical contemporary view was that the community operated under a regime controlled by 'moral schoolmasters', not least because children were the prime target of Owen's reforming impulse.[44] More generally, in the early nineteenth century encouragement of a workplace social life was often cultivated by employers to reinforce discipline and quell the threat of anti-social behaviour. This was reflected in the sudden surge of support for temperance societies in Scotland during 1830. The Bairds of Gartsherrie, wealthy Lanarkshire ironmasters and prominent evangelicals, actively directed their workforce to more abstemious pursuits via the temperance movement.[45]

As the century progressed, employers increasingly made a show of their commitment to workplace welfare. Again, there was a controlling subtext in the regimented groups of workers who participated in day excursions organised and subsidised by factory owners.[46] The proprietors of Glasgow's Atlas Ironworks, Messrs Rowan & Company, typified this well-publicised form of paternalism, which also represented a form of corporate branding. In 1853 they chartered the steamer *Sovereign* and packed their workmen and families off to Garelochhead, along with the Caledonian Brass Band.[47] In fairness, friendly societies and other self-help organisations shared a similar fondness for organising 'away day' recreation, which helped to advertise their members' good-humoured solidarity. However, there can be no doubt that employer interest in amenity welfarism did start to yield tangible returns, especially as there was growing anxiety about the moral effects of a corrosive working environment. For instance, the Early Closing Movement quite literally took an evangelical approach to the shorter-hours campaign, as radical clergy, like the Rev. Dr James Begg of the Free Church, were outspoken supporters. In 1856 one of the movement's pamphleteers stressed how the nation's health was being undermined by excessive working: 'In these cases where the hours are long, the young men are neither able nor willing to take a stroll in the country. Their mental faculties and bodily powers are alike exhausted.'[48] An enervated labour force made bad business sense, while there was obvious value in channelling workers' energies more rationally. Moreover, welfarism muted workplace divisions and helped to affirm company loyalty.

Understandably, larger industrial enterprises were most preoccupied with promoting welfare strategies, partly because of the economies of scale involved and partly as a disincentive to trade union organisation. For the Paisley thread manufacturer Thomas Coats there was a symbiotic relationship between himself and the labour force, which he explained at the J & P Coats annual soirée in 1880: 'We are all connected in one way or another, and we have but one interest, namely the prosperity and success of the Ferguslie Threadworks.'[49] Coats' Liberalism gave political resonance to this statement; class cohesion, he implied, was the generator of economic progress. Historian Catriona MacDonald has discussed the nature and form of Coats' paternalism, which included canteen facilities as early as the 1880s, assorted pension and sickness schemes, and a range of social and educational clubs.[50] In Scotland generally after World War I, paternalism was professionalised through the creation of dedicated workplace welfare departments, and J & P Coats acquired theirs in the 1930s. James Templeton & Company, the carpet manufacturer based in Glasgow's East End, set up a department in 1918 to cater especially for female employees.[51] The Templeton Club was opened as a recreation centre, with much ceremony in 1926. The building was large enough to include a grand hall, where dances, concerts and other social gatherings were held.

The mining industry constituted one of the most comprehensive examples of the changing approach to workplace welfare. The problem of scarce amenities had been addressed to some extent from the 1890s, when temperance campaigners along with coal employers attempted to make positive efforts to change the corrosive nature of drinking culture in mining districts. The phenomenon of the Gothenburg public house, or 'goth', was an innovation imported from Sweden that came to be associated especially with Fife and the Lothians. Described as 'a mixed commercial/cooperative system of liquor licensing', goths were intended to be more than just establishments where patrons indulged in drinking.[52] Social activities were encouraged as a counter-attraction to alcohol, and goths were usually housed in spacious buildings complete with games and reading rooms, as well as restaurants. A controlled and wholesome environment was deemed vital, as this would enhance overall wellbeing. At a wider level, goths were run as trusts, where profits were ploughed back into the community for the purpose of creating recreational and other public amenities. In twenty-first century Scotland a handful of the original goths still remain, based in Armadale, Newtongrange and Prestonpans.[53]

In some ways, goths were the precursors of miners' welfare institutes, although their sphere of operation was narrower and more localised. Significantly, central government had been compelled to take the welfare initiative during the 1920s, in an effort to improve conditions in mining communities across Scotland. That the postwar period was a time of bitter political tensions in the industry also made it desirable for government leaders to be seen as responding to social needs. The Miners' Welfare Fund arose out of legislation in 1920 and derived its income from a levy of one penny per ton of coal produced in Britain. The money was used to provide essential amenities like pit-head baths, but funds were also allocated for leisure facilities. Thus in Loanhead, Midlothian, the local welfare club oversaw the laying-out and civic opening of a bowling green in 1930, at a cost of £1,000.[54] By the early 1940s nearly 300 schemes in Scotland had been facilitated by the fund.[55] In particular, miners' welfare institutes were erected to serve as community centres; some were imposing buildings that could accommodate meeting halls or even swimming baths. The official history of the Coltness Iron Company states that the fund 'revolutionised life in the mining villages', a telling statement about conditions prior to 1920.[56] Miners' welfare institutes and recreation grounds certainly encouraged participation in convivial, cultural and sporting pursuits, and despite the late twentieth-century demise of Scottish mining they continue to play a pivotal role in communities. In 2006 Loanhead Miners' Welfare was still going strong, claiming half the town's population of 8,000 as members.[57]

CREATIVE ASSOCIATIONAL CULTURE

It has already been seen that the boundaries between work and leisure are not always clear-cut. Nor is work easy to define, as modern concepts of managerial hierarchies, job specification and division of labour fit uncomfortably with less rigid traditional work patterns. Cultural expression was sometimes an integral part of the day-to-day work process. A well-known example is the 'òrain luaidh', the rhythmic women's work songs of the Scottish 'Gàidhealtachd', which helped to focus the communal effort needed for the efficient 'waulking' or fulling of handwoven cloth.[58] As John McInnes has shown, the 'òrain luaidh' were originally more widespread, accompanying a variety of work activities, including milking cows and the reaping and grinding of corn.[59] In Scotland, from the nineteenth century, social and work relationships were often difficult to disentangle, especially in smaller agricultural settlements and single-industry 'company' towns. Whether by accident or design, cultural expression was harnessed to workplace identity. Participation was both sought and encouraged in a variety of outlets, from choral music to competitive sport. This could have benefits all round, reinforcing worker esprit de corps and highlighting employer beneficence. It also gave cohesion to newly created, sometimes inchoate, communities, which were striving to create a civic identity. As will be seen, the Govan Police Pipe Band was a formidable asset from the 1880s in promoting the public image of this fast-growing shipbuilding town.

During the early nineteenth century, music had polemical as well as publicity value. The Airdrie Union Band was formed in 1819 at a time

Figure 26.1 Govan police pipe band, *c*1890.

of economic depression and political unrest in the industrial districts of Scotland. According to one Airdrie historian in 1921, the term 'union' referred to the radical reform societies that flourished in what was then predominantly a weaving town.[60] The band gave musical expression to weavers' grievances, but over time it metamorphosed into a local institution (as the Airdrie Old Union Band) that was still performing in the 1950s. The band was reputedly the oldest of its kind in Scotland, having started with flutes before switching to brass.[61] However, this longevity reflects the ability to track history through record-keeping. Instrumental bands were by no means a novelty in the early nineteenth century and had an established place in Scottish customary celebrations. For instance, the January annual holiday of Handsel Monday was a buoyant occasion, the last of the 'Daft Days', as the twelve days of Christmas were known in Scotland.[62] In the village of Stanley in Perthshire it was traditional for ploughmen to parade on Handsel Monday with a flute band. Yet by the 1830s the transformation of Stanley into a cotton textile town meant that the character of the occasion was changing. Workers had formed a singing club and instrumental band whose performances were welcomed by the local newspaper as much more sober, disciplined and 'rational' than their predecessors.[63]

Instrumental bands proliferated across Scotland during the nineteenth and early twentieth centuries. Their origins could be diverse, but many were connected with large industrial enterprises, especially collieries, iron foundries and steelworks. The *Third Statistical Account* for Lanarkshire commented that 'almost every [mining] village had its band', which would be maintained by voluntary deductions from colliers' wages.[64] Of thirteen Lanarkshire brass bands still functioning by 1960, eight were from mining communities. Sponsorship was not necessarily employment-based. The Lesmahagow and District Public Miners' Welfare Band started in 1927 and had origins in the temperance movement. However, as its name suggests, it came to be funded by the local miners' welfare institute.[65] The temperance connection indicated that such bands were still keen to promote what music historian Trevor Herbert has called 'the ethos of social harmony and cooperation that the rational recreationists held in such high regard'.[66] This quintessentially Victorian ideal actually intensified during the twentieth century as a distinctive yet coherent band culture developed, notably through the Scottish Amateur Brass Band Association. Band formation was encouraged by the vogue for competition that took off from the 1880s. This in turn reflected increased leisure time, better transport communications and the availability of relatively inexpensive instruments, often imported from Germany.[67]

Brass bands pre-dated the popularity of pipe bands in Scotland, with the latter developing from the second half of the nineteenth century. Piping ensembles had been given a major boost when the Volunteer Force was formed in 1859. A civilian defence movement that sported the trappings of militarism, pipe bands injected an element of glamour and manly Scottish patriotism to the Volunteer image.[68] From this example the popularity of pipe bands spread. As William Donaldson has put it, 'exotic, unregulated and sometimes disturbingly plebeian pipe bands sprang up in towns and villages all over Scotland'.[69] The pioneer, and for many the role model, was the Burgh of Govan Police Pipe Band, formed in 1883 by public subscription. Govan, as a municipal entity, had only come into being in 1864, and part of the new burgh's rationale was to provide effective policing for its mushrooming population.[70] The pipe band was a reassuring symbol of order and security, as well as representing the civic aspirations of the community; it was also musically dynamic, usually winning the top prizes at Scotland's first pipe band competitions in the 1890s.[71] When Govan was absorbed into Glasgow in 1912 the burgh band lived on as the prestigious City of Glasgow Police Pipe Band. Thereafter, the tradition was steadily consolidated, with the Central Scotland Police becoming one of the newest piping (and drumming) forces in 2006.[72]

Of course, workplace pipe bands were not just confined to the policing realm. Wallacestone in Stirlingshire, a focal point for Free Colliers' ceremonial, had its own colliery band as early as the 1890s.[73] Colliers even transferred the tradition beyond Scotland's boundaries. In 1936 migrant Ayrshire miners to the West Midlands of England formed the Binley Colliery Pipe Band. Much-needed sponsorship from Coventry's Triumph Motor Company later brought about the band's change of name.[74] Funding for workplace bands could generally be a vexed question. Ian MacDougall has shown in his oral history of the Newtongrange miners that not all workers appreciated the need to maintain a musical profile. The unilateral practice of 'offtakes' from miners' wages was ingrained in the Lothian Coal Company. One interviewee recalled of the interwar period that: 'The coalowners were great at startin' up bands and buyin' ambulances and doin' this and that. But it always came from the wages o' the men.'[75] James Reid worked as a wages clerk for the company, and while he remembered the Newtongrange Pipe Band as a workers' initiative from the 1930s, the autocratic general manager Mungo Mackay was determined to make his personal mark on the new band's appearance. He insisted that bandsmen should wear the yellow Mackay tartan; a choice he later rescinded in favour of less bilious Stewart colours.[76]

During the 1930s bandswomen started to make an appearance in this hitherto male-dominated sphere, although the phenomenon of the 'girl piper' long retained a curiosity value. Moreover, Scotland's heavy industries had an overwhelmingly male labour force, which fixed much of the character of organised workplace social life throughout the nineteenth and early twentieth centuries. Yet there were areas where women could become involved. Participation in choral music developed in Scotland after tonic sol-fa notation became widespread from the 1850s.[77] A striking late-Victorian phenomenon was the Newhaven Fishergirls' Choir, founded in 1896, which was followed in 1927 by the Fisherwomen's Choir.[78] East-coast 'fishwives' had acquired a rough reputation during the early nineteenth century. The married and unmarried women of Fisherrow, near Musselburgh, even confounded gender stereotypes by forming competing football teams every Shrove Tuesday.[79] The Newhaven choirs reversed this image by embracing contemporary notions of self-improvement and rational recreation through the uplifting, harmonious qualities of choral music. However, the women did keep an important link with the past. In performance they wore the traditional costume that had been made famous by the evocative Hill and Adamson calotype photographs of the 1840s.[80] And despite (or maybe because of) the brisk pace of social and economic change in Newhaven, the choirs survived well into the twentieth century. The Fisherwomen's Choir eventually folded in 1977, while the Fishergirls sang for the last time in 1995.

While choirs and bands could have considerable staying power, organised work-based recreational pursuits were often more ephemeral. Nearly thirty types of club and society have been identified as operational in the two great Paisley thread mills from the 1880s, and some survived the closure of the Anchor mills in 1993.[81] From the list of Paisley activities, sport had an enduring appeal and took the lion's share of worker attention, while other pursuits were products of their time, such as the Film Society and Automobile Club. The J & P Coats Amateur Dramatic Society flourished during the 1950s, reflecting the revival of interest in community theatre after World War II. Drama was an ambiguous part of the original rational recreation agenda; as one Victorian entertainer remarked about his childhood in the 1830s: 'When the name of the theatre was mentioned in our house, you would really believe you smelt brimstone.'[82] And there was a subversive edge to twentieth-century community theatre, as the career of miner-turned-playwright Joe Corrie exemplifies. Already a poet and author by the early 1920s, Corrie began to produce dramatic material for the Bowhill Players, a group of miners and their wives from Cardenden, Fife, whose employment base was the Bowhill Colliery.[83] Corrie's most famous play, *In Time o' Strife* (1927), centred on the bitter experience of the 1926 General Strike, while his poetry provided wry commentary on contemporary employer–worker relations:

> The next boss I had was a musical hand,
> He stood like a sodger and waggled the wand,
> So I learned the cornet and played in his band,
> A' to keep in wi' the gaffer.[84]

ROBUST ASSOCIATIONAL CULTURE

Corrie may have had the legacy of the Volunteer Force in mind when he wrote his poem. As has been seen, Volunteering was far from subversive, given that its rationale was national defence. Launched in 1859, at a time of public anxiety about the threat of French invasion, members supplied their own rifles, equipment and uniforms. To maintain recruiting momentum for what could be an expensive pastime, Volunteering projected a direct appeal to local patriotism – loyalty was identified as much with the community as with the nation. However, the movement was more than just a manifestation of citizenship in action. Historian Hugh Cunningham has explained that at a time when working hours were contracting, Volunteering provided young men with new outlets for robust recreation, notably drilling and rifle shooting.[85] Because of the movement's character-forming qualities, employers actively encouraged recruitment. Wylie & Lochhead Ltd, Glasgow cabinet makers and chair manufacturers, claimed to be the first firm in the United Kingdom to raise and equip a Volunteer corps from its workforce.[86] In Edinburgh, at the height of Volunteer recruiting fever in 1860, several occupation-based corps and companies were formed, focusing on skilled groups such as engineers, railway workers and print workers in the newspaper industry.[87] The range of Volunteer recruitment was broad, including business and professional activists as well as lowlier 'artisans'. By 1868 Edinburgh had eight artisan companies, whose members constituted the mainstay of the city's Rifle Volunteer Brigade until 1908, when the Volunteer Force became the more authentically militaristic Territorial Army.[88]

Like other work-based social activities that took a high profile in the public sphere, there could be peer or employer pressure for Volunteers to come forward. In 1859 the directors of the North of Scotland Bank agreed to supply their clerks with rifles in response to an exuberant appeal from an Aberdeen branch manager, who claimed his staff 'had the stuff in them to make good soldiers'.[89] This was positive publicity, too, for the directors' public spirit – their decision was well-reported in the local press. Yet while employers could set the initial pace, there were undoubted attractions for the workforce. Volunteering exuded a culture of masculine clubbability, from tactical war games to pipe bands and assorted competitive sporting

Figure 26.2 North British Locomotive Works sports day, 1950s.

activities. This was at a time when organised sport was generally becoming a focus of public interest. Shorter working hours and the advent of Saturday half-day holidays also played a significant part in cultivating enthusiasm. So too did rising real incomes across industrial Scotland, along with dietary and public health improvements.[90] By the late nineteenth century healthy lifestyles were associated with the channelled, constructive use of physical energy. This was a campaigning concern of Glasgow's influential medical officer of health, James Burn Russell, who was determined to boost the well-being of the city's population. In 1886 he endorsed the work of the National Physical Recreation Society, which promoted exercise for the working classes through its magazine, the *Gymnasium News*.[91]

There were occupations where physical culture was taken very seriously. Police forces, which were ever conscious of the need to interact socially with the community, encouraged annual sports days where the men could demonstrate athletic prowess. Govan police initiated its first such gathering in 1881, and the tug-of-war contests with Glasgow teams were for years the highlight of proceedings.[92] This modern form of combat suggested much about Govanite determination to demonstrate the burgh's independence from it larger, predatory neighbour. Edinburgh's police sports day commenced in 1911 and was similarly projected as an inclusive, civic-orientated affair.[93] The *Recorder* magazine, produced monthly from 1903 by the Ayrshire Constabulary Mutual Improvement Society, reported on a range of social activities. These included a talk by an expert on 'muscular development', whose advocacy of strict vegetarianism and teetotalism as the route to perfect health may not have been universally appreciated.[94] At a wider level, sports days became an annual event for many Scottish workers in large enterprises, especially after World War I. Some were intended as family occasions, involving children as well as adults; others, like the Glasgow meetings of the North British Diesel Locomotive Recreative Association, were large-scale events in which outside athletes competed.[95]

Inevitably, types of sports clubs varied and changed over time. Working women, for instance, were more likely to be sporting participants as the twentieth century unfolded, in pursuits such as hockey, netball or bowls. However, it is revealing that the earliest clubs formed by the J & P Coats Ferguslie Threadworks during the 1880s – the bowling, cricket and lawn tennis clubs – were among the most enduring. Ferguslie Cricket Club, which at the time of writing (2008) competes in the Scottish National Cricket League Premier Division, commenced in 1887 as a team solely for Coats' employees.[96] Nor was this an unusual example of a workplace club becoming a significant institution. Countless association football clubs, senior and junior, emerged from an industrial background. For instance, journalist Ron Ferguson has written about his own family's role in Cowdenbeath Football Club's formation in 1881.[97] Migrant workers from Ayrshire, including the Fergusons, came to work for the Fife-based Cowdenbeath Coal Company during the 1870s, and brought with them the remarkable soccer skills of their home county. In the mid-twentieth century the amateur boxing club of the North British Locomotive Company in Glasgow became famous for

Figure 26.3 City of Glasgow police quoits club, 1929.

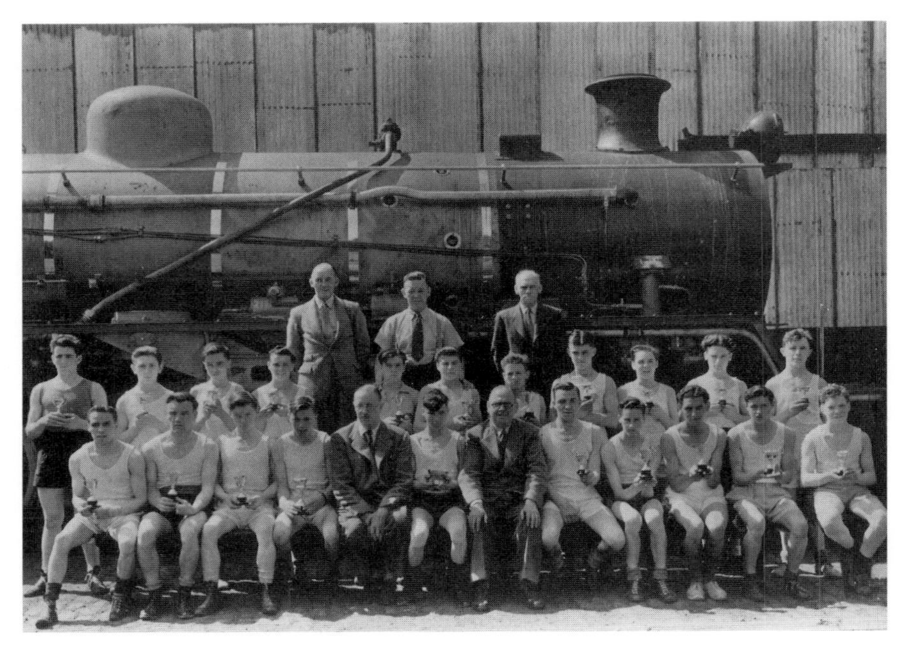

Figure 26.4 North British Locomotive boxing club, 1950s.

nurturing young talent. One of its protégés was John 'Cowboy' McCormack, who in 1956 was the British and Scottish light-heavyweight champion, as well as an Olympic medallist.[98]

There are dangers in too readily attributing sinister social control motives on the part of employers in encouraging amenity welfarism.[99] The rapid rise of organised workplace sports clubs during the late nineteenth century developed in tandem with the general growth of popular interest in competitive sport. There is no doubt that employers harnessed the leisure pursuits of their labour force to ensure that the company brand – and company authority – was secured. Yet as one study has shown, the carefully cultivated image of workplace harmony could mask tensions. Pullars of Perth was a family-run enterprise, specialising in dyeing and dry cleaning. Its owners exemplified nineteenth-century Liberal paternalism. While much was made in the 1890s about the recreational provision the company afforded, on close examination this proved to be grudging. Sports club members were even refused time off to travel to Saturday away-fixtures.[100] At other times Scottish workers could display distinctly inharmonious qualities in their sporting preferences. The Reds Quoiting Club of Blantyre, Lanarkshire, was a prize-winning team of communist miners active in the 1920s.[101] And the old, unregulated pastimes could remain doggedly resilient, even in the twentieth century. A journalist's investigation of 1936 showed how cock-fighting, and its heavy betting ethos, still attracted followers in Scotland's mining communities, despite its illegality.[102]

CONCLUSION

This analysis has focused on the links between work, community and sociability from the structured perspective of urban associational culture. The nineteenth century was the growth era of Scotland's clubs and societies, which were partly an expression of group identity at a time of social and economic change. The tendency of friendly societies, in particular, to root themselves in remote history was evidence of contemporary insecurities about the onset of industrialisation. These societies also exemplified their members' independence. They were not employer-sponsored, unlike many of the works-based social and welfare clubs that emerged at the turn of the twentieth century. A number of these organisations were of their time, notably the friendly societies, which eventually declined as a result of growing welfare-state provision. Others continue to thrive, despite the impact of deindustrialisation on communities from the 1970s. Indeed, their survival often helps to reaffirm local identity, especially in the highly visible context of customary festivals and civic and sporting events. In addition, the drastic process of deindustrialisation has reinforced the reasons for the emergence of such associational culture in the first place. For Glasgow, Scotland's largest city, the erosion of the old manufacturing base has been blamed for the city's prominent place in Britain's league table of unhealthy blackspots.[103] By 2000 a lack of amenities, especially sports facilities, characterised the most deprived areas. Whatever the complex rationale behind workplace associational culture as it evolved from the nineteenth century, it did contribute to community cohesion and wellbeing.

NOTES

1. Quoted in Morris, 1990, 416.
2. Campbell, 1925, 125.
3. Campbell, 1925, 234.
4. Campbell, 1925, 234.
5. Robertson et al., 1929, 272.
6. Colston, 1891, xlix.
7. Keir, 1966, 774–9.
8. Robertson and Wood, 1928, 182.
9. Bogle, 2004, 32.
10. McGowran, 1985, 203–4.
11. Brotchie, 1938, 138–57.
12. McNeill, 1968, 101–2.
13. Cooper, 2006, 162.
14. Cooper, 2006, 164–5.
15. Mitchell, 1883, 57.
16. Knox, 1921, 102.
17. *Scotsman*, 16 December 1826.
18. Cooper, 2006, 169–70.
19. Cordery, 2003, 17–18.

20. Campbell, 1925, 241. The Burns poem in questions is 'Epistle to a Young Friend' (1786).
21. Cordery, 2003, 17.
22. SHELF website on friendly societies, http://www.historyshelf.org/shelf/friend/08.php, accessed 31 October 2006.
23. McNeill, 1968, 222–3.
24. Neat, 2002, 34–5. The speaker was Sandy Moncrieff.
25. Bremner, 1969, 23.
26. Scott, I. The Free Colliers. On the website http://www.falkirklocalhistorysociety.co.uk/home/index.php?id=98, accessed 29 November 2006.
27. King, 2007, 134.
28. Neat, 2002, 62.
29. Adams, 1991, 63.
30. Neat, 2002, 62; Buchan, 1984, 226.
31. Stevenson, 1988, 130.
32. Cordery, 2003, 17; Finn, 1990, 167.
33. Willox, 1898, 61.
34. Harvie, 2000, 126.
35. MacDougall, 1995, 44–5.
36. Knox, 1999, 116.
37. MacRaild and Martin, 2000, 130.
38. Ó Catháin, 2007, 32–40.
39. Walker, 1991, 50.
40. Denny, 1989, 32.
41. See Maver, I. Dame Jean Barr McDonald Roberts, in the *Oxford Dictionary of National Biography*.
42. Glasgow Trades House website at http://www.tradeshouse.org.uk/, accessed 21 November 2006.
43. Smiles, 1986, 19.
44. Donnachie, 2000, 159.
45. Maver, 2006, 166.
46. Fraser, 1990, 249.
47. *Glasgow Herald*, 27 June 1853.
48. 'Justus', 1856, 4.
49. *Paisley and Renfrewshire Gazette*, 7 January 1880.
50. Macdonald, 2000, 62–7.
51. Young, 1939, 77.
52. Mulhern, 2006, 118.
53. Mulhern, 2006, 123.
54. *Scotsman*, 12 May 1930.
55. Duncan, 2005, 210.
56. Carvel, 1948, 178.
57. Loanhead Miners' Welfare website, http://www.loanheadmw.org.uk/home.htm, accessed 27 October 2006.
58. Borsay, 2006, 3.
59. McInnes, 2007, 419–20
60. Knox, 1921, 56.
61. Scobbie, 1954, 217–18.
62. McNeill, 1961, 122–5.
63. *Perthshire Courier*, 28 January 1836. Quoted in Cooke, 2003, 132.

64. Thomson, 1960, 141.
65. Extinct Brass Bands, available online at: http://www.harrogate.co.uk/harrogate-band/hb-links.htm, accessed 27 October 2006.
66. Herbert, 2000, 49.
67. Eydmann, 2007, 199–200.
68. Donaldson, 2005, 17–18.
69. Donaldson, 2000, 359.
70. Brotchie, 1938, 170–4.
71. Donaldson, 2005, 19.
72. Central Scotland police website, http://www.centralscotland.police.uk/homepages/pipe_band/history.php, accessed 21 December 2006.
73. Donaldson, 2005, 19.
74. Triumph RBL Pipe Band, Coventry website, http://www.triumphpipeband.co.uk/history/history.htm, accessed 29 February 2008.
75. MacDougall, 1995, 144. The speaker was Pat Flynn.
76. MacDougall, 1995, 69.
77. Farmer, 1947, 439.
78. McGowran, 1985, 65–6.
79. Miller, 1999, 12–13.
80. McGowran, 1985, 156.
81. Project Team, 1995, 68.
82. Findlay, 2007, 542. The quote came from James Houston, writing in 1889.
83. Mackenney, 1985, 9–10.
84. Mackenney, 1985, 139. From the poem 'It's Fine to Keep in wi' the Gaffer'.
85. Cunningham, 1975, 111–13.
86. Cunningham, 1975, 21.
87. *Scotsman*, 16 January, 2 February and 1 March 1860.
88. Gray, 1976, 103.
89. Carnie, 1902, 326.
90. Fraser, 1990, 251.
91. Russell, 1905, 319.
92. *Govan Press*, 27 June 1885.
93. *Scotsman*, 28 November 1911.
94. *Recorder*, 21 March 1907.
95. *Scotsman*, 17 May 1920.
96. Project Team, 1995, 64–5.
97. Ferguson, 1993, 22.
98. Donald, 1988, 145.
99. Tranter, 1998, 36–9.
100. Davies, 1993, 32–3.
101. Hutton, 1997, 69.
102. Beattie, 1936, 213–15.
103. Cunningham, 1999.

BIBLIOGRAPHY

Adams, D G. *Bothy Nichts and Days: Farm Bothy Life in Angus and the Mearns*, Edinburgh, 1991.

Beattie, G. The Scottish miner and his game-cock, *Scots Magazine*, 25 (December 1936), 213–17.

Bogle, K R. *Scotland's Common Ridings*, Stroud, 2004.
Borsay, P. *A History of Leisure: The British Experience since 1500*, Basingstoke, 2006.
Bremner, D. *The Industries of Scotland: Their Rise, Progress and Present Condition*, Newton Abbot, 1969; first published 1868.
Brotchie, T C F. *The History of Govan*, 1938.
Buchan, D. The expressive culture of nineteenth-century Scottish farm servants. In Devine, T M, ed., *Farm Servants and Labour in Lowland Scotland, 1770–1914*, Edinburgh, 1984, 226–42.
Campbell, C. *Provident and Industrial Institutions*, Manchester, 1925.
Carnie, W. *Reporting Reminiscences*, Aberdeen, 1902.
Carvel, J L. *The Coltness Iron Company: A Study in Private Enterprise*, Edinburgh, 1948.
Colston, J. *The Incorporated Trades of Edinburgh*, Edinburgh, 1891.
Cooke, A. *Stanley, from Arkwright Village to Commuter Suburb, 1784–2003*, Perth, 2003.
Cooper, R L D. *Cracking the Freemason's Code: The Truth about Solomon's Key and the Brotherhood*, London, 2006.
Cordery, S. *British Friendly Societies, 1750–1914*, Basingstoke, 2003.
Cunningham, H. *The Volunteer Force: A Social and Political History, 1859–1908*, Hamden, Connecticut, 1975.
Cunningham, J. Gap in areas of good and ill-health widening, *The Herald*, 2 December 1999.
Davies, J M. Social and labour relations at Pullars of Perth, 1882–1923, *Scottish Economic and Social History*, 13 (1993), 27–42.
Denny, N. Self-help, abstinence and the voluntary principle: The Independent Order of Rechabites, 1835–1912, *Journal of the Scottish Labour History Society*, 24 (1989), 24–46.
Donald, B. *The Fight Game in Scotland*, Edinburgh, 1988.
Donaldson, W. *The Highland Pipe and Scottish Society, 1750–1950*, East Linton, 2000.
Donaldson, W. *Pipers: A Guide to the Players and Music of the Highland Bagpipe*, Edinburgh, 2005.
Donnachie, I. *Robert Owen: Owen of New Lanark and New Harmony*, East Linton, 2000.
Duncan, R. *The Mineworkers*, Edinburgh, 2005.
Eydmann, S. Diversity and diversification in Scottish music. In Beech, J, Hand, O, MacDonald, F, Mulhern, M A and Weston, J, eds, *Oral Literature and Performance Culture* (Scottish Life and Society: A Compendium of Scottish Ethnology, vol. 10), Edinburgh, 2007, 193–212.
Farmer, H G. *A History of Music in Scotland*, London, 1947.
Ferguson, R. *Black Diamonds and the Blue Brazil: A Chronicle of Coal, Cowdenbeath and Football*, Ellon, 1993.
Finn, G. In the grip? A psychological and historical exploration of the social significance of freemasonry in Scotland. In Walker, G and Gallagher, T, eds, *Sermons and Battle Hymns: Protestant Popular Culture in Modern Scotland*, Edinburgh, 1990, 160–92.
Findlay, B. Theatre and drama. In Beech, J, Hand, O, MacDonald, F, Mulhern, M A and Weston, J, eds, *Oral Literature and Performance Culture* (Scottish Life and Society: A Compendium of Scottish Ethnology, vol. 10), Edinburgh, 2007, 531–55.
Fraser, W H. Developments in leisure. In Fraser, W H and Morris, R J, eds, *People and Society in Scotland, Volume II: 1830–1914*, Edinburgh, 1990, 236–64.
Gray, R Q. *The Labour Aristocracy in Victorian Edinburgh*, Oxford, 1976.
Harvie, C. *No Gods and Precious Few Heroes: Twentieth-Century Scotland*, Edinburgh, 2000; first published 1981.

Herbert, T. Nineteenth-century bands: Making a movement. In Herbert, T, ed, *The British Brass Band: A Musical and Social History*, Oxford, 2000, 10–67.
Hutton, G. *Lanarkshire's Mining Legacy*, Catrine, 1997.
'Justus', *The Advantages of Early Closing to the Employed*, Glasgow, 1856.
Keir. D. *The Third Statistical Account of Scotland: The City of Edinburgh*, Glasgow, 1966.
King, E. The material culture of William Wallace. In Cowan, E J, ed., *The Wallace Book*, Edinburgh, 2007, 117–35.
Knox, J. *Airdrie: A Historical Sketch*, Airdrie, 1921.
Knox, W W. *Industrial Nation: Work, Culture and Society in Scotland, 1800–Present*, Edinburgh, 1999.
Macdonald, C M M. *The Radical Thread: Political Change in Scotland, Paisley Politics, 1885–1925*, East Linton, 2000.
MacDougall, I. *Mungo Mackay and the Green Table: Newtongrange Miners Remember*, East Linton, 1995.
McInnes, J. *Òrain luaidh* and other work songs. In Beech, J, Hand, O, MacDonald, F, Mulhern, M A and Weston, J, eds, *Oral Literature and Performance Culture*, (Scottish Life and Society: A Compendium of Scottish Ethnology, vol. 10), Edinburgh, 2007, 412–26.
McGowran, T. *Newhaven-on-Forth: Port of Grace*, Edinburgh, 1985.
Mackenney, L ed. *Joe Corrie: Plays, Poems and Theatre Writings*, Edinburgh, 1985.
McNeill, F M. *The Silver Bough, Volume III: A Calendar of Scottish National Festivals, Hallowe'en to Yule*, Glasgow, 1961.
McNeill, F M. *The Silver Bough, Volume IV: The Local Festivals of Scotland*, Glasgow, 1968.
MacRaild, D M and Martin, D E. *Labour in British Society, 1830–1914*, Basingstoke, 2000.
Maver, I. The temperance movement and the urban associational ideal: Scotland, 1820s to 1840s. In Morton, G, de Vries, B and Morris, R J, eds, *Civil Society, Associations and Urban Places: Class, Nation and Culture in Nineteenth-Century Europe*, Aldershot, 2006, 159–74.
Miller, J. *Salt in the Blood: Scotland's Fishing Communities Past and Present*, Edinburgh, 1999.
Mitchell, J M. *The Duthie Park: A Descriptive and Historical Sketch*, Aberdeen, 1883.
Morris, R J. Clubs, societies and associations. In Thompson, F M L, ed., *The Cambridge Social History of Britain, 1750–1950: Volume 3, Social Agencies and Institutions*, Cambridge, 1990, 395–443.
Mulhern, M A. 'A bridge from the worse to the better': The Gothenburg public-house in Scotland, *Review of Scottish Culture*, 18 (2006), 114–27.
Neat, T. *The Horseman's Word: Blacksmiths and Horsemanship in Twentieth-Century Scotland*, Edinburgh, 2002.
Ó Catháin, M S. *Irish Republicanism in Scotland, 1858–1916: Fenianism in Exile*, Dublin, 2007.
Paisley and Renfrewshire Gazette.
Project Team [Anderson, P et al.], *Six Cord Thread: The Story of Coats' and Clarks' Paisley Thread Mills*, Paisley, 1995.
Robertson, D and Wood, M. *Castle and Town: Chapters on the History of the Royal Burgh of Edinburgh*, Edinburgh, 1928.
Robertson, D, Wood, M and Mears, F C. *Edinburgh, 1329–1929*, Edinburgh, 1929.
Russell, J B. The children of the city: What can we do for them? In Chalmers, A K, ed.,

Public Health Administration in Glasgow: A Memorial Volume of the Writings of James Burn Russell, Glasgow, 1905, 301–23.
Scobbie, J K, ed. *The Book of Airdrie, Being a Composite Picture of the Life of a Scottish Burgh by its Inhabitants*, Glasgow, 1954.
Smiles, S. *Self-Help, with Illustrations of Conduct and Perseverance*, Harmondsworth, 1986; first published 1859.
Stevenson, D. *The Origins of Freemasonry: Scotland's Century, 1590–1710*, Cambridge, 1988.
Thomson, G ed. *The Third Statistical Account of Scotland: The County of Lanark*, Glasgow, 1960.
Tranter, N. *Sport, Economy and Society in Britain, 1750–1914*, Cambridge, 1998.
Walker, G. The Protestant Irish in Scotland. In Devine, T M, ed, *Irish Immigrants and Scottish Society in the Nineteenth and Twentieth Centuries*, Edinburgh, 1991, 44–66.
Willox, D. *Poems and Sketches*, Glasgow, 1898.
Young, F H. *A Century of Carpet Making, 1839–1939*, Glasgow, 1939.

PART SIX

Public Service and Industry

27 The Working Lives of Scottish Schoolteachers

JANET DRAPER

INTRODUCTION

There are many ways of looking at the working life of teachers. In this chapter, which focuses on those who teach in schools, two main paths will be followed that explore the experience of being and working as a teacher. The first path will examine changes in the work of teachers, especially those changes which have occurred since the middle of the twentieth century. The second section will consider diversity in the work of teachers.

At first glance, teaching appears to be very public work, a 'performance' involving a group of learners (learners in schools have traditionally been called pupils, but in secondary school in particular the use of 'student' is increasingly common). But this element of teachers' work is only a part, albeit a very important part, of a much wider role which involves a cycle of planning and preparation prior to performance, and subsequent assessment and evaluation which shape planning for the next performance. In addition, while 'performance' time is mainly spent with children, other work time is spent alone or with other adults, for example, teacher colleagues and school managers, support staff in schools, parents and external agencies, such as social workers, police, church or educational psychologists.

Teaching in Scotland is predominantly public-sector professional work. Teachers comprise between one quarter and one third of the public sector workforce[1] and constitute the second largest public-sector group after health-service workers. There is a small independent school sector, but most teachers (97 per cent) work as employees of local authorities. Teachers' work experiences are not, however, unitary: they work with learners of varying ages; in institutions of different sizes; in varying policy contexts; and in significantly different communities, within which differing attitudes to education prevail. These differences set the immediate tasks of teaching learners as a specific range of challenges where the focus may vary significantly: for example, between motivating reluctant learners and satisfying a powerful thirst for learning; or between facilitating the mastery of key basic skills and dealing with complex and abstract or contested ideas.

Teachers' working lives should also be understood against a broader canvas than solely that of the schools where they work. Teachers carry substantial responsibility for the development and preparation of young people as workers, as individuals, as parents, as members of their communities. Studies of Scottish teachers' views have shown them to be more in line with popular views than, for example, the views aired by politicians,[2] and they have a pronounced tendency to identify themselves as Scottish before British. Gatherer[3] argues that teachers are a talented and educated group with vitality and a wide range of interests. Those who leave teaching to move to other areas of responsibility in wider education and beyond, in politics, business, the arts and so on, and the profession is enriched by those who have worked in other areas and then come into teaching, seeking new satisfaction and bringing wide experience. Teachers are a diverse group.

TEACHING AS WORK

The experience of working as a teacher changed significantly during the twentieth century. In many ways this experience is little different from other types of work in which shifts in expectations of work and career have also changed, especially since the 1980s, summarised as changes in:

- The type of work, with a shift in emphasis from manufacturing and heavy industry to the service sector, including the public sector and financial services;
- The constitution of the labour force (with more women working);
- The nature of contractual work (with more part-time and short-term work; more casualisation);
- Expectations of career (from lifelong careers to a series of different types of work);
- The range of work opportunities over the period.

TEACHING AS PROFESSIONAL WORK

Teaching has long been perceived as a profession in Scotland. This perception entails expectations of teachers; for example, that they will have a knowledge and skills base, a period of professional preparation, a degree of professional autonomy, a commitment to improvement, and a regulatory body which will oversee standards on entry and of conduct. Teaching has also been affected by changing perceptions of professionalism, which emphasise professional development and a focus on lifelong learning, partly to keep up with new developments and partly to support innovation. This has combined with a more critical appraisal of professional work as a service and increased governmental shaping of public-sector work.

CHANGES IN TEACHERS' WORK

Over the past century there have been significant shifts in what is taught, in how it is taught and whose decisions these are, and these have changed the nature of teachers' work. The laying down of more detailed and formal curriculum and assessment guidelines for schools from 1987 onwards heralded a more overt form of central control over the educational process. This has been reflected in some areas by attempts to prescribe the way teaching is carried out: direction on the 'how' as well as the 'what' of teaching. During the second half of the twentieth century there were major shifts in the relationships teachers had with their employers, government and the parents of those they taught, associated with changes in school governance and accountability. There were also major changes in the relationships between teachers and their pupils and in their ways of working. There have therefore been significant changes in teachers' work.

From well before the establishment of the Scottish parliament in 1999, Scottish teachers had understood themselves to be working in a different system from their English counterparts. The system had always been distinct, having been protected to an extent in the Treaty of Union. Concerns and frustrations were often tolerated because of the shared belief that it was 'worse over the border'. The politicisation of education in England with the reassertion of direct central control[4] by government from the late 1980s, however, led to considerable pressure to make similar changes in Scotland, some of which were resisted successfully. For example, when the government sought to introduce national testing of children at the ages of seven, eleven and fourteen, the move was resisted in Scotland jointly by teachers and parents, and a more educationally acceptable solution was found. While English teachers struggled with an unwieldy, imposed national curriculum, Scottish teachers worked with curriculum guidelines – although these guidelines shaped the nature of school inspection and were not perhaps as advisory as was sometimes suggested.

As public-sector accountability increased, so teachers found they were regarded less as autonomous professionals and more as accountable employees. Government policy shaped curriculum and some aspects of direct teaching itself, including the assessment of learners and their learning. The specified curriculum for learners aged five to fourteen, first introduced in 1987[5] but with a development programme running to 1994, led to the identification of levels within the five curriculum areas (language, mathematics, environmental studies, expressive arts, and religious and moral education) in primary, against which attainment could be assessed. This formalised the structure of teachers' work and reshaped their role more towards being deliverers rather than designers of the curriculum. The Munn and Dunning reports (both published in 1977) reshaped the secondary school curriculum with, over time, Standard Grade and Higher Still replacing the old Ordinary (O) grade and Higher Grade public examinations. While teachers have been involved in development groups for these changes, the role of teacher as curriculum leader has diminished. The inspection process was also enhanced, standards set by those outside classrooms and the outcome of assessments were made more publicly available. The process also incorporated an element of self-evaluation. The role of parents was strengthened in various ways, such as allowing parental choice of schools, greater demands for contact with, and fuller reporting from, teachers and parental influence on school governance through school boards. While it is difficult to generalise about teachers, since their situations and roles are so diverse, there is powerful evidence to support a claim that teachers' work in Scotland has been shaped by the changes in perceptions of professionalism and by shifts in the governance and control of education, as well as by more generic changes in work and conceptions and realities of career. While changes in the work of Scottish teachers show similarities to changes in teachers' work outwith Scotland, there are also ways in which teachers' work has been uniquely shaped by the Scottish context. At times, change in Scotland has been not as great as that encountered by teachers elsewhere, for example in England.

CHANGES IN TEACHERS' WORK: SOME EXAMPLES

Alongside changes in the curriculum and expectations of teacher behaviour and performance in the classroom, there have been many other ways in which teachers' work has changed. A selection of examples follows.

Teaching as work is varied but always challenging since, as is now understood, individual learning needs are varied. Similarity of age and/or stage does not mean similarity of response to learning opportunity, hence learning opportunities must be varied to foster learning in all learners. Indeed one of the major shifts in teachers' work has followed from the acknowledgement that differences between learners are significant and require to be accommodated. A 'one size fits all' form of teaching is now seen as inappropriate. Earlier views that teachers taught and learners accommodated and complied or, alternatively, failed to use the opportunity, have been replaced by expectations that teachers will differentiate learning opportunities for different learners, by taking account of differing starting points in knowledge and understanding, past experience, attitudes and approaches to learning, personalities and more. These expectations signal a real shift in the focus of teachers' work from teaching in one's own way to supporting the learning of others. The most obvious sign of this shift was the Primary Memorandum published in 1965,[6] which encapsulated the new focus on child-centred learning. The publication of this memorandum was not the start of the process but rather a reflection that expectations of teachers and education were already moving that way. A further focus on learners' needs followed in 1978, when the Warnock report[7] encouraged the education of as many children as possible in mainstream schools. This led eventually to broad policies that argued for the inclusion of all children rather than the total segregation of those with special needs. This has been controversial, as it placed extensive new demands on teachers, who have often argued that they do not have the skills, time or resources to attend adequately to pupils with very distinct needs.

Running alongside these changes in teachers' work was the major shift to comprehensive schooling in the late 1960s. The process took some time to roll out completely (Munn suggests around ten years[8]) – but it altered the landscape of schools as places of work and as places where pupils were taught, as well as the hierarchy of school and teacher roles. It removed the role of primary schools as the test bed for subsequent allocation to secondary schools. Before comprehensivisation, there had been selective entry to a ladder of secondary schools: junior and senior secondary schools for those identified through a test process as more academically able, which were often called academies or high schools, and secondary modern and technical schools for others. The different types of school had different expectations of learners and of the destinations of pupils upon leaving. They offered different curricula, and a clear hierarchy had developed, with those covering higher academic work at the top of the ladder. Comprehensivisation was to change all that and was ideologically rooted in egalitarian values. It offered a promise of children being educated in comparable schools across the country and moving as established groups from primary to secondary schools, rather than being sorted and selected for different educational experiences which would firmly shape their future prospects. In practice, however, schools carried their previous histories with them, and in spite of various strategies, including name changes, Adler's work on parental choice of school showed that the old selective senior secondary schools (the top of the ladder) remained schools of choice. Teachers thus continued to work in an informal, or covert but understood, hierarchy of schools. The range of ability in classes was perceived as being significantly greater and much more challenging for teachers, who had previously taught 'streamed' classes (the 'A' class, 'B' class, etc) in what were, in effect 'streamed' schools. The division of pupils into 'sets' of 'similar ability' pupils within subjects in the new schools represented one attempt to cope with this perceived range. While this approach was less narrow than 'streaming', it still applied labels and their associated expectations to learners. It overlooked, as streaming did, the unreliability inherent in assessments of 'ability' and the tendency to assume that differences between individuals remained constant over time, in spite of the fact that both childhood and adolescence are periods of rapid and individually very varied development. Mixed-ability teaching is now more common, especially in the earlier years of secondary schooling. Comprehensivisation did proceed, however, unlike in England, where it was less successful, and it is now a hallmark of the Scottish system and shapes teachers' work in that primary schools are not now focused on competitive testing in their final year. Secondary schools, while not seen as wholly interchangeable, have the same basic function.

Another significant change in teachers' working situations follows from the age at which young people leave school and where they go next.

The school-leaving age rose from fourteen in the immediate postwar years to fifteen, and then to sixteen in 1972, occasioning concern from teachers about working with unwilling learners who were unable to leave school. A related significant further shift occurred during the latter quarter of the century as the number of young people staying on at school beyond the statutory leaving age of sixteen gradually exceeded those who left. This has resulted in changes to the curriculum to take account of changing post-school destinations. The range of courses in the middle and later years of secondary schooling had been designed to support the few who were staying on with a view to taking up either Higher Grade work or further study after leaving school. The shift to higher levels of 'staying on' and mass higher education has radically altered demands placed on teachers, who now prepare more of their charges for subsequent study and/or training than for the workplace.

A further major change in the relationship between teachers and learners was finally formalised in 1986. The removal of the strap or tawse brought an end to the use of corporal punishment in school as a form of discipline. In reality, many teachers had already given up this form of disciplining and the profession was generally supportive of its removal whilst remaining concerned at the absence of clear alternatives for sustaining a positive working environment in the classroom. Complex collaborative policies and systems followed to support the creation and maintenance of disciplined working environments. It is clearer now than perhaps it was before the decision to give up the strap that effective and harmonious classrooms are achieved more by co-operation between teachers and learners, accommodation to negotiated classroom rules and positive encouragement than by force or hierarchical dictat. However, that managing behaviour in classrooms remains a challenging part of teachers' work is attested to by its having been the focus of a special inquiry in 2001, resulting in a published report entitled *Better Behaviour, Better Learning*.[9] The end of corporal punishment aided developments in the relationship between teachers and learners and a perception of schooling as being more egalitarian, this approach being further reflected in changes in the processes for evaluating schools, whereby views are sought from teachers, pupils and parents.

TEACHERS AS AN OCCUPATIONAL AND PROFESSIONAL GROUP

Scottish teachers are a highly unionised group. The strength of unionisation and the narrow union structure established teachers as a distinct and powerful occupational group. The OECD[10] identified teachers as being able to shape the direction of educational reforms through structures that create the opportunity for strong collective bargaining. The Scottish teachers' unions have considerable influence. Joint action over such issues as testing, mentoring and appraisal have stood as testament to the power of teachers as a collective to impact on policy. After over 100 years in existence as a teachers' association (since 1847), members of the Educational Institute of

Scotland (EIS) went on strike for the first time in 1961 in response to the diminishing status of teachers, reflected in relatively low levels of pay.[11]

The strike led to the Wheatley Commission, which recommended the establishment of a General Teaching Council (GTC). After the strike in 1961, major industrial action occurred again in a long and bitter dispute during the mid-1980s, again raising the issue of pay levels, which had fallen relatively, and reflecting a concern that the only way to a decent salary was to leave the classroom and take up managerial posts in schools or elsewhere. The agreement which ended this strike included significant changes in teachers' contracts of work and the creation of a new post, that of 'senior teacher', which was intended to recognise excellence in teaching and reward those who remained in the classroom in contrast to those who moved out of class to manage schools. But the role of 'senior teacher' was short-lived. It quickly became absorbed into the management structure in schools, especially in primary schools, and its intended function was unfulfilled. Further major change took place following the McCrone inquiry into teachers' pay and conditions of work.[12] The inquiry was set up to address teacher resistance to participation in new developments in response to unresolved long-term workload issues, and concerns about teacher supply and retention in the light of problems identified in England and elsewhere. In the run-up to the inquiry, concern was expressed by the EIS[13] that the inquiry should not confine itself to pay and conditions but should consider the circumstances in which teachers worked, including 'being forced to spend more time out of the classroom', the 'unchecked and under-resourced' pace of educational reform, class sizes and the undervaluing of teachers.

The professional identity of teachers has been clear for some time, especially since the establishment by statute in 1965 of the professional body, the General Teaching Council, following the Wheatley Commission. The GTC is recognised as one of the first of its kind, and others have followed across the globe – for example, in Ontario, Canada; Victoria, Australia; New Zealand; and England, Northern Ireland and Wales. The title of the GTC is important, as the council is not a *teachers'* but a *teaching* council, with a wide remit to maintain and enhance the status of the profession, to advise on the supply of teachers, to protect professional standards, partly through its regulatory work on behalf of the profession and, more recently, to oversee the continuing professional development (CPD) of teachers. The council exercises its powers not only on behalf of the profession but also in the public interest, since the public has a clear interest in the conduct of education. The GTC is therefore the professional body which represents and speaks for the profession on professional matters, but not on matters of pay and conditions of service, which are the preserve of the unions or professional associations. One consequence of the establishment of the council was the removal of unqualified teachers from schools. It also played a significant role in achieving all-graduate entry to teaching by the late 1980s, leading to an all-graduate and professionally qualified profession. For forty years, therefore, teachers have had an independent guardian of professional standards in Scotland. Few workers laud their regulatory bodies, given their regulatory function, but the fact that the teaching profession has a distinct and independent professional body reinforces the public perception of teaching as a profession. By its sheer existence, as well as through its growing range of responsibilities, the council occupies space as an independent professional body. Different arrangements are found elsewhere. For example, in England these responsibilities are divided between a government agency (the Teacher Training Agency), established in 1993 and renamed the Training and Development Agency for Schools in 2005, and the General Teaching Council for England, which was subsequently established in 2000.

Teachers' professional status is partly protected through the control that is exercised over entry by the GTC. There continue to be a limited number of routes into teaching: a four-year Bachelor of Education degree; a four-year (or four-and-a-half-year) concurrent degree, both of which combine academic and professional study and qualification; and a one-year programme (two years part-time) for graduates. All involve substantial periods of school experience. Teachers qualify to teach in a particular school sector (primary, secondary) or further education. Those who teach in secondary or further education gain a qualification in a specific subject area. School-based and flexible entry routes to teaching have been developed elsewhere in response to teacher shortages. The Scottish profession has rejected these, taking the view that prospective teachers should be professionally qualified before assuming responsibility for children's learning. Prior to 2002, beginning teachers (probationers) had very varied employment experiences, and the provision of support for development was markedly uneven.[14] Some teachers began their careers in unsupportive settings, and some on short-term contracts. Since 2002, following McCrone, probationer teachers have been given a post for one school year, coupled with support for professional development. By the end of the year they are expected to meet the required standard (Standard for Full Registration), after which they register with the GTC as full members of the profession.

RECENT CHANGES IN TEACHERS' WORK: THE MCCRONE INQUIRY

The final report of the McCrone inquiry[15] offered an analysis of issues related to teachers' work and made strong recommendations for change, highlighting the economic significance of teaching, the importance of having a developed professional and satisfied workforce in schools, and public attitudes to teaching.

> Scotland's future prosperity depends crucially on the skills of its people, the people educated in Scotland's schools[16]
>
> We need high quality, trained, professional, motivated and contented teachers; and we need to restore public esteem for the teaching profession.[17]

Not all of the recommendations were accepted by the employers, but many featured in the final agreement. Endorsed by teachers, the new agreement heralded major change in teachers' work, including the formal establishment of a 35-hour working week and a reduction in class contact time to relieve workload pressures and make space for planning, preparation and CPD. The spirit of McCrone to make teachers more autonomous and better regarded (and paid) professionals remained. There were to be new arrangements for new teachers, and a new, unpromoted high-status career route – chartered teacher status[18] – was to replace the hijacked 'senior teacher' role. 'Chartered teacher' was to be a status not a role, to recognise excellent classroom practice unencumbered by managerial responsibilities.

The changes finally agreed to teachers' work were numerous and significant. They included changes:

- To the structure of schools (with some flattening of the long hierarchy of posts);
- To the nature of the working day, week and year;
- To pay scales;
- To administrative support for teachers;
- To support on discipline;
- To career routes through teaching;
- To arrangements for newly qualified teachers;
- In the purpose, nature and control of CPD as an entitlement and as an obligation.

THE ROLE OF PROFESSIONAL DEVELOPMENT

These changes are currently in train, and many are progressing well partly because they were ideas whose time had come. Teachers' professional development is a case in point. While one dimension of professional behaviour is held to be commitment to improvement, there had been limited opportunities for teachers to develop their knowledge, skills and understanding, except during time out of school. There was a rhetoric of staff development as an important activity, but often little appetite for what was on offer. This was not helped in the 1980s by a significant investment in training for school managers but much less provision for classroom teaching.

Two changes were important in bringing professional development more to the fore in the working lives of teachers. One was a change in the way professional development was conceived, and the other was the demarcation of time within the job to pursue development. Models of teacher learning had tended to assume attendance at courses and at schoolwide in-service days. Evening and weekend courses had been on offer, along with a few summer schools. Hard-pressed, tired teachers would struggle to attend and participate in courses which in earlier years had somewhat indirect links with professional action. Courses run by local authorities tended to focus on system-wide issues and, like in-service days in schools, were often targeted at providing updates on new policies and system- or schoolwide issues. Individual teacher needs for development were less likely to be acknowledged or met. At the same time, teachers described important professional development taking place within their work as they evaluated their practice and sought to improve it – 'the most developing thing I do is my job'. The identification of teacher development needs was seen as important in ensuring that useful professional development opportunities would be offered, but there were skirmishes over the identification of needs through appraisal. Although the centre of 'appraisal' is praise, teachers concluded that, in an era of growing managerialism and accountability, appraisal had sinister potential, and they resisted participating.

New forms of professional development were needed. Educational thinking was developing to emphasise the value of more active learning and the value of learning embedded in day-to-day work situations.[19] Both work-based learning and action research were seen to offer development activities which focused on actual teacher practice but which could still be rooted in a wider context of understandings about teaching. To think carefully and critically about one's own teaching was to take account of what was known, thought and written about teaching by others. Engagement with colleagues and with the ideas and reflections of others was understood to deepen teachers' reflections about their own practice and to offer a more fruitful route for developing their actual practice and professional actions. New programmes like the development programme for chartered teachers and the Scottish Qualification for Headship both use active learning methods and are rooted in critical reflection on participants' own professional actions.

TEACHERS AS PARTICIPANTS IN THE POLICY PROCESS

McCrone had acknowledged that motivation and contentment were important, and this suggested a return to teachers being viewed as significant players in the system. In 1969, James Scotland had stated that the teacher was the 'most important person in the school',[20] but in the 1980s others[21] argued that the authoritarian methods of policy making in Scotland at the time contributed to a democratic deficit and limited the contribution of teachers to policy making. The McCrone changes suggest a greater acknowledgement of the contribution that teachers are able to make to education and the consequent need for them to be able to participate fully in the policy processes which shape their work.

Opportunities for teachers to contribute their views at an individual level have been more notable recently, since the establishment of the Scottish parliament. The process of consultation during the McCrone inquiry actively sought the views of individual teachers as well as those of teachers as a whole through standard representational mechanisms. The development process for the proposed high-status unpromoted career route to chartered teacher status similarly actively canvassed views of teachers and others. The 'national debate on education' invited participation from teachers and others. Opportunities have therefore existed for teachers to contribute to the policy process, and teachers as experienced professionals now have channels to make their own distinct contribution to policy development in education, alongside the contributions of learners themselves, parents, future employers, communities, teacher-employers and others.

Since teachers vary in their work and views, the message may not be consistent, and representation of their views as a whole is not straightforward. Ways in which their work varies will now be explored.

VARIATIONS IN TEACHERS' WORK

Gatherer, in his eloquent summary of teachers,[22] distinguishes three early types of teacher in the eighteenth and early nineteenth centuries: the dominie, the burgh teacher and the women. Each of these played a different role in the system and accrued different status in their respective communities. The burgh schoolmasters were an élite group of scholars who worked in the larger towns and cities and were paid well. The dominies were local parish schoolmasters, erudite and committed, though poorly paid and housed, who played a significant role in educating and raising the expectations of the young. The women taught womanly crafts and some acquaintance with the scriptures, and it was only in the latter part of the nineteenth century, as pupil numbers soared, that women were allowed to teach core academic areas and were given access to training. Women were paid less for the same work until 1962, and had to resign when they married until 1945. So from early on there is evidence that not all teachers were the same: they were not doing the same work nor playing the same role. There are many areas where teachers differ. In this section we will explore three main variables: sector (and gender), region and post, the latter taking account of career patterns and aspirations.

Sector

While some variations in teachers' work relate to the age and stage of learners, much can also be attributed to the nature of the institutions that have evolved to provide education or schooling. There are four clear sectors in Scottish schoolteaching: nursery/pre-school, primary, secondary and special and each has its own way of working. Most teachers (95 per cent) work in the primary and secondary sectors. The size of the special sector has been relatively static, in spite of the major commitment to include children with special educational needs in mainstream schools. Over the past twenty years the pre-school sector has been growing. Day-to-day experience of work is very varied. Patterns of staffing in primary and secondary schools have been changing with the increasing involvement of teaching assistants in classrooms, some as specific supporters of children with special needs and others as extra support for teaching. The private classroom of teacher and class behind a closed door is thus increasingly unusual.

Over the last fifty years there have been major changes in the pattern of primary and secondary schooling. In the 1950s, primary school classrooms bulged with the children of the postwar 'baby boom', and class sizes were running at a level which would now be considered wholly unacceptable. Classes of fifty were not unusual, and part-time education operated in some areas, most notably in Glasgow. In larger schools, classes were streamed by ability. As mentioned above, secondary schools were differentiated by selective entry: the qualifying examination (or 'qually') dictated access to educational opportunity, as well as representing 'satisfactory completion of the primary stage'.[23] Young people were dispersed to a range of schools which offered different provision and quite different career prospects.

Variations in teachers' work by sector

While there are clear similarities in teachers' work across sectors, there are also many differences. Examples of difference discussed below highlight size and structure, gender split, the nature of the working week, generic and specialist roles, and involvement with parents. A further key aspect of difference is status. Secondary teachers were measured as having higher status than primary teachers as far back as 1954.[24] Secondary teachers were paid more, had higher qualifications, taught older (and thus more advanced) pupils and were specialists. They were more likely to be graduates and more likely to be men. All of these factors contributed to the higher status of the secondary teacher and to divisions between the two main school sectors.

Primary schools are generally smaller than secondary schools, and the experience of teachers within them is rather different. Staff numbers in primary are commonly small enough to enable teachers to operate as a single large group. Secondary school staffs, however, are larger and develop sub-groupings. They often have a tighter and more territorial structure based on subject departments, and they also have a larger and more distinct senior management team. A further and major difference between primary and secondary settings is the gender split. In Scotland, 92 per cent of primary teachers are female, while some 60 per cent of secondary teachers are female.[25]

The structure of the school week varies, with those in secondary schools working to a tight, complex school timetable, with bells ringing, teacher ownership of classrooms and much movement of learners between working spaces. In contrast, primary classes (teacher and class) have their own space and, apart from the use of shared facility space and arrangements for specialist teachers, there is considerably more freedom and flexibility in how time is used. Primary teachers are generic: that is, they teach all elements of the curriculum, although they may be assisted by specialist teachers. Most primary teachers work with one class for most or all of their teaching time. In urban areas and large schools they generally work with one age group but in smaller schools and more rural areas they may teach a class with a wide age range, for example from five to eight years, or eight to twelve years. They have responsibility for up to 1,000 hours of each child's life in a year. Secondary teachers, in contrast, commonly interact with one class of pupils for no more than one or two periods each day. They generally engage with pupils of differing ages and stages, and most secondary teachers spend their teaching time with a much larger number of learners over a week, perhaps 300–400, though some see fewer and a few see many more. This is dependent mainly on the subject being taught. Amongst secondary staff, there is also considerable variation in work undertaken. Some subjects have a larger share of the timetable and the number of different classes taught is smaller, with more time during the week with each group, while other teachers may have fewer lessons per week with more groups. Mathematics and English, for example, have a larger share of the timetable and more lessons per week. Religious Education has a smaller share in the early years in secondary, and so a teacher may teach twenty-five classes with twenty-five to thirty pupils in each. Depending on how the work of the subject department/grouping is organised some teachers may cover the whole age range of twelve to eighteen years in their teaching, while others may work only with younger or older classes. In comparison to primary teachers, responsibility for the whole learner is shared with colleagues and others with pastoral responsibility. Teaching is therefore seen to be a very diverse occupation.

Interactions with, and involvement of, parents is another dimension of difference. Parents, especially of younger primary pupils, are more likely to bring children to and collect them from school in person, although there is wide variation in practice here. For those who do so, opportunities may arise for informal and unscheduled interaction with teachers. In secondary schools, on the other hand, parents are more likely to encounter teachers in more formally arranged sessions, and there is little expectation of the involvement of parents in classroom activity. Parents are more commonly involved in helping in primary classrooms and participating in the work of the school than is the case in secondary schools. Informal parental participation is a relatively common feature in primary schools, but this is less the case in secondary schools. The culture, working relationships and working experience of teachers in schools thus vary significantly.

Robertson, writing about teachers in England,[26] argued that teachers have considerable scope, both individually and collectively, to shape events in their workplaces, but also that the teaching profession constitutes a highly stratified occupational group, separated by sector, gender and place in the organisational hierarchy, and that this stratification weakens its force as a group. Scottish teachers are little different in this regard.

Geographical variations

There are marked differences in the work of teachers in different parts of the country, as a consequence of different sizes of school and different economic situations. One difference is between urban and rural schools. The distribution of population and thus of schools in Scotland is not uniform. Urban schools are generally larger, and to an extent they operate in competition with other local schools, especially since the introduction of parental choice of school. Teachers may therefore be more involved in activities to help market their schools. Rural schools are generally smaller and more genuinely comprehensive, and they draw in most local children. Teachers in rural schools are more likely to live in the catchment area of their school and thus to be part of the community that the school serves. In small rural primary schools especially, teachers are much more likely to teach pupils of a wide range of ages, perhaps five to eight or nine to twelve, rather than a class of pupils of similar age. This alters the nature of teachers' work significantly, as do local social conditions. The number of children receiving free school meals is a commonly used (though contested) measure of social disadvantage. High levels of social disadvantage are held to militate against educational progress and attainment, resulting in more challenging situations for teachers. Clark[27] noted in 1995 that the proportion of pupils receiving free school meals varied significantly. The national average was 20 per cent, but the range varied between 6 per cent in the Borders and 40 per cent in the city of Glasgow. Teachers in their daily working lives are thus engaging with children living in very different circumstances, and the demands of teaching them will vary. The contexts in which teachers work are thus very diverse.

Post and career

Scottish schools were notable for their internal hierarchy, with a range of five different levels of post in primary school and up to seven in secondary schools. There were many more promoted posts in secondary than there were in the smaller primary schools, and in 2000, one third (32 per cent) of primary teachers and over half (56 per cent) of secondary teachers held promoted posts. However, because there were more than six times as many primary schools as there were secondary schools, 10 per cent of primary teachers were headteachers and only 1.6 per cent of secondary teachers were headteachers. A much higher proportion of primary headteachers than secondary headteachers were women. One effect of the McCrone agreement[28] was to reduce the hierarchy of posts to four (class teacher, principal teacher, depute headteacher, headteacher) and thus to change the career options available to teachers. Alongside this hierarchy, the status of 'chartered teacher' was created as an unpromoted but high-status career route for those who did not wish to leave the classroom.

How enthusiastic were teachers about seeking promotion? For a long time it was assumed that teachers, especially male teachers, would progress up the career ladder as far as they could – that they would seek promotion until stopped by failure to secure a post on the next step. However, studies of principal teachers, assistant headteachers (a role which is now defunct) and depute headteachers[29] have suggested that this is not the case. Many teachers stop by choice on various steps of the ladder, deciding either that they are happy where they are or that the next step is not enticing enough to be pursued. This has been particularly noticeable as senior posts in schools have become more managerial over the last twenty years. While there remains a good number who wish to run schools, there are also many teachers who do not see this as a career goal, but who actively prefer to stay working with learners in classrooms, where they can still determine much of what they do.

CONCLUSION

In their work, teachers have always been subject to influence from their paymasters, and for most teachers these are local authorities and the public purse. As public servants who are also professionals, their work and identities are shaped by the structures and policies prevailing in education as well as in the wider work context. The latter part of the twentieth century and the early years of the twenty-first have been times of significant change in these policies and work context, and these changes have directly affected teachers' working lives, shaping what they do. However, what they do is very diverse and it is clear that teachers continue to have scope for taking decisions about how they teach and how their careers develop, and that these choices also contribute to their sense of who they are.

NOTES

1. Growth of Scottish public sector is costing jobs, *Scotsman*, 23 January 2006.
2. Paterson, 1998.
3. Gatherer, 2003.
4. Ozga, 2000.
5. Scottish Education Department (SED), 1987.
6. SED, 1965.
7. Department of Education and Science, 1978.
8. Clark and Munn, 1997.
9. Scottish Executive Education Department (SEED), 2001b.
10. OECD, 1996.
11. Clark and Munn, 1997.
12. SEED, 2000.
13. Educational Institute of Scotland, 1999, 3.
14. Draper et al., 1991; Draper et al., 1997.
15. SEED, 2000.
16. SEED, 2000, para. 2.3, 4.
17. 17 SEED, 2000, para. 2.7, 5.
18. Kirk et al., 2003.
19. Lave and Wenger, 1991.
20. Scotland, 1969, 275.
21. Humes, 1986; McPherson and Raab, 1988.
22. Gatherer, 2003.
23. Osborne, 1966, 95.
24. Caplow, 1954.

25. Scottish Goverment Statistics. Teachers in Scotland, 2007, http://www.scotland.gov.uk/publications/2008/03/18093809/0.
26. Robertson, 2000.
27. Clark and Munn, 1997.
28. SEED, 2001.
29. Draper and McMichael, 1998; Draper et al., 1998.

BIBLIOGRAPHY

Caplow, T. *The Sociology of Work*, Minneapolis and Oxford, 1954.

Clark, M and Munn, P. *Education in Scotland: Policy and Practice from Pre-school to Secondary*, London, 1997.

Department of Education and Science. *Special Educational Needs: Report of the Enquiry into the Education of Handicapped Children and Young People (The Warnock Report)*, London, 1978.

Draper, J, Fraser, H, Smith, D and Taylor, W. *A Study of Probationers*, Edinburgh, 1991.

Draper, J, Fraser, H, and Taylor, W. Teachers at work: Early experiences of professional development, *British Journal of In-Service Education*, vol. 23, no. 2 (1997), 283–95.

Draper, J, Fraser, H and Taylor, W. Teacher careers: Accident or design?, *Teacher Development*, vol. 2, no. 3 (1998), 373–84.

Draper, J and McMichael, P. Preparing a profile: Likely applicants for primary headship, *Educational Management and Administration*, vol. 26, no. 2 (1998), 161–72.

Educational Institute of Scotland. Why solidarity in the profession is crucial, *Scottish Educational Journal* (September 1999), 3.

Gatherer, W. Scottish teachers. In Bryce, T G K and Humes, W, eds, *Scottish Education Post-Devolution*, Edinburgh, 2003, 1022–30.

Humes, W. *The Leadership Class in Scottish Education*, Edinburgh, 1986.

Kirk, G, Beveridge, W and Smith, I. *The Chartered Teacher*, Edinburgh, 2003.

Lave, J and Wenger, E. *Situated Learning: Legitimate Peripheral Participation*, Cambridge, 1991.

McPherson, A F and Raab, C D. *Governing Education: A Sociology of Policy since 1945*, Edinburgh, 1988.

Osborne, G S. *Scottish and English Schools*, London, 1966.

Paterson L. The civic activism of Scottish teachers: Causes and consequences, *Oxford Review of Education*, 24 (1998), 279–302.

OECD, *Education at a Glance*. Paris: OECD/CERI, 1996.

Ozga, J. *Policy Research in Educational Settings*, Buckingham, 2000.

Robertson, S. *A Class Act*, New York, 2000.

Scotland, J. *The History of Scottish Education*, vol. 2, London, 1969.

Scottish Education Department. *Primary Education in Scotland (The Primary Memorandum)*, Edinburgh, 1965.

Scottish Education Department. *Curriculum and Assessment: A Policy for the 90s ('5–14')*, Edinburgh, 1987.

Scottish Executive Education Department. *A Teaching Profession for the 21st Century (The 'McCrone' Report)*, Edinburgh, 2000.

Scottish Executive Education Department. *A Teaching Profession for the 21st Century: Agreement Reached Following Recommendations Made in the McCrone Report*, Edinburgh, 2001a.

Scottish Executive Education Department. *Better Behaviour, Better Learning*, Edinburgh, 2001b.

28 From Engrossing Clerks to Mandarins: Civil Servants in Scotland

EWEN A CAMERON

A brief consideration of the history of work in the government service in modern Scotland is important for four reasons. First, there is a very large public sector component to the Scottish workforce; sometimes this is adduced as one of the principal problems of the Scottish economy, reflecting a deadening dependence on the state and an aversion to a dynamic entrepreneurship which could have provided a potential salvation for the travails which the Scottish economy has encountered in the twentieth century.[1] Second, this expansion of the state has by no means been exclusive to Scotland and has involved a huge shift in emphasis over the period since the middle of the nineteenth century. In 1850 the intellectual ideal and aspiration was of a minimal state. Even, perhaps especially, in times of emergency, such as the famine conditions in the Highlands and the depressed economic state of textile towns such as Paisley in the 1840s, the task of relieving the destitute was placed in the hands of private agencies with resources generated from philanthropy. The involvement of the state, through the activities of powerful figures such as Sir John MacNeill (chairman of the Board of Supervision, 1845–68) or, in London, Sir Charles Trevelyan (Assistant Secretary to the Treasury, 1840–59), was purely administrative, and they saw their role as guardians of minimalism. Nineteenth-century administrators feared the expansion of the state not only lest populations lose self-reliance and independence but also because of the political consensus around the limitation of government expenditure and taxation.[2] The same thoughts were present at a local level: Scottish local government, outside the relatively well-developed structures of burgh government, was pretty basic. Until the establishment of elected county councils in 1889, most non-burghal and rural local government was in the hands of 'Commissioners of Supply', mostly landlords and their representatives. Since revenue was raised from 'rates' payable by the owners and occupiers of land on the basis of the value of that land, the commissioners had an obvious incentive to keep expenditure and activity to a minimum. Obviously, this contrasts markedly with the expansion of the state in the twentieth century, stimulated by the expansion of welfare after 1906, by the experience of world war from 1914 to 1918 and 1939 to 1945, and then through nationalisation and comprehensive welfare provision after 1945. Although post-1979 governments have paid lip service to the restoration of the minimal state, this has proved impossible to achieve. The third reason such a consideration is important is that the history of public administration touches on another important theme in twentieth-century Scottish history: the relationship between Scotland and the British state. This has involved a journey from a position of perceived neglect in the late nineteenth century to the current dispensation which involves a devolved parliament and a 'Scottish Executive' which have responsibility for the administration of the broad swathe of domestic policy, with the exception of social security and energy policy.[3] Until the establishment of the Scottish Office in 1885 and the steady development of its functions up to the devolution settlement of 1999, Scottish administration was set in a pattern established by the Union of 1707 and political perceptions generated in the aftermath of the Jacobite rebellion of 1745–6.[4] Since that period Scotland had undergone a social, economic and demographic revolution, but the structure of its institutions of government and administration had, by stark contrast, ossified. Scottish administrative capacity was siphoned off to national and imperial fields, while problems and difficulties mounted at home. Lastly, although the history of work has been a rich field in the historiography of modern Scotland, the experiences of white-collar workers (the focus of this chapter) and of those working in the public service have been relatively neglected, whereas the history of work in heavy industry and agriculture has been subject to extensive analysis.[5] Although the Scottish Office was at the apex of what some have argued was a considerable degree of autonomy, even before devolution, this was somewhat circumscribed. The Scottish Office, like all departments, had to plead to the Treasury for additional funds, resources and staff, and the Treasury may have been even more suspicious of the Scottish Office than it was of Whitehall functional departments with which it had a greater interchange of staff.

With these four points in mind this short chapter will examine the experience of work in the service of the state in Scotland since, roughly, the middle of the nineteenth century, although the bulk of the chapter will focus on a historical approach to the period from 1880 to 1945. In order to give the discussion further focus there will be a concentration on those who have worked for central government and for the Scottish legal infrastructure. The survival of the latter was a consequence of the guarantee of the distinctive Scottish legal system in 1707, and with the expansion of the economy, business and commerce in the nineteenth century it saw, for example, an early modern system of property transactions and land registration struggle to cope with and adapt to modern conditions. Although this naturally gives the chapter a focus on Edinburgh, where many of these workers were located, this is not exclusive of those who worked for the expanding arms of the state in other parts of Scotland, including the Highlands and Islands, where a considerable infrastructure developed to manage the tasks of agricultural support and the administration of the crofting system, including the management of estates which were in the hands of the government through the Board – later Department – of Agriculture for Scotland. This was not unproblematic for the staff who were sent to this administrative frontier. When the Congested Districts Board purchased estates in the north of Skye, they were careful to avoid using the historically loaded term 'factor', preferring the more neutral 'land manager'. This did not help the incumbent, Angus Mackintosh, when he was placed in the difficult position of having to identify to a sheriff's officer tenants who had been involved in a land raid.[6]

Although there may be an element of geographical concentration on Edinburgh, this can be defended through the recognition of Edinburgh as a 'city of government', second only in the United Kingdom to London; although prior to 1921 Dublin was also an important administrative centre. Diversity can be found in other elements of the material presented here. Even within the relatively narrowly conceived group of workers discussed, there is considerable occupational diversity: the experiences ranged from those of the engrossing clerks of the Register of Sasines – who were paid on a piece-work basis and whose early twentieth-century conditions and remuneration remained based on a statute of 1811 – to the classically educated men – and they were an almost exclusively male group – who staffed the higher echelons of the Scottish Office after 1885. A final introductory point relates to source material: although some resort has been had to the oral testimony of workers in the categories analysed in this chapter, a good deal of the evidence on which it is based has been gleaned from 'traditional' sources, such as the records of central government. It may be thought surprising the extent to which this yields information on the experiences of workers up and down the hierarchy, including the 'voice' of those near the bottom of the ladder.

A brief introduction to the structure of Scottish government is required. The creation of the Scottish Office in 1885 is the most significant event. This came at the end of a long campaign, stretching back at least to the 1850s, which sought to highlight the neglect of Scotland by the British government. For organisations like the National Association for the Vindication of Scottish Rights in the 1850s and, from 1886, the Scottish Home Rule Association, the answer to this problem was not to break the Union but to improve its operation, and administrative devolution, as it came to be called, was part of the answer.[7] The Scottish Office was located in London, at Dover House on Whitehall, for most of the first fifty years of its existence and employed very few civil servants. In Scotland there was a range of functional organisations for which the secretary (secretary of state from 1926) for Scotland had overall responsibility. Some of these predated 1885: the Board of Supervision which oversaw the operation of the new Poor Law in Scotland had been established in 1845, and the Scottish Education Department (SED; technically a sub-committee of the Privy Council in London) in 1872. The Board of Supervision was especially important, because not 'since the Jacobite rebellion, had the British government allowed a principal public department north of the Border'.[8] These functional boards proliferated in the period from

1885 to the 1920s: the Congested Districts Board (to advance land purchase in the Highlands) from 1897; the Local Government Board, which absorbed the Board of Supervision, in 1894; the Board of Agriculture (which superseded the Congested Districts Board) in 1912; the Board of Health for Scotland in 1918. These were autonomous from the Scottish Office, and their staff were appointed by patronage rather than through the system of competitive entry to the civil service that had been developed from 1869 to 1873.[9] This matter of recruitment was a major issue for the Royal Commission on the Civil Service, which issued a series of reports in the years from 1912 to 1916. Throughout the questioning of witnesses – including the most senior permanent member of the Scottish Office, the Under-Secretary for Scotland, Sir James Dodds – there was a clear sense that the commissioners were engaged in the difficult task of trying to understand a very distinctive corner of the machinery of government, and that the patronage appointees to the boards was a large part of this.[10] Dodds was defensive about the patronage appointments who ran the Edinburgh boards and explicitly tried to draw an equivalence between them and the élite civil servants who came through the highest examination in the competitive entry system. In doing this he used the evidence of their educational backgrounds.

> On the whole the men who are running the Edinburgh boards, I think, generally, are of the same type as Class I men as regards their qualifications. They are university men generally of high standing of one of the Scottish Universities; some of them have English university qualifications as well.[11]

Later in his evidence Dodds asserted the superiority of the Scottish over the English education system.[12] Sir John Struthers, Secretary of the Scottish Education Department also defended the separate identity and method of working of the Scottish administration. The SED had a separate identity even within the Scottish system; it predated the Scottish Office and often regarded itself as a superior and aloof institution. He specifically deprecated civil service generalism and argued that educational specialists, often drawn from the senior ranks of the teaching profession, were preferable to those who had emerged from the top of the competitive system.[13]

The Scottish Office was reorganised in the interwar period and began to acquire a greater physical presence in Scotland, first through the opening of an Edinburgh branch of the office at 28 Drumsheugh Gardens in 1935, and then with the opening of St Andrews House on Calton Hill in September 1939. By then, after an investigation by a committee chaired by Sir John Gilmour, a former Secretary of State for Scotland, the system had been modernised and appeared, if not more like, then at least less unlike, other Whitehall departments, although its responsibilities were territorial rather than functional.[14] Patronage had been abolished, the Secretary of State had authority over a structure of departments – initially Agriculture, Home, Health and Education. The Secretary of State retained an office with a small staff in London, but the bulk of the Scottish Office and its civil servants were not located in Edinburgh. Although further reorganisations took place in the postwar period – the creation of the Scottish Development Department in 1962, the merging of the home and health departments in the same year, and the creation of the Scottish Economic Planning Department in 1972 – the broad principles of this structure remained in place until devolution in 1999, at which time there were just over 46,000 working for the Scottish Office, agencies such as the prison service, Historic Scotland or the Students Awards Agency for Scotland.[15] The contrast with 1885, when 'leaving out messengers, cleaners and ... porters ..., the Scottish Office at the outset consisted of seven', could not be greater.[16] These Scottish Office, and later Scottish Executive, civil servants are outnumbered by their colleagues working for United Kingdom departments in Scotland: Social Security or defence, for example. An important point to bear in mind is that they are all, including those who work for the post-devolution executive, members of the home civil service: there is no 'Scottish civil service'. Indeed, in the Scotland Act of 1998 civil service matters are clearly identified as an area reserved to Westminster. Since devolution this workforce has experienced a vast increase in its workload, as the number of ministers has increased threefold and the number of parliamentarians twofold.[17]

As has been noted in the brief discussion of recruitment in the early history of the Scottish Office, there is clear evidence that those working in it at the élite level perceived its clear Scottish identity – it was seen as not just another government department which happened to have responsibility for a particular geographical area. It was embedded in the Scottish education system. This clear Scottishness of the Scottish Office went beyond its peculiarities of organisation and recruitment, and it survived beyond the modernisation of the interwar period. At the Royal Commission in 1912, James Dodds, then Under-Secretary, was asked whether there would be any obstacle to the migration of staff from 'the great public offices here in London' to the Scottish Office. Dodds replied that there would not be any obstacle but went on to qualify this point: 'Of course his experience having all been in England, and we in Scotland being accustomed to a different series of statutes in many cases, and having on many points differences from English ways of thought, *it would not be welcomed* [my emphasis].'[18] The same feeling was expressed by Sir Ronald Johnston, a senior civil servant looking back on his career in the Scottish Office from the vantage point of retirement in the late 1970s. He recounted his first entrance to Dover House, when he took up his appointment in the Scottish Office because he did not like the other offers which had come his way after his entrance examination:

> I realised on the day of my arrival at Dover House that I had been reckless, I was welcomed kindly enough by Norman Duke, Tom McQueen Walker and Bruce Fraser and billeted in room 4, but the sound of their voices showed me that I was in the wrong box. The Under Secretary for Scotland had nothing more to say to me when he heard I was English.[19]

Similarly, 'J W', a retired civil servant interviewed by this author, strongly expressed his belief in the Scottish identity of the Scottish Office based on his memory of his career, which straddled the Department of Agriculture, the Crofters' Commission in Inverness and the office of the Registrar General for Scotland, from 1949 to 1985.[20] In fact the civil servants in the Scottish Office were quite distinct from those in Whitehall departments: very few were educated at Oxford and Cambridge – Johnston was one of those – and the majority were products of the Scottish university system. Further, there was not much traffic between Edinburgh and Whitehall, the Treasury was not very successful in placing its men in St Andrews House, and few Scottish Office figures went south to serve Whitehall departments. Sir Charles Cunningham, who moved from the Scottish Home Department to become permanent secretary at the Home Office, was one of the exceptions which proves the rule.[21]

A more prosaic theme, but a crucial one in a study of work, is the sheer labour-intensity of administrative work in the late nineteenth and early twentieth century – although by the time of the Royal Commission on the Civil Service, there is some evidence of the application of the available technology, especially typewriters and rudimentary copying machines. Once again, in this matter the Scottish Office was keen to emphasise its autonomy and the fact that these new working methods were not being pressed upon them by the Treasury and the Stationery Office.[22] Any scholar working on government files in the late nineteenth century will be familiar with the multiple copies of documents in the wonderfully legible copperplate handwriting of the clerks. The massive labour-intensity of copying was most evident in the legal departments, especially the Register of Sasines. In legal departments such as this, the copying had to be done to a very accurate standard, and much of it was done by workers operating on a piece-work basis. Their story is an extraordinary one, largely hidden from history and, although an extreme example, it says much of the grim conditions for some workers in nineteenth-century administrative employment. Their job was to transcribe accurately the title to land on the occasion of every conveyance of property, and they were fined by reduction in their wages for every error they made. Further, engrossing clerks were paid 6d. per page of 200 words of handwritten transcription, a rate of pay fixed by a statute of 1811. It was not without exaggeration that the newly formed 'H.M. Register House Engrossing Clerks' Yearly Society' remarked in 1905 'that the Engrossing Clerk is a man with a grievance goes without saying'.[23] In many cases, it was alleged, the mental and physical strain (the latter taking the form of what would now be recognised as repetitive strain injury) took its toll, and 'sooner or later a man's handwriting deteriorates to such an extent that it becomes the painful duty of the Keeper to tell him that his hand is no longer fit for the Record, he will have to go'.[24] This was not an isolated complaint: there had been petitions as far back as 1884, and they would continue until at least 1909.[25] One answer that was suggested was technological: the replacement of a manuscript record by a typewritten one. The Keeper of the Register of Sasines, who did not lack sympathy with the engrossing clerks, had considered this, but produced a series of objections:

> Another consideration would be how much a Type writer could write in a day of seven hours, and would this be as much as a fast MS writer in 9 or 10 hours, the present hours of the Engrossing Staff? No doubt the speed of type writing far exceeds manual, but for how long could a young Type Writer keep it up? Can he sit steadily on for not half an hour merely, but for months and keep up a high speed? This is the kind of test I am ignorant of, and which would require to be ascertained by experiment. I am very sorry for the present Engrossing Staff who would be thrown out of employment if Type Writing were introduced, but I think that it would only be right to try how it would work and I should be prepared to accommodate say one really good Type Writer and test him with the writs of one of the smaller counties as an experiment. The information thus given would certainly be important and valuable as a guide to future arrangements.[26]

By 1915 the practical objections against typewriting were much reduced, as machines had been developed which could type directly into the bound volumes of the register, thereby reducing the worry of working with loose leaves, which could be lost with disastrous consequences.[27] The plight of the wretched engrossing clerks remained an issue when W T Ketchen, then Keeper of the General Register of Sasines, was called before the Royal Commission on the Civil Service in 1915. He did not demur from the suggestion that the engrossing clerks, even if they ought not to be employed as regular civil servants, ought to receive a pension, 'because there are a great many hardships.'[28] Another leading civil servant, Sir Kenneth MacKenzie, the King's and Lord Treasurer's Remembrancer in Scotland, was of a similar view: 'I always think the engrosser's is an extremely bad trade, and I trust this Commission is going to deal with the engrosser.'[29]

We have noted above the controversial nature of the prospective introduction of the typewriter and the potential unemployment which might result. The typewriter was controversial from another point of view: it was by the outbreak of the Great War almost exclusively operated by women, and government offices were, in the late nineteenth and early twentieth century, largely a male preserve. Nevertheless, by the 1900s the pressure of work in the Scottish Office, as in other government departments, necessitated dealing with this issue. Although women began to gain employment in the civil service from the 1880s, their role was subordinate, and the tendency for the typewriter to be at the heart of gender segregation in the workplace is a well-established theme in the literature.[30] In the Local Government Board in the early years of the twentieth century, most of the typing was done by 'girl typists', and a question of accommodation was raised. The female typists sat in a separate room from the men, a ubiquitous arrangement in the civil service,[31] and the notion of extending their duties to clerical work beyond typing was rejected because, according to the Vice-President of the Board, they 'have a different way in tackling their work.'[32] Women were not entirely confined to typing duties, but the other roles in which they were employed provide evidence for the extension of the Victorian and Edwardian concept of 'separate spheres' to the workplace. In 1935 one senior figure considered that certain types of 'menial' administrative work in the registry, such as file hunting, was 'not especially suitable for a woman'.[33] There were female workers in the census office, the post office and the prison service – the latter an unavoidable consequence of the existence of female prisoners in Scottish prisons. In a revealing comment about female inspectors of the poor, Sir George McCrae revealed classic male attitudes of the time towards the employment of women, being of the opinion that they had special qualities in matters of sensitivity and in work relating to children, in this case the children who were 'boarded out' from harmful urban environments to more healthy rural ones.[34] He remarked:

> We felt that although our general superintendents were going throughout Scotland inspecting these children, there were many things that a man could not perhaps discover, and, therefore, we were very strongly of the opinion that we ought to have a lady inspector.[35]

In the period from 1881 to 1911 the number of female clerks in the civil service expanded from 4,657 to 27,129, but they still only represented about a quarter of the total number of clerks, putting them in a very small minority of the civil service as a whole.[36] Female clerical workers had the advantage of being cheap to employ, for due to the marriage bar they were unlikely to accrue many incremental salary increases, and in the eyes of their male superiors, 'they were possessed of secondary sexual characteristics that rendered them particularly suitable for routine work. The latter argument also served to restrict the sort of work considered suitable for women.'[37] Women were employed in the civil service in temporary positions during World War I, but they were eased out after the war by the preference given to ex-servicemen and the rigorous implementation of the marriage bar.[38] This preference, extended to the widows of ex-servicemen, remained a factor as late as the mid-1930s and was even a consideration in the appointment of a cleaner to the new branch of the Scottish Office.[39] It was not until World War II that women began to enter the Scottish Office in an administrative, rather than a clerical, capacity. Sir Ronald Johnston recalled, in a comment which in itself reveals much about the male culture of the office:

> Ruth Waterfield was the first female administrator known to the SO, it was not so very long indeed that there had not been a female typist in the Office. Ruth served in turn at Dover House, Drumsheugh Gardens and St Andrews house before leaving to become Mrs Ian Bell. How far she would have got if she had not left to get married is a subject of interesting speculation, the few gaffes she made became known and were spoken about, in this she was less fortunate, or less sly than me.[40]

In this comment Johnston, as well as eliding the discrimination to which this woman was subjected, also refers to the effective marriage bar which operated in the civil service at this time; 'J W' also recalled the financial incentive offered to women to leave the service on marriage.[41] The novelist Dorothy Dunnett was another of the pioneering women in the Scottish Office during the war and, at least by her husband's account, her femininity was a constant source of amusement to her male colleagues.[42]

Although the Scottish identity of the Scottish Office was strong, as we have seen, this was not unproblematic, due to the position of the civil servants as part of the home civil service and to the structure of the Scottish Office itself. As has been noted, prior to the mid-1930s, the Scottish Office was relatively weak compared to the functional boards, and although the latter were located in various sites in Scotland, the principal site of the Scottish Office itself was in London, at Dover House. Once the functional boards had been brought into the structure of the Scottish Office, the balance of staff became remarkably skewed towards Edinburgh. In 1934 over 90 per cent of the staff worked in the Scottish capital, with only 117 being left in London.[43] The incentive for opening an office in Edinburgh was the much greater degree of contact between the Scottish people, institutions and local authorities and the Scottish Office since the experiment had last been attempted in 1912 and abandoned in 1916 because there was insufficient work to justify the commitment.[44] The occasion of the opening of an Edinburgh branch of the office in 1935 revealed a tension between the perceptions of those who worked in London and the need to expand the presence in Edinburgh. This is evident in the attempts to persuade staff to move from London to Edinburgh. The Civil Service Clerical Association was exercised by the extent to which the move was motivated by nationalist pressure, worried that 'particular regard' would be paid to the 'nationality of the staff' and required reassurance that there would be no 'differentiation by nationality' or that 'pressure of Scottish opinion' would 'block promotion of Englishmen in London and Edinburgh'.[45] The Under-Secretary for Scotland was able to provide reassurance that 'merit and suitability are, and will continue to be the determining factors for promotion', but he conceded the existence of a Scottish identity for the Scottish Office in remarking that 'while amongst successful candidates of Scottish birth or descent a proportion will naturally elect to be assigned by the Civil Service Commissioners for service in the Scottish departments'.[46] A more tangible problem was arranging the pay and conditions of those members of staff who would be required to staff the office in Edinburgh. Major problems arose when it became clear that those who agreed to move would lose the extra weighting they received by virtue of working in London. The Secretary of State, Godfrey Collins, was supportive of retention of the London weighting for those who were to transfer from London to Edinburgh, and John Jeffrey attempted to persuade the Treasury to permit the retention of this payment, especially for those outside the higher levels of the service who earned more than £500 per year. Not surprisingly the Treasury was absolutely opposed to this idea.[47] This provoked a strident response from those members of staff whom the Scottish Office hoped would staff the new branch in Edinburgh, both at an institutional and a personal level. The Association of First Division Civil Servants was up in arms, and pointed out to the Under-Secretary for Scotland:

> The constant contingency of re-transfer will involve a degree of insecurity which will mean a degree of insecurity which will mean an increased cost of living for all officers liable to transfer; moreover social conditions in Edinburgh will in themselves impose a higher standard of living in many respects than is necessary in London for administrative officers and will nullify the initial advantage, where it may exist, of relatively lower charges for a number of commodities and for certain items of expenditure ... No recruit to the Scottish Office was informed that the scale of salary of his post would be liable to a deduction of 5% in the event of transfer.[48]

The expectation of a higher standard of living is an intriguing comment, and there was a certain amount of social snobbery among the higher echelon of civil servants. Ronald Johnston records that Charles Cunningham, Secretary of the Scottish Home Department, was shocked when he summoned one of his subordinates from a dinner party late on a Friday evening to find him wearing a sports jacket and flannels, and reflected that he must keep very poor company![49] More prosaic concerns worried the rank and file of the Scottish Office staff faced with the prospect of coming to work in Edinburgh. A range of letters indicates that many of them refused on the grounds that they resented the loss of salary consequent upon departing from London, were worried that they might be disadvantaged in property transactions, and also that they would lose out in the promotion stakes.[50]

The range of roles in the Scottish Office became more diverse during the 1930s; for example, with the appointment of a press and public relations officer in 1936. This was not only designed to enhance the public profile of the Scottish Office and the Secretary of State for Scotland but was also the occasion of some soul-searching in the Treasury as to the precise nature of the responsibilities of civil servants. There was a perception that: 'It is a well recognised part of the duty of civil servants to prepare material for issue to the public through the press or otherwise, and save in exceptional cases, this work should not necessitate the appointment of a special officer for the purpose.'[51] The Treasury relented and agreed to the appointment on an 'experimental basis' for an initial period of twelve months only, but the role soon became permanent due to the pressure of work, as the press officer himself explained, possibly in the mode of self-justification, but also in recognition of the expansion of the role of the Scottish Office in the 1930s:

> The press work is heavy. Apart from the issue of notices from this office and from the depts in Edinburgh, (I arrange for issue from Edinburgh so far as possible), it involves daily contact with the newspapers to

> deal with requests for information and views. On the whole, we have what is called a good press: i.e. we get a sympathetic presentation of our news and views, and although there is – and I hope there always will be – plenty of healthy criticism, editors and correspondents tend, with growing frequency, to consult me before publishing news or articles that contain criticisms based on doubtful information.[52]

The expansion of the Scottish Office and the diversification of its role in the 1930s brought difficulties for the day-to-day operations, which required an increased level of administrative work. New responsibilities, such as the administration of the Special Areas legislation or the creation of the Scottish Special Housing Association, as well as the impact of the presence of the Scottish Office in Edinburgh, made it much more of a focus in Scottish public life. The activities of the Scottish Economic Committee and the fallout from the Local Government (Scotland) Act of 1929, which 'rationalised' (very controversially) the structure of local government, also brought extra work. This was reflected in senior civil servants working excessive overtime (sixty-five hours in one case) and the formation of a case to the Treasury for more administrative staff. The clerical staff were also under extreme pressure, and their augmentation was also urgently required.[53] Increasing legislative activity, on subjects as diverse as housing and herring, placed an equally heavy burden on the staff at Dover House, as Thomas MacQueen Walker, a senior official there, explained to the Treasury, to which he was pleading for extra support in 1935:

> Experience during the present session shows that some special provision is urgently required. The typing work involved by the Housing Bill is exceptionally heavy and numerous calls for dictation and typing (and retyping) of briefs and memoranda are made mostly at short notice entailing great strain on our typing resources. As examples of rush work I may say that on Friday last two of our typists were kept until 7.30 and one until 9.30 for work to be completed early on Saturday morning; and yesterday a 19 page Second Reading Brief required to be retyped in quadruplicate for the seventh time in 50 minutes beginning at 1.10pm, and for the eighth retype the Secretary of State's personal secretary was pressed into service at 10 pm and was here again at 8.15 this morning. The work will be still heavier in Cttee stage when notes on clauses (of which at least 15 copies will be required) will have to be prepared under pressure; and the difficulties will be increased by the work involved by the Herring Bill which will become law shortly.[54]

Thus laborious procedures and extreme labour-intensity was not confined to the engrossing clerks. In addition, the increasing political advocacy undertaken by the Secretary of State and the improved structure of public relations meant that sixty or seventy copies of an increased number of speeches had to be typed and distributed to the press.[55]

In addition to the strong Scottish identity of the Scottish Office civil servants, there was also a very powerful departmental identity. Perhaps this was a hangover from the days when their predecessors, the boards, had a greater degree of autonomy. This was perhaps most notable in the Scottish Education Department, which regarded itself as the senior department. This manifested itself in the relative infrequency with which staff moved from one department to the other and in clear identification of 'departmental' working practices. Sir Charles Cunningham told a colleague in 1935 that he would prefer an inexperienced clerk to one who would have to 'unlearn' the procedures of another department.[56]

Thus, even this brief and narrow study suggests a number of points about work in the government service in Scotland, the most important being that, although these workers were members of the home civil service and their role was no different from their counterparts in the Home Office, they were recruited in such a way and affected by the history of the Scottish Office in a manner which saw the development of a strong Scottish identity. Study of the history of the Scottish Office and other aspects of Scottish administration also touches on themes which are generic to the history of white-collar work in the late nineteenth and early twentieth centuries: two of the most prominent being the controversies surrounding the implementation of new technology and the increasing presence of women in the workforce, first in a clerical, later in an administrative, capacity. The final point to make is that the history of white-collar work has been neglected in Scottish historiography. Even such an apparently narrow subject as the history of work in the Scottish Office presents rich opportunities for further research and presents abundant source material for future work.

NOTES

1. Hutchison, 1996, 46.
2. For discussion of these points see Morton, 1998; Devine, 1988; Smout, 1979, 218–42.
3. Keating, 2005, especially 96–118, provides a good introduction.
4. Hutchison, 1996, 46–63 provides a good overview of the period up to the early 1990s; see also Parry, 1993, 41–57.
5. For example, see Knox, 1999.
6. NAS, AF42/7371, Angus Mackintosh to CDB, 7 Jun. 1910; AF42/7574, Mackintosh to CDB, 23 Aug. 1910; Cameron, 1996, 120.
7. Morton, 1996, 257–79; Morton, 2001, 113–22.
8. Levitt, 1988, xi.
9. Pellew, 1982, 20–2, 33–4.
10. Mitchell, 1985, 219–31 gives the background.
11. *Parliamentary Papers* 1913, Sir James Dodds, Q18993.
12. *Parliamentary Papers* 1913, Sir James Dodds, Q19110.
13. *Parliamentary Papers* 1913, Sir James Dodds, Qs19437, 19498.
14. Hanham, 1965, 205–44; Hanham, 1969, 51–78; Mitchell, 1989, 173–88.
15. Parry, 1999, 66.
16. Gibson, 1985, 29.

17. Parry, 2000, 85.
18. PP 1913, Sir James Dodds, Qs19058, 19123.
19. NLS, Scotland's Record, Acc 7330/21a, R E C Johnston recorded in Edinburgh, 18 August 1979.
20. 'J W' interviewed by Ewen A Cameron on 1 February 2007.
21. Hutchison, 1996, 50, 53.
22. *Parliamentary Papers* 1913, Sir James Dodds, Q19087; Sir George McCrae [Vice-President of the Local Government Board], Qs19291, 19402.
23. NAS, HH1/876, Statement of the reasons which led to the formation of H.M. Register House Engrossing Clerks' Yearly Society, February 1905.
24. NAS, HH1/876, Statement of the reasons which led to the formation of H.M. Register House Engrossing Clerks' Yearly Society, February 1905.
25. HH1/876, Unto the Right Honourable The Lords Commissioners of Her Majesty's Treasury, Memorial for the Engrossing Clerks in the Office of the General Register of Sasines, Register House, Edinburgh, May 1884; Geo Henderson, Chmn of, and as authorised by, a meeting of the Engrossing Clerks held on 10th November, 1909, to Lord Pentland, 10 Nov. 1909.
26. NAS, HH1/856, Additional memorandum by the Keeper of the General Register of Sasines on the Number of the staff necessary for the work of the office, 18 Dec. 1891.
27. *Parliamentary Papers* 1914–16, Q56529.
28. *Parliamentary Papers* 1914–16, Q56372.
29. *Parliamentary Papers* 1914–16, Q56504.
30. Shiach, 2000, 117; Anderson, 1988, 6–7.
31. Zimmeck, 1984, 903.
32. PP 1913, Qs19294–300.
33. HH1/837/2, Cunningham to Thomas MacQueen Walker, 16 Feb. 1935.
34. Gordon, 1990, 206–35; MacDonald, 1996, 197–220.
35. *Parliamentary Papers* 1913, Q19373.
36. Anderson, 1988, 4.
37. Lewis, 1988, 37.
38. Zimmeck, 1988, 88–120.
39. NAS, HH1/837/2, Letter to Oliphant (Department of Health for Scotland), 10 Jan. 1935.
40. NLS, Scotland's Record, Acc 7330/21a, R E C Johnston.
41. 'J W' interviewed by Ewen A Cameron, 1 February 2007.
42. Dunnett, 1984, 40–1.
43. *Glasgow Herald*, 14 March 1934.
44. NAS, HH1/837/1, Under Secretary for Scotland to Secretary, Treasury, 15 Dec. 1933.
45. NAS, HH1/837/1, H Thrower, Civil Service Clerical Association, Scottish Office Branch, to John Jeffrey, Under Secretary of State, 20 Mar. 1934.
46. NAS, HH1/837/1, H Thrower, Civil Service Clerical Association, Scottish Office Branch, Jeffrey to Thrower, 31 May 1934.
47. NAS, HH1/837/1, H Thrower, Civil Service Clerical Association, Scottish Office Branch. Under Secretary for Scotland to Secretary, Treasury, 15 Dec. 1933; James Rae, Treasury, to Under Secretary for Scotland, 16 Jan 1934.
48. NAS, HH1/837/2, Association of First Division Civil Servants to Under Secretary for Scotland, 22 Oct. 1934.
49. NLS, Scotland's Record, Acc 7330/21a, R.E.C.Johnston; for Cunningham's own

reflections on the history of the Scottish home department from 1939 to 1957, see Acc 7330/20.

50. NAS, HH1/837/5, Margaret Macintyre to Jeffrey 22 Mar. 1934; A Crawford to Jeffrey, 20 Mar. 1934; J. Bell (?) to Jeffrey, 24 Mar. 1934.
51. NAS, HH1/838, J H McC Craig (Treasury) to John Jeffrey, 17 Jan. 1936.
52. NAS, HH1/838, Memo 9 Apr. 1937.
53. NAS, HH1/837/7, Memo, Edinburgh Office, Pressure of Work, 22 Apr. 1936.
54. NAS, HH1/837/2, TMW [T MacQueen Walker] to Trickett (Treasury) 13 Feb. 1935.
55. NAS, HH1/837/2, TMW [T MacQueen Walker] to Trickett (Treasury) 13 Feb. 1935.
56. HH1/837/2, Cunningham to Thomas MacQueen Walker, 16 Feb. 1935.

BIBLIOGRAPHY

Anderson, G. The White-blouse revolution. In Anderson, G, ed., *The White-Blouse Revolution: Female Office-Workers Since 1870*, Manchester, 1988, 1–26.

Cameron, E.A. *Land for the People? The British Government and the Scottish Highlands, c. 1880–1925*, East Linton, 1996.

Devine, T M. *The Great Highland Famine: Hunger, Emigration and the Scottish Highlands in the Nineteenth Century*, Edinburgh, 1988.

Dunnett, A. *Among Friends: An Autobiography*, London, 1984.

Gibson, J S. *The Thistle and the Crown: A History of the Scottish Office*, Edinburgh, 1985.

Gordon, E. Women's spheres. In Fraser, W H and Morris, R J, eds, *People and Society in Scotland, Vol. II, 1830–1914*, Edinburgh, 1990, 206–35.

Hanham, H J. The creation of the Scottish Office, 1880–87, *Juridical Review*, 10 (1965), 205–44.

Hanham, H J. The development of the Scottish Office. In Wolfe, J N, ed, *Government and Nationalism in Scotland*, Edinburgh, 1969, 51–78.

Hutchison, I G C. Government. In Devine, T M and Finlay, R J, eds, *Scotland in the Twentieth Century*, Edinburgh, 1996, 46–63.

Keating, M. *The Government of Scotland: Public Policy Making after Devolution*, Edinburgh, 2005.

Knox, W W. *Industrial Nation: Work Culture and Society in Scotland, 1800–Present*, Edinburgh, 1999.

Levitt, I, ed. *Government and Social Conditions in Scotland, 1845–1919*, Edinburgh, 1988.

Lewis, J. Women clerical workers in the late nineteenth and early twentieth centuries. In Anderson, 1988, 27–47.

MacDonald, H J. Boarding out and the Scottish Poor Law, 1845–1914, *SHR*, 74 (1996), 197–220.

Mitchell, J. The emergence and consolidation of Scottish central administration, 1885 to 1939. Unpublished DPhil Thesis, Oxford, 1985.

Mitchell, J. The Gilmour report on Scottish central administration, *Juridical Review*, 34 (1989), 173–88.

Morton, G. Scottish rights and 'centralisation' in the mid-nineteenth century, *Nations and Nationalism*, 2 (1996), 257–79.

Morton, G. *Unionist–Nationalism: Governing Urban Scotland, 1830–1860*, East Linton, 1998.

Morton, G. The first home-rule movement in Scotland, 1886–1918. In Dickinson, H T and Lynch, M, eds, *The Challenge to Westminster: Sovereignty, Devolution and Independence*, East Linton, 2001, 113–22.

Parliamentary Papers 1913 XVIII, Royal Commission on the Civil Service (RCCS).

Parliamentary Papers 1914–16 XII, Royal Commission on the Civil Service (RCCS).

Parry, R. Towards a democratised Scottish Office?, *Scottish Affairs*, no. 5 (Autumn, 1993), 41–57.

Parry, R. The Scottish civil service. In Hassan, G, ed, *A Guide to the Scottish Parliament: The Shape of Things to Come*, Edinburgh, 1999, 65–73.

Parry, R. The civil service and the Scottish Executive's structure and style. In Hassan, G, Warhurst, C, eds, *The New Scottish Politics: The First Year of the Scottish Parliament and Beyond*, Norwich, 2000, 85–91.

Pellew, J. *The Home Office, 1848–1914*, London, 1982.

Shiach, M. Modernity, labour and the typewriter. In Stevens, H and Howlett, C, eds, *Modernist Sexualities*, Manchester, 2000, 114–29.

Smout, T C. The strange intervention of Edward Twistleton: Paisley in depression, 1841–3. In Smout, T C, ed, *The Search for Wealth and Stability: Essays in Economic and Social History Presented to M. W. Flinn*, Basingstoke, 1979, 218–42.

Zimmeck, M. Strategies and stratagems for the employment of women in the British civil service, 1919–1939, *Historical Journal*, 27 (1984), 901–24.

Zimmeck, M. 'Get out and get under': The impact of demobilisation on the civil service, 1919–32. In Anderson, G, ed., *The White-Blouse Revolution: Female Office-Workers Since 1870*, Manchester, 1988, 88–120.

29 The National Health Service in Scotland: Past, Present and Future

CHRISTOPHER NOTTINGHAM

The National Health Service is probably the most important institution in Scotland: broader than any church, more comprehensive than the education system, more favourably regarded than the legal system, and infinitely more respected than any political body. Its most visible agents, the doctors and nurses, attract a degree of public trust four to five times greater than the politicians.[1] Even a cursory examination of the Scottish media indicates that health issues catch the public's attention in a way that others do not even approach. In terms of its presence throughout our lives, the NHS admits no rivals. The vast majority of us are born within it and will die within it. In between, its agents attend and ease our pains, illnesses and neuroses, watch and record our biological growth and deterioration, regulate and supply most of the drugs we take, rescue us from the consequences of our more dangerous activities, tell us when to go to work and when to stay at home, advise us on how to nurture our children and, ever increasingly, try to persuade us to consume more vegetables, less alcohol and no tobacco.

Some of us buy insurance so that elements of our medical care can take place outside its jurisdiction, but even this will usually be carried out by professionals trained within, and often still working part-time for, the NHS. Moreover, very few of us are rich enough to afford the range of private care that would carry us through all the biological vicissitudes of life. Insurers are as aware as the planners within the NHS that it is in our declining years that we make the most costly claims on health and medical services, and they tailor their policies accordingly so that even the better off and those protected through their working lives by private health insurance tend to find their way back to the NHS in the end. Even when we are not using its services, the NHS is frequently a focal point for public attention, as its employees are intimately involved with the ethical issues that most divide and disturb us, such as the care of the terminally ill, abortion, the screening of embryos for hereditary conditions and stem-cell research. Even when we are not using its services, the NHS remains part of our consciousness. With its core promise to provide access to the best-quality care, free at the point of delivery, the NHS remains a source of reassurance, exempting us from anxious thoughts about what might happen if we or our relatives were to fall ill and our insurance cover prove inadequate. However, this expectation can still prove a source of anxiety. We follow the stories of the performance of the NHS carefully with thought to what we might expect in our time of need. Many aspects of the relationship between citizen and state are in question in the twenty-first century and active engagement with politics is in steady decline, but the citizen rights embodied in the right to access NHS services are as jealously guarded as they ever were.

The importance of the NHS in Scotland can also be expressed in terms of its geographical ubiquity. Its network of care matches and probably exceeds other public utilities, extending from the smallest inhabited island to the bleakest urban scheme. Hospitals, surgeries, health centres, clinics and the peripatetic professionals supplemented by mobile emergency services ensure that the individual in need is never far from its grasp. NHS24 now endeavours to ensure that professional medical advice is always available at the end of a phone.

Its importance can additionally be expressed in terms of the number of people who work within it. At the latest count, NHS Scotland employs 153,996 full- and part-time staff, around 7 per cent of the total employed population.[2] In an age when government agencies joyfully celebrate inward investment producing a few hundred jobs, the economic significance of this figure needs no emphasis. The 'Whole Time Equivalent' figure as of 30 September 2005 was 129,425, made up of nearly 55,500 nurses and midwives, with just under 40,000 in the registered grades; around 14,000 doctors, of whom 4,000 were consultants; 2,700 dentists; 8,600 allied health professionals; 3,200 technical staff; 2,900 ambulance staff; 25,000 administrative clerks and senior managers; and 12,500 works, trades and ancillary staff.[3] It employs many others indirectly – for example, the people providing cleaning, catering and laundry services, now largely contracted out – and exercises an even wider economic influence through the goods and services it buys. When one considers the sociological range of the NHS workforce, from the highest paid and best educated to the worst paid and least secure, one gets a sense of a society within a society.

Inevitably, another of the ways in which the importance of the service impinges upon popular and political consciousness is through costs. The numbers are so big as to be almost incomprehensible. Currently (2004–5) expenditure on NHS Scotland runs at £7.8 billion per annum[4] – that is, nearly 11 per cent of GDP – and represents a rise of 11 per cent on the previous year, for we are in a period of substantial expansion. It is of great significance that this represents nearly 35 per cent of what the Scottish Executive spends in total. Just under £7 billion of this is spent by the fifteen health boards, of which around £4.2 billion goes on hospitals, £900 million on community services, and £1.9 billion on family health services. Yet even this is not the limit, as the NHS has an impact on services beyond its formal boundaries. Current ambitions to provide 'seamless' care and 'joined-up' services in the health and social sectors extend NHS influence into community care. This has led to the recognition that staff are in a good position to assume new responsibilities – for example, in the field of domestic violence.

A recent health department report pointed out that 'health care staff are very often the first point of contact for women who have experienced domestic abuse', and guidelines to staff are now in place.[5] Child protection is another area in which the responsibilities of staff are being stretched beyond purely healthcare issues. A further zone of influence is geriatric care. Long-term care of elderly people has been transferred from the geriatric wards of the NHS into the local authority sector with enhanced community services and the use of private nursing homes operating under the scrutiny of the care commission. But if the NHS has lost an area of exclusive responsibility, it has gained in terms of influence. The 'Joint Futures' policy of the Scottish Executive, and indeed every white paper, emphasises the importance of co-operation between the health and community sectors.

We should also take into account the impact of the NHS on the education of future and current service employees in Scotland's universities and colleges. Under the 'Project 2000' plan, the NHS colleges were closed and nurse education transferred into the higher and further education sectors. Entrants to many other NHS professions are also now largely educated in the universities. Again, it would be difficult to see this as a diminution of NHS influence, for what has been lost is compensated for by the capacity to influence standards and curriculum in the public education sector. The new agency NHS Education Scotland ensures that those who educate future and existing staff do so in line with national clinical priorities as defined by the Scottish Executive. Recent initiatives have required educators to deal with specialist care in remote areas, waiting times, infection control, multi-disciplinary teamworking and communication technology. This agency has recently established an e-library to foster a culture of 'life long learning' among NHS professionals.[6] The NHS Education Scotland Corporate Plan 2005–6 emphasises that educators' courses should meet needs as centrally determined.[7] Therefore if one takes into account not only what NHS Scotland provides directly, but also its broader area of influence, one gets a better picture of the scope of its jurisdiction. In terms of expenditure, it suggests that health, broadly defined, must account for well in excess of 40 per cent of the total spend of the Scottish Executive.

HEALTH AND HEALTHCARE IN SCOTLAND

This chapter will attempt to address the question of the significance of the NHS for the people of Scotland in the past, now and in the future. Questions of organisation and policy will be considered, but my argument is that, although these are important, they take place in a broader context of public expectations and anxieties. The dynamic nature of health provision and health politics provides an additional complication. Change is the key to understanding health issues. Health policy has always had to, and must continue to, respond to changes in demography, technology, professional cultures and public expectation. It is, moreover, sensitive to economic fluctuations. An effective system of delivery must not only meet existing needs but must be capable of successfully responding to developing circumstances. My main focus will be on how the new Scottish institutions are currently responding to these complex challenges and how they are likely to do so in the future. The NHS in Scotland has always worked in a different way from the NHS in other parts of the UK, and the devolution settlement has provided the opportunity for an even more distinctive approach. The promise of devolution was 'Scottish solutions for Scottish problems', and given the significance of the service outlined above, it is no exaggeration to say that judgments of the whole devolution settlement will rest heavily on the ability of the Scottish Executive to demonstrate competence in coping with existing health needs while adapting to new circumstances and rising expectations. Some of these demands will be generated within Scotland, but others will not. Policy and performance will remain subject to popular judgements conditioned by developments in the UK, European and international contexts. Similarly, the Scottish Executive's options will be restricted by broader economic and political factors over which they will have limited influence.

HEALTH AND POPULAR CULTURE

One simple way of understanding the contexts in which health policy is made is to look at media coverage of health stories in Scotland. Health articles are the staple of all types of media. Every week throws up a new crop of issues. Whatever else we say about NHS Scotland, we have to take account of the fact that it now operates in a society whose interest in health amounts to something akin to obsession.

One type of article can be found in the 'lifestyle' sections of almost all newspapers and magazines. No section of the population is seen as exempt from the tide of advice; women are seen as most likely to be interested, but helpful advice is also regularly directed towards children, teenagers, the old, the middle-aged and even now young men, a group often characterised as notoriously indifferent to the health consequences of their actions. The focus is mostly on exercise and diet, and the emphasis often slips from health to aesthetics. Advice ranges from overall healthy eating and exercise plans to specific benefits to be gained from particular products. Newspaper readers cannot escape advice to ward off bladder problems with cranberry juice, keep joints supple with cod liver oil or glucosamine tablets, or reduce the threat of heart disease with broccoli, white meat and whole grains. Many purchasing decisions are now invested with health implications. Even the manufacturers of biscuits and drinking chocolate feel impelled to assure us that their products are not as unhealthy as we might have supposed. This indicates a growing association of health and consumption.

There is also discussion of the benefits of prescription drugs, for example, those designed to reduce cholesterol, inhibit appetite, improve the attention span of children and alleviate the infirmities of ageing. When the NHS was founded, one of the expectations of its supporters was that it would put an end to the trade in patent medicines and quack remedies. Health was to be removed from the market and lodged in the zone of professional expertise. An expert would determine whether an individual needed false teeth, spectacles, ointments or treatments, and these would be issued accordingly. One can only imagine the amazement of the founders if they could return to the Scotland of the present day. After fifty-eight years of free access to steadily improving care, Chinese herbalists operate in shopping centres and a large section of every Boots is devoted to patent remedies and supplements.

GPs have identified a category of patient they call the 'worried well'. Some of these might be victims of our growing obsession with, though poor understanding of, risk. At one time or another doubts have been raised about the safety of every type of food and drink. Farmed salmon, red meat, chicken, fizzy drinks, bottled water, tap water, and even apples and lettuce (because of their pesticide residues) are just a few of the products which have enjoyed their moment of notoriety. Perceptions of risk associated with health interventions also abound. The MMR (measles, mumps and rubella) panic was one notable example, but anxieties have been expressed about prescribed drugs such as antidepressants and antibiotics. Fears of diseases picked up as a result of or in the course of medical intervention, such as MRSA (methicillin-resistant staphylococcus aureus), are but one side of a contradictory popular discourse about medicine, which veers from fear and suspicion to unrealistic expectation. The latter is fed by a constant diet of stories of new treatments – often prompted by drug companies – destined to radically alter the prognosis of serious diseases.[8] The human genome project has served as a major stimulus to this type of speculative enthusiasm. Consider also the regularity with which new cures for cancer are announced. Even though many reports owe much to the competitive market for research funding and qualifications can often be found in the small print, each new report increases popular expectation.

In many accounts of the development of health services, this side of the equation is ignored. Patients, current and potential, are assigned a passive role, or it is assumed that their views can be contained within formal channels. However, while it is much easier to deal with the influence of the politicians, administrators and professionals, there is every reason to suppose that turbulent lay attitudes, reflected in and stimulated by the media, will have a major bearing on the development of NHS Scotland in the post-devolution era. In the past, when the public was more quiescent, life was easier for policy makers. Douglas Jay once offered this opinion: 'In the case of nutrition and health, just as in the case of education, the gentlemen of Whitehall really do know better what is good for the people than the people know themselves.'[9] Whether the proposition is true or not, it was once possible to act as if it was. The National Health Service as constructed in the 1940s assumed that patients would be passive. There is much to suggest that the patients, conditioned as they were by the social discipline of the war years, with vivid memories of unmet need and grateful for a new service, accepted this role. If that was the case, it certainly no longer is. Deference tends to be confined to the older generation. One of the more dramatic expressions of the new relationship is the fact that on weekend nights in accident and emergency departments in the main Scottish cities, professionals have to be protected by police officers. This is the extreme end of a broad tendency towards patient assertiveness. Professionals now recognise that the modern patient – or, at least, many of them – is a more active ingredient: questioning, conscious of rights, and deeply disinclined to take no for an answer. It is perhaps more in the role of critical consumer than active citizen that the patient should now be understood.

The precise impact of this is somewhat elusive, but we cannot understand the world of NHS Scotland, nor would those in charge be able to run it successfully, if some account were not made of the complex requirements of the modern patient. At the core of the NHS is the promise to cater for every medical 'need'. Need is a notoriously elastic concept, and in practice the ability of the NHS to contain costs has rested on the fact that professionals defined patients' needs. The GP acted as 'gatekeeper', determining whether the patient needed medicines or access to other services. The current inability, or unwillingness, of professionals to perform this role in the old manner carries important consequences. In practice, it has become difficult to make the distinction between what the patient needs and what he or she wishes.

Aspects of this new culture surface in the Scottish media on an almost daily basis. Many health stories are negative: scandals of professionals botching or exploiting their duty of care, or units failing to provide the expected service. Also familiar are stories of individuals denied treatments or 'wonder' drugs such as Zyban, Beta Interferon or, currently, Herceptin. Such stories are usually cast in the 'human interest' mould with 'heartless' politicians always apparently denying comfort to deserving individuals. A similar one-way reaction is produced by unit or hospital closures, real and rumoured. Here, too, media coverage, and public opinion, is almost invariably set against any closure.[10] Concerns about hospital services led to the victory of a single-issue MSP in the 2003 parliamentary elections.[11] In the case of other familiar negative stories, such as those involving uncomfortable waits on hospital trolleys or patients picking up illnesses while in hospital, it would be too much to expect disinterested coverage, even if some attention to the complexities of dealing with such matters might be fairer to those responsible. One final type of story to be included in this catalogue is the apparent failure of the NHS to make much impression on the long-term health problems of the poorer sections of the Scottish population. After almost sixty years of comprehensive healthcare free at the point of delivery, there are few international measurements of health in which Scotland does not perform badly. Over the past few years countless stories have suggested that poor health and premature death remain, for a large minority, an inescapable feature of the Scottish way of life. Naturally enough, all of the above negative health stories are compounded by the attention they receive from opposition parties, from the many and various unions and quasi-unions representing professionals and other employees, by patients' and carers' organisations, and the countless other groups active in the hinterland of the NHS.

There are, of course, positive stories concerned with remarkable recoveries and devoted professionals, but these do not balance the negative material. Even when the health minister recently announced that waiting-time targets had been met, he was met with the reaction that his figures were misleading and in any case things were not as good as they should be.[12] In many ways this is inevitable: routine operations taking place within an acceptable timescale with successful outcomes will always be taken for granted. 'Woman reasonably satisfied with outcome of treatment' will never make a good headline, even though all evidence and reason suggest that it would be the best representation of the daily reality of NHS Scotland. Policy makers themselves may also be contributing to the negative configuration. The constant stress on improvement, the enhanced scrutiny of professionals (often carried out in a way that implies mistrust) and the eternal escalation of 'targets', the contemporary reality of management in the public sector, imply that existing performance is inadequate. When professionals or managers point to things that are actually working quite well, they are often accused of complacency.

There are two conclusions that we can draw from the constant and intense interest in health and healthcare stories. The first is that the Scottish public is anything but apathetic. Interest in many areas of public life may be in sharp decline, but this is anything but the case with regard to the NHS. The second is the remarkable fact that, no matter how negative the stories, they have little effect on the overall commitment to the NHS. It is difficult to think of another area where a diet of stories about poor service would not produce speculation about system failure and a discussion of alternative ways of running things. Which of us when faced with a delayed train can resist a thought about restructuring the whole rail system? In health, public opinion always maintains a distinction between the service itself and the custodianship of those in charge of it at a particular moment. The recognition of deficiency produces only a strengthened commitment to core principles. This tells us a great deal about the place occupied by the NHS in the Scottish national consciousness.

NHS SCOTLAND AS AN INSTITUTION

At the beginning of the chapter the word 'institution' was used to describe the NHS. The word is appropriate, as the NHS is an institution in the fullest sense. It enshrines and reinforces a set of beliefs about the nature of social obligation and individual entitlement. It embodies the benefits of citizenship. It serves as an expression of social continuity, linking our sense of past, present and future, and offering security in a changing world. It does, of course, promise to provide a service, and if it were seen to fail in this respect the other layers of meaning would slip away, but our attachment goes beyond the utilitarian. The NHS, as all debates testify, occupies a place where sentiment and emotion are at least as important as calculation and reason. To depict it simply as a set of hospitals, clinics and surgeries, a major employer of labour, the costliest responsibility of the Scottish Government, or even as the means of delivering a vital service, would be to miss its deeper significance for Scottish society and politics. I have argued above that popular support for the NHS transcends worries about day-to-day inefficiencies or mistakes, rather in the way that strong relationships are unaffected by quarrels. Another way to describe this might be to ask what we mean when we wear those 'Defend the NHS' badges which appear at moments of crisis? The implication is of a golden past when all was well. If, however, we could be miraculously transported to a 1950s' hospital, the glow would soon disappear. We would be awakened at some desperate hour on a ward with at least thirty other patients, routinely denied information about our condition, expected to keep silent while doctors talked over us, and used as teaching material for medical students without our consent. Moreover, our stay in hospital would be three or four times longer, the pain and discomfort greater, and the treatment less likely to prove effective. It is not the day-to-day realities that campaigners defend, but something more fundamental.

Aneurin Bevan described his creation of the original UK NHS as 'a triumphant example of the superiority of collective action and public initiative applied to a segment of society where commercial principles are seen at their worst'. Its significance went well beyond health: 'Society becomes more wholesome, more serene, and spiritually healthier, if it knows that its citizens have at the back of their consciousness the knowledge that not only themselves, but all their fellows, have access, when ill, to the best that medical skill can provide.' As Rudolf Klein put it, the NHS is 'the only service organised around an ethical imperative'.[13] The interesting question is why the NHS became embedded in popular sentiment while so much else that the Attlee governments created stuttered and failed? Other nationalised industries proved easy to dispense with. The public housing sector developed and declined. Other aspects of the 'cradle to grave' security promised in the Beveridge report survive but are no longer as important to the majority of the citizens as they were to those of the 1940s. However, any attempt to make changes in the NHS has always had to be accompanied by the claim that reform was in the interests of preservation, from Margaret Thatcher's insistence that it was 'safe with us' to Tony Blair's and Jack McConnell's more recent denials that introducing private service provision undermines core principles.

Perhaps the totemic status of the NHS rests on a capacity to mean different things to different people. On the left, it remains an institutional demonstration of core egalitarian and communitarian values, a working model of an ideal society, and one that is all the more important as other social democratic creations have melted away. Yet even for the many who do not think in such terms, the NHS remains a national achievement to be proud of, and this in an age that is largely sceptical of institutions and where respect for anything connected with government is rare. Support, of course, is not purely altruistic and depends on the capacity of the NHS to meet individual needs with an acceptable standard of service, but this does not detract from the ability of the service to act as a vehicle for broader values. As I will argue below, popular support for the NHS played a very special part in producing a clear 'yes' vote in the devolution referendum. The NHS was a UK creation, but there was and is a widespread feeling in Scotland that support for its key values runs deeper here, and that this creates a possibility of a health policy in line with its original intentions and structures.

HEALTH AS POLITICS

At the centre of the debate about the NHS is the proposition that health is qualitatively different from other issues. To deny treatment to an individual in need of healthcare is seen as a very different matter from denying satisfaction of other kinds of need. The health service is regarded as a zone in which market principles should not apply. In most other areas, majority opinion accepts that richer people will have easier access to goods and services and enjoy superior provision, but this is rejected in respect of healthcare. No doubt the position is somewhat illogical, but it constitutes a political fact. Some would argue that the same principle should apply to other needs, but it is only in health that egalitarian argument commands broad acceptance. Access to healthcare should not depend upon ability to pay: need should be the only criterion determining access. Similarly, the quality of healthcare provided should be unaffected by the social or economic status of the patient, or where he or she lives.

These principles are bound to produce difficulties for policy makers. Although the NHS must be seen to run on these ethical principles, it is impossible in practice to guarantee the same treatment for every patient, given the geographical variety of its jurisdiction and the organisational complexities of delivery. Problems will always arise, such as the 'postcode' anomalies which produce such damaging publicity. Moreover, the NHS operates in a society and an economy that work on principles different from its own. Doctors and nurses have to be paid at rates which guarantee there are enough of them. Hospitals need to be built and maintained, and drugs must be bought in the international market. The NHS operates almost entirely on tax revenues, and the enthusiasm of the electorate is not always matched by an eagerness to pay taxes. There are, moreover, effective limits on the proportion of national income which any government can allocate to public services and, within the public sector itself, limits on how much can be directed towards health. In many ways the NHS is an industry like any other, with limits on overall spending and a consequent requirement that resources should be allocated in the most effective manner. Such requirements do not always sit easily with public expectations. Closing a hospital or an A&E unit might make perfect sense in the interests of the service as a whole, but the decision will not please those who live nearby. Given that 'human interest' (that is, 'individual' interest) is the guiding media principle, and that responding to need is the promise of the NHS, protesters will usually attract favourable coverage. There are good reasons why oppositions tend to be more comfortable with health issues than governments.

It is largely for such reasons that health came to assume the importance that it did in the campaign for devolution, and why the defence of the core principles of the NHS became a foundation stone for the new Scottish political identity that developed between 1979 and 1997, for during this period the majority of Scots were in a state of permanent opposition.

HEALTH AND THE CAMPAIGN FOR DEVOLUTION

In order to plot the way the NHS came to play such an important role in the politics of devolution it is necessary to go back to the 1980s. This was a period in which policy makers in all developed countries became obsessed with rising healthcare costs. The first cause for concern was demographic change; Citizens over sixty-five cost a health service about five times more than those under sixty-five, and for those over eighty-five the figure rises above ten times; and, of course, these were the sections of the population which were growing while the revenue-producing sections of the population were shrinking. A second major concern was technological development. Every year more could be done. Transplant and open heart surgery moved from science fiction to daily reality in a very short period of time. Hip and knee replacement operations were becoming routine. New drugs for cancers came along with increasing frequency and were often spectacularly costly. Sometimes the hope was expressed that new treatments would reduce costs, but examples were rare. Pharmaceutical companies became increasingly adept at appealing directly to patients as consumers in creating demand for their new and, almost invariably, more expensive products; Prozac was but one of the successes. As health budgets rose, governments were under increased international pressure to contain public spending. However, it proved difficult to make savings in an industry where labour costs accounted for around 70 per cent of the total. Health work could be organised so that the efforts of the more highly skilled were reserved for more complex tasks, but there are always limits to possible savings. In any case, absolute numbers of doctors and nurses, along with waiting times, are always used as indices of success in public battles over health.

It was not that UK health costs were comparatively high. For the first fifty years of its existence the NHS kept labour costs down by using low-paid junior doctors and even lower-paid student nurses to keep its hospitals working. The NHS was also infamous for the very low rates of pay of ancillary workers. During the 1980s, when such issues came to the fore, health expenditure in the UK ran at under 7 per cent of GDP, compared with 11 per cent in France and 14 per cent in the United States. Yet financial regulators are always more concerned with marginal rises than comparisons.

Rising costs have always been seen as a problem. The founders of the NHS had assumed that there would be an initial surge of demand created by a residue of unmet need, but this, they anticipated, would level out. It quickly became clear that this would not be the case, and the Labour government was forced into imposing charges for prescriptions, false teeth and spectacles within two years of the NHS' foundation. Demand constantly rose, and there were few mechanisms for cost control within the service. The NHS as it appeared in 1948 was less a rational design than the outcome of a battle. In essence it rested on a political bargain, sometimes described as a 'gentlemen's agreement', between the state and the doctors. Administrative structures were designed to keep politicians, both national and local, at arm's length. GPs retained their independence and contracted their services to the NHS, and within the hospital sector senior consultants enjoyed a high degree of autonomy. The price of autonomy was to accept some responsibility for cost containment. The GPs accepted the role of 'gatekeeper', regulating (in effect, rationing) access to virtually all NHS services. The consequent queues had proved an effective way of restraining cost rises in the early decades of the service, partly because of patient docility, but by the 1980s it was widely regarded as insufficient.

The Griffiths report of 1983 can be seen as a watershed between the original and the current politics of NHS.[14] Where previous reports following inquiries had been complex, informed, dull and balanced, this was snappy, glossy and argued a very specific case. Moreover, the message was in the membership. Previous experience of the NHS was positively undesirable. Roy Griffiths himself was managing director of Sainsbury's, with other members drawn from British Telecom, United Biscuits and Television South West. The report produced one memorable sentence: 'In short if Florence Nightingale were carrying her lamp through the corridors of the NHS today she would almost certainly be searching for the people in charge.'

What the NHS needed was management of the sort that enabled private businesses to control personnel and ensure that cost-effectiveness informed their day-to-day decisions. In the jargon, clinical autonomy was to be replaced by clinical governance. Margaret Thatcher moved cautiously. The values of the NHS did not appeal to her, but she could recognise a political landmine. General management and new executive structures were introduced in both England and Scotland, but it was not until 1987 that radical change was made. When the NHS was faced with a financial crisis, as it frequently was (and is), it usually proved impossible for government to resist appeals for more money. Doctors, nurses and patients in need make popular advocates, and health service crises express themselves in poignant 'human interest' stories. Just as Tony Blair was later persuaded that extra money had to be found in the wake of the winter flu crisis of 1999–2000, after stories of old people spending many hours on trolleys, so Margaret Thatcher capitulated in the face of postponed operations which had had tragic consequences. The difficulty, though, both for Blair and Thatcher, was to ensure that the extra money was spent in ways that would prevent future crises. For Thatcher, the priority became ensuring that the new management structures were fully empowered. The result was another glossy white paper, Working for Patients.[15] Alain Enthoven, an influential adviser, captured the essence. The work of the professionals within the NHS was excellent, but the 'command and control' systems of the 1940s no longer worked. Only a system of incentives, such as existed in private businesses, could push individuals in more productive directions.[16]

The Thatcher government settled on the introduction of the 'internal market'. The units of the health service were divided into providers and purchasers of services. Individual hospitals, the providers, were removed from the direct control of health boards and set up as quasi-independent trusts, with the expectation that they would compete against one another. They would be funded in proportion to the patients they attracted: 'money would follow the patient'. Efficient units would expand and others decline, and the sharper focus produced by competition would empower hospital managements. The choice of hospital was not, however, left to the patients but to their GPs, who would become purchasers on their behalf. Instead of automatically sending their patients to the nearest hospital or clinic, they would choose the one that provided the best service and represented the best value for money. They would have an incentive to choose carefully because, as fund holders with their own budgets, they could retain any surplus and use it to improve their practices.

The *Working for Patients* changes were met with almost undivided professional hostility across the UK. Doctors' campaigns in defence of the NHS were as public and combative as had been their opposition at the time of its foundation. Public opposition was assuaged to a degree by massive additional investment in the NHS and the muting of the more radical proposals. To say that the changes were unpopular in Scotland is to risk serious understatement. Hospitals duly became trusts, and some general practices took on fund-holder status, although the latter process was slower and less complete than in England. But changes were undertaken with a near unanimous lack of enthusiasm. In the first place they were seen as a step towards privatising the NHS and undermining its values. In the second they were seen as an alien imposition, entirely inappropriate to Scottish sentiments and circumstances; just another example of measures which might make sense in England but could only harm Scotland. The fact that they were wrapped up in anti-state ideological rhetoric only made things worse. The idea of state action to maintain tolerable social conditions had become so integral to the Scottish way of politics that Thatcher's attack on the state activity, as David McCrone put it, 'came to be viewed in large part as an attack on the country itself'.[17] Support for 'independence', which stood at 7 per cent in 1979, had reached 37 per cent by 1997, and Conservative party support slumped.[18] The sense of affront went across the political spectrum; even some of the dwindling band of Conservative supporters felt that Scottish traditions were being wilfully ignored. The feelings of alienation were underpinned by the contrast between the visible desolation of industrial Scotland and the prosperous south-east of England, and were multiplied by the capacity of Margaret Thatcher to antagonise the Scottish majority in almost every utterance. By the time of Blair's landslide victory in 1997, all political parties operating in Scotland, save one, had taken on a markedly Scottish identity. Thus the defence of the NHS, hostility to applying business principles to health services, and the related question of developing a health service appropriate to Scottish circumstances combined and became a key foundation stone of the campaign for devolution. As the final report of the Scottish Constitutional Convention put it, one of the first tasks of a Scottish parliament would be 'to arrange the organisation, funding and policy of health provision to deliver the sort of health service Scotland wants.'[19]

SCOTTISH SOLUTIONS?

There were thus a number of reasons to suppose that a devolved Scottish Government could deliver in the area of health. In the first place it could become the flagship policy. Given the place of health in the executive's budget and the importance that the electorate placed on health issues, nobody was going to complain if time and effort was spent on it. Moreover, it was an area in which it had a relatively free hand, at least in formal constitutional terms, and its actions could build on a broad consensus that there was scope for a distinctively Scottish approach. It was also convenient that what everyone wanted was essentially a restoration of the principles of the NHS, albeit re-energised and reapplied in the light of Scottish needs and circumstances.

Scottish difference though, did not begin in 1998. The Thatcher years were pivotal, but had there been no existing sense of Scottish distinctiveness the reaction could not have been so emphatic. Scottish differences of approach to health and welfare questions were not as obvious as they were in law and education, with their long histories of institutional separation, but there was much to build on.

As early as 1831, a major cholera epidemic found Scottish authorities 'more willing to enforce strict measures' than their English counterparts. Police turned back beggars seeking to enter Edinburgh, the sick poor were hospitalised using 'every means other than outright force' and physicians were required to report all cases.[20] In addition, Scotland's doctors had a less equivocal commitment to the principle of state service, demonstrated by the fact that Scottish teaching hospitals did not seek the autonomous status that was granted their English counterparts.[21] Pay beds were always a more peripheral issue, and indeed the proportion of Scottish patients with private insurance has always been, and remains, lower than in England.[22] We must also take some account of the fact that the monolithic myth of the NHS always concealed all manner of differences in terms of service delivery both across the UK and within Scotland. Local differences could arise as a response to unique geographical conditions or the ideas of particular individuals, such as a medical officer of health who ensured that health visitors were more numerous and more central to NHS activities in Aberdeen than in the rest of the country.[23] Whether the differences between Scotland and England rested on distinct popular attitudes is a more complex question. Claims of 'a distinct culture and a distinct set of political and moral preferences'[24] have proved easier to assert than to demonstrate, but what cannot be doubted is that if the NHS eliminated some Scottish features it did not eliminate them all. This enduring sense of Scotland's special relationship with the NHS was one of the factors which reinforced the belief of the Labour administration that took over in 1997 that it had a special mandate as regards health.

NEW DIRECTIONS

Soon after the Labour landslide in 1997, and well before devolution, *Designed to Care: Renewing the National Health Service in Scotland* announced the abolition of GP fund holding practices and an amalgamation of hospital trusts, which effectively terminated the function they were created to perform within an 'internal market'. The new large primary care trusts looked more like administrative units designed to carry out the intentions of ministers, rather than purchasing units in a market. The emphasis had shifted to collaboration rather than competition as the means of producing improvement.[25] NHS Scotland was set on a course which separated it from its immediate past and from the changes taking shape in England. While ministers insisted that they had no intention of 'turning the clock back to a time when the NHS was run by a crude command and control system', Scottish politicians showed little interest in diversity and choice, each of which were major themes in England.[26] The green paper of 1998, *Working Together for a Healthier Scotland,* identified health improvement as the new core theme. 'Good health,' it asserted 'is more than the absence of disease. It has to do with the way we live, the quality of our life and our environment.' Turning around Scotland's troubling rates of premature death was a priority. 'Restructuring the NHS as a public health organisation' was the means and 'health improvement' was the 'main aim'.

After devolution, the responsibilities of the Scottish Office were split among six main departments, of which the Scottish Executive Health Department was one. Its new minister, Susan Deacon, presented another white paper, *Our National Health –A Plan for Action, a Plan for Change* in December 2000, which ran along the same lines. The functions of existing health boards, acute hospital trusts and primary care trusts were combined and entrusted to fifteen revamped and unified boards, which came into being in October 2001. The boards were made responsible for improving population health as well as health services. Improved care standards were to be achieved by 'valuing and empowering staff and working in partnership with them to work in new, more collaborative, flexible and effective ways'. At the core was a commitment to 'restore the NHS as a national service – setting national standards of care to be delivered locally across Scotland'.

In 2003, yet another pronouncement, *Partnership for Care: Scotland's Health White Paper,* spoke of 'removing unnecessary organisational and legal barriers' and asserted that 'the public, patients and staff expect the NHS at local level to be a single organisation with a common set of aims and values and clear lines of accountability'. NHS boards were to provide 'strategic leadership and performance management', and the paper highlighted 'the important role of managers in working with clinicians to enable service change'.

In March 2003 a new health minister, Malcolm Chisholm, announced the final abolition of NHS trusts, removing the last vestige of the internal market. In subsequent legislation, health boards were required to co-operate with one another, and ministers were empowered to intervene if they failed to do this. Ministers and health boards were placed under a statutory duty to take action to promote health improvement.[27]

Decisions about the drugs and treatments to be made available to clinicians have always been at the core of NHS politics. In Scotland, formal responsibility is part of the remit of the Scottish Medicines Consortium. This agency takes note of the decisions of the National Institute of Health and Clinical Excellence in England but is not bound by them. Its decisions so far have been consistently more permissive than its English counterpart. There is acceptance of the private finance initiative, politically irresistible as it guarantees new units but transfers much of the cost to future governments, but an explicit rejection of the concept of foundation hospitals which has been embraced in England and which allows efficient hospitals to operate autonomously. This, perhaps, is seen as weakening the control that ministers can exercise. This same instinct is apparent in quality issues, where centralisation and managerialism rule. Responsibility is vested in the new agency, NHS Quality Improvement Scotland: 'We set standards, monitor performance, provide advice, guidance and support to NHS Scotland on effective clinical practice and service improvements.'[28] Improvement will flow from a firm application of clinical governance, 'on understanding the scientific evidence, the needs and preferences of patients' and 'collecting and publishing clinical performance data, assessing the clinical and cost effectiveness of health interventions, clinical guidelines, set[ting] clinical and non-clinical standards, review[ing] and monitor[ing] performance through self-assessment and external peer review'. The organisation also promises to identify and investigate 'serious service failures'.

On the issue of patient involvement, Scottish ministers have abolished the old health councils and placed the emphasis on responding to patients as consumers rather than citizens. The Patient Focus and Public Involvement Strategy, unveiled in December 2003, promised an independent complaints procedure, a new NHS Scotland forum to bring together various groups, including patients and community bodies, to identify possibilities for improvements in services, and a new Scottish health council to monitor the service's response to patients' views. The policy document *Involving People* promised a patient-focused approach in professional training, and required boards to demonstrate 'a diverse range of modern and appropriate methods for communicating with their local communities'.[29]

The activities of the Scottish parliament have reinforced the significance of health within the new Scottish politics. In its first session there were five major debates and 1,100 parliamentary questions on health issues in the first six months of 2000.[30] The health and community care committee has instituted a number of enquiries and asserted its right to look into the affairs of individual NHS trusts. It has also considered health legislation – for example, proposing changes to the Community Care and Health (Scotland) bill, published in 2003.[31]

In substantial ways, then, the devolved institutions are building on separate Scottish traditions and responding to priorities while eliminating perceived deficiencies and accommodating current demands. There is greater emphasis on central responsibility and, in spite of protestations of partnership, on establishing management structures and agencies to ensure that both the professionals and the reinforced health boards sing the tune prescribed by ministers. There are distinctive policies: the swift and enthusiastic embrace of the ban on smoking in public places, for example, which was a perfect device for demonstrating a vigorous approach to health improvement at little direct cost. There was the acceptance of the Sutherland Report, which recommended that charges for both the nursing and social care of the elderly should be provided free to the user, a useful means of asserting 'welfarist' priorities and a distinctly Scottish approach. Former first minister Henry McLeish, on the passage of the Community Care and Health (Scotland) bill, made the point explicitly: 'The satisfaction is that devolution is making a difference.'[32] There is a general impatience with anything that gets in the way of problem-solving, exemplified in one minister's statement: 'I do not think there is any contradiction in saying that more decision-making should be devolved to front line staff and saying that as a last resort if things are really going wrong and not being sorted out that government must have the powers to intervene.'[33]

Scotland has two major distinct health characteristics: a poor record of population health substantially if not entirely related to the large minority of the population which suffers economic and social deprivation; and significant geographical disparities. Population density is 67 persons per sq km, as opposed to 365 in England, with 200,000 people living on 130 islands. As a result the costs of maintaining similar standards of service are far higher in some areas than others. In the Western Isles for instance, family healthcare services cost £506 per head per year, whereas the comparable figure for the Borders is £327. Net operating expenditure per head between April 2004 and March 2005 ranged from £2,173 for the Western Isles to £1,243 in Fife, with Greater Glasgow at £1,538. So far the potential for conflict has been easy to handle, for three main reasons. First, the overall increases in UK health expenditure this century have muted competitive instincts. Second, while the comparative requirements of remote areas are great, the totals are small. The Western Isles for example, has a population of 26,260 and net operating costs of £57 million out of a Scottish budget of over £7 billion. Third, there is widespread popular support for the remote areas. The National Review of Resource Allocation in 2000, in recommendations endorsed by the Scottish Executive, placed high priority on ending the 'tyranny of distance' by accepting 'the costs of remoteness'. There are, however, ongoing disputes over proposals to rationalise hospital services in some less-populated areas.

The Scottish Executive's commitment to deal with Scotland's poor health record by addressing class differences in health outcomes presents difficulties of a higher order. Few dispute the significance of the issue. The National Workforce Planning Framework calculated that 200,000 Scottish households – 10 per cent of the total – suffered multiple deprivation, that 52 per cent of Scots lived in deprived communities and that 80 per cent of those that live in the two lowest social category areas live within the Greater Glasgow area. The connection between this endemic poverty and ill-health is now broadly accepted. Overall figures on the health of Scotland's population indicate that cancer and heart disease account for 25 per cent each of all deaths, putting Scotland at the bottom of the Western European league, and that 128 per million of Scotland's residents will die of heart attacks compared to 31 in France. However, these overall figures actually conceal a far worse situation in particular sections of the population.[34] The same class differences underlie the marked regional differences. The national average of people reporting their health 'not good' is 10.15 per cent, yet this rises to 15.64 per cent in the Greater Glasgow health board area, with North and South Lanarkshire, Inverclyde and West Dunbartonshire not far behind. Glasgow has 12.35 per cent of its citizens aged 16–74 permanently sick or disabled, as compared to the national average of 7.44 per cent.[35]

Few people would want to take exception to the aim of creating a 'health producing health service' that recognises the importance of social and economic factors and individual lifestyle choices, as well as medical intervention, in producing good population health. However the aim of bringing the health of the poorest sections of the population up to the standards of the better-off could prove impossible. The former are being targeted, but experience suggests that many may not be particularly receptive to health advice and, in any case, they will find it difficult to alter their behaviour. The experience of multiple deprivation, the sense of being at the bottom of the heap, does not dispose individuals to prioritise their own health. Given that the better-off and more healthy are likely to adjust behaviour, the net effect of NHS agencies and professionals putting out health advice might well be to increase inequalities, even in the context of an overall general improvement.

Moreover, recent research commissioned by the Public Health Institute of Scotland has indicated that Scotland's poor health record is a more complex problem than is often imagined, and improvements may only appear in the very long term. It seems that childhood experiences rather than current behaviour have a more potent influence on predispositions to adult ill-health and premature death.[36] The concern must be that a well-intentioned initiative might produce yet more material for the critics.

CONCLUSION

When the devolved authorities faced up to the task of creating a distinctively Scottish approach to the NHS they had foundations on which to build. There was a tradition of co-operation on health and social questions in the close community of the Scottish Office and its clients, and a developed capacity to identify national priorities. There were habits of co-operation between politicians, administrators and professionals, and an impatience with niceties of dogma which got in the way of practical remedies. Moreover, after the Thatcher and Major years, there was a popular sense that Scottish instincts were closer to the true principles of the NHS. However, there are limitations on what can be achieved within Scotland. Although the Scottish Government enjoys unqualified formal control over health policy, its ability to attain its more ambitious aims, such as making inroads into Scotland's poor health record, rests as much upon broad economic and social security policy as upon health policy itself, and these are reserved to Westminster. Similarly, it must be remembered that although Scottish ministers are free to allocate resources as they wish, the global total of their expenditure is still determined in the UK treasury.[37] Rising UK health expenditure and politically compatible governments in Edinburgh and London, both content with the somewhat generous Barnett formula, mean that conflict on the expenditure issue is currently unlikely, but future changes could restrict the Government's capacity to follow its priorities.

The Scottish Executive cannot, of course, simply assume that the Scottish public will automatically support any endeavour to develop distinctive health policies. John Curtice has suggested that Scots, like other citizens across the UK, do favour a 'fairly generous' social policy.[38] We can assume a degree of public sympathy for efforts to target the health needs of the poorest sections of the community, and a general conviction that policy made in Edinburgh is likely to meet needs better than policy made in London. But as support for the NHS involves a mixture of altruism and self-interest, the public is unlikely to support any policy which compromises their access to the best available healthcare. Work in the public health sphere would become unpopular if it was seen as diverting significant funds away from healthcare provision itself. A further restriction on the actions of Scottish ministers is that public judgments of what constitutes an adequate level of provision will continue to be influenced by what they know of standards elsewhere, and particularly standards of care in England. Two examples highlight this imposition on Scottish ministers. When John Reid was briefly English health secretary, he took the opportunity of a speech in Scotland to suggest that the Scottish Executive should follow more consumer-orientated policies so that his constituents in Hamilton and Bellshill might enjoy the benefits he was conferring on English patients.[39] When the Scottish Executive decided, no doubt for the best of reasons, that it would not fund the provision of the anti-smoking drug Zyban, which was available in England, it was met with banner headlines such as 'Scots denied life saving treatment'.[40] The decision was immediately reversed. The Scottish Executive is fortunate that in the present circumstances it has sufficient funds and a Barnett cushion to develop distinctive policies while delivering healthcare which is popularly perceived as being on an equivalent level to England. However, changed circumstances could produce harder political choices, and much in the future will depend on popular perceptions of the performance of the distinct models that are taking shape on different sides of the border. This point of comparison will continue to be reinforced by the professionals whose representative bodies continue to be organised on a UK-wide basis.

However, the future of NHS Scotland will also be influenced by what seem to be inevitable changes in the broader devolution arrangements themselves. The one possible future that can be ruled out with confidence is a simple reversal of the devolution project: even though the Scottish public have no particular affection for their new institutions, it seems inconceivable that they would vote to reverse the referendum decision. Otherwise there is uncertainty. As Ron Davis, former Welsh secretary, once said, devolution should be seen as a process rather than an event. The current arrangements in Scotland owe more to the political balance at the time of their creation – a Labour government in London desperate to maintain support in Scotland, and a Labour-dominated coalition in Edinburgh – than constitutional rationality. John Curtice has argued there may be some virtue in current arrangements: however logically incoherent the three devolution settlements are, they represent what the public wanted.[41] However, the anomalies, such as the 'West Lothian' question and the Barnett formula, contain a potential for disruption if they become the currency of serious power struggles. Whether this will happen is impossible to predict, but change of some sort is inevitable. Recently the Liberal Democrats announced their support for fiscal autonomy for Scotland, thereby opening the possibility of cooperation with the SNP.[42] David Cameron, leader of the Conservative party, has indicated that he is 'relaxed' on the issue: if the Scottish Conservatives want it, he is prepared to support it.[43] This, of course, could leave Labour as the only party still wedded to the world of Barnett. Richard Parry has recently detected a similar 'moving on' among the administrators: allegiances in the home civil service are beginning to fragment on territorial lines.[44] Though no one can say where the political momentum released by devolution will take us, it is certain that NHS Scotland, as the major responsibility of the Scottish Executive, would be affected.

What then can be said about the future of the NHS in Scotland? Prediction in such an area is hazardous. There is the technological dynamism of health provision. We have, for instance, just left the century when the hospital ruled as the dominant healthcare institution. Scottish patients of the future will receive much more of their healthcare in the community. Operations which thirty years ago required a three- or four-week incarceration can now be performed in an ambulatory care unit, with recovery at home supported by community nurses. The pace of technological change which has transformed the nature of healthcare during the first fifty-eight years of NHS Scotland will increase rather than slacken. While it is always dangerous to predict any decline in the power of the medical profession, new patterns of professional work are certainly emerging to cope with new demands and new modes of delivery. The simpler hierarchies of the past will not return: patients will continue to become better informed and less deferential. Most of them will be healthier and live longer, but on past experience this will not diminish the demands they make on the healthcare system.

From the political perspective it is likely that the prospects for NHS Scotland will fall somewhere between the regenerative dreams of the keenest supporters of devolution and the direst warnings of those who foresee inevitable decline. The politicians' room for manoeuvre is likely to be limited; much of their activity will be reactive, trying to match popular demands with limited resources, and this may become tougher in Scotland as the political balances shift. As a consequence the shape of the NHS may change considerably, but there is no reason to suppose that the core principles are inevitably compromised by either shaving off some services or by asking for direct payments from those that can afford it.[45] Neither paying for prescriptions nor, for instance, providing the bulk of eye care in the private sector were envisaged in Bevan's original NHS, but it would be difficult to argue that any fundamental principle has been breached by their introduction. Change is not always indicative of decay.

The politicians have a number of assets. The size of NHS Scotland makes it a relatively 'small pond', and that much easier to run. There is effective consensus on a number of broad issues, a long-standing tradition of purposeful co-operation and, perhaps above all, there are professions with long traditions of service. The arguments for maintaining a national service on a universal basis apply to England as well, but the principle may be easier to defend in Scotland. There are, of course, many forces that will continue to make it difficult to run the service in the twenty-first century: citizens who are used to consumer sovereignty in other areas of their lives will demand something similar of their health services, and obvious failures may precipitate the 'middle class walk out'. The first eight years of post-devolution NHS Scotland have been something of a honeymoon period. The politicians have been able to provide all that is on offer in England and a little more. If, as seems likely, choices do become tougher, this will present a new challenge to the Edinburgh politicians. Instead of asserting that Scottish attitudes to the unfortunate are more generous, they will be faced with the task of arguing that they ought to be and persuading the better-off to accept the costs. In short, the further development of the NHS in Scotland in the future may demand an increased engagement between the new institutions and the public. To guard the health of the population and provide good healthcare is the greatest responsibility of the devolved institutions, but in spite of the many difficulties, it could prove the means of mobilising a new sense of political community.

NOTES

1. See, for example, the ESRC report 'Social Capital in the UK', available online at: http://www.esrcsocietytoday.ac.uk/ESRCInfocentre/facts/uk/. Between 1983 and 2003, a period of some turbulence within the NHS doctors' trust ratings rose from 82 per cent to 91 per cent. Politicians were on 18 per cent in both years.
2. See Scottish Executive, 2006.
3. Information and Statistics Division, 'Overall Summary Trend (Whole Time Equivalent and Headcount) to 30th September 2005', available online at: http://www.isdscotland.org/workforce.
4. Information and Statistics Division, 'Costs 2005', available online at: http://www.isdscotland.org/isd/information.jsp?pContentID=3632&p_applic=CCC&p_service=Content.show&.
5. Scottish Executive, 2003.
6. NHS Education for Scotland. *Annual Report 2004–2005*, available online at: http://www.nes.scot.nhs.uk/FOI/class_b/.
7. NHS Education for Scotland, 'Corporate Plan 2005–6', available online at: http://www.nes.scot.nhs.uk/FOI/class_b/.
8. For an authoritative discussion of the activities of drug companies, see House of Commons Health Committee, 2005.
9. Douglas Jay in 1937, quoted in Hennessey, 1993, 132.
10. For instance, Two West Highland hospitals look like being saved from downgrading in another victory for people power in support of health services: People power saves two more hospitals from downgrading, *Scotsman*, 2 October 2004.
11. Dr Jean Turner, MSP for Strathkelvin and Bearsden.
12. How can Kerr boast about a sickly NHS?, *Scotland on Sunday*, 26 February 2006.
13. Klein, 1983.
14. House of Commons Social Services Committee, 1984.
15. Department of Health.*Working for Patients*, London: HMSO (Cm 555) 1989.
16. Enthoven, 1985.
17. McCrone, 1994, 198.
18. McCrone and Paterson, 2002.
19. Scottish Constitutional Convention. 1995.
20. Baldwin, 1999, 116.
21. For further discussion of the attitudes of the Scottish medical profession, see Dupree, 2000.
22. Stewart, 2003, 409.
23. See Curnow, J. The provision of health care in remote communities. In Nottingham, 2000, 124–37; see also Stewart, 2003, 397.
24. Paterson, 1997, 95. Jeffrey, for example, has found strong support for devolution but also that the public around the UK had 'very similar values and policy preferences' Jeffrey, 2005, 10–28.
25. See for an evaluation of the first changes, Bruce and Forbes, 2001.
26. For an interesting statement and development of this view see Stewart, 2004.
27. National Health Service Reform (Scotland) Act 2004.
28. 'The new "quality" regime in Scotland', available online at: http://80.75.66.189/nhsqis.

29. See http://www.show.scot.nhs.uk/organisations/orgindex.htm.
30. Jervis and Plowden, 2003, 34.
31. Jervis and Plowden, 2003.
32. Free care for elderly bill passed, *Scotsman*, 7 February 2006.
33. Chisholm defends NHS plan, *Scotsman*, 28 January 2003.
34. Annual Conference of NHS Confederation. Scotland in Profile.
35. 2001 census data, available online at: http://www.gro-scotland.gov.uk/files/key_stats_chareas.pdf.
36. Leon, Morton, Cannegieter and McKee, 2003.
37. There are, of course, the additional tax-raising powers passed with the devolution legislation, but it is currently difficult to foresee the political circumstances in which these could be activated.
38. Curtice, 2006.
39. Memories fire Reid's desire for top-level treatment, *Herald*, 27 January 2005.
40. *Evening Times*, 27 June 2000.
41. Curtice, 2006.
42. Lib Dems open door to coalition with SNP, *Scotsman*, 7 March 2006.
43. Cameron says Scotland can have fiscal autonomy if Tories want it, *Scotsman*, 21 December 2005.
44. Richard Parry. The home civil service after devolution, Devolution Policy Papers 6, available online at: http://www.devolution.ac.uk/Policy_Papers.htm.
45. This latter currently applies to NHS dental services and has been proposed as a method by which the modern, highly expensive cancer drugs could be provided. Nicholas Bosanquet and Karol Sikora, *The Economics of Cancer Care*, Cambridge, 2006.

BIBLIOGRAPHY

Baldwin, P. *Contagion and the State in Europe, 1830–1930*, Cambridge, 1999.

Crowther, M A. Poverty, health and welfare. In Fraser, W H and Morris, R J, eds, *People and Society in Scotland, Volume 2, 1830–1914*, Edinburgh, 1990, 265–89.

Bosanquet, N and Sikora, K. *The Economics of Cancer Care*, Cambridge, 2006.

Bruce, A and Forbes, T. From competition to collaboration in the delivery of health care: Implementing change in Scotland, *Scottish Affairs*, 34 (Winter 2001), 107–24.

Curtice, J. A stronger or weaker union? Public reactions to asymmetrical devolution in the United Kingdom, *Publius: The Journal of Federalism* 36, 1 (2006), 95–113.

Dupree, M. Towards a history of the NHS in Glasgow and the West of Scotland. In Nottingham, 2000, 138–49.

Enthoven, A. *Reflections on the Management of the National Health Service*, Oxford, 1985.

Finlay, R J. Scotland in the twentieth century: In defence of oligarchy?, *Scottish Historical Review*, lxxiii (1994), 103–12.

Gibson, J S. *The Thistle and the Crown: A History of the Scottish Office*, Edinburgh, 1985.

Hennessey, P. *Never Again: Britain*, London, 1993.

House of Commons Social Services Committee. *Griffiths NHS Management Inquiry Report*, London, 1984.

House of Commons Health Committee. *The Influence of the Pharmaceutical Industry*, London, 2005.

Jeffrey, C. Devolution and divergence: Public attitudes and institutional logics.

In Adams, J and Schmuecker, K, eds, *Devolution in Practice 2006: Public Policy Differences within the UK*, London, 2005, 10–28.
Jenkinson, J. *Scotland's Health Policy, 1918–48*, Bern, 2002.
Jervis, P and Plowden, W. *The Impact of Political Devolution on the UK's Health* Services, London, 2003.
Klein, R. *The Politics of the National Health Service*, London, 1983.
Leon, D A, Morton, S, Cannegieter, S and McKee, M. *Understanding the Health of Scotland's Population in an International Context*, Edinburgh, 2003.
Levitt, I. *Poverty and Welfare in Scotland, 1890–1948*, Edinburgh, 1988.
McCrone, D. Towards a principled élite: Scottish élites in the twentieth-century. In Dickson, A and Treble, J H, eds, *People and Society in Scotland. Volume 3, 1914–1990*, Edinburgh, 1994, 174–200.
McCrone, D and Paterson, L. The conundrum of Scottish independence, *Scottish Affairs*, 40 (Summer 2002), 54–75.
National Health Insurance Commission (Scotland). Report on the hospital and nursing services in Scotland, Cmd. 699, *Parliamentary Papers* 1920, vol. XXII.
NHS Scotland. Strategic review of health and care statistics in Scotland, available online at: http://www.healthstatsreview.scot.nhs.uk/, accessed 15 December 2005.
Nottingham, C. *The NHS in Scotland: The Legacy of the Past and the Prospect of the Future*, Aldershot, 2000.
Parry, R. The role of central units in the Scottish Executive, *Public Policy and Management* (April–June 2001), 39–44.
Paterson, L. Scottish autonomy and the future of the welfare state, *Scottish Affairs*, xix (1997), 95.
Scottish Constitutional Convention. *Scotland's Parliament, Scotland's Right*, Edinburgh, 1995.
Scottish Executive. Responding to domestic abuse in NHS Scotland: Guidelines for health care workers in NHS Scotland, available online at: http://www.scottishexecutive.gov.uk/Publications/2003/03/16658/19403, accessed 10 March 2003.
Scottish Executive. Employment by industry and public/private split Scotland and UK 1996–2003, available online at: http://www.scotland.gov.uk/Publications/2006/01/12105647/0, accessed 13 January 2006.
Stewart, J. The National Health Service in Scotland, 1947–74, *Historical Research*, 76, 193 (August 2003), 389–410.
Stewart, J. *Taking Stock: Scottish Social Welfare after Devolution*, Bristol, 2004.

30 Scottish Coal Miners

ANDREW G NEWBY

INTRODUCTION

Although coal had been mined on a small scale in Scotland since at least the twelfth century, it was the industrial revolution that instigated the mass exploitation of this natural resource.[1] During the seventeenth century, extraction of coal in Scotland was mainly concentrated along the Forth, its growth stimulated both by export and by increased domestic use.[2] It has been estimated that production in the 1690s reached 225,000 tons per annum, and expansion before 1707 exhausted many pits, especially those within reach of the major ports.[3] Towns such as Prestonpans in East Lothian, Brora in Sutherlandshire and others had developed as bases for both coal and salt works. By the start of the seventeenth century, increased coal exports allowed for notably greater importing of French and Dutch wares.[4] It is claimed that Scottish output rose by eight or ten times during the eighteenth century, making Scotland one of the 'fastest-growing-coal producing regions'.[5]

The start of the nineteenth century witnessed a shift in emphasis to the western coalfields, and output reached approximately 3 million tons by 1830.[6] The application of the 'hot blast' smelting process after 1828, invented by Shettleston-born J B Neilson, permitted the smelting of blackband ironstone, which was particularly abundant in Lanarkshire.[7] As industrialisation accelerated, production increased. By 1854 there were 33,000 miners working in 368 Scottish collieries, and their output of 7.5 million tons of coal represented more than 12 per cent of the British total.[8] With demand for pig iron on the increase, particularly for railways and general engineering projects, the 'hot blast' linked the fortunes of iron and coal in Scotland.[9] Something of a distinction could be drawn during this period between the large ironmasters of west-central Scotland – companies such as Bairds', or Merry and Cunninghame, who also held extensive coal workings – and the eastern fields of Fife and Lothian, which were overwhelmingly owned by small local concerns. Campbell has identified a further phase in the development of the Scottish coalfields, lasting from 1870 to World War I, during which time output and workforce roughly doubled and the proportion of output began to shift eastwards once more.[10]

In 1913, Scottish coal output reached a record 42.5 million tons, and 7.5 per cent of the nation's entire labour force was employed in mining and quarrying.[11] After a brief period of optimism, Scottish fields were hit hard in the years following World War I. A downturn in exports badly affected the Fife and Lothian collieries, and a lack of orders for new Clyde-built ships hit the west-central field.[12] As a result, Lanarkshire lost half of its mining workforce, Fife around a quarter and Lothian a fifth between 1920 and 1938, and production fell from the 1913 zenith to an average of about 30 million tons per annum.[13]

The most mechanised region in the British coal industry by 1939, Scottish output actually fell during World War II, but after nationalisation in 1947 a new dawn was promised. Scotland, however, would be responsible for half of the National Coal Board's deficits during the 1960s, and Payne notes that 'there was no alternative but to embark upon a programme of closures: a labour force of over 83,000 on vesting day was to dwindle to a few hundred when the mines were returned to private enterprise'.[14] According to figures published in the *Scottish Miner*, the industry in Scotland contracted from 82,700 employees in 1956, producing 21,000,000 tons of coal, to only 23,616 employees in 1976, producing 10,000,000 tons.[15] The miners' strike of 1984–5 highlighted the local importance of the industry but also the tensions which existed within it, and the end of deep-shaft mining in Scotland was so dramatic that it barely survived into the twenty-first century.

THE DAYS OF COLLIER SERFDOM

Many of those involved in the early extraction of coal in Scotland were serfs in the coal mines and salt pans, a group which has received a great deal of attention from historians. The traditional, pessimistic, view is perhaps best exemplified by Smout's assertion that these workers suffered 'degradation without parallel in the history of labour in Scotland'.[16] This has helped to perpetuate the idea that miners were a 'caste apart', and in spite of recent revisionism the 'pessimistic' view has been an enduring one.[17] Barbara Freese recently compared Scotland's mining practices of using legally sanctioned 'slave labor' unfavourably with those of seventeenth-century China.[18] Economic forces, however, seem to have ensured that the traditional image of collier serfdom did not always correspond to reality. Whatley has argued that the serfs were no 'brutish sub-sect' of lower-class society, and using evidence from the Rothes estate has demonstrated that mobility and pay could be higher than has previously been acknowledged.[19]

Acts passed in the seventeenth century prevented anyone from working as a collier, coal-bearer or salter unless they were able to prove they had been released by their previous master.[20] Landlords were then given further powers that allowed them to arrest vagabonds and force them to work in their mines or salt pans.[21] Children could also become the property of the local master through the custom of 'arling', a promise that a child would become a miner in return for a baptismal gift. At this stage, entire families could be implicated in the business of coal extraction. Often the man hewed the coal, and his wife, working as the 'bearer', took baskets of coal up to the pit-head.[22] Boys and girls, from the age of six upwards, worked either for their own parents or for hewers who did not have children of their own.[23] The scale of female participation in certain pits, however, indicates that the women were not always there simply to assist their own menfolk. In Loanhead in the 1680s and Bo'ness in the 1760s, for example, women outnumbered men by two to one.[24]

Landowners were keen to exploit the economic potential of the increased demand in coal, but increased supply could only be brought about by employing more workers: men, women and children.[25] There was, aside from steam-powered drainage of some mines, little in the technological innovations that could replace the hewing of coal by traditional tools. By the end of the eighteenth century serfdom no longer made economic sense and the practice was outlawed by parliamentary acts in 1775 and 1799, leading some to argue that the acts emancipated the coal owners every bit as much as it emancipated the miners.[26] It is likely that increasing combination among miners also precipitated the end of serfdom, something which would have important implications for the development of the notion of the 'independent collier' which prevailed in the nineteenth century.[27]

As coal production increased throughout the 1700s, Lanarkshire grew in importance to emerge as the largest single district.[28] Although much of the demand for coal in Scotland remained domestic, patterns differed throughout the country, with export to Ireland increasing – and therefore bolstering the industry in Ayrshire – and Holland declining in importance as a market.[29] Although coal did not play as large a role in the early days of the industrial revolution as it would later, it was still vital for fuelling the new steam-powered machines that were the driving force behind the burgeoning cotton- and linen-spinning industries.[30] Outwith the central coalfields of Lanarkshire, Ayrshire, Fife and Lothian, coal also played a part in local economies such as at Canonbie and Sanquhar in Dumfriesshire and Macrihanish in Argyll. On the duke of Sutherland's lands, coal was seen as a component of estate reorganisation, one of the possible ancillary occupations for those cleared from the inland straths to coastal crofts. At Brora, the workforce was augmented by imported miners in order to provide fuel for the local manufacture of tiles and bricks, although by 1825 this was no longer considered viable.[31]

In the aftermath of the Napoleonic War, thanks in part to an increase in imported (often Irish) labour, coal owners were able to enjoy relatively full employment in their collieries. Danger remained ever present, however, with the following 'melancholy incident' from Midlothian only one of many examples:

> On Monday forenoon, as Samuel Robertson was at work in Whitehill Colliery, near Lasswade, along with his son, in pushing a wagon laden with coal from the working to the shaft, it gave way, and upset, when the father was unfortunately crushed to death. He has left a widow and nine children to deplore his loss, five of whom are at present lying ill with fever.[32]

Women had all but ceased to work underground in western coalfields by the turn of the nineteenth century, and this had the effect of improving aspects of the home life of a family.[33] In the east, however, women continued to undertake extraordinary, if not brutal, feats of labour underground.[34]

Unions also began to develop, usually aiming at defending earnings and, sometimes, trying to ensure a 'closed shop' which would prevent foreign labour.[35] Peaks in union activity tended to coincide with troughs in wages, such as 1817, 1825–6 and 1837, when living standards were threatened.[36] Although the Combination Acts of 1799 had effectively outlawed union activity, these were repealed in 1824, leading to an upsurge in industrial action. Colliery workers had become stalwarts of the trade union movement by the 1880s, but their involvement in organised labour had been developing over several decades, with several peaks and troughs, including involvement with Chartism in the 1830s and 1840s.[37] The evidence of continuing quasi-masonic ritualism in the early unions suggests a natural progression from the combinations that existed before the abolition of serfdom.[38]

FROM THE CHILDREN'S EMPLOYMENT COMMISSION TO WORLD WAR I

The revelations of the Children's Employment Commission, which reported in 1842, caused consternation among the wider Scottish public.[39] In his history of the Scottish working classes, Tom Johnston highlighted several examples:

> Children of 11 years work from 5am to 5 pm, 'work all night on Fridays, and come hame at twelve in the day' – children carry one hundred weight at a time, in water 'coming up to their calves'. 'I have no liking for the work,' says one little witness at Sheriffhall, Midlothian, 'but faither makes me like it ... Another girl who can carry two hundred weight of coal on her back described how one day she was up to the neck in water. Another girl, Ellison Jack, of Loanhead Colliery, began at the age of eight to work from 2 am till 1 pm or 2 pm in the afternoon. 'I gang to bed at six to be ready for work the next morning ... I have had the strap when I didna do my bidding. I am very glad when my task is wrought as it sair fatigues.' ... Females at Arniston have been down the pits working to the 'last hour of pregnancy'.[40]

The wide dissemination of the findings of the commission in the press, however, led to improvements in the position of the miners and their families. In August 1842, the Coal Mines Act passed into law, prohibiting female labour underground and the employment of boys under ten, and enforcing inspections of conditions. There was also a general point relating to alcohol, noting that the 'practice of paying wages to workmen at public houses is found to be highly injurious to the best interests of the working classes'.[41]

As well as increasing the need for imported workers, the restrictions on labour imposed by the 1842 legislation changed the role of mining women in particular. Some felt that financial considerations outweighed their welfare, and they sought to remain underground, on occasion having to be driven away by force from the pits.[42] Although women were able to find some work in and around the collieries, their role generally became one of support to the male breadwinners.[43] This applied to daughters as well as wives, and the servitude was not only to their husbands or fathers, but also to the lodgers they took in to supplement their income.[44]

The 'paradox of Scottish emigration', noted by Devine and other historians, has been that Scotland, as an industrialised nation, may have attracted immigrants, but it also suffered from greater out-migration than all European countries other than Ireland and Norway.[45] In many respects, the mining industry in the mid- to late-nineteenth century reflected Scottish society at large, and whilst various groups of immigrants arrived as colliery labour, Scottish miners themselves left in great numbers for various destinations in the empire and beyond. These movements had long-term implications for Scottish society.

Inward migration had been a part of life in the mining communities long before the mid-nineteenth century, not only from Ireland but also with the arrival of experienced colliers from Wales and England.[46] However, a combination of circumstances came together to accelerate the growth in migrant labour after the 1840s. The new legislation meant that, in the absence of child and women labour, there was a shortage of pit workers. In addition, the expansion of the coal and iron industries in central Scotland created more opportunities for employment. Simultaneously, economic conditions in the Scottish Highlands and in Ireland, which culminated in the disastrous famines in the mid-1840s, meant that men from these areas would come to the Scottish coalfields to find work and fill the gaps created.[47] The presence of Irish labour was seized upon as a reason for the continuing low wages which brought destitution to the mining communities, especially in Lanarkshire. Assaults on blacklegs were common, and anti-Catholic rhetoric was used to demonise the Irish workers. The Irish were also blamed for starting a process of general 'degradation' amongst mining communities. In his report for 1848, mines inspector Seymour Tremenheere highlighted the poor condition of the colliers, but only attributed blame to 'natives' inasmuch as their propensity to strikes allowed imported Irish labour to enter the job market:

> They remain in the country as competitors with the Scotch population for the lower kinds of employment; and their presence is greatly felt in this respect the moment trade becomes dull. Mixed up as they are with the Scotch in the mining villages, their habits have an injurious effect upon their neighbours, and make it more difficult for well-disposed and decent families to preserve order and cleanliness about them.[48]

Tremenheere had not revised his opinion six years later, when his report blamed the Irish for the severe overcrowding and attendant cholera outbreaks to be found in the mining villages.[49]

As well as having an impact on housing, immigration began to spark trouble in the ironmasters' new Lanarkshire mines – notably around Coatbridge, Airdrie and Larkhall. Initially, this was portrayed as a clash between the 'honourable men' of the unions and 'degraded slaves' among non-unionised Irish.[50] Although the Irish would, in a short period of time, become intimately connected with the labour movement themselves, the issue of religious strife proved a long-term, and more intractable, problem.[51]

The problem of housing congestion was one that greatly exercised contemporaries. Although the legislation of 1842 may have given some grounds for optimism, inadequate housing remained a feature of mining communities. As Campbell has noted, the areas in which the iron industry expanded were the least able to house the expanding population. This resulted in the rapid construction of many single-storey dwellings close to the new collieries and ironworks.[52] Whilst the provision of housing could be portrayed as patrician benevolence on the part of the coal owners, it was often perceived as a 'source of social control – and hence of growing class antagonism'.[53] Laslett describes the Lanarkshire rows thus:

> The rent in these miners' rows was low, sometimes no more than fifty pence per month, but it was deducted automatically from the miner's pay packet, and he and his family could be – and often were – evicted for breaking the company's rules or for going out on strike. In fact, the threat of eviction became one of the most powerful weapons in the mine owners' arsenal ... Usually, miners' rows in Lanarkshire coal towns consisted of one- or two-room houses set end-to-end and built around a central yard containing an ash pit, a water pump, and a communal toilet known as a 'middens'. Sometimes larger two story apartment buildings were erected, as in the Craigneuk area between Wishaw and Motherwell, where miners and low grade ironworkers lived side by side. In most cases the single-story rows had dirt floors, bare whitewashed walls, and almost no windows. The middens were supposed to be emptied daily by an agent of the coal company, but they rarely were. As the mining population increased, the miners' rows became fearfully overcrowded.[54]

The overcrowding caused by an influx of labour was exacerbated by market forces, with lodgers being taken in to supplement incomes, or indeed serve as a primary income for many landladies. In 1842, the Children's Employment Commission heard how some families in Coatbridge, with two-roomed houses, took in fourteen single men as lodgers.[55] By the 1870s, miners' housing was considered such an urgent social problem that the Conservative *Glasgow Herald* published a series of articles on the subject, and union leaders such as Alexander MacDonald sought to keep the issue in the public eye.[56] As Campbell notes, however, there were considerable regional differences: 'In Ayrshire even in the older hamlets, the [*Glasgow Herald*] reporter did not find "anything approaching filth and squalor of several districts within twenty miles of Glasgow" ... The best housing was in Midlothian, while in Fife and Clackmannan the internal condition of the houses was "better kept than in similar dwellings in the west."'[57] Lodging was still a feature of mining life by 1901, when Kellog Durland was making his investigations in Fife. Durland's comments, however, serve as a reminder that there were great differences between the various mining villages through time and place, with his assertion that 'squalor' was not an adjective that could be used for the rows of Kelty:

> It was a typical miners' house, one of a brick row with triangular roofs. There was a parlour and a kitchen on the ground floor and an attic above. When we were all at home there was little spare room as, together, family and boarders, we made up a company of fourteen ... The place was recommended to me as a good representative miners' home, and when I called upon the mistress she was perfectly willing to take me in, but did not make mention of the fact that I was to have a few room-mates. It was not until well into the first night that I learned that five of us were sharing the attic. As to the terms, they were the same as if I had had the room to myself. 'I charge twelve shillings, but them that wants to gie's me thirteen,' my landlady had said to me. This was the average price in the village. The twelve shillings included board, washing, mending and any incidentals that might be needed. In some places thirteen shillings was charged but that always included black twist tobacco and clay pipes. Since then I have been in a good many houses and I have every reason to believe that it was thoroughly representative of its class.

There seems little doubt that mining families, many of whom came from rural districts of Scotland and Ireland, retained an attachment to the land beyond the mere fact that their livelihood depended on Scotland's geology.[58] Anti-landlordism as well as hunger can be attributed to the Irish immigrants' practice of hunting game on lands belonging, for example, to the duke of Hamilton, but poaching in this manner had already been undertaken for generations by native Scots miners.[59] Similarly, many rural superstitions were brought over to the new industrial context in which many workers found themselves.[60] The land was also one of the stimuli behind migration from Scotland to mines overseas. Although systematic recruitment of men for American mines was widespread in the 1860s, many of these emigrants moved on quickly to become owners of their own farms in the New World.[61]

The general point relating to temperance made in the 1842 Coal Mines Act provided a rare meeting of interests for trade unionists and coalmasters. The popular image of miners, as with dockers and other labourers, was of men who worked hard and drank hard.[62] The consumption of alcohol was partly related to workplace customs, as noted by the temperance reformer

Figure 30.1 Miners' houses, Lumphinnans, Fife, 1925. Source: Courtesy of the Scottish Mining Museum Trust.

John Dunlop in 1839.[63] On the other hand, miners' paydays often witnessed alcohol-fuelled mayhem. In Lanarkshire:

> there was approximately one public house for every twenty adult males in the Airdrie and Coatbridge area, and on pay nights the 10,000 or so people who flocked into Airdrie from the surrounding mining villages totally overwhelmed the town's five man police force and 'scenes of uncontrolled license ensue which there are no means of either preventing or punishing.'[64]

In spite of restrictions on where miners could be paid, their perceived predilection for strong liquor continued to cause concern amongst union leaders, who considered alcohol to be one of the major tools of social control in the coalmasters' armoury. When the Miners' National Association was founded in Leeds in 1863, William Brown called for the men to be 'sober and industrious' in order to gain respect, and to save money in the event of strikes.[65] Although Tremenheere, in his report of 1854, had identified some progress amongst proprietors, notably Messrs Baird and Messrs Houldsworth, in helping to break the 'pernicious habit' of their employees' drinking, riots still broke out with alarming frequency.[66]

A slightly different perspective can be found in the writing of Kellogg Durland, who took a daytrip from Kelty to Inverness in 1901 as part of a group of 3,000 Fife miners and their families. Much to his surprise (an indication of his own middle-class prejudices), Durland witnessed a day full of drink-induced merriment and bonhomie, rather than violence.

I had been warned that these excursions were frightful crushes, always very late, and that excessive drinking was the most pronounced feature. So, with a stout heart, prepared to endure all manner of discomforts, I joined the throng awaiting the train at Kelty station ... The singing lasted all the way to the Highlands and back, and at night when the trip ended ... I was able to testify that drunkenness was conspicuous for its absence. Drinking there had been, but for a trip of that kind there was little intoxication.[67]

In general, however, it seemed that the temperance advocates, be they trade union leaders or Christian missionaries, were fighting a losing battle.[68] Drink remained a part of everyday life. Efforts to control alcohol's deleterious effects on society were made, for example, through the development of Gothenburg pubs in many mining districts.[69]

By the 1880s the Irish in Scotland, whilst maintaining a political distinctiveness, had become a firmly established emigrant grouping, and indeed generations of Irishmen had by then been born in Scotland.[70] In 1888, when miners' agent J Keir Haride stood for parliament, the support of the Irish in mid Lanark was keenly sought. Many miners of Irish descent were by now fully involved in trade unionism, but in the midst of the Irish Home Rule crises of the 1880s onwards, Scotland witnessed an influx of a new group of workers. Between then and the outbreak of World War I, approximately 4,000 of the 7,000 Lithuanians who had come to the United Kingdom, initially escaping the repression of Tsarist Russia, had settled in Lanarkshire, especially around the Carnbroe mines of Merry and Cunninghame.[71] In some respects the Lithuanians suffered the same reaction as had the Irish in the early 1800s – hostility from 'natives' (notably from Hardie himself) and accusations of being strike-breakers.[72] Before long, however, many Lithuanians had also become involved in the Scottish labour movement. As a practical step, and as an acknowledgment of their strong presence, the Miners' Federation of Great Britain began to print its rules in Lithuanian and allowed the immigrants full union benefits.[73]

In spite of the increase in urgency and organisation brought about by Hardie and his colleagues, underperformance on the electoral front remained a problem. Hardie blamed the 1906 election disappointments on the complacency of the Mining Federation, and as Hutchison has noted, 'the miners – potentially the most important body – were either organisationally inept or politically indifferent for almost the entire pre-war period'.[74] The impact of World War I on the Scottish coalfields was far from uniform. The Lothian and Fife fields were hit by the closing-off of formerly profitable Baltic and German markets, as well as the requisition of the Forth coal ports by the admiralty.[75] On the other side of the country, the Clyde basin's economy boomed, with its heavy industry becoming the engine of the British war machine.[76] Enlistment for the war deprived the mining industry of more than a quarter of its workforce, causing a reduction in productivity.[77]

For many in mining communities, and for workers throughout Scotland, the most pressing post-World War I issue was the much anticipated provision of 'homes fit for heroes'. The Royal Commission on Housing in Scotland reported that working-class housing in Scotland was 'much, much worse' than that of England, showing little, if any, improvement since the 1850s.[78] Shortly after the publication of this report, in March 1919, the Sankey Commission condemned contemporary mining practices and called for the nationalisation of the coal industry.[79] Many miners still viewed tied housing with suspicion, a reminder of the days of serfdom when the coalowner controlled almost all aspects of their lives.[80]

Some of this suspicion is encapsulated in the assertion by miner Bob Selkirk that his political beliefs and reputation as a trade unionist led to difficulties in finding both employment and accommodation. Against the backdrop of 'Red Clydeside', the growth of socialism and communism in mining communities was a constant source of anxiety to the authorities and coal owners:

> Before leaving the Lothians in 1924, I was promised a house in Cowdenbeath. When my flitting arrived at the Railway Station, I called on my prospective landlord only to be told he had decided not to give me the house. My wife and three children had also arrived and we only got a roof over our heads when late that night, Annie Storione, who lived in an already overcrowded house in Lumphinnans Rows, gave me one of her two rooms.[81]

Thanks mainly to investment in new technology, productivity in Scottish pits was well above the national average during the interwar period. Notwithstanding this improved productivity, in hindsight it is clear that heavy industry was in terminal decline.[82] Employment in the collieries continued to fall, from 155,000 in 1920 to 81,000 in 1933, as new, rival energy sources – oil, gas and electricity – made their presence felt.[83] Although the Sankey Commission's call for nationalisation had shocked the coal owners, the government in fact did not take up the recommendation and, having overseen the industry since the war, handed it back in 1921. Having regained control, the owners bemoaned wage levels, which they claimed had risen out of control, and demanded that they be reduced, precipitating a three-month lock-out.[84]

The lock-out saw many clashes between blacklegs and police, but eventually the miners returned to work after support from other unions waned.[85] Although the lock-out prefigured an even more serious industrial dispute in 1926, some optimism was engendered by the 1922 election, in which Labour made gains in mining towns such as Airdrie, Coatbridge, Hamilton and Motherwell, as well as parts of Fife.[86] Yet, whilst the French occupation of the Ruhr increased coal prices and wages for Scottish miners, the re-entry of Great Britain into the gold standard in 1925 hit heavy industry, forcing up the price of exports.[87] When coal owners attempted once more to reduce wages, the TUC announced it would stand with the miners, leading the Conservative administration to grant a subsidy to bring wages back to their previous levels. A temporary success, the government announced that such a subsidy would only last for nine months, and it established a commission under the leadership of Sir Herbert Samuel to inquire into the condition of the industry. Although the Samuel Commission (reporting in March 1926) did suggest some improvements to the working conditions of miners, its immediate recommendation was that wages be reduced. On top of this, the owners demanded a longer working day.[88] With the battle-cry of 'not a penny off the wages, not a minute on the day', the miners entered into a stand-off with the employers, leading to a lock-out on 1 May and a general strike, brokered by the TUC, which lasted for just over one week. The miners remained out, however, and it was not until October 1926 that dire hardship started to force men back to the pits. By late November, most had returned, and the result was a bitter defeat: longer hours and lower wages.[89]

The strike and its aftermath had a huge impact on Scottish miners, with Fife and Midlothian witnessing some of the most violent incidents of the whole dispute.[90] William McIlvanney wrote of the effect he believed that the strike had on his fathers' generation, suggesting that 'the General Strike of 1926 was for him what Midian was for Moses. He was a miner at the time and the pain of that cataclysmic event never fully left him.'[91] T C Smout, similarly, considers the failure of the strike to have had a devastating impact on 'socialist idealism'.[92] As A J P Taylor wrote, however, in discussing the British coal industry in general, the victory was a pyrrhic one for the coal owners, sealing the eventual destruction of what he termed 'the least worthy element in the British community.'[93]

The 1926 reverse was quickly followed by the Great Depression, leaving over a quarter of Scottish miners unemployed and highlighting the value of recently instigated welfare provisions.[94] The Miners' Welfare Fund, established in 1920 as a result of recommendations made in the Sankey Report, supported recreation, health and education in the mining districts from a 1d. levy on every ton of coal produced.[95] The money raised was distributed by local welfare funds, with additional money being raised by public subscription. Communities throughout Scotland benefited from the scheme, building facilities such as libraries, putting and bowling greens, tennis courts, and bandstands in which colliery bands could perform. As well as leisure facilities, the welfare fund helped establish practical improvements to miners' conditions, notably baths at pits around the country. Football teams had long been a staple of mining communities, with clubs such as Cowdenbeath drawing crowds and playing at a level which nowadays seems incomprehensible.[96] The specific contribution made by the welfare organisations is remembered in the names of teams like Lochore Welfare (1934) and Whitehill Welfare (1953), which still survive and, at times, flourish.[97]

As well as more organised activities, miners continued to enjoy what had been traditional pursuits. John McArthur, a prominent figure in Fife radical politics, spoke of his childhood in Buckhaven in the early twentieth century, and of how the miners 'went in for football, cycling, quoits, and running, tended their gardens and kept dogs'.[98] Making the most of the environment surrounding the pit villages is also a prominent theme in memoirs, the attachment to land noted amongst the nineteenth-century mining families still very much in evidence in the twentieth. John McArthur's recollection of his early years presents an almost idyllic image of running on the beach, scrambling on rocks and bird-nesting.[99] In another part of Fife, Mary Docherty – who, like McArthur, would grow up to be a leading communist in the 'kingdom' – was also exposed to the natural world in her youth.

> My father took us different roads sometimes coming out at Crossgates other times nearer Cowdenbeath. All the way he would point out things that were of interest, tell us names of birds and the names of all the wild flowers and trees. At a certain time of year he would say 'Stop and listen' and sure enough we would hear a cuckoo in the distance singing 'Cuckoo! Cuckoo!' Another bird call I remember was that of the pee-weep. My father would stop and we would listen to the bird and he would point it out to us.[100]

The pre-existing 'industrial militancy' of the miners, along with the prevailing economic climate, tended to harden political attitudes in some of the mining districts. In 1935, the communist William Gallagher was elected MP for West Fife, his headquarters based in the 'Little Moscow' of Lumphinnans.[101] At Polmaise, east of Stirling, the ten-month strike of 1938 was the longest action at a Scottish pit prior to 1984–5.[102] There are also stories of Scottish miners joining up with the International Brigades against Franco, something the country could scarcely afford three years later when war broke out against Nazi Germany.[103] Miners were essential workers on the home front, with World War II temporarily halting the slow death of Scottish heavy industry.[104] As Duncan notes, however, 'the Scottish coalfield had [by 1939] become the most heavily mechanised region within the entire British mining community; and it was also the most dangerous in which to work'.[105] Therefore, the pre-war tradition of radicalism persisted, and even deepened, meaning that between one half and three-quarters of all strikes in British mining between 1939 and 1945 took place in Scotland. Most notable were strikes of over 10,000 men in response to miners being prosecuted for illegal strike action in 1943 and, alongside this, demands for nationalisation grew.[106]

The immediate postwar period, and the attendant Atlee administration, oversaw one of the most significant events in the history of mining in Britain: the nationalisation of the industry in 1947.[107] Not only did this signal the death-knell for the long-declining aristocratic control of the mines, it was believed that, given the right supervision, mining – especially in Lothian and Fife – had a bright future.[108] In retrospect, these plans were never likely to come to fruition, with many pits coming to the end of their working lives, particularly in Lanarkshire, and consumers increasingly seeing electricity, oil and gas as superior alternatives for their energy requirements.[109] By insisting that electricity was produced by coal power and providing subsidies for the National Coal Board, the government attempted to ensure a future for coal mining in Scotland. As Finlay has argued:

> Most governments offered some type of subsidy to the coal and energy industries, yet what is striking about the Scottish case is the extent of this. Had it not been for the policy of tying the electricity industry to the supply of coal so that by the early seventies it was taking about 40 per cent of total production, the Scottish coal industry would have probably died out in the late fifties.[110]

Nevertheless, the official view post-nationalisation was one of tremendous optimism for coal. With Lanarkshire in apparent decline, efforts were made to relocate skilled workers from these collieries to Fife, with the new town of Glenrothes a primary destination. It is significant that the NCB's own *Short History of the Scottish Coal-Mining Industry*, published in 1958, features a full-colour plate of the brand new Rothes colliery opposite the title page.[111] The promotional video, *New Day*, produced in 1959 to showcase the Rothes project and encourage people to settle in the new town, demonstrates the great faith that was placed in new technological advances having the ability to revive the coal industry.[112] Alongside the obvious attractions of the locale – convivial pubs, fresh new housing and schools, days at the beach and watching Raith Rovers – the commentary gives a perfect example of the 'white heat of technology' zeitgeist: 'It's odd to think that the first machines in industry were attacked by howling mobs of frightened men ... it was too early to see that the machine marked the end of an era, an era of back-breaking, muscle-tearing, bitter physical labour.'[113]

Rothes, however, was dogged by geological problems and failed to hit a seam.[114] Although £200 million was invested in 1958 in efforts to produce cost-effective coal, this effectively came to nothing and productivity was maintained only by a programme of drastic pit closures: by the late 1960s only 47 collieries remained active in Scotland, a decline from 166 in 1958. The number of workers employed in the industry almost halved over the same period.[115] Much of this was because of rationalisation and the consolidation of smaller collieries into new 'super-pits': Fife saw the creation of Seafield, in Kirkcaldy, alongside Rothes. Elsewhere, Bilston Glen and Monktonhall opened in Midlothian, and Killoch was established in Ayrshire. In a stinging critique of the post-nationalisation industry, Alex Conner wrote of the failure of the planned migration schemes, highlighting worries that the NCB were prepared to write off Lanarkshire in favour of Fife – as the 'eldorado of the industry' – and claiming, presciently, that more problems lay ahead.[116] If the new technology was intended to improve both profitability and safety, its success was, at best, questionable. In September 1967, nine miners were killed in a fire at the redeveloped Michael Colliery near East Wemyss, and five men perished after a roof collapsed at Seafield in 1973.[117]

Archie Campbell, of Kelty, summed up what many miners felt about the industrial disputes of the early 1970s, stating: 'The '72 and '74 strikes were hard, but we won and, I'll tell you, we won handsomely. Tae think we had tae wait a' that time tae get a decent rise on our wages.'[118] Just as the 1926 strike sowed the seeds of the coal owners' destruction, however, so the apparent victory of 1974 would prove costly for the miners. Margaret Thatcher saw the unions bring down Edward Heath's government and resolved never to let it happen again.[119] Several years of tension between unions and government followed Thatcher's election as prime minister in 1979, until the final crisis was reached in 1984.[120]

Devine has summarised the strike and its consequences in the following terms:

> The miners' strike in 1984 ended in complete victory for the government and the humbling of the once-all-powerful National Union of Mineworkers. But the miners' cause evoked great sympathy in Scotland, despite the uneconomic nature of much of the industry, and the government was accused of seeking to achieve efficiency while indifferent to the resulting human costs.[121]

As in other parts of Britain that were affected by the strike, however, this was more than a battle between the rulers and the ruled: bitter divisions re-emerged within the ranks of the miners themselves, and there was tension with other groups of workers, especially the steelworkers of Ravenscraig.[122] Having already seen several pits threatened with closure, notably Polmaise, organisation in Scotland was slightly in advance of the rest of the UK, with men like Michael McGahey (deputy leader of the NUM) to the fore. This was also an occasion when women, always a vital part of the mining community, once more made their presence felt in a direct way.[123]

The BBC's reporter in Scotland, Colin Blane, recalled the events of the strike on its twentieth anniversary:

> In the course of the year, there was also hardship and despair. I met strikers who had sold cars and houses and strike-breakers who had gone back to work – some broken, some defiant. It was a battle of ideas. These were the early years of Margaret Thatcher but many in Scotland were still resisting. The miners' union had defeated governments in the past; this time Mrs Thatcher was ready to take them on. When the strike ended I went to Frances colliery in Fife to watch the men march back to work. It was an emotional occasion made even more difficult for those returning because more than 200 miners across Scotland had been sacked during the strike.[124]

At the end of the bitter struggle the industry faced a situation in which, less than two decades later, all Scottish deep mines would be closed. In the 1980s, the number of active pits fell from fifteen to two. In the wake of the strike, West Lothian lost Polkemmet, and Fife saw Comrie, and Seafield/Frances, the great hopes of the 1970s, closed down.

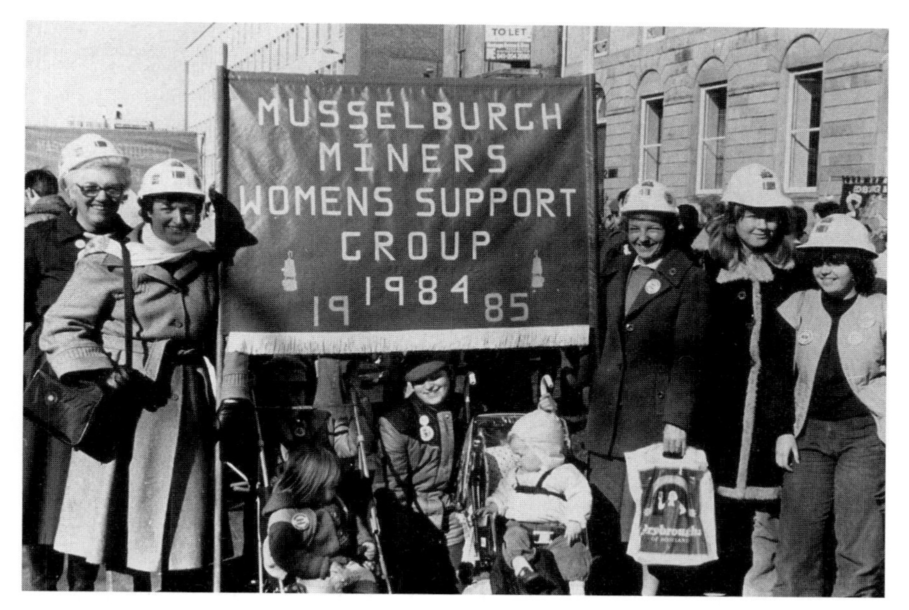

Figure 30.2 Women of the Musselburgh Miners Women's Support Group, 1985. Source: East Lothian council, courtesy of Mrs K MacNamara.

Production finally stopped at Polmaise in 1987, and the mighty Bilston Glen complex, in Midlothian, followed two years later. When the possibility of a workers' buyout to save Monktonhall fell through in 1997, the Longannet complex in Fife became the sole representative of deep-shaft coal mining in Scotland.[125]

Having survived the economic turmoil of the later twentieth century, the end for Longannet, when it came, was sudden. In March 2002, an estimated 17 million gallons of water flooded the old workings, affecting ventilation and rendering the coalface inaccessible.[126] Liquidation of the pit followed soon afterwards, and it was announced that there would be no rescue plan.[127] As one commentator observed:

> It is a great mercy that no miners' lives were lost in the weekend's sudden, devastating flooding of the Longannet mine near Kincardine. Down the years too many men have made that ultimate human sacrifice in their efforts to exploit Scotland's extensive coal reserves. However, it looks as if this latest catastrophe is a death sentence on the Scottish coalfield's last working deep mine.[128]

In spite of passionate interventions on behalf of Longannet, it was recognised that a return of any deep-shaft operations in the near future would be impossible.[129] Those who lost their jobs were advised to try 'new' sectors, such as call centres and companies operating from nearby Dunfermline or Rosyth.

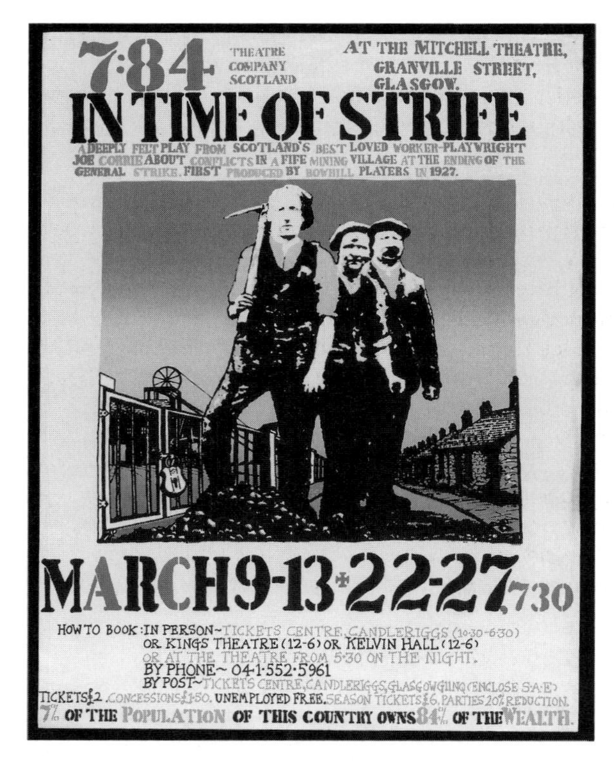

Figure 30.3 Poster of the play *In Time of Strife* by Joe Corrie, 1984. Source: Glasgow University Library, Department of Special Collections.

Since the closure of Longannet, the deep-shaft mines and the communities they employed have returned sporadically to public discourse. Former pit towns, especially in Scotland and northern England, have become occasional objects of curiosity for newspaper editors obsessed with London house prices. A study in metropolitan aloofness, the *Observer* ran a piece entitled 'For Misery go to Lochgelly', in which the Fife town was compared to Beirut and social problems listed which, in fact, would have struck a chord in many small towns the length and breadth of the country. The fact that it was possible to buy seven houses in Lochgelly for just one in Henley-upon-Thames was given as proof that it was 'the last place in Britain people want to live – and that's official'.[130] Commuters, it is argued, may be the saviours of this town, and the concluding paragraph sums up the position of many former mining communities:

> Yet it is that very road [the A92] that might just prove to be Lochgelly's salvation. For this forgotten corner of Fife is less than 40 minutes' drive from the heart of one of Britain's most prosperous cities. Edinburgh and its burgeoning financial sector is just a short hop over the Forth Road Bridge … Eileen McKenna, of Lochgelly Community Regeneration Forum, is confident the town is poised to rocket out

of the doldrums. 'There's a man living down in the Beaches who is commuting to London,' she told *The Observer*. 'For years, we were a forgotten town,' added McKenna. 'But in the last four years, a real strong spirit has developed. And Lochgelly has changed a lot for the good.'[131]

If the communities are in a state of flux, the industry itself is now firmly located within the heritage sector, in spite of the continuation of open-cast mining in Scotland.[132] Where Ladybird once published children's books on *The Miners* alongside policemen, doctors, nurses, builders and postal workers, school courses now deal with mining in a historical context.[133]

The thousands who flock to Lochore Meadows to enjoy boating, orienteering, or simply sunbathing, do so in the shadow of the Mary Colliery's winding gear – a monument to a bygone era.[134] Novels and plays, such as those by William McIlvanney and Joe Corrie, and the music of performers like Dick Gaughan, open a window to the past, and the image of the miner as representative of a dogged Scottish identity is an enduring one.[135] Gordon Brown, for example, recently articulated a popular theme in citing a mining background as a factor behind the success of certain football managers. Discussing Jim Baxter's time at Raith Rovers, Brown explained:

> We snapped him up while he was working as a miner in Fife: many of the Rovers players came from mining backgrounds. Becoming a footballer was seen as a way out of the mines. I think that's why so

Figure 30.4 Mary Colliery winding gear, Lochore Meadows, Fife. Source: Andrew Newby.

many players, and also managers such as Jock Stein and Bill Shankly, came from mining areas like Lanarkshire and Ayrshire.[136]

The work of local heritage groups in almost all of the former mining regions is supplemented by larger attractions, such as Summerlee Heritage Park at Coatbridge and Prestongrange Industrial Heritage Museum in East Lothian. The Scottish Mining Museum, located in Newtongrange on the site of the Lady Victoria Colliery, employs former miners as guides and is surely one of the most poignant visitor centres in the UK.[137] Nevertheless, museums portray miners very much as a 'caste apart', the revisionism relating to serfdom, for example, little in evidence. Mining is now part of the heritage and tourism sector, which increasingly occupies the position once held by heavy industry in the Scottish economy. Almost symbolically, the site of Seafield Colliery now forms a sprawling southern outskirt of Kirkcaldy, rebranded as 'Forth View' to be peopled, hope the developers, by aspirational young families working on the other side of the Forth Bridge. If this stirs up contradictory emotions, it is worth recalling T C Smout's warning of how:

> The social historian has to be careful to avoid the sentimentality that has dogged the economic history of Scotland, where there has been much lament over the failure of the old heavy industries of the nineteenth century and less appreciation of the far better standard of living that the bulk of the population has come to enjoy since their final demise after the Second World War.[138]

To appreciate this fully, it is necessary to look outwards. Although deep-shaft mining in Scotland barely survived into the twenty-first century, there are many parts of the world where miners face daily perils to extract the black diamonds from the earth. Even while writing this article an underground explosion killed five in Lviv, in an accident which brought to 4,300 the number of miners who have perished in post-Soviet Ukraine.[139] Almost simultaneously with the Lviv explosion, four were killed in a blast in Zhejiang, China, an incident which was itself dwarfed by the February 2005 calamity at Sunjiawan mine, Liaoning, which claimed over 200 lives. In 2004, over 6,000 people perished in China's mines.[140] Lessons learned over centuries from places like High Blantyre, Udston and Seafield can still, perhaps, be learned outwith Scotland.

NOTES

1. Smout, 1970, 170; Cochrane, 1985, 29; Hatcher, 1993, 97.
2. Duncan, 2005, 10, 16.
3. Duncan, 2005, 16; Whatley, 1987, 26–7.
4. Smout, 1970, 103; Withers, 1988, 309.
5. Cooke and Donnachie, 1998, 146.
6. Whatley, 1994, 7.
7. Campbell, 2000, I, 20; *Scotsman*, 23 January 1865.

8. Cooke and Donnachie, 1998, 146.
9. Cooke and Donnachie, 1998, 148.
10. Campbell, 2000, I, 21.
11. Payne, 1998, 80; for an extensive set of statistics for Scottish coal output by region, 1875–1935, see Campbell, 2000, I, 22–3.
12. Duncan, 2005, 201.
13. Duncan, 2005, 204.
14. Payne, 1998, 88.
15. *Scottish Miner*, April 1976. Graphic reproduced in Duncan, 2005, 246.
16. Smout, 1970, 168.
17. Hair, 2000, 136–51; Ashton and Sykes, 1929, 74. For the 'optimistic' view of the period of serfdom, see Whatley, 1991, 3–20; Whatley, 1995a, 66–79.
18. Freese, 2003, 211.
19. Whatley, 1995b, 28.
20. Houston, 1983, 3.
21. Smout, 1970, 168; Campbell, 1979, 9; Johnston, 1929, 79.
22. Campbell, 1979, 26.
23. Smout, 1970, 169.
24. Houston, 1989, 121.
25. Devine, 1999, 119.
26. Whatley, 1987, 141.
27. Houston, 1983, 3.
28. Whatley, 1994, 2–23; Duckham, 1971, 29–32; Cochrane, 1985, 27; Nef, 1932, II, 355.
29. Cochrane, 1985, 27–8; Hume and Butt, 1966, 160–183.
30. Devine, Lee and Peden, 2005, 46; Devine, 1999, 114.
31. Sutherland, 1825, 100; Richards, 1973, 227; Withers, 1988, 309; National Library of Scotland, Sutherland Papers, Acc. 10225, Brora Colliery Extracts. I am grateful to Annie Tindley for references to Brora. See also *Scotsman*, 23 August 1872.
32. *Scotsman*, 20 December 1828.
33. Smout, 1970, 406.
34. Smout, 1970, 409.
35. Smout, 1970, 406–7; Campbell, 1979, 15.
36. Campbell, 1979, 23; Johnston, 1929, 330–49.
37. Campbell, 1979, 1; Arnot, 1955; Youngson Brown, 1953, 35–50; Fraser, 2000, 61.
38. Campbell, 1988, 144–7; Campbell, 1995, 230.
39. *Scotsman*, 18 May, 1842.
40. Johnston, 1929, 323–33.
41. *Times*, 2 and 5 August 1842; *A Collection of the Public General Statues, Passed in the Fifth and Sixth Year of the Reign of Her Majesty Queen Victoria*, London, 1842; 5º & 6º Victoria, Cap. XCIX, An Act to prohibit the employment of women and girls in mines and collieries, to regulate the employment of boys, and to make other provisions relating to persons working therein, 10 August 1842, para. X.
42. Johnston, 1929, 334.
43. Humphries, 1981, 1–33; Mark-Lawson and Witz, 1988, 151–74.
44. Laslett, 2000, 38–40. Durland noted that the 'subjection of women' was one of the most striking features of home life in a mining village. Durland, 1904, 116–19.

45. Devine, 1992, 1–13.
46. Campbell, 1979, 156–57.
47. Johnston, 1929, 335; Haddow, 1888, 365; Campbell, 1979, 295.
48. *Scotsman*, 13 Sep. 1848.
49. *Scotsman*, 14 October 1854
50. Campbell, 1979, 156–7; Smout, 1987, 18–19; Devine, 1999, 260.
51. Campbell, 1979, 110, 117, 157; Mitchell, 1998, 33–40.
52. Campbell, 1979, 103.
53. Laslett, 2000, 38.
54. Laslett, 2000, 38–9. This type of housing was not, however, limited to Lanarkshire, and was not easily improved. The Fife miners' leader, Bob Selkirk, described a similar scene in relation to his own childhood in Midlothian. Selkirk, 1967, 3–5; Docherty, 1992, 14–15.
55. Campbell, 1979, 129.
56. Campbell, 2000, I, 214.
57. Campbell, 2000, I, 214. For a detailed breakdown of overcrowding in mining counties, see Campbell, 2000, I, 215.
58. It was even hoped that rows could be made more habitable by encouraging gardening among the miners. *Scotsman*, 23 August 1859.
59. Laslett, 2000, 39.
60. Campbell, 1979, 36–7.
61. Harvey, 1969, 18; Campbell, 1979, 268–9; Page Arnot, 1955, 64; Roy, 1905, 52.
62. Taplin, 1985, 19–20.
63. Dunlop, 1839, 81; Campbell, 1979, 35–6, 42.
64. Campbell, 1979, 104.
65. Campbell, 1979, 271.
66. *Scotsman*, 14 October 1854; Campbell, 1979, 130, 227.
67. Durland, 1904, 129.
68. For missionary activity, see Smout, 1987, 199–200.
69. Durland, 1904, 149–75; Anderson, 1986; Rowntree, 1901, 63–86; Prestoungrange, 2002.
70. Haddow, 1888, 363.
71. Millar, 1998, 7–12. There was also considerable settlement in Midlothian. See MacDougall, 2000, 107.
72. Lunn, 1980, 311.
73. Devine, 1999, 508–9; Millar, 2005, 514–34.
74. Hutchison, 1999, 39.
75. Devine, 1999, 266; Finlay, 2004, 64.
76. Finlay, 2004, 8.
77. Finlay, 2004, 5.
78. Harvie, 1981, 29; MacDougall, 2000, 70.
79. Page Arnot, 1955, 147; Lyall, 2000, 25–6; Finlay, 2004, 67.
80. Melling, 1982, 87.
81. Selkirk, 1967, 17.
82. Devine, 1999, 270; Payne, 1996, 17.
83. Devine, 1999, 267.
84. MacDougall, 1981, 51.
85. MacDougall, 2000, 111. For an employer's perspective, see Muir, n.d., 42–46.
86. Finlay, 2004, 57.
87. Finlay, 2004, 68–9.

88. Taylor, 1965, 243.
89. For the strike from the perspective of communist miners, see MacDougall, 1981, 94–7; Selkirk, 1967, 19.
90. Campbell, 2000, II, 225–7; MacDougall, 2000, 112.
91. Quoted in Finlay, 2004, 279.
92. Smout, 1987, 274.
93. Taylor, 1965, 249.
94. Finlay, 2004, 83.
95. John Morgan, 1990, 201.
96. Scottish Screen Archive, *Years Ago in Cowdenbeath* (Kay, E. and Brown, J.W., dirs, 1923–4). *Years Ago in Cowdenbeath* shows footage of a game versus Hearts at Central Park. On the banked terracing – made up of slag from the local pits – large crowds gathered to witness Cowdenbeath's first game in the top division of Scottish football. See also Ferguson, 1993, 30.
97. Smith, 1996.
98. MacDougall, 1981, 3. The scene in Midlothian was very similar. See MacDougall, 2000, 98. For the communist miners, sport had a particularly important role to play in society, with Bob Selkirk writing of how 'the Fife YCL [Young Communists' League] had a mass membership and support which made possible a Football League, which competed for, amongst other things, a Cup presented by Saklatvala, the Communist MP. The Fife YCL also organised successful Physical Culture Clubs in most big centres and ran popular boxing tournaments.' Selkirk, 1967, 23.
99. MacDougall, 1981, 2–3.
100. Docherty, 1982, 4.
101. Macintyre, 1980, 74; MacDougall, 1981, 146; Campbell, 2000b, 395.
102. See, for example, *Scotsman*, 14 May, 14 June and 31 August 1938.
103. Selkirk, 1967, 37.
104. Finlay, 2004, 178–9.
105. Duncan, 2005, 206.
106. McIlroy and Campbell, 2003a, 27–72; McIlroy and Campbell, 2003b, 39–80; *Scotsman*, 29 June, 27–30 September, 1 October 1943; 23 September 1944.
107. MacDougall, 1981, vi; Finlay, 2004, 216; Devine, 1999, 557.
108. Devine, 1999, 456, 548; Finlay, 2004, 206.
109. Devine, 1999, 571.
110. Finlay, 2004, 258–9.
111. National Coal Board (Scottish Division), 1958.
112. Scottish Screen Archive, *New Day* (Templar Film Studios, 1959).
113. Scottish Screen Archive, *New Day* (Templar Film Studios, 1959).
114. Halliday, 1990, 49–77; Finlay, 2004, 259.
115. Finlay, 2004, 259.
116. Conner, 1962, 45–8.
117. For a view of the industry after World War II, often from a miner's perspective, see McIvor & Johnston, 2002; McIvor and Johnston, 2007; Smout and Wood, 1991, 114.
118. Owens, 1994, 69.
119. Thatcher, 1993, 339–78.
120. Michael McGahey claimed that the strike was unavoidable, and that Thatcher had been 'cute enough' to wait until the right moment to crush the unions. Thatcher stated that she had ordered stockpiles of coal to be built up from

1981, 'steadily and unprovocatively', to allow a strike to be endured. Owens, 1994, 80; Thatcher, 1993, 341.

121. Devine, 1999, 602–3.
122. See, for example, Blane, C. Miners' strike was a battle of ideas, *BBC News Online*, 5 March 2004, available online at: http://news.bbc.co.uk/go/pr/fr/-/1/hi/scotland/3502855.stm.
123. See, for example, Margot Russell's account of the Dalkeith Women's Support Group. MacDougall, 2000, 140–1.
124. Blane, C. Miners' strike was a battle of ideas, *BBC News Online*, 5 March 2004, available online at: http://news.bbc.co.uk/go/pr/fr/-/1/hi/scotland/3502855.stm.
125. *Herald*, 9 February 1998; *Hansard*, 3 May 1989, Cols. 161–2; Devine, 1999, 592.
126. Holme, C and Macleod, C. Massive Flood Hits Scotland's Last Pit, *The Herald*, 22 March 2002.
127. Community mourns death of Longannet, *Evening Times* (Glasgow), 30 March 2002; Flooded Longannet set for closure, *Herald*, 4 April 2002.
128. Young, A. Seam of sentiment does mining no favours, *Herald*, 28 March 2002. The same author noted, a few days later, that 'the black stones will no longer be dug from Longannet, Scotland's last deep mine, and that fringe of gold now fears the tarnish of economic inactivity and community depression'. Young, A. It's time to accept that the black stones should be left alone, *Sunday Herald*, 31 March 2002.
129. Scott, K. Last Scots coalmine beaten by flooding, *Guardian*, 30 March 2002; Ross, S. Longannet miners enter jobs market, *Herald*, 9 April 2002; Wazir, B. A Call centre job? That's the real pits, *Observer*, 18 February 2001.
130. Khan, S. For misery go to Lochgelly, *Observer*, 25 January 2004.
131. Khan, S. For misery go to Lochgelly, *Observer*, 25 January 2004.
132. Open-cast mining continues, with the Scottish Coal Company planning to extend environmentally controversial operations in South Lanarkshire at Brokencross and, it hopes, open a new site at nearby Poneill.
133. Havenhand and Havenhand, 1965; Millar, 1998, 3.
134. Mason, 1982; *Birdwatching*, September 2005.
135. Corrie, 1985.
136. *The Observer*, 6 August 2000. The story of the Fife-born miner Tom 'Tuck' Syme, who became a star in North American ice hockey, evokes similar rhetoric. See Gordon, D. Tuck, pucks & big bucks, *Scotland on Sunday*, 24 April 2005.
137. Lingstadt, 2003, 99–111.
138. Smout, 1987, 275.
139. Five coal miners killed in explosion in Ukraine, *International Herald Tribune*, 26 June 2005, available online at: http://www.iht.com/articles/2005/06/26/europe/web.0626ukraine.php.
140. Deadly pits claim 200 more lives, *Times*, 16 February 2005, available online at: http://www.timesonline.co.uk/tol/news/world/article514818.ece; McCurry, J. 200 miners killed in China blast, *Guardian*, 16 February 2005.

BIBLIOGRAPHY

Books

Anderson, A. *The Dean Tavern: A Gothenburg Experiment*, Newtongrange, 1986.
Arnot, R P. *A History of the Scottish Miners from the Earliest Times*, London, 1955.
Ashton, T S and Sykes, J. *The Coal Industry in the Eighteenth Century*, Manchester, 1929.

Campbell, A B. *The Lanarkshire Miners: A Social History of their Trade Unions, 1775–1874*, Edinburgh, 1979.
Campbell, A B. *The Scottish Miners, 1874–1939, Vol. 1: Industry, Work and Community*, Aldershot, 2000.
Campbell, A B. *The Scottish Miners, 1874–1939, Vol. 2: Trade Unions and Politics*, Aldershot, 2000.
Challinor, R. *Alexander Macdonald and the Miners*, London, 1968.
Challinor, R. *The Miners' Union: A Trade Union in the Time of the Chartists*, London, 1968.
Cochrane, L E. *Scottish Trade with Ireland in the Eighteenth Century*, Edinburgh, 1985.
Conner, A. *Coal In Decline*, Glasgow, 1962.
Corrie, J. *Plays, Poems and Theatre Writings*, Edinburgh, 1985.
Devine, T M, ed. *Scottish Emigration and Scottish Society*, Edinburgh, 1992.
Devine, T M. *The Scottish Nation 1700–2000*, Harmondsworth, 1999.
Devine, T M, Lee, C H and Peden, G C, eds. *The Transformation of Scotland: The Economy since 1700*, Edinburgh, 2005.
Docherty, M. *A Miners' Lass*, Cowdenbeath, 1992.
Duckham, B F. *A History of the Scottish Coal Industry, 1700–1815*, Newton Abbot, 1971.
Duncan, R. *The Mineworkers*, Edinburgh, 2005.
Dunlop, J. *The Philosophy of Artificial and Compulsory Drinking Usage in Great Britain and Ireland*, London, 1839.
Durland, K. *Among the Fife Miners*, London, 1904.
Ferguson, R. *Black Diamonds and the Blue Brazil*, Ellon, 1993.
Finlay, R J. *Modern Scotland 1914–2000*, London, 2004.
Freese, B. *Coal: A Human History*, Cambridge, Massachusetts, 2003.
Finlay, R. *Modern Scotland, 1914–2000*, London, 2003.
Fraser, W H. *Scottish Popular Politics: From Radicalism to Labour*, Edinburgh, 2000.
Halliday, R S. *The Disappearing Scottish Colliery*, Edinburgh, 1990.
Harvey, K A. *The Best Dressed Miners: Life and Labor in the Maryland Coal Region, 1835–1910*, Ithaca and London, 1969.
Harvie, C. *Few Gods and Precious Few Heroes: Scotland Since 1914*, London, 1981.
Hatcher, J. *The History of the British Coal Industry, Volume 1: Before 1700: Towards the Age of Coal*, Oxford, 1993.
Havenhand, I and Havenhand, J. *The Miner*, Loughborough, 1965.
Hollis, P. *Jennie Lee: A Life*, Oxford, 1997.
Johnston, T. *A History of the Working Classes in Scotland*, Glasgow, 1929.
Laslett, J H M. *Colliers Across the Sea: A Comparative Study of Class Formation in Scotland and the American Midwest, 1830–1924*, Chicago, 2000.
Lyall, A. *Model Housing for Prestongrange Miners*, Prestonpans, 2000.
MacDougall, I, ed. *Militant Miners*, Ilkley, 1981.
MacDougall, I, ed. *Voices from Work and Home*, Edinburgh, 2000.
Macintyre, S. *Little Moscows*, London, 1980.
Mason, A J. *Lochore: A Playground for Fife*, Edinburgh, 1982.
McIvor, A and Johnston, R. *Miners Lung: A History of Dust Disease in British Coal Mining*, Aldershot, 2007.
Millar, J. *The Lithuanians in Scotland*, Colonsay, 1998.
Miller, A. *The Rise and Progress of Coatbridge*, Glasgow, 1864.
Mitchell, M J. *The Irish in the West of Scotland, 1797–1848*, Edinburgh, 1998.
Muir, A. *The Fife Coal Company Limited: A Short History*, Leven, n.d.

National Coal Board (Scottish Division). *A Short History of the Scottish Coal Mining Industry*, Edinburgh, 1958.
Nef, J U. *The Rise of the British Coal Industry*, London, 1932.
Oglethorpe, M K. *Scottish Collieries: An Inventory of the Scottish Coal Industry in the Nationalised Era*, Edinburgh, 2006.
Owens, J, ed. *Miners 1984–1994: A Decade of Endurance*, Edinburgh, 1994.
Prestoungrange, G. *The Prestoungrange Gothenburg: The Goth's First 90 Years and the Coming Decade*, Prestonpans, 2002.
Richards, E. *The Leviathan of Wealth*, London, 1973.
Rowntree, J. *British 'Gothenburg' Experiments and Public House Trusts*, London, 1901.
Roy, A. *A History of the Coal Miners of the United States*, Ohio, 1905.
Selkirk, B. *The Life of a Worker*, Dundee, 1967.
Sinclair, J, ed. *Statistical Account of Scotland*, 21 vols, Edinburgh, 1791–9.
Smith, A C. *A History of Whitehill Welfare*, n.p., 1996.
Smout, T C. *A Century of the Scottish People, 1830–1950*, London, 1987.
Smout, T C. *A History of the Scottish People, 1560–1830*, London, 1970.
Smout, T C and Wood, S. *Scottish Voices 1745–1960*, London, 1991.
Sutherland, A. *A Summer Ramble in the Highlands*, Edinburgh, 1825.
Taplin, E. *The Dockers' Union: A Study of the National Union of Dock Labourers, 1889–1922*, Leicester, 1985.
Taylor, A J P. *English History, 1914–1945*, Oxford, 1965.
Thatcher, M. *The Downing Street Years*, London, 1993.
Withers, C J W. *Gaelic Scotland: The Transformation of a Culture Region*, London, 1988.

Chapters and articles in books

Campbell, A. 18th century legacies and 19th century traditions: The labour process, work culture and miners' unions in the Scottish coalfields before 1914. In Westermann, E, ed., *Vom Bergbau – zum Industrierevier*, Stuttgart, 1995, 217–38.
Cooke, A and Donnachie, I. Aspects of industrialisation before 1850. In Cooke, A, Donnachie, I, MacSween, A and Whatley, C A, eds, *The Transformation of Scotland, 1707–1850* (Modern Scottish History 1707 to Present, vol. 1), East Linton, 1998, 130–54.
Houston, R A. Women in the economy and society. In Houston, R A and Whyte, I D, eds, *Scottish Society 1500–1800*, Cambridge, 1989, 118–47.
Hutchison, I C G. The impact of the First World War on Scottish politics. In Macdonald, C M M and McFarland, E W, eds, *Scotland and the Great War*, East Linton, 1999, 36–58.
Lunn, K. Reactions to Lithuanian and Polish immigrants in the Lanarkshire coalfield, 1880–1914. In Lunn, K, ed, *Hosts, Immigrants and Minorities*, Folkestone, 1980, 308–42.
Melling, J. Scottish industrialists and the changing character of the Clyde Regior, *c*1880–1918. In Dickson, D, ed, *Capital and Class in Scotland*, Edinburgh, 1982, 61–142.
Millar, J. The Lithuanians. In Beech, J, Hand, O, Mulhern, M A and Weston, J, eds, *The Individual and Community Life* (Scottish Life and Society: A Compendium of Scottish Ethnology, vol. 9), Edinburgh, 2005, 514–34.
Payne, P L. The economy. In Devine, T M and Finlay, R J, eds, *Scotland in the Twentieth Century*, Edinburgh, 1996, 13–45.
Payne, P L. Industrialisation and industrial decline. In Cooke, A, Donnachie, I,

MacSween, A and Whatley, C A, eds, *The Modernisation of Scotland, 1850 to Present* (Modern Scottish History 1707 to Present, vol. 2), East Linton, 1998, 73–94.
Whatley, C A. New light on Nef's numbers: Coal mining, c1700–1830. In Cummings, A J G and Devine, T M, eds, *Industry, Business and Society in Scotland since 1750*, Edinburgh, 1994, 2–23.

Articles in journals
Brown, A J Y. Trade union policy in the Scots coalfields, 1855–1885, *Economic History Review*, 2nd Ser., vi (i) (1953), 35–50.
Haddow, R. The miners in Scotland, *The Nineteenth Century* (September 1888), 360–71.
Hair, P E H. Slavery and liberty: The case of the Scottish colliers, *Slavery and Liberty*, xxi (2000), 136–51.
Houston, R. Coal, class and culture: Labour relations in a Scottish mining community, 1650–1750, *Social History*, viii (1983), 1–18.
Hume, J R and Butt, J. Muirkirk, 1786–1802, *Scottish Historical Review*, xlv (1966), 160–83.
Humphries, J. Protective legislation, the capitalist state and working-class men: The case of the 1842 Mines Regulation Act, *Feminist Review*, vii (1981), 1–33.
John Morgan, W. The miners' welfare fund in Britain, 1920–1952, *Social Policy & Administration*, xxiv (1990), 199–211.
Lingstadt, K. Scotland's black diamonds: Heritage, education, entertainment and commercialism, *Scottish Economic and Social History*, xxiii (2003), 99–111.
McIlroy, J and Campbell, A. Beyond Betteshanger: Order 1305 in the Scottish coalfields during the Second World War, part 1: Politics, prosecutions and protest, *Historical Studies in Industrial Relations*, xv (2003), 27–72.
McIlroy, J and Campbell, A. Beyond Betteshanger: Order 1305 in the Scottish Coalfields during the Second World War, part 2: The Cardowan Story, *Historical Studies in Industrial Relations*, xvi (2003), 39–80.
Mark-Lawson, J and Witz, A. From 'family labour' to 'family wage'? The case of women's labour in nineteenth century coal mining, *Social History*, xiii (1988), 151–74.
McIvor, A and Johnston, R. Voices from the pits: Health and safety in Scottish coal mining since 1945, *Scottish Economic and Social History*, xxii (2002), 111–33.
Whatley, C A. A caste apart? Scottish colliers, work community and culture in the era of 'serfdom', c1606–1799, *Journal of the Scottish Labour History Society*, xxvi (1991), 3–20.
Whatley, C A. Collier serfdom in mid-eighteenth century Scotland: New light from the Rothes colliery MS, *Archives*, xxii (1995), 22–33.
Whatley, C A. Coal, salt and the union of 1707: A revision article, *Scottish Historical Review*, lxvi (1987), 26–45.
Whatley, C A. Scottish 'collier serfs', British coalworkers? Aspects of Scottish collier society in the eighteenth century, *Labour History Review*, lx (1995), 66–79.
Whatley, C A. 'The fettering bonds of brotherhood': Combination and labour relations in the Scottish coalmining industry, c1690–1775, *Social History*, xii (1987), 139–54.
Youngson Brown, A J. Trade union policy in the Scots coalfields, 1855–1885, *Economic History Review*, 2nd Ser. vi, i (1953), 35–50.

31 Scottish Dock Trade Unionism, 1850 to Present

WILLIAM KENEFICK

This chapter defines and describes the development of modern dock trade unionism in Scotland from the mid-nineteenth century. This will be accomplished by charting the rise and decline of dock unionism in Scotland from the early phase of emergent dock unionism in the mid-nineteenth century through the decades of decline and collapse towards the end of the twentieth century. It is ironic indeed that when the Scottish Transport and General Workers' Union (STGWU) finally rejoined the TGWU in 1972, it did so on the advent of the collapse of the port transport industry in Scotland. From the 1970s onwards the number of dockers across Scotland dwindled to insignificant numbers, and before the turn of the twentieth century the Scottish port transport sector (with a few exceptions, such as the port of Aberdeen) was virtually non-existent.

In the Scottish context, Glasgow and the Clydeside ports were clearly influential. After a brief discussion of sources and the absence of academic research into Scottish docks and dockers, there will follow two brief context-setting sections, looking first at the importance of Glasgow to the Scottish economy, then at the pivotal role played by the Irish and the Scottish Highlanders in Glasgow in influencing and directing developments in dock trade unionism in Scotland and Britain from the mid-nineteenth century onward.

As a guide, it may prove useful to briefly detail some of the main organisations which are the subject of this chapter:

Union	*Abbreviation*	*Founded*
Glasgow Harbour Labourers' Union	GHLU	1853
Glasgow Harbour Mineral Workers' Union	GHMWU	1887
National Union of Dock Labourers	NUDL	1889
Dock, Wharf, Riverside General Workers' Union	DWRGWU	1889
Scottish Union of Dock Labourers	SUDL	1911
National Transport Workers' Federation	NTWF	1910

Union	Abbreviation	Founded
National Joint Council for the Port Transport Industry*	NJC	1920
Transport and General Workers' Union	TGWU	1922
Scottish Transport and General Workers' Union	STGWU	1932
National Amalgamated Stevedore and Dockers	NASD	1923
South Side Labour Protection League	SSLPL	1872

* National Joint Council (NJC) actually refers to the National Joint Industrial Council for the Port Transport Industry. It was one of the recommendations to emerge from the Shaw Enquiry (Enquiry into the Wages and Conditions of Dockers and Waterside Labour) 1920, and was established jointly by the National Council for Port Labour Employers and the National Transport Workers' Federation (NTWF). Later the Transport and General Workers Union (TGWU) took over the NTWF role when it was formed in 1922.

A PAUCITY OF SOURCES

There are, without doubt, some major problems associated with research into trade unionism generally in Scotland, not least amongst those bodies such as the dockers' unions, who were considered to represent 'unskilled' workers. This may explain why it is only recently that a comprehensive history of dock workers in Scotland has been produced.[1] Records of early Scottish trade unions are so fragmentary that any in-depth analysis of certain groups of workers is extremely difficult. To William Marwick, writing in the mid-1930s, the reason was simple:

> Scottish trade unionism has never found its Webbs [Sydney and Beatrice], and it may be that the opportunity has passed. Many old minute books and other records have perished, few if any odd copies survive of such 'Labour' periodicals as the *Liberator* and the *Sentinel*, references in ordinary periodicals are slight, often uninformed, and uninformative. Most of the existing bodies are of recent origins, and are merely branches of an English body.[2]

Ironically, Marwick was writing at a time when the Glasgow dock labour force had decisively rebelled against their status as a 'mere branch of an English body' in seceding from the TGWU to STGWU in 1932.

Clearly, the STGWU was of recent origin, yet its records, like those of earlier Scottish unions, are still largely fragmentary.[3] For the few references that are made to other early general Scottish unions we must, ironically, thank the Webbs and the collection they left behind. However, there are other sources which can be exploited, such as newspapers and parliamentary papers, as well as the scattered remnants listed in Ian MacDougall's excellent catalogue of labour records in Scotland – an invaluable resource to any researcher working in this field.[4]

It is clear, however, that despite Clydeside's overwhelming importance in terms of Scottish trade and the central function of Glasgow as Scotland's premier port, remarkably little has been written about the activities of this geographical area of trade, and still less is known about the labour force which handled the goods that passed through the Clydeside ports.[5] What has been written in relation to port development and dock life in general refers to the English experience (mainly Liverpool and particularly London), and while some reflect on certain aspects of dock life in Scotland, they amount to little more than passing references and occasional footnotes.[6]

Moreover, until recently there has been a general lack of interest in the position of the unskilled worker in Scottish historiography, and such omissions, argue Bill Knox and Conon Fischer, 'have distorted the historical record'.[7] Recent evidence demonstrates that the dockers of Glasgow and Clydeside were overwhelmingly Irish and Catholic, and were generally considered unskilled. Viewed in this context a study of the Glasgow dockers helps to set this particular record straight. But before turning our attention to this matter it is important to consider briefly the economic impact of Glasgow and the Clydeside ports (the main West of Scotland ports), and how the right conditions were to emerge to help cultivate emergent dock trade unionism in the mid-nineteenth century; the formation of national and general dock unionism in the late nineteenth century; survival and consolidation, and disintegration and resurgence before World War I; and the modern development of dock trade unionism from 1914 to the last decades of the twentieth century in Scotland.

THE ECONOMIC ROLE OF GLASGOW AND THE WEST OF SCOTLAND

There is little disagreement that the modern history of the port transport industry in Scotland focuses upon the activity of Glasgow and the west-coast ports. Ports such as Aberdeen, Leith and Dundee (and over twenty smaller east-coast ports) may have had longer and perhaps more illustrious trading histories than Glasgow, but in terms of their economic impact they pale considerably when compared with the Clydeside ports. For this reason it is appropriate to consider briefly something of Glasgow's trading pre-eminence and its relationship to the Scottish economy in the nineteenth and twentieth centuries.

In general, the nineteenth century proved a lucrative period for the Scottish ports and the Scottish port transport industry as a whole. Between 1820 and 1913, for example, the growth in Scottish trade was truly remarkable. Ports such as Aberdeen, Dundee and Leith fared particularly well from this profitable expansion in trade, but none more so than the port of Glasgow and the other west-coast ports. In 1913, for example, Glasgow's total value of trade was almost £55 million (accounting for over 56 per cent of total Scottish figures), whereas the combined value of both Leith and Dundee was £32.5 million. By 1918 Glasgow had increased its share to £96.5 million, while

Leith and Dundee had fallen to just under £20 million (Leith experienced the greatest loss, falling from just under £23 million to around £14 million *c*1913–18).[8] This was part of the period of 'great prosperity' experienced before World War I, and while in relative terms this sector of the Scottish economy did less well after the war than before, Glasgow and the Clydeside ports were still the prime movers within the Scottish port transport industry thereafter.

After the boundary changes of 1912, Glasgow's population expanded to over one million and accounted for a fifth of the total population of Scotland.[9] Coupled with significant demographic change, an immense productive capacity, the increase in shipping and trade, and the city's vast and growing distribution services, Glasgow clearly deserved its well-earned reputation as Britain's 'Second City of Empire'.[10] Likewise, when investigating the growth and development of dock trade unionism in Scotland it is clear that Glasgow and the other west-coast ports were also the largest employers of dock labour.

The rapidly expanding port of Glasgow employed ever greater numbers of dock workers from the 1860s onwards, until this reached a high point just after World War I. On a decennial basis, based on census returns, 1911 is traditionally considered the peak year (although other evidence points to this as being 1920), and that after this date official numbers began to fall. However, with figures falling for Scotland as a whole, as a percentage of Scottish totals, Glasgow and Clydeside dock labour force was still the largest in Scotland. Because of the casual nature of the work, however, it may never be possible to enumerate adequately the actual numbers employed in dock work before 1914, and this means that official statistics, extracted from the census figures, offer only a general impression.[11]

However, it is possible to say something of significance when using these figures. For example, in 1861 there was a total of 3,487 dockers in Scotland, rising to 13,147 by 1911. Over the same period the number of Glasgow dockers increased from 1,197 to 5,586 – or from 35 to 42 per cent of Scottish totals. By the time of the 1931 census the port of Glasgow still accounted for around 40 per cent of the Scottish dock labour force, and with a slight expansion of trade across the Clydeside port transport sector in general, around half of the Scottish dock labour force was still to be found in the West of Scotland.[12] This was to remain the case well into the second half of the twentieth century. With such a consistent and concentrated number of workers within one geographical area it would come as no surprise that the NUDL were to emerge from the Clydeside region by the late 1880s.

Glasgow had seen the emergence of other dock unions, in particular the GHLU, but only the NUDL was to develop a presence beyond the confines of Glasgow harbour, from west to east Scotland, north-west England – including Liverpool and Birkenhead – and across the Irish sea and the eastern seaboard of Ireland. This was in no small part due to the presence of a great number of Irish who settled in ports and harbours along the western seaboard of Britain stretching from Liverpool to Glasgow.

The number of Irish immigrating to Scotland in the nineteenth century peaked at 207,000, or 7.2 per cent of the total Scottish population in 1851, and steadily declined thereafter to level out at around 3.3 per cent by 1921.[13] In west-central Scotland, however, the picture was very different, and in Glasgow the Irish percentage of total population was consistently and notably higher than Scottish averages. In 1881, for example, the Irish constituted 5.9 per cent of total Scottish population, but in Glasgow they accounted for 13.1 per cent of the total population. Forty years later, when the Irish formed 3.7 per cent of the Scottish population, they accounted for 7.1 per cent of the population of Glasgow.[14]

Glasgow was also a great employer of Irish labour and it is clear that the impact of the Irish at Glasgow was significant, but in terms of waterfront employment – as will be discussed in the next section – their impact was of a greater order of magnitude. Combine the Irish and the next significant and influential section of waterfront workers, the Scottish Highlanders, and the cause of dock trade unionism would be advanced beyond the aspirations of even the most hopeful mind by the latter decades of the nineteenth century.

THE IRISH, THE SCOTTISH HIGHLANDERS AND GENERAL DOCK TRADE UNIONISM

It is evident that the ethnic composition of the Clydeside and Glasgow dock labour force was overwhelmingly Roman Catholic Irish and Protestant Scottish Highlanders. The Irish in particular were crucial to the emergence of the NUDL as a major trade union force in Scotland and Britain. It is also probable that the Irish played a leading role in forming the GHLU (or 'Old Union') in 1853, and it is also probable that the Irish were pivotal in the formation of the two other early dock unions at Liverpool in 1849 and London between 1849 and 1853. In London, by the 1850s, for example, it was estimated that around two-thirds of all those engaged in coal portering (loading and discharging of coal cargo) were Irish, and they were to dominate this trade and the London dock unions well into the first half of the twentieth century. Likewise, in Liverpool and New York from the 1880s and 1890s, we see a similar picture develop as the Irish came to dominate dock work in those locations.[15]

Glasgow's development was comparable, but the main difference was the addition of the Highlanders. Indeed, in terms of dock work, as far back as the 1830s Glasgow's harbour master estimated that out of the '1,000 or so men labouring at the harbour, the great majority were either Irishmen or Highlanders'.[16] A study of the Irish and the Highlanders in Glasgow between 1830 and 1870 estimated that around two-thirds of the quay labourers in 1851 were Irish, mostly from the north and in particular Donegal.[17] By the time of the 1891 census the Irish and the Highlanders had consolidated their hold on dock work, and recent research suggests that something in the region of 63 per cent of the Glasgow dock labour force at that time were

Irish-born, and just under 20 per cent were Highlanders. A full survey of the dock workers in the Clydeside port of Ardrossan on the Ayrshire coast showed a similar pattern to that of Glasgow, demonstrating a labour force composition of 58 per cent Irish and 15.5 per cent Scottish Highlander. In addition, at both Ardrossan and Glasgow almost 10 per cent of the total dock labour force was English.[18]

It is important to recognise that the ethnic composition of any group of workers can play a fundamental role in developing their attitude and stance towards industrial relations. In the context of this discussion, this is particularly so with the Irish. According to John Lovell, over the longer term: 'The Irish element brought stability and cohesion to wide areas of waterside employment. In particular, the tradition of the son following father into dock work did much to raise the calibre of the labour force.'[19] David Wilson also notes similar links between the Irish and dock labour, and that their early influx into dockwork helped to engender 'a significant sense of community'.[20] Indeed, it was this sense of community that accounted for the solidarity of sections of the dock labour force during labour disputes, and aided the process and development of trade unionism.[21]

This is precisely what happened at Glasgow and the other Clydeside ports. The process of building up a dock union in Glasgow was intimately linked with Irish politics. This factor played a significant part in the formation of the first general and non-sectional dock union at the port. It was an Irish-born leadership that forged a spirit of trade unionism among the dockers at Glasgow between 1889 and 1893, and led directly to the formation of the NUDL.[22] As discussed in greater detail in later sections, the Glasgow Trades Council and the National Amalgamated Seamen and Firemen's Union (NASFU, or the 'Seamen's union') in particular gave great encouragement and assistance towards the formation of the NUDL. This important role remains to be documented.[23] As fellow members of the waterfront community, however, the seamen and the dockers shared common interests and bonds that aided trade union developments within both sectors of the port transport industry.

Ports and port towns and cities are part of a great network of distribution and exchange, and as maritime cosmopolitan trading centres they are a conduit for the exchange of ideas as well as goods. Given the evident and large concentration of Irish and Highlanders in the West of Scotland, the potential for the exchange of political ideas across the Glasgow and Clydeside waterfront was considerable. Before, during and after the period of formation, familiar names in Irish political history – particularly Daniel O'Connell and Henry Grattan – were used to further the cause of trade unionism and help strengthen the fledgling NUDL in its first days and weeks.[24] The other substantial group to be involved with dock work were the Scottish Highlanders, and both they and the Irish shared the experience and memory of life and work on the land before their migration to urban, industrial and maritime centres of Scotland. Here we find another common political dimension that could unite Irish Catholics and Highland

Protestants. Of course, this also holds true for Lowland Scots who left the land to make a living in urban employment. The common experience of relocation to an urban environment was as bewildering to a Lowland cottar as a Highland crofter.

It is no accident that the Irish leaders of the NUDL were also firm adherents of land reform. They argued for free access to land, the ending of the monopoly by landlords and the end to what they considered 'rent robbery'. These and other issues constituted what was known as the 'Land Question', which brought together Irish and Highland political issues through a coalition with Irish and Scottish Nationalists, Irish and Scottish Land Leaguers, and urban Scottish political radicals during the 1880s and 1890s. As Bernard Aspinwall has argued: 'Irish Nationalism, support for the Land League and Catholicism were part of a complex mixture that could unite with Scottish Highland tenants in a critique of the existing order.' The ideas of the American radical Henry George and the Land League, therefore, could and did act to bind together the Irish and Highland elements of the Glasgow dock labour force.[25]

It is clear that ethnicity and religion are important in the history of dock trade union organisation at Glasgow, and that this had an impact across Scotland. Given the high proportion of Irish Catholic and Highland Presbyterian dockers who worked at the Glasgow ports, religion and ethnicity were notable features of this workforce. Despite the possibility of sectarian tension, there existed a broader proletarian culture at Glasgow, where sectarian loyalties were tempered by mutual political interests and increasing occupational class solidarity. It would appear that the Irish never entered into competition with Lowland Protestant Scottish workers on the docks, and they co-existed with few problems between them and their Protestant and Celtic Highland cousins. Taking into account the influential nature of Glasgow dock labour organisation, this ability to avoid tensions arising from ethnicity and religion is suggestive of an important aspect of the history of dockers across Scotland as a whole.

The Irish built the new docks and wharves of Glasgow in the 1860s and 1870s, and it was they who remained to work these docks and wharves once they were completed. There was therefore no tradition of Scots-born Protestants working the docks, and this left the field open to the Irish – and they made it their own. The Irish and the Highlanders thereafter shaped the custom and practice in the workplace, influenced developments in the area of industrial relations, and inculcated these within their grand vision for organised dock trade unionism across Scotland. They became 'an occupational community' with a national perspective, and it was this factor that helped the NUDL to become the first truly national and general dock union in Scottish and British history, the growth and development of which will be charted in the following sections.

THE ROOTS AND EARLY EVOLUTION OF SCOTTISH DOCK TRADE UNIONISM: THE ARTISANAL PHASE

This section is concerned principally with the evolution of dock trade unionism in Scotland and the roots and development of this movement in Clydeside and the port of Glasgow in particular. In tracing the roots of that development we can also consider the manner in which dock union organisation on Clydeside and Scotland compared with or differed from the growth of dock unionism elsewhere in Britain. By studying the history of the British waterfront we add to the understanding of Glasgow, Clydeside and the Scottish docks. In studying the Scottish port transport industry in this comparative framework we also add to our understanding of British port transport industry.[26] What becomes clear in adopting such a comparative approach is the pivotal role played by the Scottish ports in the development of dock trade unionism across Britain as whole.

The early phase of dock unionism developed from within the ranks of shipworkers such as the Liverpool South End Dock Labourers Association formed in 1849, and the Glasgow Harbour Labourers' Union formed in 1853. The GHLU were highly exclusive, sectional in outlook and primarily concerned with defending their higher skilled 'craft' status in the face of growing competition from quayside workers. The GHLU imposed an entrance fee of £5 in the 1860s – later increased to £8 – and when compared to the 16 shillings charged by the Greenock Harbour Labourers' Union in the late 1880s, it can be seen that this levy was intended to maintain exclusivity. Moreover, it provided an effective barrier to dockers from the lower levels of the labour force from entering the ranks of the GHLU, and proved an obstruction to the growth of dock union expansion at the port.[27]

London developed a similar system by the 1870s, with the formation of the South Side Labour Protection League (SSLPL) and the Stevedores' Union, although other smaller societies of corn porters, coaltrimmers, watermen and lightermen were in evidence at London and other locations before then. Quayside workers in Scotland also formed trade unions such as the Aberdeen Shore Labourers' Union in the early 1880s, the Greenock Dock Labourers' Union, in Glasgow the Harbour Mineral Workers' Union in 1887, and for a brief time the American Knights of Labour organised dock workers at the port of Ardrossan on the Clyde in the late 1880s.[28] But with the exception of the Knights of Labour, these trade unions, or trade societies, functioned as mutually exclusive, sectional and highly localised bodies of workers.

The GHLU made little attempt to organise the quay workers at the port until after 1872 when, during an extensive period of strike activity, they attempted to recruit other dock workers, presumably assisted by a reduction in the entrance fee. The strike action ended in defeat for the GHLU, and was broken in the time-honoured tradition by the importation of outside, or scab, labour by the port and shipping employers. The immediate result of the dispute was a reversal in the fortunes of the GHLU, with the consequence that membership levels fell dramatically between 1872 and 1889. After the strike, those who filled the places of the strikers remained employed at the docks, and this further weakened the GHLU. Indeed, this may perhaps account for their distrust of general dock unionism and the NUDL between the late 1880s and early 1890s.[29]

It is also clear that waterfront industrial relations were changing as the influence of the large steamship companies increased: they would not tolerate any combination of workers interfering with their business. This was a further formidable barrier against the emergence of general dock unionism, and with ever-increasing numbers of steamship companies located and operating out of Glasgow and the Clydeside ports, the employers became even more anti-union than was generally the case at other major British ports.[30] The Glasgow port employers had single-handedly destroyed the Mineral Workers' Union, and had inflicted serious damage on the GHLU by the late 1880s. It was only with the emergence of the NUDL in 1889 that there emerged any serious challenge to their authority, when it became clear that if Glasgow was organised the rest of the country would quickly follow.

THE 'NEW UNIONISM' – 1889 AND ALL THAT!

The main phase of organised dock unionism across Britain essentially dates from the 1870s, but it did so on the basis of a highly developed separatism. For example, London employers came to recognise the Stevedores' Union (considered a 'skilled' body of workers), but would not recognise the great mass of general dock labourers. At Glasgow the GHLU was recognised by the sailing ship owners, but not the steamship owners. After a series of militant strikes across Britain between 1889 and 1890, however, beginning at Glasgow in February 1889, there emerged a palpable and growing solidarity among dock workers, which gradually spread to take root in other ports around Scotland and Britain.

It was during an extensive strike at Glasgow in June 1889 that the NUDL began to extend its influence more widely. They led by example, and in taking on the powerful and autocratic Glasgow shipping employers they were sending a message to dockers across the country that they could do likewise. Apart from Aberdeen and Greenock, and to a lesser extent Ardrossan harbour, no other dock workers were organised at that juncture (except for some small local societies), but this was set to change when Dundee joined the Glasgow strike action, followed by Greenock, Leith, Aberdeen, Derry and Liverpool, and a little later at Grangemouth and Burntisland. With the exception of Aberdeen and Greenock, branches of NUDL were quickly established in all these ports.

As noted previously, the link between the Glasgow Trades Council, the Seamen's union (NASFU) and the Glasgow dockers was an important one. The Seamen's union membership had swelled across the country by 1888, and after several protracted strikes during the following year they had recruited 7,500 members across Scotland.[31] The seamen's union encouraged others to organise, and at Glasgow, working closely with the Glasgow Trades Council, they played a central role in organising the Glasgow dockers. Indeed, Glasgow was following the example of Aberdeen, where the Aberdeen Trades Council in league with the Aberdeen Shore Labourers helped to set up a branch of the Seamen's union in 1887.[32]

It was reported with great pride to the annual conference of the Glasgow Trades Council in 1889 that the recently formed NUDL had a membership of nearly 4,000 and was affiliated to the council.[33] In terms of affiliate membership, only the Railwaymen's Union could claim to have more members – and only marginally so – than the NUDL.[34] As the second largest affiliated union to the Glasgow Trades Council, the influence of the NUDL would have been considerable. Glasgow was by far the most influential of the Scottish Trades Councils, and by the end of the nineteenth century this influence was growing considerably.

It is true that trade union density in Scotland was considerably less than in England and Wales. Moreover, the Scottish trade unionism structure was federal and quite different to the 'model union' framework more typical of England. This federal structure – compounded by the fact that there were more local and fewer national unions in Scotland – was thus argued to have weakened trade unionism in Scotland. However, this view ignores the fact that almost every organised trade union that existed in Scotland – large or small – affiliated to one of the sixteen regional trades councils in the country. Indeed, some unions and most trades councils were also affiliated to the Glasgow Trades Council. With two-thirds of all trade unionists in Scotland residing in Glasgow or greater Clydeside, Glasgow Trades Council was one of the most influential bodies in the Scottish and the British trade union movement, central to the growth, expansion and progress of the Scottish labour movement before 1914.[35]

The NUDL had a similar impact in relation to the emergence of general dock unionism in Scotland and Britain. Within a year of its formation, the NUDL had branches in many major ports across the country, and by 1890 the Glasgow-based union had an estimated membership of 50,000 across Britain and Ireland. Indeed, the NUDL almost matched the London-based Dock, Wharf, Riverside General Workers' Union (DWRGWU), with an estimated membership of 57,000 for England and Wales.[36] As will be seen below, the early 1890s were to prove the highpoint of the New Unionism, but the impact of this movement was considerable, and in the Scottish and British context it was fundamentally important to the future of dock unionism.

Because of the New Unionist gains, employers across various sectors of the British economy came together in the early 1890s to form strong and formidable trade associations (see Chapter 25 of this volume), and in the employer counterattacks that followed, no association became more feared than that of the Shipping Federation. In the aftermath of the Shipping Federation-orchestrated counterattack that took place between 1890/1 and

1893, many of the new unions were destroyed, including NASFU in 1894. While large dock unions such as the NUDL (now headquartered in Liverpool) and the London-based DWRGLU managed to survive, they were severely weakened. This brought to a close the first and dramatic phase of general dock unionism.[37]

By 1894 the membership of both the NUDL and the DWRGWU had fallen to around 10,000 nationally, only recovering slightly before the turn of the nineteenth century. With the amalgamation of the GHLU into the NUDL at Glasgow in 1899, giving a combined membership of 3,500, the survival of two smaller branches at Bo'ness and Burntisland, and the formation of a new branch at Aberdeen (following the total collapse of the Shore Labourers' Union there in 1893), the future of Scotland's 'pioneering organisation for new unionism' seemed guaranteed.[38] But by the early twentieth century the NUDL was virtually moribund in most ports in Scotland. Nevertheless it did survive, and in time built a bridge between the New Unionism and the labour unrest in the years before 1914.

However, with the relocation of the NUDL headquarters to Liverpool during 1891, a dangerous split had emerged between the Scottish and English dock labour forces, which in 1910 resulted in another body blow to the NUDL when what was left of the Glasgow and West of Scotland dock labour force left the union. In the aftermath of this event, during 1911, a stronger – and what would prove more effective – Scottish Union of Dock Labourers was formed, and this guaranteed the long-term survival of dock unions across much of the Scottish waterfront. But it also marked a fundamental east/west-coast split and created tension between Scottish and English dock trade unionists, which in effect remained the case (although in another form) through to the 1970s. Despite this split, however, the maritime trade unions and dock unionism in particular were once more a force to be reckoned with by the time of the labour unrest of 1910–14.[39] They would never again suffer the setbacks they experienced in the aftermath of the New Unionist success.

A DIVIDED WATERFRONT?: DOCK TRADE UNIONISM FROM THE LABOUR UNREST TO THE FORMATION OF THE TGWU, c1910–22

The strike activity of the new unionism (*c*1889–93) and the later labour unrest (*c*1910–14) did help to strengthen the bonds of solidarity between the Scottish dockers and their historic allies, the seamen. In the interim, however, the maritime unions were more or less moribund apart from occasional, sporadic activity in some of the larger ports such as Belfast, Dublin, Glasgow, Liverpool or London, and much of the solidarity evident in the earlier period began to dissipate. Moreover, an intense period of inter-union rivalry was also experienced across many ports, despite the best efforts of the National Transport Workers' Federation (NTWF), which was formed in 1910 (from an earlier organisation known as the International Federation of Ship, Dock and

River Workers, formed in 1896) and attempted to heal division and partition within the waterfront unions.

Religious and ethnic differences between Protestant and Catholic maritime workers, and the Irish and local workers at ports such as Liverpool, Leith and London, was a cause of considerable tension.[40] But, as was discussed in the previous section, there were no such sectarian or sectional divisions apparent at Aberdeen or Glasgow – despite Glasgow's potential for sectarian division – between the late 1880s and the years after World War I. Indeed, despite heightened sectarian tension and conflict during the interwar years, there was little evidence to suggest that this was a problem for the Scottish waterfront community.[41]

The split that did emerge within the Scottish dock unions in 1910 was based neither on sectarianism nor sectional differences but simply on politics. During the first decade of the twentieth century the fortunes of the NUDL in Glasgow changed considerably, and across Scotland dock unions were finding it difficult to maintain their position, let alone form new branches. Due largely to the activities of Jim Larkin, the NUDL had halted an employer attempt in Aberdeen in 1906 to convert it to a 'free labour' port, although by then it was too late for Dundee, which became a free labour port in 1904.[42] But despite the best efforts of the NUDL, by 1910 the Glasgow branch was lost, and thereafter the influence of the NUDL in Scotland rested entirely with the east-coast ports of Aberdeen and Bo'ness.[43]

James O'Connor Kessack temporarily resigned as Scottish organiser of the NUDL after losing Glasgow, but he returned to support the cause of dock unionism and remained leader of the NUDL until he joined the forces at the outbreak of World War I. He did have some considerable success in reorganising the east-coast ports, increasing the number of branches from two (Aberdeen and Bo'ness) to eight (adding Alloa, Burntisland, Grangemouth, Leith, Methil and Montrose), and the membership from around 500 to over 3,500 between 1909 and 1914.[44] The NUDL was not the power it had been prior to the secession of Glasgow and the earlier loss of the west-coast ports, but it proved formidable in improving the lot of the east-coast dock labour force, as well as in its impact on future dock union developments in England.

There is considerable debate and discussion as to why the Glasgow branch of the NUDL failed.[45] Within six months, however, and with the assistance of Glasgow Trades Council and the Seamen's union, dock trade unionism made a spectacular comeback with the formation of the Scottish Union of Dock Labourers in July 1911. The dockers of Glasgow and Clydeside flocked into the ranks of the new union in their thousands, and before the year was out they were in a series of disputes across Clydeside and one in Dundee (where a branch was formed for the first time since becoming a free labour port in 1904). In the interim, the SUDL affiliated with the National Transport Workers' Federation (NTWF), and through the NTWF to the Triple Industrial Alliance.[46] The ground was now being prepared for the extensive general strike that was to take place during January and February 1912.

It was noted in the board of trade *Gazette* that the challenge to the Clyde port employers had been threatened since December 1911, when they first began to enforce new working conditions. As a result, on 29 January 1912 some 7,000 dock workers across Clydeside went on strike.[47] Indeed, in order of magnitude only the strike at the Singer sewing machine factory at Clydebank during March and April 1911 and the Dundee Mill workers' strike during February and April 1912 involved more workers before 1913.[48] The strike ended in something of a compromise but resulted in official recognition of the SUDL and the beginning of joint collective bargaining – the first time this had occurred at the port of Glasgow. Moreover, the strike wave that swept across Clydeside, and the bond of solidarity that emerged from this, far exceeded anything that had been achieved during the glory days of New Unionism.[49]

Once again the role of the powerful and influential Glasgow Trades Council was important, as was the invigorated relationship with the seamen and the growing political influence of the NTWF, which was constantly arguing the case for industrial unionism. Indeed, the NTWF helped extend the bond of unity and solidarity by forging a pan-national industrial alliance with the seamen and railwaymen, and at the same time acted to bring together the membership of both the SUDL and the NUDL. This was further strengthened through the affiliation of both unions to the TUC and the STUC. On the eve of war, in August 1914, the SUDL had become firmly established along the entire west coast of Scotland, Dundee and Bo'ness on the east, and in several west-coast English ports, including Workington and Whitehaven.

THE IMPACT OF WORLD WAR I

The impact of World War I had two main effects on the dock labour force at Glasgow. The first was the intervention of the state in port affairs. The second, leading on from the advances made during the labour unrest, was an enhancement of the status of the dockers within the broader trade union movement and British society generally. The docklands was virtually an industrial relations battlefield before World War I, but it became pacified during the war.[50] By the end of the war the dockers had benefited greatly through the formal setting-up of courts of arbitration and conciliation, and because of their links with the NTWF and the Triple Industrial Alliance, they had more influence in labour affairs in Britain than at any other time in their history.

By 1920 the SUDL had a membership of 10,000, and its members proudly, with some justification, proclaimed it 'the best organised little trade union in Britain'.[51] Except for Dundee and Bo'ness, the east-coast dockers were still represented by the NUDL, which had by 1920 significantly increased its Scottish membership from around 3,500 to over 5,000. Once again the NTWF played a crucial role in this development insofar as it helped unite the great majority of transport workers within one central organisation and worked hard to develop a port industry-wide perspective. When the first national agreement for the British port transport industry was signed in 1920 – to which the small but effective SUDL was a signatory – the future looked good. With this agreement came increased calls for a permanent and independent trade union body to be organised to represent all of the unions federated in the NTWF. This became reality in 1922, when the Transport and General Workers' Union (TGWU) was formed.

But, as befits the troubled and chequered history of Clydeside dock unionism, there was to be yet another twist in the tale, one that would highlight once again the Glasgow dockers' stubborn and independent nature. The dockers were generally considered not to have been too intimately involved in the militant campaigns and disputes that became associated with Red Clydeside, although some recent evidence suggests that this position might well have to be reappraised.[52] However, when they struck in sympathy with the miners in response to the expected general strike of the Triple Industrial Alliance in April 1921, they clearly demonstrated their militant credentials. It was an episode in the history of the SUDL, however, that was to have significant implications for the future of Clydeside dock unionism.

The miners' strike of 1921 was fought over the issue of decontrolling the industry on the part of the government. The miners had secured from the government a guarantee that they would negotiate with the mine owners to establish a national wages board by the end of March 1921. As a result of this promise the miners called off the strike scheduled to take place during October 1920. However, by March 1921 the government showed it had no intention of honouring its previous promise. The miners struck on 1 April 1921, and as part of the Triple Industrial Alliance it was agreed that a general strike would take place on 15 April. The sympathetic action never materialised, however, and the general strike was called off. This episode became known as 'Black Friday' and it marked the end of the Triple Alliance as an effective industrial weapon. Key sections of the trade union movement were to be defeated in the confrontations that followed, and it was to spell disaster for the SUDL.[53]

At the STUC conference held on 20–23 April 1921, the SUDL demonstrated its commitment to the miners by calling upon all workers in Scotland to support it, and also identified those whom they blamed for the failure of the Triple Alliance.[54] Its resolution was not supported. Nevertheless, on 6 May 1921 the SUDL downed tools, first at Glasgow, then at Ayr, Ardrossan and Dundee, with other branches quickly following.[55] But there was little support for this industrial action, and with the exception of some railwaymen who had refused to transport 'blackleg' coal from Glasgow harbour, the SUDL was more or less on its own.[56] The NTWF had called an embargo on all incoming foreign coal, and although the SUDL considered this an inadequate response, it noted too that the NUDL members had been instructed to ignore the order. Indeed, James Sexton, leader of the NUDL, had personally ordered striking dockers at Leith to go back to work on shipments of Belgian coal entering Scotland.

The SUDL felt it had been 'deserted' and 'betrayed': deserted by the NTWF, which failed to endorse a national stoppage, and betrayed by Sexton and the membership of the NUDL, who together undermined the SUDL strike action. The SUDL was out on a limb, and on 23 May 1922 the executive recommended that the men went back to work.[57] Somewhat reluctantly, most dockers did so, but the Clydeside dockers held out until 6 June.[58] The strike was finally over, but with this began the slow yet sure disintegration of the SUDL. The rank-and-file dockers were already retreating from the close involvement they shared with the wider labour movement after June 1921. More importantly, they came to distrust the type of 'corporate' trade unionism symbolised by the NTWF and the leadership provided by Ernest Bevin, and specifically the actions of NUDL leader, James Sexton, whose union they rejected in 1910. Indeed, they became deeply distrustful of the actions of the SUDL leadership. The dockers' 'honeymoon' with the Scottish trade union movement was over.[59]

THE FORMATION OF THE TRANSPORT AND GENERAL WORKERS UNION: FROM GLASGOW DOCKS BRANCH TO INDEPENDENT TRADE UNION, 1923–32

The events outlined above must also be viewed against the backdrop of the proposed formation of the TGWU and how this was occurring at a time when a third request was placed before the membership of the SUDL to amalgamate with the proposed new union. Two previous ballots had already failed to secure a merger and although there was a large majority in favour, in neither case did the ballot produce the five-sixths majority of the total membership which, according to the constitutional rules, was the required number needed to officially close the union.[60] Put simply, the SUDL was not part of the initial formation of the TGWU.

By November Ernest Bevin informed the executive council of the SUDL that the membership would have to transfer to the TGWU, as the cost of balloting the 300,000 TGWU members would be exorbitant. It was stressed, however, that the TGWU would clear all of the SUDL's debts and accept all legal liabilities should it decide to join, and that Joseph Houghton, the leader of the SUDL, would become Scottish organiser for the TGWU. Bevin also stated that no more affiliation fees would be paid to the NTWF. Thereafter the TGWU more or less assumed the role previously taken up by the NTWF.[61] The SUDL was all but wrapped up by the end of December 1922, and finally laid to rest when the certificate of registration was cancelled by request of the leadership on 9 November 1923.[62]

During the amalgamation discussion it was stressed to TGWU officials that there was a strong minority in Glasgow who might mount a legal challenge to the transfer procedure over the failure to meet the five-sixths majority clause contained in the SUDL constitution. It was also known at this time that a committee from the Glasgow docks branch had independently approached the employers to draw up a 'new agreement', the first clear indication that Glasgow's dockers were once again intent on deciding port affairs by and for themselves.[63] John Veitch – the area secretary of the TGWU – dismissed such reports, arguing that this minority would come over to the TGWU and by doing so 'bring unity where unity was so much desired'.[64] Unity, however, was the last thing that this strong and influential minority wanted, and unity was something that the TGWU would ultimately fail to achieve at Glasgow.

Many of the problems that were to beset the Transport and General Workers' Union in its dealings with the dock workers of Clydeside were already well established before 1923. First, there was a hard core of Glasgow dockers unhappy with the transfer of their membership to the TGWU in December 1922. Second, a faction representing the dockers of Glasgow had already approached employers there to draw up their own agreement as to wages and working conditions. Third, the Glasgow branch members of the SUDL had shown on numerous occasions that they had an independent streak and continually questioned the rulings of the SUDL's executive council.

A fourth factor was that the dockers of Glasgow exhibited a great sense of their own history and their place in the development of dock unionism nationally. They always felt that the base for such an organisation should be in Glasgow – the antecedents of which stretched back to the New Unionist period of the late 1880s and the formation of the NUDL. Fifth, their own traditions had shown that they would not function well as a 'mere branch of an English trade union', as illustrated by the disintegration of the NUDL at Glasgow in 1910 under James Sexton's leadership. In short, they would always want to have control of their own affairs, retain their headquarters in Scotland and at the same time proclaim that they should not suffer under 'an English dictatorship', whether it was that of James Sexton or Ernest Bevin.

When considering the 'breakaway' that was to take place at Glasgow in January 1932 the above factors need continually to be borne in mind. There are, however, other factors to consider. It was shown through Joseph Houghton's evidence before the Shaw Enquiry of 1920 that the Glasgow dockers would not accept the principle of decasualisation and the necessary prerequisite of registration.[65] It was also reported widely in the press that Glasgow would have 'divided the Federation' (the NTWF) in 1920 rather than accept registration. Ernest Bevin noted at that juncture that 'Glasgow had opposed a great many reforms until they were shown to be good', and concluded 'that this was perhaps part of the Scottish temperament'.[66] This clearly illustrates Bevin's thoughts on trade union democracy – evident in the use of the phrase 'shown to be good'. This could quite easily translate into 'forced to accept what he perceived to be good'. Indeed, this was exactly what he, and the TGWU, attempted to do in Glasgow.

Perhaps the biggest problem with registration was that it was pushed vigorously by the leadership of the TGWU, and by Ernest Bevin in particular. Bevin was associated with authoritarianism, and with the climb-down of the Triple Alliance in 1921. This led directly to the first docks revolt at London, resulting in an irreconcilable split between the dock unions in Scotland and the planting of bitter seeds of dissent at Glasgow.[67] But dockers' registration, perhaps more than any other issue, was the main cause of division within the ranks of the British dock labour force in the 1920 and 1930s.

To the rank and file, registration was a pact entered into between union officials and the employers in order to control the workforce. By 1925 Glasgow's dockers had arranged meetings around the docks and voted strongly 'against the adoption of registration at Glasgow',[68] arguing strongly that registration was potentially 'destructive of union power'.[69] Events at the port of Ardrossan were shown to be a prime example of this rationale. The dockers at Ardrossan had accepted the register in 1919 but quickly demanded its suspension because it allowed non-unionists to participate in dock work.[70] This was an attack on dock custom and tradition and on dock trade unionism.

Registration, in the minds of the dockers of Glasgow, was seen as a principal method of control over them and their work. In this perceived struggle for control the dockers of Glasgow formed the Anti-Registration League, along with dockers at the port of Aberdeen and various other Scottish ports.[71] The Scottish Area minutes of the TGWU show that the league was up and running in Glasgow by April 1928.[72] The saga continued for many years, and the existence of the Anti-Registration League in Glasgow proved to be the focus for every complaint that could be raised against Bevin and the TGWU leadership. The events unfolding at Glasgow created much interest in trade union circles and the local and national press, and in December 1931 the *Glasgow Herald* reported extensively on what Bevin saw as the increasing 'rebelliousness and contrariness' of the Glasgow dockers. In January 1932 it all came to a head when Glasgow finally seceded from the TGWU to form the Scottish Transport and General Workers' Union.[73]

THE SCOTTISH TRANSPORT AND GENERAL WORKERS UNION, 1923–72

There is perhaps no greater evidence of the commitment of the dockers of Glasgow to the democratic principals of trade unionism than the fact that the STGWU came into existence and remained an independent dock union for over forty years. When Glasgow broke from the TGWU it caused a real stir, with the *Glasgow Herald* noting that the STGWU manifesto was issued under the title 'Scotland Forever'. It was stressed that the STGWU was not a 'breakaway' but a manifestation that the Glasgow dockers 'were loyal to the principles of trade unionism'. At least, concluded the report, they now had the satisfaction of 'appointing their own officials'.[74]

The STGWU did not become the great power in the Scottish trade union movement as suggested by the initial euphoria that surrounded its formation. It had set up a branch at Grangemouth, but quickly lost it to the TGWU, although it did create a more permanent foothold at the small port of Campbeltown. Despite their earnest desire to organise all Scottish transport workers, however, Campbeltown was as far as the organisation went (although it did represent galvanisers at Lochaber, and agricultural workers around Campbeltown itself).

STGWU reports of the 1930s testify to the great efforts made by the TGWU to exclude the STGWU from the wider trade union movement in Scotland and the National Joint Council for the Port Transport Industry. But by this time the STGWU was already negotiating directly with the Glasgow shipping employers on working conditions and wages – an arrangement the TGWU was attempting to break up[75] – and had been admitted to the Glasgow Trades Council, and latterly the STUC. By 1935, the STGWU had agreed favourable conditions with employers, but this was 'wrecked' because of the procrastination of the NJC in London, and through the interference of Bevin and the TGWU.

Despite further attempts by the TGWU to exclude the STGWU from joining the NJC, by 1945 the STGWU was a co-signatory to the National Agreement alongside the TGWU, the National Union of General and Municipal Workers, and the National Amalgamated Stevedores and Dockers.[76] The events leading to secession in 1932, however, cannot be understood without some appreciation of the historical developments of dock unionism at Glasgow from the time of the SUDL, or the early days of the NUDL. Indeed, to fully understand future developments in Scottish dock unionism by the second half of the twentieth century, we must again look to the influential and important role played by the dockers of Glasgow.

The defining characteristic of Glasgow dockers is perhaps best summed up in a report in the *Glasgow Herald* that the rank and file had historically 'been in closer touch with authority at Glasgow', and that this was a 'privilege' they were not willing to give up. This was evident in the formation of the STGWU, and while Ernest Bevin vowed that the Glasgow dockers would 'eat grass', they were to outlive him and that threat and go on to defend and maintain their independent Scottish union for over forty years.[77]

CONCLUSIONS

This chapter has stressed the importance of Glasgow to developments in Scottish dock unionism stretching back to the formation of the GHLU in 1853, and as the birthplace of general and national dock unionism in Scotland and Britain in the early months of 1889. Glasgow's strength and resilience lay in its determination to develop principled trade unionism that was sympathetic to local traditions and customs, but this brought a peculiarly Scottish view of trade unionism into direct conflict with the type of corporate trade unionism that was developing in England in the twentieth century. Thus the NUDL, the NTWF and finally the TGWU became, to general semi-skilled and unskilled dock labour force in the twentieth century, what the English model trade unions were to the skilled and artisan labour force in the nineteenth century. Indeed, they were a natural consequence of the nineteenth-century developments. This situation did not prevail in Scotland, where the trade union movement developed along different lines.[78]

When the organisation of work was being threatened by outside interference, whether by the employers, the trade unions or the state, the dockers would rise to the defence of that system of organisation. This can be seen through the mass rejection of any form of decasualisation or registration, at Glasgow. When coupled with the desire for greater autonomy and democracy on the part of the Glasgow dockers, the central insistence of the Glasgow branch on electing its own branch officials and an extreme awareness of what was right and proper in defining trade union constitutionality, it is clear that a direct appeal was being made to the affections of the Glasgow dockers, and this helped to create the best conditions for a united front.

Other ports rejected schemes for registration, most notably Aberdeen which, with Glasgow, was one of only two ports not to be registered by 1939. But Aberdeen stayed to fight within the TGWU and showed no inclination to follow Glasgow's example in seceding from the union. Indeed, there were no secessions elsewhere except London, where the National Association of Stevedores and Dockers (NASD), formed in 1923, refused to have any dealings with the TGWU and remained a separate union. Its numbers, however, were not as great as at Glasgow, and the NASD did not involve the entire casual dock labour force of the port of London.

What occurred at Glasgow was therefore unparalleled elsewhere along the British waterfront, and in that sense it made Glasgow unique. While the dock labour force shared much with other dockers, it was sufficiently different in composition and attitude to fuel a fierce and stubborn streak of independence which was not evident to anywhere near the same degree outwith the port of Glasgow. This determination and self-assurance were to serve and sustain Glasgow's dockers well in their many conflicts with the TGWU during the 1920s and 1930s.

The question of trade union democracy was one the dockers of Glasgow had been addressing since the early period of New Unionism. Twenty years elapsed before the Glasgow dockers decided to reject the NUDL. It took them just ten years to decide to leave the TGWU. In total it had taken the Glasgow dock labour force over forty years to find a solution to their problems. The Glasgow men 'were out and they were staying out', stated the *Glasgow Herald* when it reported the dockers' secession in January 1932. They were to remain out for forty years, until the STGWU was officially closed down in 1972. That the Scottish Transport and General Workers' Union came into existence at all is an illustration of the patience and determination of Glasgow's dock labour force: that the organisation survived independently for four decades is proof of this.

It is ironic that Bevin, who was the cause of so much disquiet among waterside workers, was responsible for the introduction of the Dock Labour Scheme in 1947, a scheme that Glasgow and dockers from all over Britain would fight to maintain and defend forty years later. The Dock Labour Scheme was the first viable attempt to bring order to the casual system of employment – although full decasualisation, which forced time discipline on the dockers in return for guaranteed wage earnings, was not finally achieved until 1967. By the 1960s, however, containerisation was the greatest problem faced by an industry in which modernisation was long overdue. If implemented, it would ultimately lead to a drastic reduction in the number of dockers employed. It was the final great struggle against the implantation of technology stretching back to the days when steam replaced sail. In essence it sounded the death knell for Scottish and, to a lesser extent, British dock unionism.

The last great hurrah came in the 1980s in the fight against the Conservative government's anti-trade union legislation. This marks a clear watershed in the history of dock trade unionism in particular, and in maritime and British trade unionism generally. There were strikes across the docklands during the 1980s, including a national strike in 1984. But the national dock strike of April 1989, in response to the government's abolition of the National Dock Labour Scheme, was to prove a decisive turning point for dock trade unionism. For over forty years the scheme protected dockers' wages and conditions of labour, and it was expected that the strike would have widespread support. But with the strike scarcely effective beyond London and Liverpool, the union leaders admitted defeat and the dockers returned to work.

According to R B Oram, it was suggested that it was easier to pass a camel through the eye of a needle than pass some machine parts through the ports of Britain. Containerisation changed this and the industry. The combination of anti-union legislation in the 1980s, or any hope of a sympathetic hearing from the Blair Labour administration in the 1990s (as manifest at Liverpool docks since 1995), demonstrates clearly that the late twentieth and early twenty-first centuries mark the end of effective dock unionism in Britain. The industry had been too defensive, failed to invest and modernise against increased competition from both British and European docks and, when faced with the spectre of containerisation, the Scottish dockers were unwilling to change or compromise. The Glasgow dock labour force was broken up and transferred to other ports, such as Hunterston on the Ayrshire coast. By 2000 the Clyde was a quiet river. and the great age of the Glasgow and Clydeside docks – once a proud metaphor for the success of the Scottish port transport industry – was now simply a memory, as were the men who once worked her docks, harbours and wharves.

NOTES

1. Kenefick, 1996a; Kenefick, 1997b; Kenefick, 2000a; Kenefick, 2000b, 19–22; Kenefick, 2007a; Kenefick, 2007b.
2. Kenefick, 2000b, 19; see also Marwick, 1935.
3. No thought was given to the preservation of the documentary materials held in the former offices of the STGWU after they rejoined the TGWU in 1972. Much of this material was thrown out or succumbed to weather damage as the former offices deteriorated. Thus the records of one of Scotland and Britain's last independent trade unions were lost.
4. MacDougall, 1978.

5. Jackson, 1983, Preface, 10–11.
6. Wilson, 1972; Lovell, 1969; Taplin, 1986.
7. Knox and Fischer, 1991.
8. Kenefick, 2000b, 56–73; see also Board of Trade, 'Summaries', 1913.
9. Kenefick, 2000b, 56–8.
10. Jackson, 1983, 134.
11. Kenefick, 2000a, 323–4; Kenefick, 2000b, 109–16; see also 'Home Office Statistics', 1912, 9. It was stated categorically that numbers were 'perhaps hardly capable of exact calculation'.
12. Kenefick, 2000b, 109.
13. Kenefick, 2003.
14. Kenefick, 1997a, 22–3.
15. Kenefick, 2000b, 112–6: see also Lovell, 1969, 57–8; Lovell, 1977; Lovell, 1987; Wilson, 1972, 25; Davis, vol. I, 2000.
16. Kenefick 2000b, 109; *New (Second) Statistical Account of Scotland*, Edinburgh, 1845, Appendix G, 117.
17. Sloane, 1987, 43.
18. Kenefick, 1997a, 23; Kenefick, 2000b, 126–31.
19. Lovell, 1969, 57–8.
20. Wilson, 1972, 50.
21. Lovell, 1977, 16–8.
22. Kenefick, 2000b, 116–8; see also Taplin, 1986, 27–30.
23. Kenefick, 2000b, 184–5.
24. Kenefick, 2000b, 116.
25. Kenefick, 2000b, 121–2; Kenefick, 2007a, 15–18; Devine, 1994, 225; Aspinwall, 1991, 103 and 98. Whilst it is true to say that the Land Question was a common feature of the politics of Ireland and the Highlands and Islands, this commonality only went so far. See Cameron, 1996, 199; Dewey, 1974, 56–68.
26. Kenefick, 1996, 51–71.
27. Kenefick, 2000b, 168 and 189; Lovell, 1987, 232.
28. Kenefick, 1993, 1–27; Kenefick, 2000b, 185–92.
29. Royal Commission on Labour. 2nd Report, Group B, Minutes of Evidence, ii. 1892; pt III, C 6795–II, Q, 13,417.
30. Lovell, 1969, 85.
31. Kenefick, 2000b, 193.
32. Buckley, 1950, 101.
33. Glasgow Trades Council. *Annual Report 1888–1889*, Glasgow, 8.
34. Kenefick, 1997b, 18–25; see also Fraser, 1978, 1–2.
35. Kenefick, 1997b, 42; Kenefick, 2007a, 9–13.
36. Board of Trade, 1899, Cd 422; see also Coates and Topham, 1991, Table 4.1, for figures for Membership of the New Union in 1890, 95.
37. Kenefick, 2007b, 153–4.
38. 38 NUDL, *Annual Report*, Fifth Annual Congress, Liverpool, 1893, 16.
39. Kenefick, 2000b, 210–12, for an in-depth explanation of the NUDL's collapse at Glasgow.
40. Kenefick, 1997a, 25–7; Kenefick, 2000b, 118–20; Taplin, 1986, 24; Lovell, 1977, 16–7.
41. Kenefick, 1997c, 23–5.
42. Royal Commission on Poor Laws, 1910, *Parliamentary Papers*, Cd 5086: Evidence

given by Mr John Malloch – the Clerk to the Harbour Trustees of Dundee, 64–7.

43. Glasgow Trades Council Minutes, 15 February, 1911.
44. STUC *Annual Reports*; see affiliation lists for reports 1909 to 1914.
45. Kenefick, 2000b, 200–9.
46. Kenefick, 2000b, 216.
47. Board of Trade labour *Gazette*, February, 1912, 68.
48. Duncan and McIvor, 1992, 83 (see Table 1 for list of principal disputes); Kenefick, 2007a, 108–13.
49. Kenefick, 2000b, 14.
50. Schneer, 1982, 103.
51. Kenefick, 2000b, 211.
52. There is evidence that shows that the SUDL were part of a 'Peace March' during the May Day celebration in 1915, but the extent of their involvement in this movement is far from clear. It does, however, suggest that we may need to look more closely at this particularly issue.
53. Hinton, 113–5.
54. Scottish Trades Union Congress, Aberdeen 1921. Debate on the Call–Off of the Triple Industrial Alliance General Strike, 15 April 1921.
55. SUDL Emergency Committee (EC) 6, 9 and 12 May; and Special Executive Committee Meeting on 14 and 18 May 1921.
56. Glasgow Trades Council minutes, 14 May 1921.
57. SUDL EC Emergency Meeting, 23 May 1921.
58. SUDL Executive Minutes, 6 June 1921.
59. Kenefick, 2000b, 233–6; Kenefick, 2007a, 186–90.
60. SUDL Executive Minutes, 4 January 1922: SUDL constitutional rules, Rule XIX Clause 1, states: 'The Union may be dissolved by the vote of five-sixths of the whole of the financial membership', Glasgow, 1914, 27.
61. SUDL Executive Minutes, 18 and 25 November 1922.
62. NAS, Registrar of friendly Societies in Scotland (FS. 8. 18).
63. SUDL Executive Minutes, May 24.
64. Executive Meeting of the Scottish Union of Dock Labourers and Transport Workers, 19 December 1922. Typed minute signed by John Veitch – area secretary of the Transport and General Workers' Union – on accepting the transference of the liabilities and membership of the SUDL.
65. *Parliamentary Papers*, Cmd 936, Cmd 937.
66. The case for Glasgow, *Glasgow Herald*, 19 February 1920.
67. Wilson, 1972, 80.
68. This was most definitely stated through the minutes of Area 7 Committee of the TGWU (which covered Clydeside as well as the central belt of Scotland) when it was noted that the National Joint Industrial Council for the Port Industry adopted the principle of decasualisation of the industry in 1924. Noted in Area 7 Committee of the Transport and General Workers' Union Quarterly Minutes, 14/7/24, 3.
69. Phillips and Whiteside, 1985, 225.
70. Phillips and Whiteside, 1985, 222.
71. Phillips and Whiteside, 1985, 225.
72. Area 7 Committee Minutes, 17/4/28, 18.
73. Kenefick, 2000b, 236–9.
74. *Glasgow Herald*, 12 January 1932; see also *Forward*, 2, 16 and 23 January 1932,

and 6 February 1932, where it was stated that the 'cardinal issue in Glasgow was the appointment of officials' and that the breakaway had secured this overriding principle.

75. TGWU, Executive Report, 1932/1933, 3.
76. National Joint Council Agreement, amended 21 December 1945, 54: edition of agreement between the co-signatories and the National Association of Port Employers, which included the STGWU for the first time.
77. The term 'eat grass' is noted in Wilson, D. *Dockers: The Impact of Industrial Change*, 1972, and was referred to on many occasions in the oral testimonies of Glasgow dockers. The phrase is part of the Glasgow dockers' history, and the fact that Bevin never made good his threat was to prove an abiding source of pride and achievement. Discussion with ex-dockers (all from Glasgow) Charles Ward, 25 June 1992; Gordon Banders, 16 June 1992; and Tom O'Connor, March 1993.
78. Kenefick, 1997b, 18–24.

BIBLIOGRAPHY

Aspinwall, B. The Catholic Irish and wealth in Glasgow. In Devine T M, ed., *Irish Immigrants and Scottish Society in the Nineteenth and Twentieth Centuries*, Edinburgh, 1991, 91–115.

Board of Trade. Sixth annual abstract of labour statistics, 1898–99, *Parliamentary Papers*, Cd 119, 1900.

Board of Trade. Report on trade unions in 1899 with comparative statistics for 1892–1898, *Parliamentary Papers*, Cd 422.

Board of Trade. Board of trade summaries, *Parliamentary Papers*, 1913, Cmd 758.

Buckley, K D. *Trade Unionism in Aberdeen, 1878–1900*, Edinburgh, 1955.

Cameron, E A. *Land for the People? The British Government and the Scottish Highlands, c1880–1925*, East Linton, 1996.

Coates, K and Topham, T, eds. *The Making of the Transport and General Workers' Union 1870–1922*, vols I and II, Oxford, 1991.

Devine, T M. *Clanship to Crofters' War*, Manchester, 1994.

Dewey, C. Celtic agrarian legislation and the Celtic Revival: Historicist implications of Gladstone's Irish and Scottish Land Acts, 1870–1886, *Past and Present*, 64 (1974), 30–70.

Duncan, R and McIvor, A eds. *Militant Worker: Labour and Class Relations on the Clyde*, Edinburgh, 1992.

Fraser, W H. Trade councils in the labour movement in nineteenth century Scotland. In MacDougall, I ed., *Essays in Scottish Labour History*, Edinburgh, 1978, 1–28.

Hinton, J. *Labour and Socialism: A History of the British Labour Movement, 1867–1974*, Brighton, 1983.

Home office statistics of compensation under the Workmen's Compensation Act (1906) and Employers Liability Act (1880), 1911, *Parliamentary Papers*, 1912, lxxv, Cd 6493.

Inquiry into the wages and conditions of employment of dock labour, Cmd 936, *Parliamentary Papers* xxiv, 1920; Cmd 937, *Parliamentary Papers*, xxiv, 1920.

Jackson, G. *The History and Archaeology of Ports*, Tadworth, 1983.

Kenefick, W. *Ardrossan: The Key to the Clyde, with Particular Reference to the Ardrossan Dock Strike, 1912–1913*, Irvine, 1993.

Kenefick, W. A Struggle for control: The importance of the great unrest at Glasgow

harbour, 1911–1912. In Kenefick, W and McIvor, A, eds, *The Roots of Red Clydeside 1910–1914: Labour Unrest and Industrial Relations in West Scotland*, Edinburgh, 1996a, 129–52.
Kenefick, W. An historiographical and comparative survey of dock labour *c*1889–1920 and the neglect of the port of Glasgow, *Scottish Labour History Journal*, 31 (1996b), 51–71.
Kenefick, W. Irish dockers and trade unionism on Clydeside, *Irish Studies Review*, 19 (1997a), 22–9.
Kenefick, W. The Scottish trade union movement *c*1850 to 1914, *History Teaching Review Year Book*, vol. 2 (1997b), 18–25.
Kenefick, W. 'Quixotically generous … economically worthless': Two views of the dockers and the dockland community in Britain in the 19th and early 20th centuries, *The Historian*, 56 (1997c), 12–16.
Kenefick, W. A Struggle for recognition and independence: The Growth and development of dock unionism at the port of Glasgow, *c*1853–1932. In Davies, S, Davis, C J, de Vries, D, van Voss, L H, Hesselink, L and Weinhauer, K. *Dock workers: International explorations in comparative labour history, 1790–1970*, 2 vols, Aldershot, 2000a, 319–41.
Kenefick, W, *Rebellious and Contrary: The Glasgow Dockers c1853 to* 1932, East Linton, 2000b.
Kenefick, W. Demography. In Cooke, A, Donnachie, I, McSween, A and Whatley, C, eds, *Modern Scottish History: The modernisation of Scotland, 1850 to the present*, vol. II, 2003, 95–118; first published 1998.
Kenefick, W. *Red Scotland! The Rise and Fall of the Radical Left, c.1872 to 1932*, Edinburgh, 2007a.
Kenefick, W. The Shipping Federation and the free labour movement: a comparative study of waterfront and maratime industrial relations, c.1889–1891. In Govski, R, ed. *Maritime Labour: Contributions to the History of Work at Sea, 1500–2000*, Amsterdam, 2007b.
Knox, W and Fischer, C. Shedding the blinkers: German and Scottish labour historiography from *c*1960 to present, *Scottish Labour History Society*, 26, 1991, 21–44.
Lovell, J. *Stevedores and Dockers: A Study of Trade Unionism in the Port of London, 1870–1914*, London, 1969.
Lovell, J. The Irish and the London docker, *Society for the Study of Labour History Bulletin*, 35 (1977), 16–8.
Lovell, J. Sail, steam and emergent dockers' unionism in Britain, 1850–1914, *International Review of Social History*, xxxii, 3 (1987), 230–49.
MacDougall, I. *A Catalogue of Some Labour Records in Scotland and Some Scots Records outside Scotland*, Edinburgh, 1978.
Marwick, W H. Early trade unionism in Scotland, *Economic History Review*, v/2 (April 1935), 87–95.
Oram, R B. *The Dockers' Tragedy*, London, 1970.
Phillips, G and Whiteside, N. *Casual Labour: The Underemployment Question in the Port Transport Industry, 1880–1970*, Oxford, 1985.
Schneer, J. The war, the state, and the workplace: British dockers during 1914–1918. In Cronin, J E and Schneer, J eds, *Social Conflict and Political Order in Modern Britain*, 1982, 96–112.
Sloane, W. Assimilation of the Highland and Irish migrants in Glasgow, 1830–1870. Unpublished MPhil Thesis, University of Strathclyde, 1987.
Taplin, E. *Liverpool Docker and Seamen, 1870–1890*, Hull, 1974.

Taplin, E. The *Dockers' Union: A Study of the National Union of Dock Labourers, 1889–1922*, Leicester and New York, 1985.
Wilson, D. *The Dockers: The Impact of Industrial Change*, London, 1972.

Index